Nanocarbons

This book provides a practical platform to the readers for facile preparation of various forms of carbon in its nano-format, investigates their structure–property relationship, and finally, realizes them for a variety of applications taking the route of application engineering. It covers the preparation and evaluation of nanocarbons, variety of carbon nanotubes, graphene, graphite, additively manufactured 3D carbon fibres, their properties, and various factors associated with them. A summary and outlook of the nanocarbon field is included in the appendices.

Features:

- Presents comprehensive information on nanocarbon synthesis and properties and some specific applications
- Covers the growth of carbon nanoparticles, nanotubes, ribbons, graphene, graphene derivatives, porous/spongy phases, graphite, and 3D carbon fabrics
- Documents a large variety of characterizations and evaluations on the nature of growth causing effect on structure properties
- Contains dedicated chapters on miniaturized, flat, and 2D devices
- Discusses a variety of applications from military to public domains, including prevalent topics related to carbon.

This book is aimed at researchers and graduate students in materials science and materials engineering, and physics.

Emerging Materials and Technologies

Series Editor:
Boris I. Kharissov

The *Emerging Materials and Technologies* series is devoted to highlighting publications centred on emerging advanced materials and novel technologies. Attention is paid to those newly discovered or applied materials with the potential to solve pressing societal problems and improve quality of life, corresponding to environmental protection, medicine, communication, energy, transportation, advanced manufacturing, and related areas.

The series takes into account that, under present strong demands for energy, material, and cost savings, as well as heavy contamination problems and worldwide pandemic conditions, the area of emerging materials and related scalable technologies is a highly interdisciplinary field, with the need for researchers, professionals, and academics across the spectrum of engineering and technological disciplines. The main objective of this book series is to attract more attention to these materials and technologies and invite conversation among the international R&D community.

For more information about this series, please visit:
www.routledge.com/Emerging-Materials-and-Technologies/book-series/CRCEMT

Nanocarbons
Preparation, Assessments, and Applications

Ashwini P. Alegaonkar and Prashant S. Alegaonkar

CRC Press
Taylor & Francis Group
Boca Raton London New York

CRC Press is an imprint of the
Taylor & Francis Group, an **informa** business

First edition published 2023
by CRC Press
6000 Broken Sound Parkway NW, Suite 300, Boca Raton, FL 33487-2742

and by CRC Press
4 Park Square, Milton Park, Abingdon, Oxon, OX14 4RN

CRC Press is an imprint of Taylor & Francis Group, LLC

Library of Congress Cataloging-in-Publication Data
Names: Alegaonkar, Ashwini P., author. | Alegaonkar, Prashant S., author.
Title: Nanocarbons : preparation, assessments, and applications / Ashwini
P. Alegaonkar, Prashant S. Alegaonkar.
Description: First edition. | Boca Raton : CRC Press, [2023] |
Series: Emerging materials and technologies | Includes bibliographical
references and index. |
Identifiers: LCCN 2022039572 (print) | LCCN 2022039573 (ebook) |
ISBN 9781032329000 (hbk) | ISBN 9781032329017 (pbk) | ISBN 9781003317258 (ebk)
Subjects: LCSH: Carbon. | Carbon nanofibers. | Carbon nanotubes.
Classification: LCC TA455.C3 A44 2023 (print) | LCC TA455.C3 (ebook) |
DDC 620.1/93–dc23/eng/20221110
LC record available at https://lccn.loc.gov/2022039572
LC ebook record available at https://lccn.loc.gov/2022039573

ISBN: 9781032329000 (hbk)
ISBN: 9781032329017 (pbk)
ISBN: 9781003317258 (ebk)

DOI: 10.1201/9781003317258

Typeset in Times
by codeMantra

Dedicated

to

Prof. V. N. Bhoraskar

and

Prof. (Mrs.) S. V. Bhoraskar

Contents

About the Authors

Dr. Ashwini P. Alegaonkar is working in the areas of Materials Chemistry. She obtained her M.Phil. degree in the year 2014 and PhD degree in Chemistry in 2019 from the Department of Chemistry, SP Pune University. Her area of research includes spin transport and magnetic correlations in nanocarbon systems, adatom doping in graphene, graphene derivatives, and 2D heterojunctions for energy storage and device applications. She has a number of papers in the journals of international repute, and book chapters.

Dr. Prashant S. Alegaonkar is working in the areas of Physics with an emphasis on Applied Physics. He obtained his PhD from the Department of Physics, SP Pune University, in 2004, and postdoc from SAINT, SKKU, Korea, in 2008. He served as a faculty of physics in College of Engineering Pune (CoEP) and as an Assistant Professor in Defence Institute of Advanced Technology, Pune, and is currently working as an Associate Professor of Physics in Central University of Punjab at Bathinda. His research interest includes CVD synthesis and super-growth of carbon nanotubes, graphene, reduced graphene, and graphene oxide for targeted applications of military and civil origin. He has more than 170 international papers, 75+ conference proceedings, and 2350+ Google Scholar citations with h- and i-10 indexes, respectively, of 25 and 50.

Preface

This book provides a practical platform to the readers for facile preparation of various forms of carbon in its nano-format, investigates their structure–property relationship, and finally realizes them for a variety of applications. The route of this book is application engineering. Moreover, it's a story of consistent, perseverant, and painstaking journey we have taken in the past 20+ years to establish insights into research and development of *Nanocarbons: Preparation, Assessments, and Applications*. The content of this book is a rich source of information published by the authors in the form of *100+ research publications* during 2002–2022 and compiled in 500+ manuscript pages. For readers, this book is arranged in 11 sections and three appendices. The sections are integrated with 100+ most relevant formulae and equations; 200+ attractive figures, schematics, and concept diagrams; and 20+ data tables.

The word **nanocarbon** is repeatedly used throughout the text and has a *notional* sense. It has a generalized meaning in the context of C_{60} (fullerene)-like spherical nanoparticles (CNS), carbon nanotubes (CNTs: think/thin, multi-/single-walled, capped with/without catalyst, ultralong, medium, short tube), graphene (single-/few-/multilayered), graphene-like carbon compounds (GNCs), reduced graphene oxide (rGO), graphene-like nano-flakes (GNF), graphene-like nano-ribbons (GLNR), graphene nano-ribbons (GNR), porous nanocarbons (PNCs), spongy graphitic phases/matrixes, graphite, 3D carbon fabric, etc., as and when referred. There are a number of recipes charted for the readers. Most of them are single step, one pot, facile, involved trivial starting material, cost, time, and large-scale effective. They are assimilation of literature review, self-discovered, optimized with an intuitive experience implemented. There are a couple of remedial exceptions too! The presented synthesis approach has a blend of laboratory as well as industry scale. A large variety of characterization probes such as UV–Vis, FTIR, Raman, microwave, relaxation, thermo-magnetic, thermo-mechanical, thermo-elastic, ion transmission spectroscopies, ellipsometry, XPS/ESCA, D-SIMS, positron annihilation, RBS, NDP, SIMS, PIXE, VBDOS, scanning electron, transmission electron, atomic force, and tunnelling microscopies, including fractographic analysis, are presented and discussed in depth. The nature of growth causing an effect on the structure and, consequently, on the properties has extensively been discussed. Such preparation and assessment strategies enabled to target applications of nanocarbons in strategic as well as in public domains branched to shock absorbent armour blocks to radar absorber/stealth coating, rocket propulsion nozzles to optical gas sensors, energy storage to field emission devices, spintronic to ion track engineering, meta-materials to radiation-optimized opto-electronic polycarbon techniques. The details are presented.

Acknowledgements

First and foremost, PSA wishes to express his deep sense of gratitude towards his grandfather (Late) Mr. Jagannath P. Alegaonakr, father (Late) Mr. Sudhir J. Alegaonkar, and mother Ms. Nanlini S. Alegaonkar and expresses special thanks to his daughter Ksheeraja for her love and support!! You made our life…alive!!!

We are thankful to Prof. Raghavendra P. Tiwari, Vice Chancellor, Prof. R. K. Wusrika, Dean Acad., Prof. Anjana Munshi, Dean Research, Prof. V. K. Garg, Dean student's affairs, Dr. Eswara Prasad, Director, Dr. S. M. Abbas, Dr. Alok Dixit, Dr. Himangshu Baskey, Mr. Ashish Dubey, Scientist, Dr. K. C. Tripathi, technical staff of DMSRDE-DRDO, Kanpur, the Director, ISRO-DRDO, SPPU Cell, the Director, UGC-DAE-CSR, German Academic Exchange Program (DAAD), Germany, Dr. Dietmat Fink, Scientist, Prof. Alexander Berdinsky, National Academy of Sciences, Russia, Dr. Alexander Petrov, Research Scholar, Hahn-Meitner-Institute, Berlin, Germany, Dr. Manjeet Singh, Ex-Director, Prateek Kishor, Director, Dr. D. R. Saroha, Dr. Inderpal Singh Sandhu, Suresh Kumar, A. Sharma, P. Chandel, TBRL-DRDO, Chandigarh, Dr. K. P. S. Murthy, Director, Dr. D. K. Kankane, Dr. R. K. Kalal, Dr. Balesh Ropia, Dr. Himanshu Shekher, HEMRL-DRDO, Pune, the staff and Director, ARDE-DRDO, Pashan, Pune, the Vice Chancellor DIAT (DU), Pune, Mr. Nikhil Navle, Mr. Vipul Dhongde, and Mr. Praphull Saste, the Head, Electronic Science, University of Pune, CSIR, New Delhi, Dr. Sumit Rane, Dr. Manish Shinde, Dr. B. B. Kale, C-MET, Pune, Dr. P. N. Vishwakarma, Dept. of Physics, NIT Rourkela, DRDO-DIAT programme on nanomaterial, ERIPR, DRDO, New Delhi, Dr. Alok Banerjee, UGC-DAE-CSR, Indore, University Grants Commission, New Delhi, India, Nanomission Council, Department of Science & Technology, Government of India, Department of Science and Technology, New Delhi, India, Dr. Babasaheb Ambedkar Research and Training Institute (BARTI), Pune, India, ERIP/ER/1003883/M01/908/2012/D(R&D)/1416 project, SPPU-DRDP, Pune, Mr. Gajanan Arbade, Dr. D. K. Chouhan, DIAT (DU), Pune, Dr. Pankaj Poddar, NCL, Dr. Aditya Dharmadhikari, Dr. Jayashree Dharmadhikari, TIFR, Mumbai, India, Prof. Lalit Mohan Patnaki, Ex-Vice Chancellor, Dr. Prahlada, Ex-Vice Chancellor, Dr. Surendera Pal, Ex-Vice Chancellor, DIAT (DU), Dr. Tejashree Bhave, Dr. Suwarnan Datar, Dr. K. Balasubramanian, Dr. (Late) Ashutosh C. Abhyankar, Dr. Prashant S. Kulkarni, Dr. Shaibal Banerjee, Dr. T. Umasankar Patro, Dr. H. S. Panda, Dr. S. Chandel, Dr. Sangeeta N. Kale, Dr. A. V. R. Murthy, DIAT (DU), Pune, Dr. R. K. Goyal, COEP, Pune, Dr. Mangesh Desai, Dr. Srikrishna D. Sartale, University of Pune, the Director, IUAC, New Delhi, Head, Department of Physics, University of Pune, Head, Sophisticated Analytical Instrumentation Facility, Indian Institute of Technology, Mumbai, SERB, New Delhi, DAE, Mumbai, Dr. J. S. Rawat, Dr. Seema Gautam, Dr. Neeraj Jain, the Director, SSPL-DRDO, New Delhi, Dr. Rishi B. Sharma, CSIR Emeritus Scientist, Dr. S. B. Ogale, IISER, Pune, R. V. Godbole, V. P. Godbole, University of Pune, Dr. Sunita Bhagwat, AGC, Pune, Prof. Ji-Beom Yoo, SKKU, Korea, Prof. Sumio Ijima, Director, CNNC, Korea, Prof. Yong-Hee Lee, Prof. Y. U. Kwon, Prof. D. J. Kang, Dr. J. H. Han, Dr. J. W. Nam, Dr. J. H. Park., Dr. T. Y. Lee, Mr. H. C. Lee, Mr. S. H. Lee, Mr. S. J. Park, Mr. H. C. Lee, Mr. J. S. Jeon, Mr. J. S. Jeong, Dr. S. P. Patole, Dr. Archana S. Patole, J. H. Lee, Ms. M. J. Kim, SKKU, Korea, Prof. V. N. Bhoraskar, Director, IUC-DAEF, Indore, Prof. (Mrs.) S. V. Bhoraskar, University of Pune, Prof. N. Stolterfoht, Hahn-Meitner-Institut, Berlin, Germany, Dr. Bela Sulik, Dr. B. R. Sankapal, Dr. G. Richster, Mr. P. Symcowize, and Prof. (Late) V.G. Bhide, Department of Physics, University of Pune, Dr. Murlay Sastry, Dr. S. R. Sainkar, Dr. A. B. Mandle, Dr. N. R. Pawaskar, and Dr. S. Radhakrishnan, National Chemical Laboratory, Pune, Dr. P.S. Goyal, Center Director, IUCDAEF, Mumbai Centre, Prof. P. Balaya (currently SNU, Singapore), Dr. S. K. Deshpande, and Mr. M. Venu Gopal, Dr. V. Siriguri, Dr. P. D. Babu, Dr. J. V. Joshi, Dr. R. Mukhopadhaya, Dr. Ajay Gupta, IUC-DAEF, Indore Centre, Dr. A.V. Pimpale, Dr. (Late) S.M. Chaudhari, Mr. Avinash Wadikar, Mr. Sudarshan Potdar at INDUS -1, Centre for Advanced Technology beam line, Indore, Dr. V. Ganesan, Dr. T. Shripati, and Mr. U.P. Deshpande,

Dr. D. M. Phase, Dr. Vasant Sathe, Dr. N. P. Lalla, Dr. A. M. Awasthi, Dr. Alok Banerjee, Dr. S.R. Burman, Mr. S Bharadwaj, Dr. A.K. Sinha, Centre Director, IUC-DAEF, Kolkata Centre, Dr. D. Das, Dr. M. Sudarshan, Dr. V.S. Raju, Dr. Sanjeev Kumar, Dr. G. L. M Reddy, National Center for Characterization of Composite Materials, B.A.R.C. Hyderabad, Dr. T. Som, Dr. B. R. Sekar, Institute of Physics, Bhubaneshwar, Prof. (Late) V. N. Kulkarni, IIT Kanpur, my labmates in Department of Physics, University of Pune, who became distinguished academicians in India and abroad Dr. K. A. Bogle, Department of Physics, Prof. S. K. Mahapatra, Department of Physics, CU Punjab, Bathinda, Dr. V. R. Kulkarni, MIT, Pune, Prof. V. S. Purohit, Brown University, the USA, Dr. F. D. Attar, Asso. Prof. A. I. College, Pune, Dr. A. A. Datar, Dr. S. D. Dhaiwale, University of Pune, Dr. Gisha George, Dr. C. Balasubramanyam, Dr. M. Naddaf, Dr. Shilpa Jain, Dr. P. S. Chaudhary, Prof. (Late) Indrani Banerjee Mahapatra, Dr. Harshada Nagar Babrikar, Dr. (Late) Navin Kulkarni, Dr. Ryhid Mohammad, Dr. Anand Gole, Dr. Vidya, Dr. Ashwini Kumar, Dr. D. B. Rautrai, Dr. N. L. Mathkari, Dr. G. R. Pansare, Dr. K. C. Mohite, Dr. S. D. Chakane, Dr. Ninda Shinde, Mr. A. P. Gadgil, Mr. S. Shinde, Department of Physics, Dr. Archana Jaiswal, IUC-DAEF, Indore, Dr. Anupam Sharma, Dr. Rajneet Brajpurya, Dr. Pramod Bhatt, Dr. Archana Lakhani, Dr. (Mrs.) Shilpa Tripathy, Dr. Bindu Pande, Dr. Ajay Shukla, Dr. Ram Prakash, Dr. Sandeep Kumar Chaudhary, Dr. Somsunder Mukhopadhaya, Dr. Samrat Mukharjee, IUC-DAEF, Dr. (Late) S. K. Date, Dr. K. P. Adhi, Dr. A. G. Banpurkar, Dr. S. D. Dhole, Prof. S. K. Pardeshi, Department of Chemistry, University of Pune, Pune, Mr. S. S. Shinde, and Mrs. Madhuri Joshi, Dr. V. N. Krishnnamurthy, Hon. Director, DRDO - ISRO CELL, University of Pune, and Department of Space, Bangalore, Council of Scientific and Industrial Research, Human Resources Development Group, New Delhi, German Academic Exchange Service, Bonn, Germany (DAAD Fellowship programme). I must admit my apologies, if any name is omitted.

Abbreviations and Symbols Used

α_L	Coefficient of thermal expansion (CTE)
$\bar{g}$	Degree of graphitization
$\dot{\gamma}$	Disordered velocity
γ_{e-p-c}	Electron–phonon coupling
Γ_{spin}	Spin relaxation rate
$\vec{J}$	Thermal flux
λ	Thermal conductivity
2D	Two-dimensional
3D	Three-dimensional
AAO	Anodized aluminium oxide
ADA	Areal density of aggregates
AlPO$_4$	Aluminium phosphate
BET–BJH	Brunauer, Emmett, Teller–Barrett, Joyner, Halenda
C$_{60}$	Buckyball, buckminsterfullerene
CB	Carbon black
CBP	Carbon black powder
CD	Charge–discharge
CNPs	Carbon nanoparticles
CNS	Carbon nano-spheres
CNS	Spherical nanoparticles
CNTs	Carbon nanotubes
C_P	Heat capacity@constant pressure
CQD	Carbon quantum dot
C_{SP}	Specific capacitance
CV	Cyclic voltammetry
C_V	Heat capacity@constant volume
CVD	Chemical vapour deposition
DC-PECVD	Direct current plasma enhanced chemical vapour deposition
DDSC	Dye-sensitized solar cell
DLC	Diamond-like carbon
D-SIMS	Dynamic secondary ion mass spectroscopy
E_D	Energy density
EDLC	Electric double-layer capacitor
EIS	Electrochemical impedance spectroscopy
EMC	Electromagnetic interference compatibility
EMI	Electromagnetic interference
EOS	Equation of state
EPC	Electron–phonon coupling
ESCA	Electron spectroscopy for chemical analysis
ESR	Electron spin resonance
ESR	Equivalent series resistance
ESRF	European Synchrotron Radiation Facility
FC/ZFC	Field-cooled and zero-field-cooled measurements
FE	Field emission
FEDs	Field emission devices
FN	Fowler–Nordheim
FRP	Fibre-reinforced plastic

FTIR	Fourier transform infrared spectrometer
G_{IC}	Critical strain energy release rate
GLNR	Graphene-like nano-ribbons
GMH	Graphene/MoS_2/hBN multilayers
GNCs	Graphene-like carbon compounds
GNF	Graphene-like nano-flakes
GNR	Graphene nano-ribbons
GO	Graphene oxide
H	Enthalpy
hBN	Hexagonal boron nitride
HEL	Hugoniot elastic limit
HIP	Hot isostatic pressing
HOPG	Highly oriented pyrolytic graphite
HR	High resolution
HSR	High strain rate
ID/OD	Inner diameter/outer diameter
IED	Improvised explosive device
IT	Impregnation time
ITRS	International Technology Roadmap for Semiconductors
K_{IC}	Fracture toughness
L_a	Crystalline length
L_D	Disordered length
LHM	Left-handed materials
MD	Molecular dynamics
MoS_2	Molybdenum disulphide
MWCNTs	Multiwalled carbon nanotubes
n_D	Defect density
NDP	Neutron depth profiling
NRA	Nicolson–Ross algorithm
NRW	Nicolson–Ross–Weir
PCR	Polycyclic rings
P_D	Power density
PDMS	Poly(dimethylsiloxane)
PIXE	Particle-induced X-ray emission
PMMA	Poly(methyl methacrylate)
PNCs	Porous nanocarbons
PR	Photoresist
PTFE	Polytetrafluoroethylene
PU	Polyurethane
PVA	Polyvinyl alcohol
PVD	Physical vapour deposition
RADAR	Radio detection and ranging
RAM	Radar absorbing material
RBS	Rutherford backscattering
RCS	Radar cross section
R_L	Reflection loss
S	Entropy
S	Scattering parameter
S_A	Specific area
SAED	Selected area electron diffraction
SDS	Sodium dodecyl sulphate

SE	Shielding effectiveness
SEM	Scanning electron microscopy
SHPB	split Hopkinson pressure bar
SIMS	Secondary ion mass spectroscopy
SO	Spin–orbit coupling
SRRs	Split-ring resonators
STM	Scanning tunnelling microscopy
SWCNTs	Single-walled carbon nanotubes
t	Thickness
TDAE	Tetrakis(dimethylamino)ethylene
TEM	Transmission electron microscopy
TEMPOS	Tuneable electronic materials in pores in oxide on semiconductors
TMAH	Tetramethylammonium hydroxide
TMC	Transition metal dichalcogenides
TMD	Transition metal dichalcogenides
T_{sl}	Spin–lattice relaxation
T_{SS}	Spin–spin relaxation
UHF	Ultrahigh frequency
ULSI	Ultra-large-scale integrated circuits
UN	United Nations
U_P	Particle velocity
U_S	Shock velocity
UTS	Universal tensile strength
UV–Vis	UV–visible spectroscopy
VBDOS	Valence band density of states
VDG	Variable density graphite
V_F	Fermi velocity
VISAR	Velocity interferometer system for any reflector
VNA	Vector network analyser
VSM	Vibrating sample magnetometer
WACVD	Water-assisted chemical vapour deposition
XPS	X-ray photospectroscopy
Z	Impedance
α	Attenuation coefficient
α	Thermal diffusivity
β	Mean field enhancement factor
β	Propagation coefficient
γ	Field enhancement factor
γ	Gyromagnetic ratio
γ	Helmholtz constant
γ_E	Elastic shear strain energy
δ	Skin depth
$\varepsilon(\omega)$	Dielectric function
ε–μ	Constitutive parameters in electromagnetics
ζ	Wave function
η	Characteristic impedance
θD	Debye temperature
λ	Wavelength
$\mu(\omega)$	Diamagnetic function
ρ_d	Field emission current density
σ	Poisson's ratio

σ_{ac}	AC conductivity
σ_{dc}	DC conductivity
σ_T	Areal density
$\sigma - \varepsilon$	Constitutive parameters in mechanics

1 Introduction and Survey

1.1 STATUS OF CARBON

Carbon, the sixth element of the periodic table, is one of the most versatile atoms. With its four valence electrons, carbon offers a multitude of chemical bonds. The carbon–carbon bond is the basis of life and the key parameter to organic materials. One of the reasons why living matter has such an extraordinary richness of different molecules is probably due to the bonding scheme of carbon. But besides living matter, carbon atoms form minerals, allotropes, crystals, jewels, etc.

In nature, atomic carbon forms various multi-atomic structures with different molecular configurations, called allotropes. Two allotropes were known, until recently: (a) the graphite and (b) the diamond. In graphite (Figure 1.1a), every carbon atom has three neighbours and they are in two-dimensional (2D) layer format (Figure 1.1b). In diamond, the coordination number is four for each atom and they form a three-dimensional (3D) crystal. These are very abundant forms of carbon and very much appreciated, especially diamond. As an extraordinary surprise, in recent years, scientists have been impressed more by two new allotropes: fullerenes [1] (discovered in 1985) and carbon nanotubes (1991) [2]. In these structures, the carbon–carbon bonds are in bent or curved format. In the former, a ball–shaped molecule – the so-called buckyball (C_{60}) – is formed, while in the latter, carbon atoms form a long cylinder-shaped tube. Both of them are in the nanometre range. These two discoveries provided researchers a wealth of knowledge in basic science, and they serve in many applications. For example, C_{60} is used in creating new medications such as antioxidant drugs, while carbon nanotubes are basic constituents of transparent conducting films, dye-sensitized solar cells, and flat-panel screens. They seem to be very promising for application in compact electronic circuits.

1.1.1 INCREDIBLE NANOCARBONS

In the history of carbon research, nanotubes and graphene have been of major interest, both from fundamental point of view and for applications. The most attractive features of these nanostructures are their electronic, mechanical, optical, and chemical characteristics, which opened an express way for future applications. Their properties are reported to be measured down to a single tube or one layer. However, for applications to be commercialized, gram-scale quantities are required and there are varieties of nanotubes and graphene that can be produced using numerous fabrication tools. The most common techniques used nowadays are as follows: arc discharge, laser ablation, chemical vapour deposition, and flame/combustion synthesis (discussed in detail in Chapter 2). These structures itself are inert in nature and hence needed surface modifications in terms of oxidation, acid treatment, annealing, sonication, filtering, and functionalization. Economically feasible, large-scale production and modification techniques still have to be realized.

Fundamental and practical nanotube as well as graphene research demonstrated their potential use in the fields of displays to energy storage, molecular electronics to nanomechanical devices, and composite materials to nanomedicine. However, realistic implementation is still in its infant stage.

1.1.1.1 Dimensionality Effect: Buckminsterfullerenes to Nanotubes

The other popular format of nanocarbon is buckminsterfullerenes! These are supra-molecules with closed-caged structure of carbon clusters and have several special characteristics that were not identified in any class of compounds before. Hence, fullerenes form an interesting category of carbon compounds that are implemented in drug delivery technologies and applications in medical science. It was a generally accepted notion that these large spherical molecules were unstable prior to the

DOI: 10.1201/9781003317258-1

first synthesis and detection of the smaller fullerenes C_{60} and C_{70}. However, some co-workers [1,2] already predicted the existence of C_{60} in the gaseous form as a stable compound with a relatively large band gap. Fullerenes were accidentally discovered, like the case with numerous important scientific discoveries occurred in science before! In 1985, Kroto and Smalley [1] reported strange signals in the recorded mass spectra for evaporated carbon samples, proving the discovery of fullerenes and their stability in the gaseous state. This triggered the search for other classes of fullerenes, leading to the discovery of a one-dimensionally extended form in 1991 by Iijima and co-workers [2], which is termed as carbon nanotubes and had been investigated by a number of researchers globally. Such structures had extended 1D length (up to several μm) and smaller diameter (up to few nanometres), resulting in a large aspect ratio. They can be seen as the nearly one-dimensional form of fullerenes. Therefore, these materials are expected to possess additional interesting electronic, mechanic, and molecular properties. Especially in the beginning, all theoretical studies on carbon nanotubes focused on the influence of the nearly one-dimensional structure on molecular and electronic properties.

1.1.1.2   2D Graphene and Beyond

The researchers were of the opinion that not much more could come from carbon, but it has surprised us once again! The surprise was the emergence of the new form of carbon: graphene. One could not say that it was discovered in 2004 because graphene was produced by chemical methods in large quantities well before. Furthermore, its physical properties had extensively been studied theoretically and there were trials launched well before producing large surface for microelectronic applications [3]. However, the breakthrough came from the isolation of graphene from graphite by a simple mechanical exfoliation process using Scotch tape. This isolated 2D crystal showed a phenomenon called the quantum Hall effect. This was the first experimental demonstration of such a phenomenon detected in structures prepared by sophisticated technologies. For this, the Nobel Prize in Physics for 2010 was awarded to the persons who isolated graphene using fairly simple techniques involving Scotch tape and observed them by bare eye [4]. In the case of graphene, the challenge was to inherently study a 2D crystal, as well as the facility to produce it. Consequently, it launched a vast activity in the scientific community. This led to a large excitement with the hope for widespread applications of such a system. This led to the new era of other classes of 2D materials subsequently. Some of the allotropes of carbon are summarized in Figure 1.1. The lower panel of figure shows other 2D materials such as MoS_2, h-BN, and phosphine (Figure 1.2).

Since the discovery of graphene, several 2D materials have been isolated and several could be isolated in the upcoming years. The significant growth of such research field may be due to the simple techniques of mechanical exfoliation giving high-quality samples showing fascinating properties and wide applications. It does not require any sophisticated techniques and equipment. Just with a

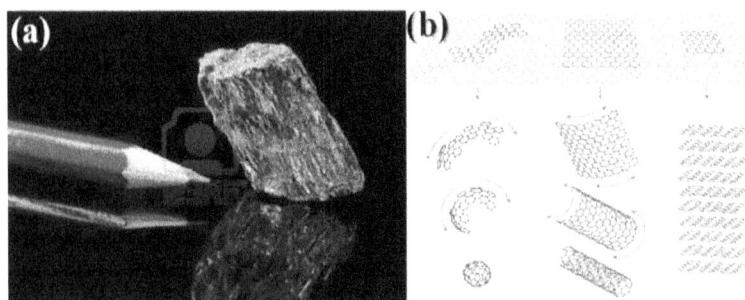

FIGURE 1.1   (a) Pure graphite is a mineral form of carbon (with courtesy of ESRF, France), for example pencil lead. It is a soft and conducting material. The building blocks are the graphene layers. (b) Top: a single layer of graphite, called graphene. The carbon atoms form a honeycomb lattice. Buckyballs (left), nanotubes (centre), and graphene layers (right) all share the same honeycomb structure.

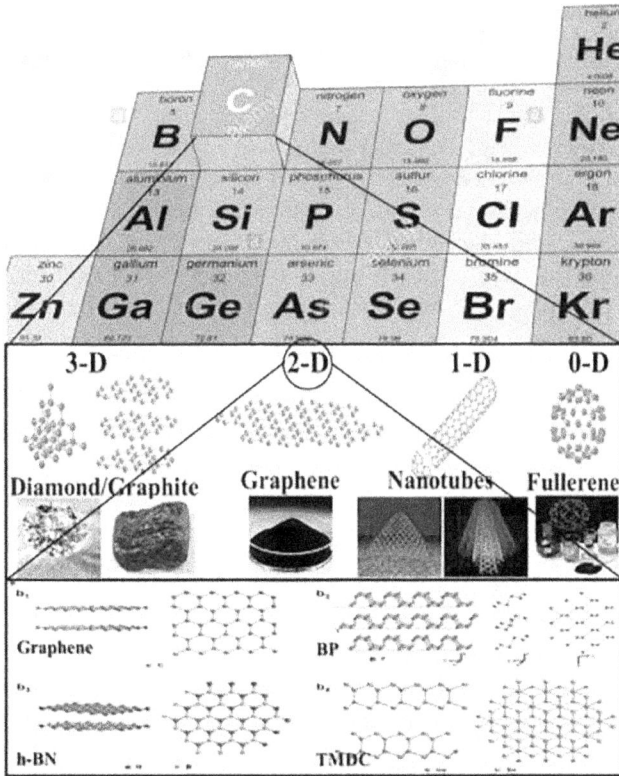

FIGURE 1.2 Top panel: position of carbon in the periodic table, middle panel: a summary of allotropes of carbon, and lower panel: structures of other 2D materials such as black phosphorus (BP), hexagonal boron nitrate (h-BN), and transition metal dichalcogenide (TMD).

piece of a bulk layered material, a roll of tape, and an optical microscope, any trained researcher can isolate atomically thin layers of many different 2D materials ranging from wide-band-gap insulators to superconductors [5]. Under the umbrella of 2D materials, graphene derivatives [6–8], mono-elemental membranes [9–12], polymer thin films [13–17], MXenes (where M = transition metal and X = C or N) [16–21], transition metal dichalcogenides (TMDs) [22–26], metal oxides, and hydroxides [27–30] have emerged.

After graphene, the well-studied 2D materials are layered transition metal dichalcogenide (TMD) compounds and MX_2, where M is a transition metal (e.g. Mo, W, Re, and Ta) and X is a chalcogen (e.g. S, Se, and Te), and TaS_2 and $NbSe_2$, group V TMDs also isolated. More than 30 stable TMDs have been isolated and studied for various applications such as catalysis and lubrication. They possess an intrinsic band gap within the visible region, which shows deviations with a change in the number of layers. Most importantly, TMDs have large spin–orbit interaction due to the heavy transition metals, leading to the splitting of the valence band that strongly affects their optical spectra. These materials with ultrathin thickness induce new properties in comparison with the corresponding bulk materials. 2D materials beyond graphene with different lattice structures and extended energy gaps further motivate the rapid development of these materials in electronic devices [31–33].

1.1.2 SPECIAL PROPERTIES OF NANOCARBONS: A BRIEF OVERVIEW

To a large extent, the structure of nanocarbons determines their electronic, molecular, and structural properties. The most important properties of such carbons and their molecular background are highlighted below.

1.1.2.1 Electrical Conductivity

Theoretically, for a single nanotube, chirality vector decides the conducting properties of the system, and with a small diameter, such tubes could be found either semiconducting or metallic in nature. In tubes, the difference in the conducting properties originates due to the molecular structure that leads to the different band structure and, thus, offers a distinction in the band gap. Fundamentally, the difference in the conductivity is based on the properties of the graphene sheet [34]. As per the group theory of molecules, it can be seen that a (n, m) nanotube is metallic with $n=m$ or $(n-m)=3i$, where i is an integer and n, m are the chiral indices defining the nanotube. The conduction is determined by the quantum mechanical aspects and was proved to be independent of the nanotube length [35].

On the contrary, C_{60} fullerene is an electrical insulator, in its nascent state, though comprised of the sp^2 state of hybridization like graphene. So, fullerene is considered to be a bad conductor of electricity although the molecular environment is similar to graphene. The mechanism that makes C_{60} a bad conductor is that it has a shorter range of continuity than graphene. In graphene, the carbon is made of sheets that can be extended as long as the sample dimension. These sheets have high conductivity as electrons can have ballistic transport to continue their conducting path; hence, the behaviour of graphene is metal-like. Now, extending this transport model to fullerene, within supra-molecular network the high conductivity is estimated to be high. However, to have large percolation conductance, electrons have to perform an intra-C_{60} jump that offers resistivity to fullerene as compared to graphene. For more information and other transport details, readers are referred to Ajayan and Ebbesen [36].

1.1.2.2 Optical Activity

For tubes, theoretical studies have revealed that for larger tubes, the optical activity of the chiral nanotubes disappears [37]. So, other physical properties have an influence on optical parameters also. The manipulation of optical activity might result in the setting up of optical device in which C_{60}, CNTs, and graphene may play a crucial role. Interestingly, C_{60} fullerene exhibits exceptional optical properties that have potential applications on wideband photo-optical devices. C_{60} suspension in non-polar organic solvents such as toluene, xylene, and trichloroethylene (TCE) displayed excitation bands at 625, 591, 570, 535, and 404 nm that correspond to $A_g \rightarrow T_{1u}$ and $A_g \rightarrow T_{1g}$ optical transitions. Notably, these bands differ from solid C_{60} recorded using diffuse reflectance UV–Vis spectroscopy. In suspension state, two emission band energies for toluene and xylene were reported: ~1.78 and 1.69 eV (i.e. nearly the same), whereas for TCE, a shift was reported to 1.72 and 1.65 eV. This is mainly due to the polarity of TCE that is higher than that of toluene and xylene. By and large, the report underlined the fact that the optical activity in C_{60} exists due to polarity difference between organic solvents and fullerene.

For graphene, the optical properties vary by the number of layers; that is, a single layer only absorbs 2.3% of light with 97.7% of transmission through a single layer and ~0.1% reflected from its initial trajectory. However, increasing graphene stack enhances light absorption by lowering optical transparency; however, the relationship is linear. For each layer absorbing 2.3% of light, a sample comprised of five layers would have an absorption of 11.5% and an optical transparency of ~88%–88.5%, including small degree of light reflection. Graphene is a wideband optical absorber due to the response of Dirac's fermions to photons causing variations in visible to near-infrared to far-infrared region. This is mainly due to the absence of discrete energy band levels like most materials. Graphene shows gate-dependent optical transition due to the existence of Dirac point in band, near Fermi level. The low density of states near Fermi level to shift under the applied electrical field which is used in electronics to modulate the current. The shift in Fermi level changes the conductivity, as well as transmission tuning for an optical source. However, this is valid for a single-layer graphene, as more than one layer can significantly alter the absorption properties and the ability to tune the absorbance.

Graphene can also exhibit some form of photoluminescence. Due to its zero band gap structure, it can't form the relaxed states, which is, in general, the photoluminescence mechanism. In such a

TABLE 1.1

Mechanical Parameters of Stand-Alone Nanocarbons [35]

S. No.	Material	Mechanical Properties			
		Young's Modulus (TPa)	Tensile Strength (GPa)	Elongation at break (%)	Ref.
1	C_{60}[a]	0.92–0.105	0.380	25	
2	Single-walled CNT[b]	≈1 (from 1 to 5)	13–53	16	
	Armchair SWCNT[a]	0.94	126.2	23.1	
	Zigzag SWCNT[a]	0.94	94.5	15.6–17.5	
	Chiral SWCNT[b]	0.92	-	-	
	MWCNT[b]	0.2–0.8–0.95	11–63–150	-	
3	Stainless steel[b]	0.186–0.214	0.38–1.55	15–50	
4	Kevlar[b]	0.06–0.18	3.6–3.8	≈2	
5	Graphene[b]	2.4-2.0	130	39.8	

[a] Theory.
[b] Experiment.

phenomenon, before releasing a photon an electron is excited to a higher-energy state and returns to its ground state. However, a virgin single-layer graphene emits light upon its excitation by a near-infrared laser, which is due to the transient temperature rise by the femtosecond laser photons which hit the graphene layer, as they are known to emit in the visible light spectrum. For doped graphene, photoluminescence is observed due to the creation of the band gap via functionalization or through cutting graphene layers into smaller pieces to generate quantum confinement compounded with local disorders. These zones are responsible for opening the band gap, which enables conventional photoluminescence mechanisms to take place [37–41].

1.1.2.3 Mechanical Strength

In general, tubes have axial and radial Young's modulus, which is very high, i.e. ~250–950 GPa, in the axial direction. Nanotubes, thus as a whole, are very flexible due to their extended structure. Such carbon compounds are potentially suitable for their applications in reinforced composite materials that need anisotropic properties. C_{60} being a point particle has only theoretical predictions on mechanical parameters, whereas graphene has layer-dependent characteristics. Table 1.1 shows the compilation of mechanical parameters for stand-alone C_{60}, CNTs, and graphene [35].

1.1.2.4 Chemical Reactivity

All carbons are neutral in their perfectly ordered (i.e. unreflective) state, and hence, chemical reactivity is zero for such nanocarbon systems. However, they can be made functional by introducing chemically active groups. The functionalization strategy depends upon the end application, such as composites for mechanically reinforced hybrid matrix, field emission, solar/photovoltaic, shielding, and shock mitigation. Further, the chemical reactivity of a CNT when compared to a graphene sheet is high as a direct consequence of the curvature effect in CNT. For nanotubes, the reactivity is associated with the π-orbital mismatch caused by an increased curvature. This makes a distinction between the sidewall and the end caps of a nanotube. For the same reason, a smaller nanotube diameter results in an increased reactivity. Covalent chemical modification of either sidewalls or end caps is possible to enhance chemical reactivity. For example, the solubility of CNTs in different solvents can be controlled this way. However, the direct investigation of chemical modifications on nanotube behaviour is difficult as the crude nanotube samples are still not pure enough.

1.1.2.5   Functionalization

A number of co-workers has used −COOH functionalization technique [42,43]. In this, CNTs obtained as a raw material were, initially, subjected to acid treatment using a mixture of concentrated sulphuric and nitric acids (95% H_2SO_4:65% HNO_3 = 3:1 volume ratio) to induce the formation of various functional groups such as −COOH, −OH, v–CO, and C–O on the side walls of the nanotubes [44]. The acid treatment was carried out under mild ultrasonic conditions for a period of ~3–7 hours. The acid treatment time and ultrasonic power conditions were optimized to minimize the damage caused by the acid treatment and/or sonication. After the acid treatment, the solution was filtered by an evacuation technique using a polytetrafluoroethylene (PTFE) membrane; the pore size of the membrane was ~0.2 μm. The nanotubes collected on the membrane were extensively washed four to five times in deionized (DI) water to neutralize the walls of the nanotubes. In the final stage, the nanotubes were washed with acetone to remove the water and dried at a temperature of ~90°C under atmospheric conditions for ~1 hour. This process was carried out to avoid the agglomeration of the nanotubes.

In another study, −SO functionalization was reported as follows: Purified CNTs were mixed with 1 wt% sodium dodecyl sulphate (SDS) solution as 1 mg/mL concentration and subjected to 30 minutes of sonication and 1 hour of centrifugation at 10,000 RPM. The well-dispersed suspension was gathered and diluted in deionized water as 100 times. Various amounts of the diluted CNT solution were filtered through the cellulose nitride membrane that has 200 nm pore diameter [45].

Peroxide functionalization was reported in a study carried out by Lee et al. In this, multiwalled CNTs (MWCNTs, supplied by ILJIN Nanotech), synthesized by the thermal chemical vapour deposition (thermal CVD) technique, were used. The raw powder contains MWCNTs of diameter 25 nm, amorphous carbon, and carbon-encapsulated metal nanoparticles. MWCNTs were oxidized in a hydrogen peroxide (H_2O_2) solution under ultrasonication condition for 24 hours at the temperature of 50°C to produce finely dispersed MWCNTs terminated with carboxylic acid groups. The resulting solution was filtered by a PTFE membrane with a pore size of 1 μm. At this step, the carbonaceous impurities were removed from the as-grown MWCNTs. Raman spectrometer and Fourier transform infrared spectrometer (FTIR) were used to identify the formation of carboxylic acid groups on MWCNTs [46].

1.1.2.6   Doping, Decoration, and Ad-Atom

In a study carried out by Alegaonkar et al., nitrogenized graphene and their derivatives were investigated for their spin transport and magnetic correlation. The doping of nitrogen (N) was carried out using the tetrakis(dimethylamino)ethylene (TDAE) compound. Initially, the suspension of GNCs/graphene was prepared in 25 mL of tetrahydrofuran and 0.1–0.5 mg of TDAE was added in GNCs/graphene suspension. After adding TDAE, the suspension was sonicated for 30 minutes followed by room temperature stirring for 8–10 hours. The suspension was allowed to settle for about 5 hours. The vacuum filtration was carried out using PTFE filter (pore size ~1.2 μm). Among five samples from 0.1 to 0.5 mg, the highest concentration (0.5 mg) was used for further studies. The N-doped graphene and GNCs were designated, respectively, as N-graphene and N-GNCs.

In another study, the work has been carried out on the improvement of catalytic activity of Au nanoparticles incorporated in carbon nanostructures for glucose sensing. For this, gold chloride ($HAuCl_4.3H_2O$), trisodium citrate ($Na_3C_6H_5O_7$), sodium dodecyl sulphate (SDS, $CH_3(CH_2)_{10}CH_2OSO_3.Na$), potassium dihydrogen orthophosphate (KH_2PO_4), hydrogen peroxide (H_2O_2, 30%), sodium hydroxide (NaOH), aniline, ammonium peroxydisulphate, hydrochloric acid (HCl, 35.4%), and ethanol (C_2H_5OH) were readily available sources, while GOD and glucose ($C_6H_{12}O_6$, 99.5%) were made available from Sigma-Aldrich. All chemicals were used as received without further purification. Deionized water (DI water, ~18 MΩ) was used in the experiment. MWCNTs of 5–7 nm in diameter and 2–5 graphene walls were synthesized by the water-assisted CVD method. For the synthesis of AuNPs:MWCNT:PANI composite, an aqueous solution of $HAuCl_4.3H_2O$ (100 mL of 1 mM) with (10 + 10) mg of MWCNTs and sodium dodecyl sulphate (SDS) was boiled under the reflux condition for half an hour. Afterwards, 5 mL of 38.8 mM $Na_3C_6H_5O_7$ was added rapidly into the above solution for reduction of gold ions. Twenty-five milligrams of the synthesized PANI was also added to the above solution just after the addition of $Na_3C_6H_5O_7$. Further reaction was continued for an additional 20 minutes and then allowed to cool down to room temperature. This solution was used for the characterization and electrode preparation. Finally, the solution was filtered and washed with DI water. The powder was dried at 45°C in a vacuum oven for 30 minutes, and this powder was also used for characterization [47].

In a study carried out to investigate the electrochemical properties of tellurium-reduced graphene oxide hybrid system, GO was prepared using natural graphite as a precursor by modified Hummers' route. In situ (1–11 w/w%) incorporation of Te metal was carried by on set reduction using hydrazine hydrate [48,49]. It is interesting to note that the subsequent enhancement in Te wt% (such as 3, 5, 7, 9, and 11) has no added effect on the structure–property relationship of the composite system analysed, extensively, by FTIR and Raman spectroscopy that mainly focused on the low weight fraction effects, typically 1 wt% Te-rGO system [50,51].

However, these are representative examples. The literature is vast, and presenting a survey on all would be lengthy and out of scope from the current discussion.

1.2 SCOPE OF NANOCARBONS

In the subsequent journey, we would be looking for the scope of nanocarbons by preparing them and assessing for their quality and implementing for typical applications. The word *nanocarbon* is repeatedly used throughout the text and has a *notional* sense. It has a generalized meaning in the context of C_{60} (fullerene)-like spherical nanoparticles (CNS), carbon nanotubes (CNTs: think/ thin, multi-/single-walled, capped with/without catalyst, ultralong, medium, short length tube), graphene (single-/few-/multilayered), graphene-like carbon compounds (GNCs), reduced graphene oxide (rGO), graphene-like nano-flakes (GNF), graphene-like nano-ribbons (GLNR), graphene nano-ribbons (GNR), porous nanocarbons (PNCs), spongy graphitic phases/matrixes, graphite, and 3D carbon fabric including nano-diamond and carbon quantum dots, as and when referred. There are a number of recipes charted for the readers. Most of them are single step, one pot, facile, involved trivial starting material, cost-, time-effective, and could be realized at large-scale. They are assimilation of literature review, self-discovered, optimized, and with intuitive experience implemented. There are couple of remedial exceptions too! The presented synthesis approach has a blend of laboratory as well as industry scale. Large variety of characterization probes such as UV– Vis, FTIR, Raman, microwave, relaxation, thermo-magnetic, thermo-mechanical, thermo-elastic, ion transmission spectroscopies, ellipsometry, XPS/ESCA, D-SIMS, positron annihilation, RBS, NDP, SIMS, PIXE, VBDOS, scanning electron, transmission electron, atomic force, tunnelling microscopies including fractographic analysis are presented and discussed in depth. The nature of growth causing effect on the structure and, consequently, on the properties has extensively been discussed. Such preparation and assessment strategies enabled to target applications of nanocarbons in strategic as well as in public domains. They are used in many ways such as armour blocks, radar absorber/stealth coating, rocket propulsion nozzles, optical gas sensors, energy storage, field emission devices, spintronic, ion track engineering, metamaterial, and radiation-optimized opto- electronic poly-carbon techniques [52,53]. The details are presented.

Towards the end of the book, the summary and outlook is presented, including Appendices A–C. Appendices are supplemental or additional information to Chapters 3–8.

REFERENCES

[1] Kroto, Harold W., James R. Heath, Sean C. O'Brien, Robert F. Curl, and Richard E. Smalley. "C_{60}: Buckminsterfullerene." *Nature* 318, no. 6042 (1985): 162–163.

[2] Iijima, Sumio. "Helical microtubes of graphitic carbon." *Nature* 354, no. 56 (1991): 56–58.

[3] Sprinkle, M., J. Hicks, A. Tejeda, A. Taleb-Ibrahimi, P. Le Fèvre, F. Bertran, H. Tinkey et al. "Structure and electronic properties of epitaxial graphene grown on SiC." arXiv preprint arXiv:1001.3869 (2010).

[4] Novoselov, Kostya S., Andre K. Geim, Sergei Vladimirovich Morozov, Dingde Jiang, Michail I. Katsnelson, Irina V. Grigorieva, S. V. Dubonos, and Alexander A. Firsov. "Two-dimensional gas of massless Dirac fermions in graphene." *Nature* 438, no. 7065 (2005): 197–200.

[5] Castellanos-Gomez, Andres. "Why all the fuss about 2D semiconductors?" *Nature Photonics* 10, no. 4 (2016): 202–204.

[6] Gómez-Navarro, Cristina, Jannik C. Meyer, Ravi S. Sundaram, Andrey Chuvilin, Simon Kurasch, Marko Burghard, Klaus Kern, and Ute Kaiser. "Atomic structure of reduced graphene oxide." *Nano Letters* 10, no. 4 (2010): 1144–1148.

[7] Voiry, Damien, Jieun Yang, Jacob Kupferberg, Raymond Fullon, Calvin Lee, Hu Young Jeong, Hyeon Suk Shin, and Manish Chhowalla. "High-quality graphene via microwave reduction of solution-exfoliated graphene oxide" *Science* 353, no. 6306 (2016): 1413–1416.

[8] Lin, Dingchang, Yayuan Liu, Zheng Liang, Hyun-Wook Lee, Jie Sun, Haotian Wang, Kai Yan, Jin Xie, and Yi Cui. "Layered reduced graphene oxide with nanoscale interlayer gaps as a stable host for lithium metal anodes." *Nature Nanotechnology* 11, no. 7 (2016): 626–632.

[9] Pumera, Martin, and Zdeněk Sofer. "2D monoelemental arsenene, antimonene, and bismuthene: beyond black phosphorus." *Advanced Materials* 29, no. 21 (2017): 1605299.

[10] Liu, Han, Yuchen Du, Yexin Deng, and D. Ye Peide. "Semiconducting black phosphorus: synthesis, transport properties and electronic applications." *Chemical Society Reviews* 44, no. 9 (2015): 2732–2743.

[11] Zhang, Zhuhua, Sharmila N. Shirodkar, Yang Yang, and Boris I. Yakobson. "Gate-voltage control of borophene structure formation." *Angewandte Chemie* 129, no. 48 (2017): 15623–15628.

[12] Sharma, S., S. Kumar, and Udo Schwingenschlögl. "Arsenene and antimonene: two-dimensional materials with high thermoelectric Figs of merit." *Physical Review Applied* 8, no. 4 (2017): 044013.

[13] Xu, Jie, Sihong Wang, Ging-Ji Nathan Wang, Chenxin Zhu, Shaochuan Luo, Lihua Jin, Xiaodan Gu et al. "Highly stretchable polymer semiconductor films through the nanoconfinement effect." *Science* 355, no. 6320 (2017): 59–64.

[14] Shi, Hui, Congcong Liu, Qinglin Jiang, and Jingkun Xu. "Effective approaches to improve the electrical conductivity of PEDOT: PSS: a review." *Advanced Electronic Materials* 1, no. 4 (2015): 1500017.

[15] Cai, Guofa, Peter Darmawan, Mengqi Cui, Jiangxin Wang, Jingwei Chen, Shlomo Magdassi, and Pooi See Lee. "Highly stable transparent conductive silver grid/PEDOT: PSS electrodes for integrated bifunctional flexible electrochromic supercapacitors." *Advanced Energy Materials* 6, no. 4 (2016): 1501882.

[16] Eklund, Per, Johanna Rosen, and Per O. Å. Persson. "Layered ternary M n+ 1AX n phases and their 2D derivative MXene: an overview from a thin-film perspective." *Journal of Physics D: Applied Physics* 50, no. 11 (2017): 113001.

[17] Chaudhari, Nitin K., Hanuel Jin, Byeongyoon Kim, Du San Baek, Sang Hoon Joo, and Kwangyeol Lee. "MXene: an emerging two-dimensional material for future energy conversion and storage applications." *Journal of Materials Chemistry A* 5, no. 47 (2017): 24564–24579.

[18] Yoon, Yeoheung, Keunsik Lee, and Hyoyoung Lee. "Low-dimensional carbon and MXene-based electrochemical capacitor electrodes." *Nanotechnology* 27, no. 17 (2016): 172001.

[19] Xu, Chuan, Libin Wang, Zhibo Liu, Long Chen, Jingkun Guo, Ning Kang, Xiu-Liang Ma, Hui-Ming Cheng, and Wencai Ren. "Large-area high-quality 2D ultrathin Mo2C superconducting crystals." *Nature Materials* 14, no. 11 (2015): 1135–1141.

[20] Cassabois, Guillaume, Pierre Valvin, and Bernard Gil. "Hexagonal boron nitride is an indirect bandgap semiconductor." *Nature Photonics* 10, no. 4 (2016): 262–266.

[21] Weng, Qunhong, Xuebin Wang, Xi Wang, Yoshio Bando, and Dmitri Golberg. "Functionalized hexagonal boron nitride nanomaterials: emerging properties and applications." *Chemical Society Reviews* 45, no. 14 (2016): 3989–4012.

[22] Voiry, Damien, Aditya Mohite, and Manish Chhowalla. "Phase engineering of transition metal dichalcogenides." *Chemical Society Reviews* 44, no. 9 (2015): 2702–2712.

[23] Zhang, Y. J., M. Yoshida, R. Suzuki, and Y. Iwasa. "2D crystals of transition metal dichalcogenide and their iontronic functionalities." *2D Materials* 2, no. 4 (2015): 044004.

[24] Kannan, Padmanathan Karthick, Dattatray J. Late, Hywel Morgan, and Chandra Sekhar Rout. "Recent developments in 2D layered inorganic nanomaterials for sensing." *Nanoscale* 7, no. 32 (2015): 13293–13312.

[25] Onga, Masaru, Yijin Zhang, Toshiya Ideue, and Yoshihiro Iwasa. "Exciton Hall effect in monolayer MoS_2." *Nature Materials* 16, no. 12 (2017): 1193–1197.

[26] Zhang, Zhepeng, Jingjing Niu, Pengfei Yang, Yue Gong, Qingqing Ji, Jianping Shi, Qiyi Fang et al. "Van der Waals epitaxial growth of 2D metallic vanadium diselenide single crystals and their extra-high electrical conductivity." *Advanced Materials* 29, no. 37 (2017): 1702359.

[27] Kalantar-zadeh, Kourosh, Jian Zhen Ou, Torben Daeneke, Arnan Mitchell, Takayoshi Sasaki, and Michael S. Fuhrer. "Two dimensional and layered transition metal oxides." *Applied Materials Today* 5 (2016): 73–89.

[28] Mei, Jun, Ting Liao, Liangzhi Kou, and Ziqi Sun. "Two-dimensional metal oxide nanomaterials for next-generation rechargeable batteries." *Advanced Materials* 29, no. 48 (2017): 1700176.

[29] Zhao, Mingming, Qunxing Zhao, Bing Li, Huaiguo Xue, Huan Pang, and Changyun Chen. "Recent progress in layered double hydroxide based materials for electrochemical capacitors: design, synthesis and performance." *Nanoscale* 9, no. 40 (2017): 15206–15225.

[30] Yin, Huajie, and Zhiyong Tang. "Ultrathin two-dimensional layered metal hydroxides: an emerging platform for advanced catalysis, energy conversion and storage." *Chemical Society Reviews* 45, no. 18 (2016): 4873–4891.

[31] Li, Xinming, Li Tao, Zefeng Chen, Hui Fang, Xuesong Li, Xinran Wang, Jian-Bin Xu, and Hongwei Zhu. "Graphene and related two-dimensional materials: structure-property relationships for electronics and optoelectronics." *Applied Physics Reviews* 4, no. 2 (2017): 021306.

[32] Fiori, Gianluca, Francesco Bonaccorso, Giuseppe Iannaccone, Tomás Palacios, Daniel Neumaier, Alan Seabaugh, Sanjay K. Banerjee, and Luigi Colombo. "Electronics based on two-dimensional materials." *Nature Nanotechnology* 9, no. 10 (2014): 768–779.

[33] Saito, Yu, Tsutomu Nojima, and Yoshihiro Iwasa. "Gate-induced superconductivity in two-dimensional atomic crystals." *Superconductor Science and Technology* 29, no. 9 (2016): 093001.

[34] Avouris, Phaedon. "Carbon nanotube electronics." *Chemical Physics* 281, no. 2–3 (2002): 429–445.

[35] Tans, Sander J., Michel H. Devoret, Hongjie Dai, Andreas Thess, Richard E. Smalley, L. J. Geerligs, and Cees Dekker. "Individual single-wall carbon nanotubes as quantum wires." *Nature* 386, no. 6624 (1997): 474–477.

[36] Ajayan, Pulickel M., and Otto Z. Zhou. "Applications of carbon nanotubes." In *Carbon nanotubes*, edited by Mildred Dresselhaus and Gene Dresselhaus, pp. 391–425. Springer Berlin Heidelberg, 2001.

[37] Damnjanović, Milan, Ivanka Milošević, Tatjana Vuković, and R. Sredanović. "Full symmetry, optical activity, and potentials of single-wall and multiwall nanotubes." *Physical Review B* 60, no. 4 (1999): 2728.

[38] CheapTubes: https://www.cheaptubes.com/graphene-synthesis-properties-and-applications/

[39] Zhu, Yanwu, Shanthi Murali, Weiwei Cai, Xuesong Li, Ji Won Suk, Jeffrey R. Potts, and Rodney S. Ruoff. "Graphene and graphene oxide: synthesis, properties, and applications." *Advanced Materials* 22 (2010): 3906–3924.

[40] Kavitha, M. K., and Manu Jaiswal. "Graphene: a review of optical properties and photonic applications." *Asian Journal of Physics* 7, no. 25 (2016): 809–831.

[41] Falkovsky, Leonid A. "Optical properties of graphene." *Journal of Physics: Conference Series* 129, no. 1 (2008) 012004.

[42] Ebbesen, Thomas W., and Pulickel M. Ajayan. "Large-scale synthesis of carbon nanotubes." *Nature* 358, no. 6383 (1992): 220–222.

[43] Moon, J. S., P. S. Alegaonkar, J. H. Han, T. Y. Lee, J. B. Yoo, and J. M. Kim. "Enhanced field emission properties of thin-multiwalled carbon nanotubes: Role of Si O_x coating." *Journal of Applied Physics* 100, no. 10 (2006): 104303.

[44] Shin, Jun-Ho, H. C. Lee, J. H. Lee, S. M. Park, P. S. Alegaonkar, and J. B. Yoo. "Carbon nanotube based transparent electrodes for flexible displays using liquid crystal devices." In 한국정보디스플레이학회: 학술대회논문집, pp. 897–899, 2007.

[45] Jeong, Jin Su, Jin-San Moon, Sung Yun Jeon, Jae Hong Park, Prashant Sudhir Alegaonkar, and Ji Beom Yoo. "Mechanical properties of electrospun PVA/MWNTs composite nanofibers." *Thin Solid Films* 515, no. 12 (2007): 5136–5141.

[46] Lee, Tae Young, Prashant S. Alegaonkar, and Ji-Beom Yoo. "Fabrication of dye sensitized solar cell using TiO_2 coated carbon nanotubes." *Thin Solid Films* 515, no. 12 (2007): 5131–5135.

[47] Park, J. H., P. S. Alegaonkar, S. Y. Jeon, and J. B. Yoo. "Carbon nanotube composite: dispersion routes and field emission parameters." *Composites Science and Technology* 68, no. 3–4 (2008): 753–759.

[48] Gautam, Seema, Deepak Kumar, Prashant S. Alegaonkar, Pika Jha, Neeraj Jain, and Jaswant S. Rawat. "Enhanced response and improved selectivity for toxic gases with functionalized CNT thin film resistors." *Integrated Ferroelectrics* 186, no. 1 (2018): 65–70.

[49] Gangwar, Rajesh K., Vinayak A. Dhumale, Arvind Kumar, Prashant Alegaonkar, Rishi B. Sharma, and Suwarna S. Datar. "Gold-graphene nanocomposite based ultrasensitive electrochemical glucose sensor." In *2012 1st International Symposium on Physics and Technology of Sensors (ISPTS-1)*, pp. 282–285. IEEE, 2012.

[50] Alegaonkar, Ashwini P., Arvind Kumar, Prashant S. Alegaonkar, Shobha A. Waghmode, and Satish K. Pardeshi. "Exchange interaction of itinerant electron donors of tetrakis (dimethylamino) ethylene with localized electrons in graphene." *Synthesis and Reactivity in Inorganic, Metal-Organic, and Nano-Metal Chemistry* 44, no. 10 (2014): 1477–1482.

[51] Gangwar, Rajesh K., Vinayak A. Dhumale, Kalyani S. Date, Prashant Alegaonkar, Rishi B. Sharma, and Suwarna Datar. "Decoration of gold nanoparticles on thin multiwall carbon nanotubes and their use as a glucose sensor." *Materials Research Express* 3, no. 3 (2016): 035008.

[52] Alegaonkar, Ashwini, Prashant Alegaonkar, and Satish Pardeshi. "Exploring molecular and spin interactions of Tellurium adatom in reduced graphene oxide." *Materials Chemistry and Physics* 195 (2017): 82–87.

[53] Alegaonkar, Ashwini P., Satish K. Pardeshi, and Prashant S. Alegaonkar. "Spin dynamics in graphene-like nanocarbon, graphene and their nitrogen adatom derivatives." *Applied Physics A* 124, no. 7 (2018): 1–10.

2 Preparation and Evaluation of Nanocarbons

2.1 0D CARBON SYSTEMS: CARBON NANOPARTICLES (CNP)/NANO-SPHERES (CNS)

Several research groups have studied the growth of C_{60}, CNTs, carbon onions, carbon nanohorns, DLC, and other formats of carbon [1,2]. However, the growth of CNTs has captured the attention, due to their intrinsic characteristics such as high thermal and chemical stability, high mechanical strength, superior electron emission properties, and easy methods of synthesis. A considerable amount of work has been carried out related to the templated growth of CNTs [3–5]. Shi et al. grew bead-like nanorods in the channels of mesoporous SiO_2 [6] by electro-crystallization, and Grebel et al. synthesized SWCNTs within an ordered array of nanosize silica spheres [7,8] and prevented the sintering of the catalyst nanoparticles. Huang et al. prepared laterally assembled CNTs from the parallel trenches of porous silica templates [9], whereas Zhang et al. demonstrated the diameter-controlled growth of CNTs [10] by changing the Fe loading level in the mesoporous silica substrates. In addition, a few attempts have been made to demonstrate the importance of fixed aspect ratio CNTs. The field emission characteristics of CNTs with higher local aspect ratio [11] were studied by Kuo et al., and Boccaccini et al. reported a dramatic variation in the sintering ability [12] of borosilicate glass matrix composites by using fixed aspect ratio CNTs. Iwassaki et al. grew multiwalled carbon nanotubes (MWCNTs) in anodic alumina channels [13], and Zhao et al. fabricated continuous mesoporous silica films for the synthesis of SWCNTs within an ordered array [14]. However, the studies on the subsurface region of the porous SiO_2, after the growth of the CNTs, have rarely been carried out so far. This is the topic of this section.

Here, we present the attempts to grow carbon nanoparticles (CNPs) in the subsurface region of porous SiO_2 templates. Morphological studies were carried out after separating the grown CNPs from the porous templates. In general, the CNPs were observed in the soot of the carbon; however, in the present study, we observed the growth of the CNPs in the porous subsurface region, simultaneously with that of the over-layer MWCNTs. Interestingly, the growth of the CNPs was found to be reproducible. This section centres on the CNPs in the porous SiO_2 template, particularly on their growth mechanism and characterization [15].

2.1.1 TEMPLATED GROWTH USING DC-PECVD

2.1.1.1 Template Preparation

The Co-catalyst-loaded porous SiO_2 templates were fabricated [15] and subjected to a DC-plasma enhanced chemical vapour deposition (DC-PECVD) process to obtain the CNPs. The plasma reactor was heated to 650°C in an argon (flow rate 70 sccm) environment and evacuated to a base pressure of 10^{-3} Torr. A mixture of C_2H_2 and NH_3 gases (flow rate 120/60 sccm) was then introduced into the reactor. The plasma energy was tuned to ~77 W, with a chamber pressure of 3.25 Torr. The plasma treatment was carried out for a period of 30 minutes, for all the templates.

2.1.1.2 Separation of the Over-Layer MWCNTs from the Porous Templates

The MWCNTs grown in templates were subjected to the sonication process to remove the over-layer MWCNTs grown in the surface region of the template. Initially, a trial experiment was carried out by immersing one of the templates in ethanol. This template was sonicated for a period of 5 minutes.

DOI: 10.1201/9781003317258-2

After the sonication was over, the template sample was removed from the ethanol solution and allowed to dry. The surface morphology of the sample was examined by the SEM technique. The process of sonication and morphological examination of the sonicated sample was repeatedly carried out, until the complete removal of the over-layer MWCNTs from the template surface was confirmed. SEM analysis showed that the over-layer MWCNTs had been completely removed from the template surface after 20 minutes of sonication. In this way, the other templates were subjected to the sonication process for an integrated time period of 20 minutes. Following this, the surface morphology of each sonicated template was examined by the SEM technique [15].

2.1.1.3 Morphological Studies on the Porous Subsurface Region and CNPs

Figure 2.1a shows a typical TEM micrograph recorded for the Co-catalyst-doped porous SiO_2 template. The inset shows the electron diffraction pattern of the template surface. The in-depth analysis of the porous SiO_2 templates is reported elsewhere [15]. Figure 2.1a shows that a large number of the pores in the templates were embedded with the Co-catalyst (a few of them are marked by circles) and many of the pores were empty (marked by arrows). The average size of the Co-catalyst particles embedded in the pores was 20 ± 5 nm. The embedded Co-catalyst in each of the nanopores contributes to the catalytic activity during the PECVD process used to grow the MWCNTs. We observed that the MWCNTs were grown by the tip growth mechanism.

Figure 2.1b shows a typical cross-sectional SEM micrograph recorded for the MWCNTs grown in the porous SiO_2 template. It can be seen that the thickness of the template is 150 ± 5 nm. The MWCNTs were grown in the surface region, normal to the surface of the template. In addition, a large number of voids were observed, mainly at the subsurface region of the template (indicated by arrows). It seems that these voids in the subsurface region play an important role in the growth of CNPs. Figure 2.1c shows a typical TEM micrograph recorded for the MWCNTs grown from the porous SiO_2 shells. It shows that, after the plasma treatment, the spherical shape of the porous shells changed and they became ellipsoidal.

However, it seems that the electric field has no direct effect on the deformation of the pores. One can see from Figure 2.1c that the pores protruded in the direction perpendicular to the template surface, which could be attributed to the axial stress exerted, by the growing MWCNTs, on the pores. The MWCNTs grown by the plasma enhanced CVD process are more vertically aligned because the entire height of the growing MWCNTs is submerged inside the plasma sheath where a large electric field exists in the direction normal to the template. The MWCNTs grew like individual, free-standing tower structures with a catalyst particle at the top. As a result, the electrostatic force, F, creates a uniform tensile stress across the entire particle/CNT interface, regardless of where the particle is located (tip of base). The electrostatic force, F, could produce an axial stress at the bottom

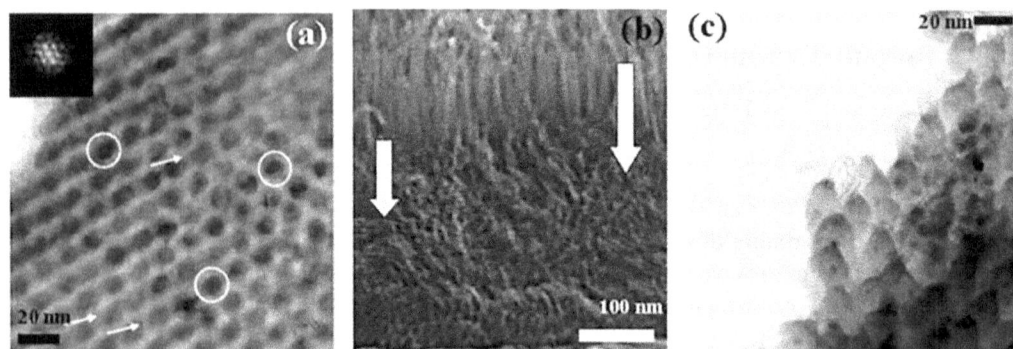

FIGURE 2.1 Typical micrographs for (a) a porous SiO_2 template (arrows show voids, and circles show encapsulated catalyst particles), recorded by TEM, (b) a subsurface region indicated by arrows (cross-sectional SEM view), and (c) the growth of CNTs from the pores of SiO_2, recorded by TEM.

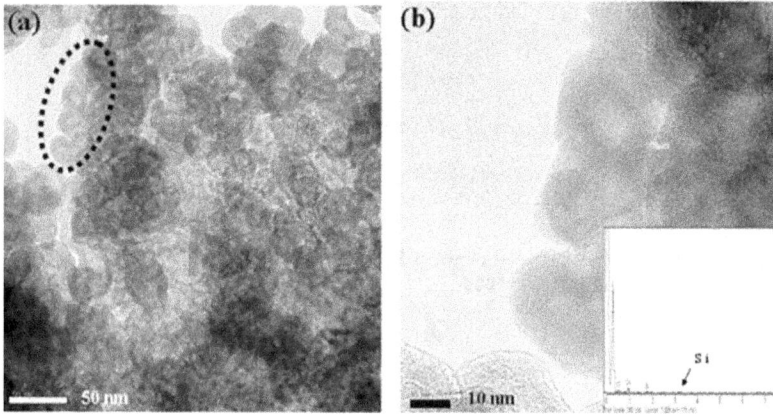

FIGURE 2.2 Typical TEM micrographs recorded for the CNPs: (a) low-magnification image and (b) details of the shaded portion in micrograph (a). The inset shows the recorded EDX profile.

of the MWCNT near the pore, and as a result, the pores may change their circular shape. Due to the multilayered and interconnected structural morphology, the top layer deformation could act as a cascade process, which in turn changes the shape of the pores in the subsurface region.

Figure 2.2a shows a typical low-magnification TEM image of the CNPs obtained from the porous SiO_2 powder. Figure 2.2b shows a high-magnification micrograph of the shaded portion in Figure 2.2a. It seems that these CNPs are grown in the voids that exist in the subsurface region of the template. Moreover, the nanoparticles grew simultaneously with the over-layer MWCNTs. The TEM micrographs show that the porous shells are completely separated from the CNPs. This could be attributed to the process adopted for obtaining the CNPs, as mentioned in Section 2.2. It seems that, during the sonication process, the porous shells might have ruptured, settled at the bottom of the solution, and thus become separated from the CNPs. As a result, we could not observe the porous shells in the recorded TEM micrographs. However, a typical EDX pattern shows a small peak associated with Si, which could be attributed to the physisorption of Si atoms within the CNPs. However, in most of the cases, the concentration of Si in the spaghetti of the CNPs was found to be negligibly small.

Figure 2.3 shows the micro-Raman spectra recorded for (a) the CNPs and (b) the over-layer MWCNTs. The intensity of both profiles was normalized with respect to the high-energy G-mode, the laser power, and the integration time. One can see that the Raman spectrum for the CNPs is distinctly different from that of the Raman spectrum recorded for the over-layer MWCNTs. For the CNPs, a broad peak was observed at ~1552 cm^{-1}, which could be attributed to the high-energy G-mode. The peak associated with the D-mode seems to be submerged with the G-mode and appeared as a shoulder at ~1385 cm^{-1}. For the MWCNTs, the peak at ~1350 cm^{-1} is attributed to the D-band [16,17], whereas the G-mode is observed at ~1580 cm^{-1} [18]. Furthermore, the difference between the G-mode and the D-mode (i.e. the inter-band difference) is observed to be 167 cm^{-1} for the CNPs and 230 cm^{-1} for the MWCNTs. The observed decrease in the inter-band difference, for the CNPs, indicates that the high-energy G-mode is shifted down, whereas the D-mode is shifted at the high wave number with respect to the G- and D-modes recorded for the over-layer MWCNTs. The estimated full width at half maximum (FWHM) of the G-mode is ~182 and 114 cm^{-1}, whereas the FWHM of the D-mode is ~250 and 208 cm^{-1} for the CNPs and over-layer MWCNTs, respectively. This indicates that the peak width of the G- and D-modes is relatively high for the CNPs as compared to the peak width of the G- and D-modes for the over-layer MWCNTs. In general, the defects in graphitic carbon are pentagonal and heptagonal carbon rings, which subsequently affect the force constants near and around the defects [15].

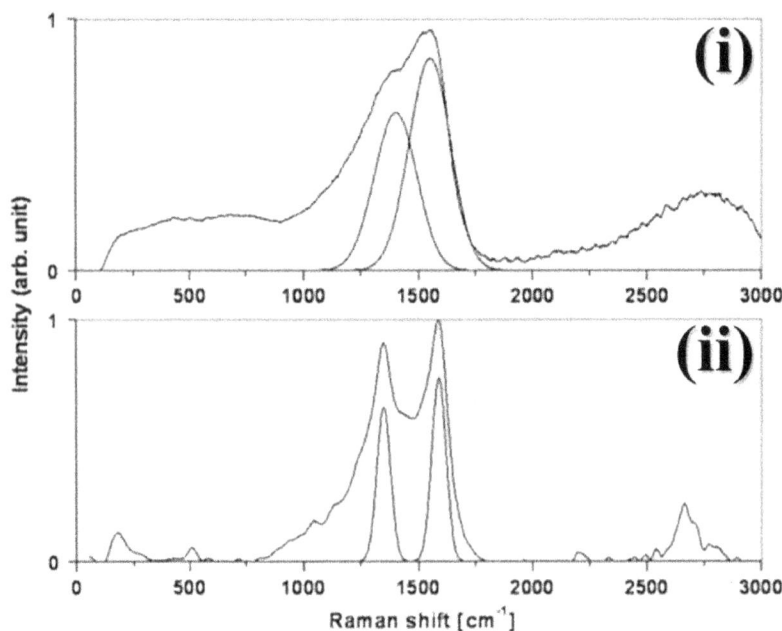

FIGURE 2.3   Recorded Raman spectra for (a) the CNPs and (b) the over-layer MWCNTs.

A small structural disorder in the graphitic planes can change the intensity, position, and width of the Raman peak [15]. Thus, the observed variations in the G- and D-modes of the CNPs and over-layer MWCNTs indicate that the CNPs have more defects than the CNTs. Furthermore, this issue could be investigated by modelling how each of the isolated tubes is transformed into a nanoparticle structure in the cavity of the porous SiO_2 templates. Their direct transformation is not possible, so the structure must go through an intermediary stage. It seems that, during the plasma treatment, carbon atoms segregate into porous shells (cavities). Initially, the hollow cavities of SiO_2 are transformed into ellipsoidal or facet shapes under an applied electric field. At this stage, the segregation of the carbon atoms takes place in the cavities. Next, the graphitization of some of the surface layers starts near the wall of the cavities and the shape of the particles becomes identical to that of the cavities, i.e. ellipsoidal or facet shaped. Gradually, the concentric shell structure becomes more and more perfect. This later process seems to proceed from the outside to inside and could be attributed to internal epitaxial growth [11–14]. The elliptical growth of the CNPs signifies that the growth occurred due to pore deformation, whereas the onset of the edges and pentagonal facets marks the appearance of some defects that disturbed the further growth of the CNPs, thus forcing them to change from elliptical to pentagonal or – in one stage – from slightly pentagonal to more pronounced pentagonal. The points of transition from one shape to another can be understood by a simple estimation. Inside the cavity, when the carbon atoms come in contact with the growing nanoparticles, they diffuse around the periphery of the particles until they occupy the lowest possible hybridized energy state (of the two possibilities sp^2 or sp^3, sp^3 hybridization is the more favourable state around the periphery of the CNPs). Usually, this will be a regular place, so that the growth of the nanoparticles can continue undisturbed [15].

2.1.2   SUBLIMATION SYNTHESIS

In Section 6.2, we demonstrate the application of nanocarbons, obtained by a facile synthesis route, as an efficient electrode medium with effective electrochemical parameters. The carbon nanospheres (CNS) are obtained from the single-step synthesis by complete atmospheric sublimation of

Formosa. It doesn't need any additional pre- and/or post-treatment. The method is cost- and time-effective. The parameters achieved using CNS are superior as compared to those reported in the literature; specifically, the specific capacitance, C_{SP}, is found to be superior. The analysis of charge storage mechanism is presented. To the best of our knowledge, there has been no report on the direct implementation of campho-carbon as a high-performance EDLC. In this section, its fabrication details are presented [19].

2.1.2.1 Single-Step Preparation of CNS

Formosa was taken as a starting material to grow CNS. The deposition was carried out on the surface of a commercially available alumina substrate (KETAO, advanced ceramic solution), under normal thermodynamic conditions. The production scheme of CNS deposition is shown in Figure 2.4. Initially, a square substrate of dimension $5 \times 5 \times 1\,cm^3$ was cut and subjected to sonication process by immersing into acetone and, subsequently, into deionized water for a period of 10 minutes at room temperature. Following this, the drying process was carried out for about 20 minutes with the help of IR heating. For deposition, a spatula of stainless steel was mounted on a fixed platform and 1 g pellet of Formosa was kept on the circular side of the spatula. The platform was a vertically moveable clamp coupled with a stand so that it can move freely in the vertical direction [19,20].

By adjusting the vertical distance between the clamp and the spatula, the deposition of CNS on the substrate was facilitated. The substrate was clamped and brought into vicinity of the precursor pellet subjected to the flash point sublimation process at 54°C. The complete process occurred in a period of 2–3 minutes or so. The powder was deposited onto the substrate and, subsequently, collected by gentle scrubbing of the surface of substrate using razor blade and collected into the crucibles/bottles as seen in Figure 2.4. In this fashion, several pellets were subjected to combustion to obtain nanocarbon. It was found that 1 g of precursor combustion yields around 70 mg nanocarbon.

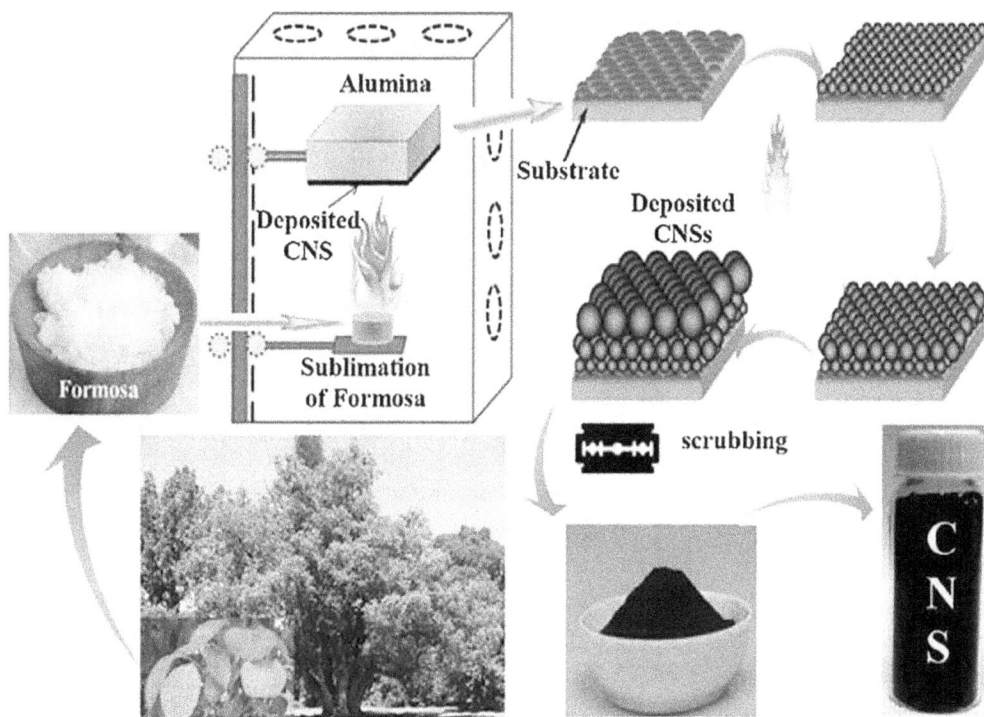

FIGURE 2.4 Ecological production route involved flash point sublimation of Formosa at 54°C, CNS deposition on substrate, scrubbing, and collection.

It is noteworthy that the adopted process is facile with no involvement of catalyst in reaction. In one step, the product is readily available and requires no pre- or post-treatment prior to application [20].

2.1.2.2 Analysis of CNS

Figure 2.5a shows a typical Raman spectrum recorded for CNS. The spectrum consisted of two peaks, G at 1597.9 cm^{-1} and D at 1374.9 cm^{-1}. The G-band corresponds to the E_{2g} phonon branch of ordered sp^2 carbon atom. The D-band is related to sp^3 + non-sp^2 fraction and attributed to the breathing mode of the k-point branch with A_{1g} symmetry [17]. In addition, the FWHM and integrated area peak ratio (I_D/I_G) indicates crystalline length, L_a, amount of amorphous carbon, and provides information about the degree of disorder. To evaluate this, curve fitting was carried out in terms of spectroscopic parameters such as peak position, peak width, line shape, and band intensity (i.e. Gaussian, Lorentzian, or a mixture of both) [19–21]. As a broad feature, the intensity of D-peak is comparable to G. This is indication of a large amount of disorder in the obtained CNS [22]. The FWHM of D-peak is quite high around 323 cm^{-1}. The contribution comes from non-sp^2 fraction in terms of topological disorder in carbon shells, Stone-Waller fraction, and sp^3 carbon phase. The FWHM of D is indicative of the disorder, dominantly. The presence G shows sp^2 fraction, rich in π-electron environment. The I_D/I_G ratio is estimated to be 1.27, resulting in L_a ~ 1–3 nm [19]. This indicates that the order of crystallinity is quite low in CNS. It seems that, the sp^2 fraction is distributed within the disordered zone. In such a heterostructured environment, itinerant electrons can have favourable participation in the charge accumulation process. This is further evident by UV–Vis spectrum obtained for CNS between 200 and 800 nm, as shown in Figure 2.5b. The sharp peak that

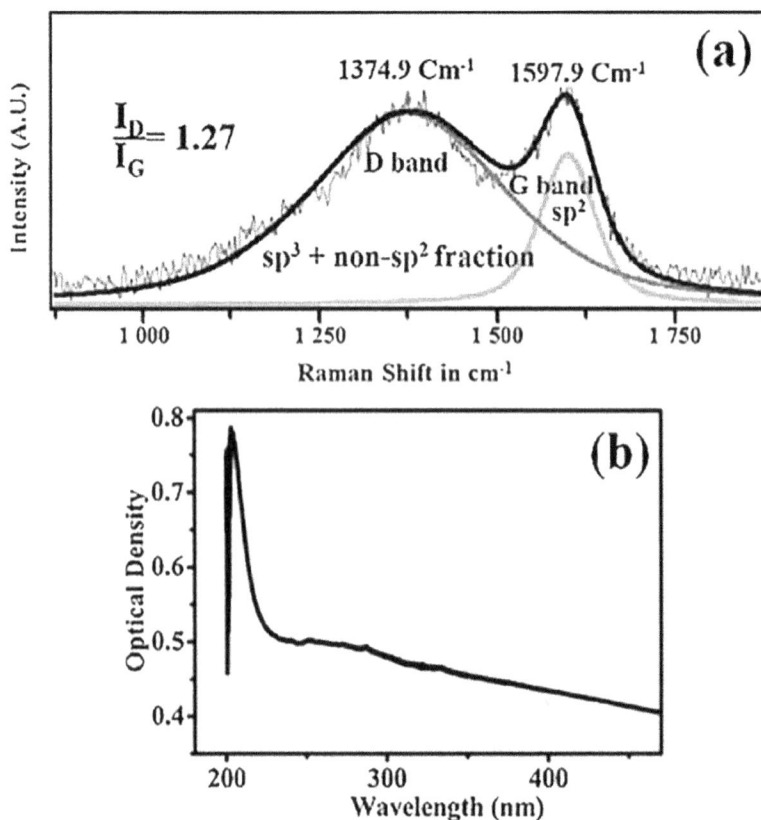

FIGURE 2.5 (a) Recorded Raman spectrum at 457 nm indicating D (1374.9 cm−1) and G (1597.9 cm−1); (b) UV–Vis spectrum showing strong π–π* electron transition.

emerged at ~210 nm is indicative of the nearly monodispersed nature of CNS. The nature of the peak indicates donor-loaded sites in CNS [21] (Figure 2.6).

The FESEM images recorded for two different magnifications are shown in Figure 2.7a and b. The morphology of the CNS shows a coagulated, interconnected 3D network of nano-spheres. The HRTEM recorded for CNS showed a more detailed structure of individual spheres, representative images shown in Figure 2.7c and d. The continuous and concentric carbon shells are rarely seen in the recorded images. The observed brightness on CNS indicates electrostatically charged surface showing electron-rich carbonaceous moieties. A large number of sites were examined to estimate the average size of nearly homogeneous nano-spheres, which was found to be ~40–50 nm.

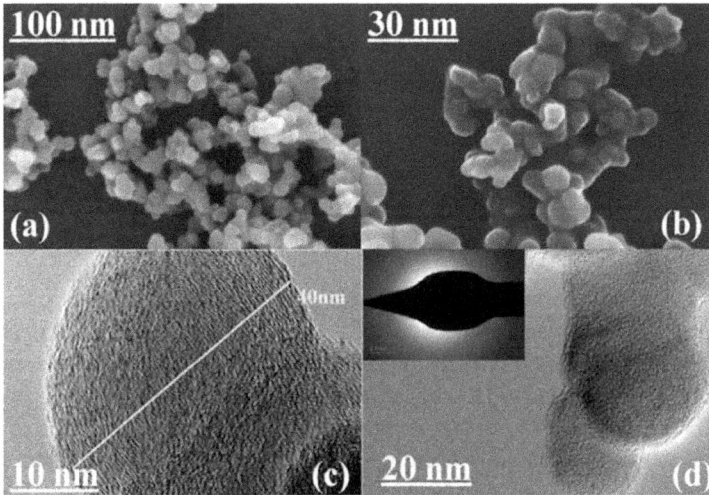

FIGURE 2.6   (a) and (b) FESEM images indicating interconnected 3D spherical carbon network; (c) and (d) details of spherical nanocarbon recorded by HRTEM (inset showing SAED pattern) for CNS.

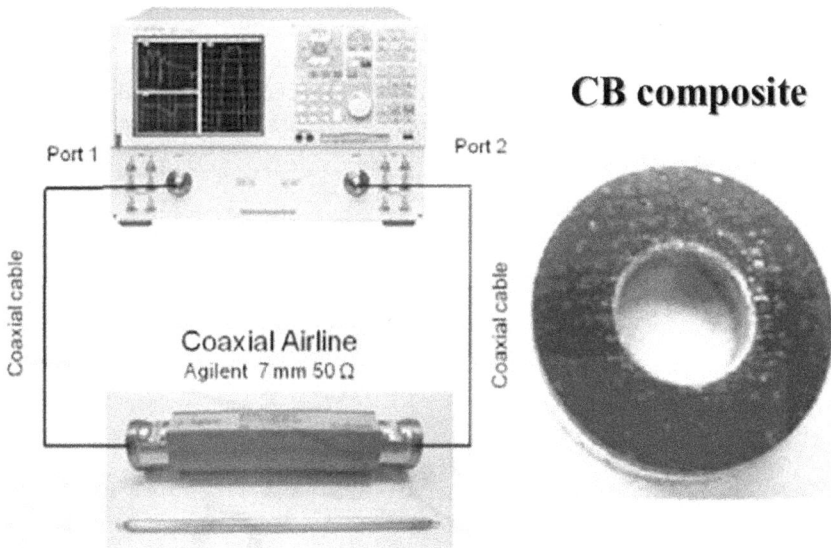

FIGURE 2.7   System set-up and prepared specimen carbon black (CB) composite for EMI measurements.

During combustion, Formosa carbon-spheres grow in a chain format with sp^2+sp^3 heterostructure nature of the spheres of nearly uniform diameter. Inset (d) shows a blur SAED pattern recorded for CNS. It seems that the presence of oxygen in precursor might be responsible for onset oxidation of the excess amorphous carbon to obtain CNS with large amount of non-crystalline zone. Such a network may provide appreciably high S_A in terms of active electrode area over the macroscopic area, thereby offering superior ultra-capacitor characteristics. The formation of amorphous zones and discontinuous shells may allow rapid access of electrolyte into the bulk phase of the electrode, expecting enhancement in specific capacitance C_{SP}. For electrochemical application, the porosity of carbon material is one of the important parameter and characterized by N_2 sorption technique at 77K. CNS at standard temperature and pressure (STP) are exposed to N_2 for about 1 hour; the average surface area is estimated to be ~791 m^2/g, which is quite higher than the previously reported values [5,10]. Further, the hysteresis curve (not shown) has been observed in sorption isotherms at a relatively low pressure (P/P0) of ~0.30, which indicates a widely distributed, heterogeneous mesophase of the material with high amount of porosity. The carbon network with a high amount of surface electrons and appreciably high S_A may provide excellent electrode parameters, thereby utilizing macroscopic area over active surface area. It is noteworthy that such a nanocarbon is obtained without any additional activation or treatment [16,21].

2.1.3 CARBON BLACK COMPOSITE

In another study for electromagnetic interference shielding (EMI), nanocomposite preparation was carried out by using carbon black powder (CBP) (Senka Carbon, India) thoroughly mixed using acetone medium in two-pack polyurethane matrix consisting of polyol-8 (Ciba-Geigy, Switzerland) and hexamethylene diisocyanate (E-Merck, Germany) mixed in a 50–50 ratio. Samples with carbon black powder (CBP) as a filler with varying weight contents (100, 150, 200, 300, and 400 mg) mixed in 1 mL polyurethane (PU) matrix have been prepared. The mixture was homogenized and then put in the mould, followed by curing under heat and pressure in a hydraulic press. The samples were prepared in toroidal shape with an outer diameter of 7.0 mm and an inner diameter of 3.0 mm to fit in a coaxial waveguide sample holder [17].

The surface morphologies have been studied with the help of SEM images. The SEM images of carbon black nanoparticles and polyurethane (PU) are shown in Figure 2.8a and b, respectively. The SEM micrograph in Figure 2.8a shows that the carbon black particles are agglomerated and form the porous structure. Figure 2.8b shows the rubberized nature of virgin polyurethane (PU) matrix. Thermogravimetric analysis (TGA) has also been carried out to study the thermal stability of the prepared nanocomposite. Figure 2.8c shows the TGA plot of the prepared nanocomposite that exhibits weight loss in several steps. But the prepared CBP/PU nanocomposite is found to have a thermal stability at least up to 300°C [17,18,22].

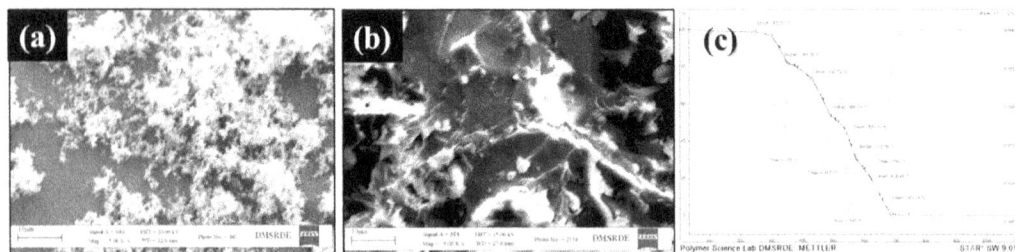

FIGURE 2.8   Recorded SEM micrographs for (a) carbon black (CB), (b) polyurethane (PU), and (c) TGA of 400 mg CB/PU nanocomposites.

2.2 CARBON NANOTUBES: 1D FORMAT OF CARBON

After the discovery of CNTs, carbon materials are broadly known as one of the most easily modified matter for various applications. However, the pace of carbon nanotechnology watershed has been halted by the efforts to synthesize CNTs at industrial standards, such as to get them surface sensitive, well ordered, vertically aligned, mono-sized, and with desired yield. A considerable amount of work related to the synthesis of CNTs has been carried out by various techniques such as direct current plasma enhanced chemical vapour deposition (DC-PECVD), catalytic CVD, rapid thermal CVD, templated CVD, as well as by patterning, selective area deposition, paste, reinforced, and metal-coated methodologies as given in [23] and references cited therein. We would be discussing a few typical methodologies in subsequent sections. However, the synthesis of CNTs by DC-PECVD is warranted [23]. In general, during the plasma synthesis of CNTs, hydrocarbon gas adsorbed onto the catalytic particle surface releases carbon upon decomposition, which dissolves and diffuses into the catalyst particle. When a supersaturated state is reached, carbon precipitates in a crystalline tubular form, with the emergence of two different possibilities: tip growth or base growth of the filament.

A large attempt has been made to grow CNTs by various techniques. In a study, mesoporous silica has been used with pore diameter ~30 nm and inter-pore distance ~100 nm to grow vertically aligned CNTs [8]. CNTs were grown in the alumina [9] nano-holes (~50 nm) by embedding Co nanoparticles electrochemically. In another study, Suh and Lee observed a reasonable amount of vertically aligned CNTs, i.e. ~1.1×10^{10} tubes/cm^2, in anodized aluminium oxide (AAO) templates [10]. The clay carbon templates [11] were prepared by Moggridge et al. for diameter-controlled synthesis of CNTs, whereas Fonseca et al. used [12] a micro-mesh-supported transition metal catalyst for the mass production of CNTs. However, it was found that such growth processes meet various difficulties such as agglomeration of catalyst particles, defected growth, and disorientation of CNTs.

In this section, we have presented the use of aluminium phosphate ($AlPO_4$-5) zeolites to grow CNTs, under which the $AlPO_4$-5 crystallites were impregnated with Fe catalyst solution, and the doping concentration was varied to obtain a variable areal density, $(\sigma_T)_{av}$, of CNTs. The analysis revealed that at $(\sigma_T)_{av}$ ~$6.24 \pm 0.19 \times 10^{10}$ t/cm^2 [24,25].

2.2.1 DC-PECVD SYNTHESIS

To grow CNTs, initially, the powder of (aluminium phosphate) $AlPO_4$-5 crystallites was calcined at ~400°C for 12 hours, under atmospheric conditions. The Fe catalyst solution (concentration ~0.5 mol/mL) was prepared by mixing iron acetate (($C_2H_3O_2$)$_2$ Fe II) in deionized water. The solution was sonicated for a period of ~20 minutes, at a temperature of ~40°C, and poured in six glass bottles. The powder of $AlPO_4$-5 crystallites of ~0.001 wt.% was added in each bottle containing the solution, and the bottles were sealed. The impregnation time was varied from one bottle to another: the solution of the first bottle was impregnated for 10 minutes, the solution of the second bottle was impregnated for 20 minutes, and in this manner, the solution of each bottle was impregnated over the range of ~10–60 minutes. These bottles were opened and kept in a furnace for drying for ~12 hours, at a temperature of ~100°C, under nitrogen atmosphere [25].

In another experiment, thin layers of Ti (thickness ~100 Å) and Al (thickness ~1000 Å) were grown, by thermal evaporation, on a silicon wafer (1 cm × 1 cm) and used as a conducting cathode substrate for field emission measurements (results not presented here). Figure 2.9a shows a sputtering set-up. The dry powder of catalyst-loaded crystallites was obtained from the bottles. The powder was mixed with ethanol solution and sprayed uniformly on the Al/Ti/Si substrates (hereafter called substrate) and used, after drying, for further experimentation. In order to obtain uniform coverage of the crystallites, the spraying was carried out under identical conditions. Various parts of the substrates were examined by SEM to estimate the coverage of $AlPO_4$-5 crystallites per unit area of the substrate. For elemental confirmation and determination of the concentration of Fe, each substrate was subjected to HRTEM and ESCA techniques, respectively [25].

FIGURE 2.9 Photograph of (a) sputter, (b) DC-PECVD, and (c) evaporation systems at Nano-materials and Device Laboratory, Sungkyunkwan University, Korea.

To grow CNTs, these substrates were employed to DC-PECVD (direct current plasma enhanced chemical vapour deposition) process as shown in Figure 2.9b. The plasma reactor was heated in the argon (flow rate ~70 sccm) environment, and as the temperature reaches ~650°C, the reactor was evacuated to a base pressure of ~10^{-3} Torr. A mixture of C_2H_2/NH_3 gases was introduced in the reactor, and the plasma energy was tuned to ~77 W, with a pressure of ~ 3.25 Torr. The plasma treatment was carried out for a period of ~15 minutes, for all the substrates. Using SEM, the number of CNTs per unit area of the crystallites was estimated on various substrates. Details of the grown CNTs were studied by FETEM [24].

Figure 2.10a shows the texture and morphology of the $AlPO_4$-5 crystallites, and Figure 2.10b and c are the recorded HRTEM-EDX and HRTEM images, which confirm the presence of Fe catalyst particles encapsulated in the $AlPO_4$-5 crystallites. Figure 2.10d is the plot of variation in the concentration of the Fe catalyst, C_{Fe}, estimated by the ESCA technique, as a function of the impregnation time, I_T, of the crystallites. Figure 2.10e is a representative SEM image of CNT-grown crystallites that were impregnated for ~10 minutes, (f) ~20 minutes, and (g) ~60 minutes. Figure 2.10h is a low-magnification SEM micrograph showing the coverage of the crystallites on the substrate. The details of the CNT grown in a crystallite, recorded by the HRTEM technique, are shown in Figure 2.10i. Figure 2.10j is the plot of the average areal density, $(\sigma_T)_{av}$, of CNTs on the crystallites as a function of the impregnation time, I_T. The plot of field emission current density, ρ_d (measured in $\mu A/cm^2$), as a function of electric field, V/μm, for a virgin Si substrate and the CNTs grown on $AlPO_4$-5 crystallites with different areal densities $(\sigma_T)_{av}$ is not shown in Figure [24].

An alternative approach was adopted to achieve the optimum field emission characteristics from the CNTs, grown in the $AlPO_4$-5 crystallites. The crystallites were loaded with Fe catalyst, and the doping concentration of Fe, C_{Fe}, was varied from ~1.71% to 8.36% by varying the impregnation time, I_T, from 10 to 60 minutes. The catalyst-loaded crystallites were sprayed uniformly on the conducting Al/Ti/Si substrates, with a constant coverage of ~10^3–10^4 crystallites/cm². These crystallites were subjected to the DC-PECVD process to grow CNTs. The results showed that the average areal density of CNTs, $(\sigma_T)_{av}$, was found to increase from ~6.24±0.19×10^{10} to ~2.04±0.61×10^{11} t/cm², with an increase in the concentration of Fe, C_{Fe}. The formation of CNTs in the $AlPO_4$-5 crystallites was attributed to several components such as the presence of Brønsted sites, the quality of crystallites, the encapsulation of hydrocarbon molecules in crystallites, and the pyrolysis conditions. In the field emission study (not presented here), the optimum value of the field emission current density, ρ_d, of ~1.78×10^3 $\mu A/cm^2$ and the turn-on electric field of ~3.69 V/μm was obtained by controlling the areal density of CNTs, $(\sigma_T)_{av}$ [24,25].

2.2.1.1 Role of Barrier/Buffer Layer: Catalytic CVD

The layer sandwiched between catalyst and substrate is termed as barrier layers (buffer layers). The role of barrier layers in the CCVD synthesis of CNTs is an important class of issue. Various materials such as SiO_2, TiO_2, TiN, Al_2O_3, Cr, and MgO could be used as the barrier layer. The details are

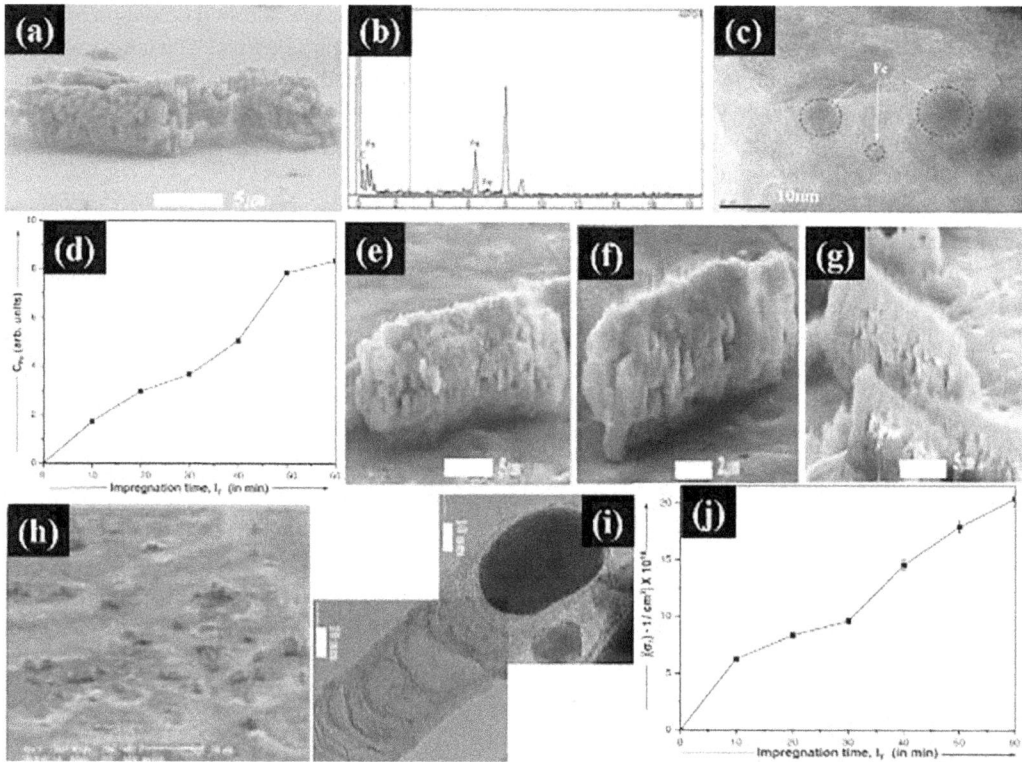

FIGURE 2.10 (a) Recorded SEM image for a virgin AlPO$_4$-5 crystallite, (b) recorded TEM-EDX image for the Fe catalyst-loaded crystallites, (c) confirmation of encapsulated Fe nano-clusters in the crystallites by HRTEM analysis, and (d) plot of variation in the concentration of Fe catalyst clusters, C_{Fe} (in arbitrary units), as a function of impregnation time, I_T, measured in minutes. Recorded SEM and TEM images for the CNTs grown in crystallites: (e) $I_T \sim 10$ minutes, (f) $I_T \sim 20$ minutes, (g) $I_T \sim 60$ minutes, (h) low-magnification image to show the coverage of the crystallites on the substrate, (i) HRTEM image of a CNT grown in the surface region of AlPO$_4$-5 crystallites, and (j) plot of variation in the estimated areal density of CNTs, $(\sigma_T)_{av}$, with impregnation time, I_T.

provided in [26–28] and references cited therein. These layers could prevent the diffusion of catalyst atoms into the substrates, improve the adhesion of CNTs to substrate, and influence the characteristics of the CNTs. A considerable amount of work related to barrier-layer-mediated growth of CNTs has been carried out that involved substrate-selective growth, influence of type of buffer layers by in situ XPS measurements, etc. So far, the systematic investigations on the diffusion of Fe atoms in multiple barrier layers have rarely been reported [26]. That is what we present in this section.

Here, we present our attempts to investigate the diffusion of Fe into multiple barrier layers. CNTs were grown on various types of barrier layers (Al, Al$_2$O$_3$, Al/SiO$_2$, Al$_2$O$_3$/SiO$_2$, and no barrier layer) coated with Fe catalyst, on Si (110) substrate. The CNT characteristics were determined by SEM and TEM techniques. The diffusion dynamics of the Fe atoms into the multiple barrier layers was investigated by the dynamic secondary ion mass spectroscopy (D-SIMS) technique, and the surface chemical analysis of the as-prepared and CNT-grown multi-barrier-layer samples was carried out by XPS [26].

2.2.1.2 Deposition of Multiple Barrier Layers

Initially, an n-type Si (100) wafer (resistivity ~132 kΩ and thickness ~450 μm) was cleaned up with the chemicals. The wafer was cut into several pieces and subjected to different treatments to obtain multiple barrier layers. SiO_2 layers were grown on the Si samples by wet-thermal oxidation process. For this purpose, a few samples were loaded into the high-temperature furnace. The temperature of the chamber was raised up to ~1300°C and maintained for a period of ~12 hours. During the thermal oxidation process, H_2O vapours were passed onto the samples. Following this process, these samples were removed from the thermal reactor and kept under dry atmosphere for the next treatment. The thickness of the SiO_2 was found to be ~550 nm.

To grow Al layers on to the samples, a few Si and SiO_2/Si samples were subjected to the e-beam evaporation technique as shown in Figure 2.9c. An aluminium pellet of purity ~99.999% was loaded into the evaporation chamber, and a high voltage of ~15 kV was applied. The chamber temperature was raised up to ~90°C; however, the substrate temperature was maintained at ~40°C. During the e-beam evaporation, the deposition rate was kept at ~0.1–0.2 Å per second. The evaporation process was carried out for a period of 5 hours to obtain ~15 nm thick Al layer. To obtain Al_2O_3 coating, a few Al/Si and Al/SiO_2/Si samples were subjected to thermal evaporation process. These samples were loaded into the annealing chamber, and the temperature was increased up to ~300°C. The thermal oxidation process was carried out for a period of 1 hour to obtain ~15 nm think Al_2O_3 layer. A few Si and multi-barrier-layer samples were subjected to e-beam evaporation process. In a similar fashion, the Fe catalyst layer of thickness ~1 nm was coated on each sample. The prepared samples were kept in a dry environment under high vacuum conditions ~10^{-5} Torr to avoid catalyst oxidation and contamination. The batch of the Fe/Si substrates are designated as sample S1, Fe/Al/Si as sample S_2, Fe/Al/SiO_2/Si as S_3, and Fe/Al_2O_3/Si and Fe/Al_2O_3/SiO_2/Si as S_4 and S_5, respectively [27].

2.2.2 THERMAL CVD GROWTH

A few S_1 and all S_{2-5} samples were subjected to the thermal chemical vapour deposition (T-CVD) treatment. Initially, samples were loaded in the thermal reactor and evacuated to a base pressure of ~20 mTorr. The reactor temperature was raised with a ramp rate of ~10°C/min, and the growth of CNTs was carried out at a temperature of ~650°C, for a period of 30 minutes. Moreover, C_2H_2/NH_3 feedstocks were used to grow the CNTs with flow rate 40/120 sccm. T-CVD process was repeated for 4–5 times to cover all category of samples. In order to simulate the diffusion conditions for the Fe atoms, a few S_{1-5} samples were subjected to the annealing process at a temperature of ~650°C, for a period of 30 minutes. These samples were characterized by D-SIMS. In another experiment, a few CNT-grown S_{1-5} samples were subjected to the sonication process to remove the over-layer CNTs grown on the surface and these samples were characterized by the XPS technique (results not discussed) [28].

2.2.2.1 D-SIMS Analysis

To investigate the diffusion of Fe atoms into the barrier layers, a few S_{1-5} samples, after annealing treatment, were subjected to the D-SIMS. One sample at a time was loaded in the mass spectroscopy chamber, and the chamber was evacuated to the base pressure of 1.4×10^{-9} mbar. The sample was exposed to the O_2^+ ions with energy 800 eV, and the beam current was varied from 9.71×10^3 pA to 1.07×10^4 pA. Spectra of the sputtering yield as a function of the sputtering time were collected for a period of ~10 minutes, for each sample. Figure 2.11a shows the variation in the sputtering yield of Fe atoms, for S_{1-5} samples, as a function of the sputtering time. It can be seen that for the Fe/Si, S_1, sample, the sputtering yield is almost constant, ~10^5 ions/ cm^2s, over the time measured for ~7 minutes. Similarly, the sputtering yield was found to be almost constant for the Fe/Al/Si and Fe/Al_2O_3/Si samples. However, for the Fe/Al/SiO_2/Si, S_3, sample, the sputtering yield of ~10^3 ions/ cm^2s decreases after a period of ~2.5 minutes. Furthermore, for the Fe/Al_2O_3/SiO_2/Si, S_5, sample, the sputtering yield of ~10^5 ions/cm^2s decreases markedly after ~0.5 minutes. This shows that the diffusion of Fe atoms is hindered by the Al_2O_3/SiO_2 barrier layer and almost all catalyst layers reside on the surface of the S_5 sample. However, it seems that the diffusion of Fe atoms is not prevented by

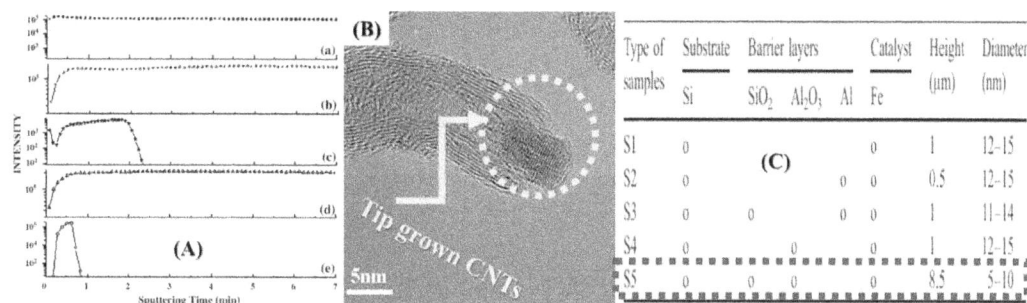

Type of samples	Substrate	Barrier layers			Catalyst	Height	Diameter
	Si	SiO$_2$	Al$_2$O$_3$	Al	Fe	(μm)	(nm)
S1	0				0	1	12-15
S2	0			0	0	0.5	12-15
S3	0	0		0	0	1	11-14
S4	0		0	0	0	1	12-15
S5	0	0	0	0	0	8.5	5-10

FIGURE 2.11 (a) Variation in the sputtering yield of Fe atoms, for S$_{1-5}$ samples, as a function of sputtering time, spectra recorded by the D-SIMS technique, (b) typical TEM micrograph of the tip-grown CNTs grown on the sample S5, (c) data table for the characteristics of the CNTs grown on the samples S$_{1-5}$, of which S$_5$ shows superior properties.

the barrier layers in the S$_{1-4}$ samples. The time of sputtering was calibrated with respect to the depth of diffusion for each sample. The analysis revealed that for the pristine Si sample, the sputtering rate is ~100 Å/min [26–28].

Figure 2.11b is a typical TEM image recorded for the tip-grown CNTs fabricated on the surface of the S$_5$ sample. It can be seen that the nanotube consists of 4–5 concentric graphene layers and the head of the CNT is decorated by the Fe cluster. It is noteworthy that the diameter of the CNTs was found to be relatively small for the S$_5$ samples as compared to the other category of samples. Figure 2.11c data table lists the height and diameter of CNTs grown on various substrates. Thus, we have observed a relatively greater height and small diameter for the Fe/Al$_2$O$_3$/ SiO$_2$/Si samples. The observed variation in the height of CNTs could be correlated to the quantity of Fe atoms available on the surface region of the barrier layer. The observed decrease in the height shows that a large number of Fe atoms diffused into the Al layer for Si/Al samples. In contrast, the diffusivity seems to be reduced for the Si/SiO$_2$/Al$_2$O$_3$ samples [28].

2.2.3 SUPER-GROWTH OF CNTS

The synthesis of super-grown CNTs is of technological importance because of their integration into fibres and sheets for lightweight and high-strength material applications. The height of CNTs synthesized by water-assisted chemical vapour deposition (WA-CVD) [28] is ~2.5 mm, whereas that of CNTs synthesized by standard CVD [26] varies in the range of 10–200 μm. In the case of WA-CVD, introducing a small and controlled amount of water enhances and retains the activity as well as the lifetime of the catalyst particles [28]. As a result, the overall efficiency of the synthesis can be increased, resulting in the formation of densely packed, defect-free, vertically aligned CNTs forests; see [28] and references cited therein. By controlling the level of water in the reaction ambience, brush-like CNT forests with a height of ~7 mm were grown in a period of 12 hours. Half-centimetre-high mats of vertically aligned single-walled carbon nanotubes were grown at 600°C by point-arc microwave plasma chemical vapour deposition. A considerable amount of work has been carried out related to the CVD synthesis of CNTs. However, in general, not much attention has been paid to the optimization of various WA-CVD parameters in order to achieve the super-growth of CNTs [29]. In this section, the optimization of various WA-CVD parameters, such as the flow rate of the reactant gas mixture, its injection temperature, and growth temperature, is discussed in an attempt to synthesize super-grown CNTs. We report that CNT forests with a height of up to 2.2±0.002 mm could be achieved in a synthesis time of about 16 minutes using the acetylene/argon ratio of 200/500 sccm, growth temperature of 810°C, and ramp time of 1 minute. The synthesized nanotubes were densely packed, were vertically aligned, consisted of 2–5 graphene walls, and were typically 5–7 nm in diameter (Figure 2.12).

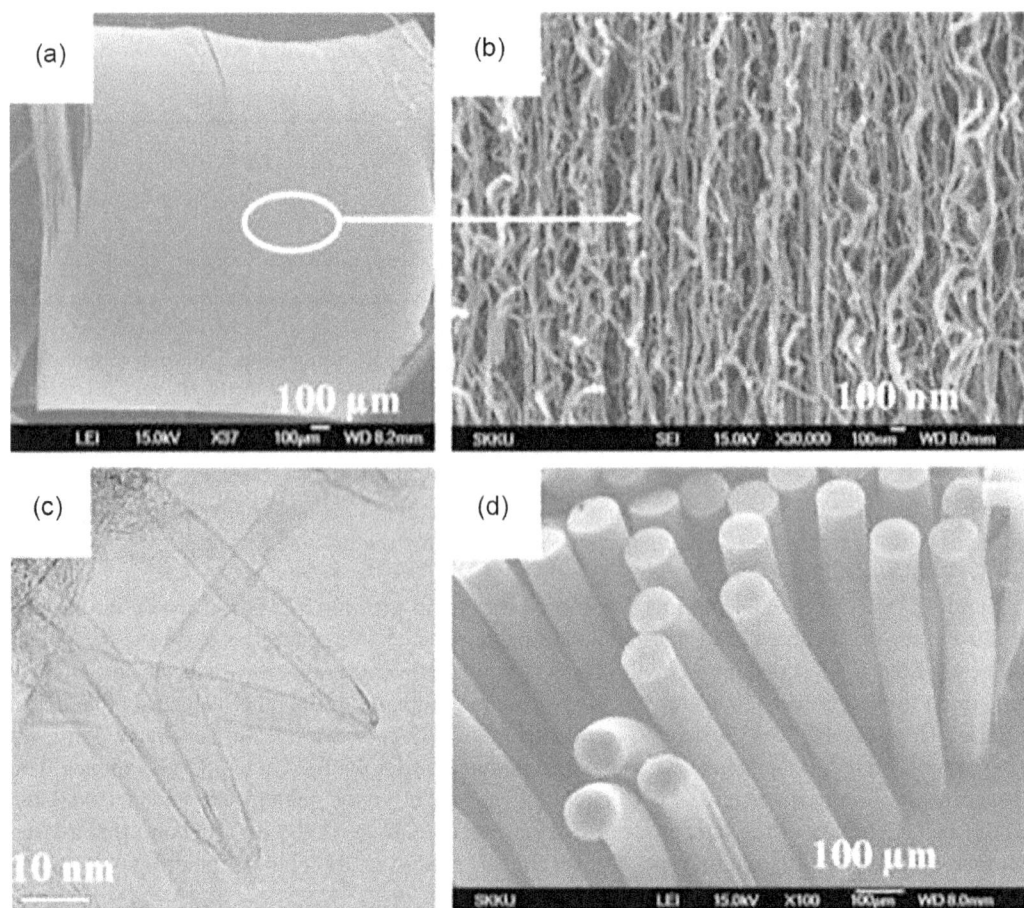

FIGURE 2.12 Recorded SEM micrographs for (a) a CNT film with a height of 2.2±0.002 mm obtained at optimized conditions, (b) morphology at the middle of the CNT film, (c) high-resolution TEM micrographs showing two graphene walls, and (d) patterned CNTs using a TEM grid as a mask [28,29].

2.2.3.1 Rapid Thermal CVD

In another study, the effect of the partial pressure of acetylene on the height of carbon nanotube films synthesized by the WA-CVD was studied as shown in Figure 2.13 [29]. Initially, Fe (2 nm)/ Al_2O_3 (15 nm) bi-layers were deposited onto a silicon substrate and subjected to the WA-CVD to grow carbon nanotubes. The growth was carried out at 800°C for a period of ~10 minutes using a mixture of acetylene/argon gases with a water bubbling system. The partial pressure of acetylene was varied from ~5% to 95% while keeping the total pressure of the gas mixture constant. The height of the carbon nanotube film was measured using the scanning electron microscopy. The analysis showed that the height of the film gradually increased from 800±15 nm to 1.4±0.002 mm as the partial pressure of the acetylene feedstock was sequentially increased from 5% to 40%. Thereafter, the height decreased gradually to 0.77±0.002 mm as the partial pressure of acetylene was further increased to 95%. These samples were subjected to X-ray photoelectron spectroscopy after removing the nanotube film from the surface. The analysis revealed that at a lower acetylene content, surface oxidation is predominant, whereas in the higher acetylene content regime carbonization influenced the height of the carbon nanotube film. It is briefly presented in Figure 2.14 [28,29].

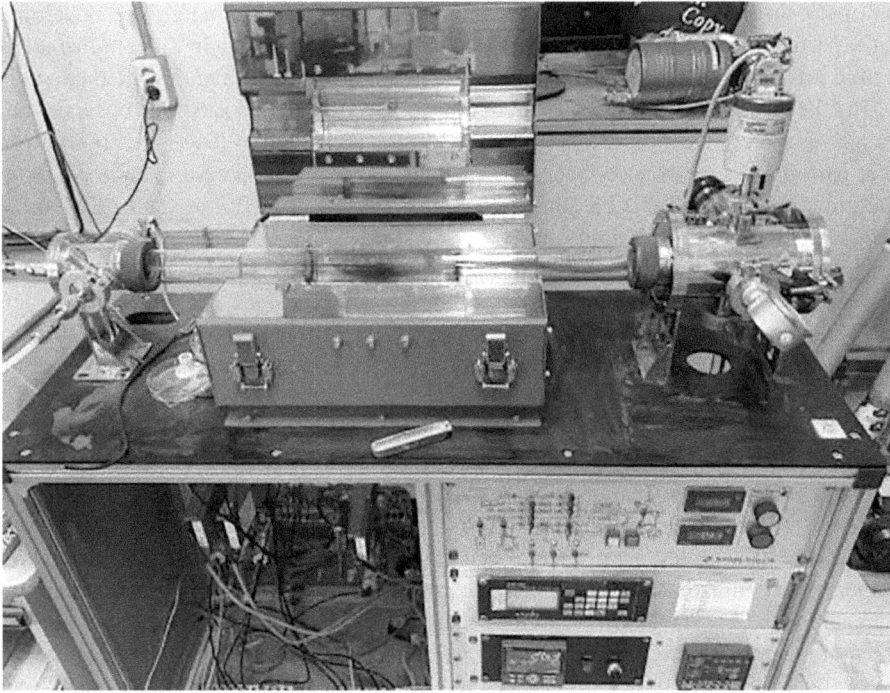

FIGURE 2.13 Rapid thermal CVD synthesis hardware for ultralong CNTs grown by catalytic effects.

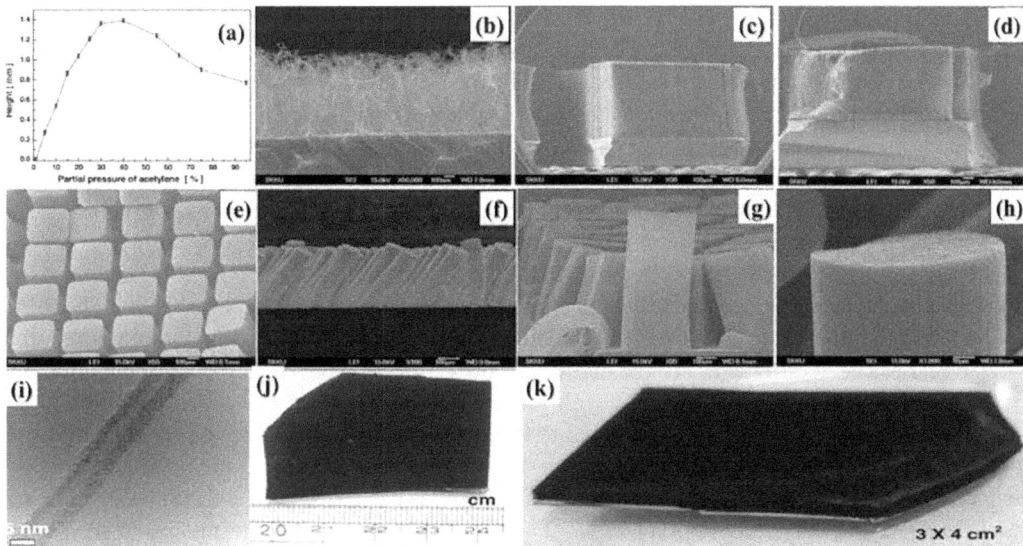

FIGURE 2.14 (a) Variation in the height of the CNT film with the partial pressure of acetylene (in %), (b)–(d) acetylene partial pressure: 5%, 40%, and 95%, respectively, (e)–(h) SEM images of the CNTs deposited onto the $Fe/Al_2O_3/SiO_2/Si$ substrate using a mask pattern, (i) typical high-resolution TEM image of the thin multiwalled carbon nanotubes grown by the WA-CVD technique, and (j), (k) top and side view digi-cam photographs of the CNT film grown on the substrate with dimensions of ~$3 \times 4\,cm^2$ [29].

In summary, the height of the CNT film was tailored from 800 ± 15 nm to 1.4 ± 0.002 mm by controlling the partial pressure of acetylene gas. The growth was carried out in water ambience, and the growth time was kept constant at ~10 minutes. The diameter of the thin multiwalled nanotubes was 5–7 nm, and they contained a small amount of amorphous carbon deposits. The observed deviation in the height of the CNT film towards the lower limit is attributed to the domination of oxidation at a lower partial pressure of acetylene, whereas at a higher partial pressure, carbonization dominates the growth process. Our XPS analysis supports this finding. The amount of minority oxide phase was larger for the virgin samples, whereas the surface concentration of carbon was larger for the samples synthesized at a partial pressure of acetylene of 95%. The growth properties, in turn, depended critically on the particular partial pressure of the feedstock. Interestingly, the tuning of the partial pressure seems to allow the synthesis of a CNT film with the desired height. Moreover, the height of the CNT film is observed to be uniform at the scalable limit [28,29].

2.2.3.2 Templated CVD
Herein, we describe a systematic approach to synthesize MWCNTs using the WA-CVD technique. Initially, the Fe/Al/Si substrate has been fabricated using the routine electron beam evaporation technique (Figure 2.9c). These samples were subjected to the standard as well as WA-CVD synthesis to grow MWCNTs [30]. The temperature has been varied from 650°C to 900°C, and the growth time is kept constant ~10 minutes. Following this, the samples have been subjected to the scanning electron microscopy (SEM). The analysis revealed that the introduction of controlled and small amount of water dramatically affected the growth of height. Furthermore, at two temperature regimes, 700°C and 800°C, the growth of carbon nanotubes has been studied at variable growth times. The growth time was varied from 10 to 60 minutes. The SEM analysis revealed that the overall height of nanotubes has been increased at the high-temperature regime [10–14]. The growth rate has been estimated. The WA-CVD facilitates the growth of densely packed, vertically aligned MWCNT forests. In our case, the optimum height of MWCNTs (>500 μm) has been achieved at 800°C in 30 minutes, using the WA–CVD process. To obtain MWCNT patterns, initially, two p-type silicon wafers have been taken and one of them was used to construct a porous mask. The pattern growth has been achieved using such masks. The brief results are presented here (Figure 2.15).

In summary, the water-assisted synthesis of long, densely packed, and patterned MWCNTs has been carried out [28–30]. Initially, a few Fe/Al/Si samples were subjected to the standard and water-assisted chemical vapour deposition (WA-CVD) techniques to grow the MWCNTs. The synthesis of nanotubes has been carried out under standard as well as WA-CVD conditions over the temperature range 650°C–900°C. The SEM results revealed that the height of nanotubes increases with increasing temperature and, at a temperature of 800°C, the maximum height of ~533 μm has been achieved by the MWCNT film. Furthermore, the growth time has been varied from 10 to 60 minutes and the growth of nanotubes has been carried out at two temperature regimes 700°C and 800°C. The SEM analysis showed that the overall height of nanotubes is increased. At this temperature regime, the growth rate has been estimated. For the regime of 700°C, the growth rate decreases monotonically from 12 to 2.69 μm/min, whereas the overall growth rate is observed to be increased at the high-temperature regime of 800°C. The growth rate for this regime also decreases monotonically from 40.5 to 8.017 μm/min; however, the decrease in the growth rate is drastic as compared to the lower-temperature regime. The patterned growth of MWCNTs has been achieved on the Si wafer by using the shadow mask catalyst deposition technique. The result offers insight into the use of such long, densely packed, patterned nanotubes as vias in layer-by-layer electronic interconnects [30].

2.2.3.3 CNT Patterning
There have been several examples of CNTs being used to realize devices and composites [31,32]. However, several hurdles remain to be overcome before CNT-based technology can be deployed on a commercial scale. Among these hurdles are locating and patterning technologies. So far, films of CNT networks have been patterned in the micrometre regime by a variety of techniques, including using a

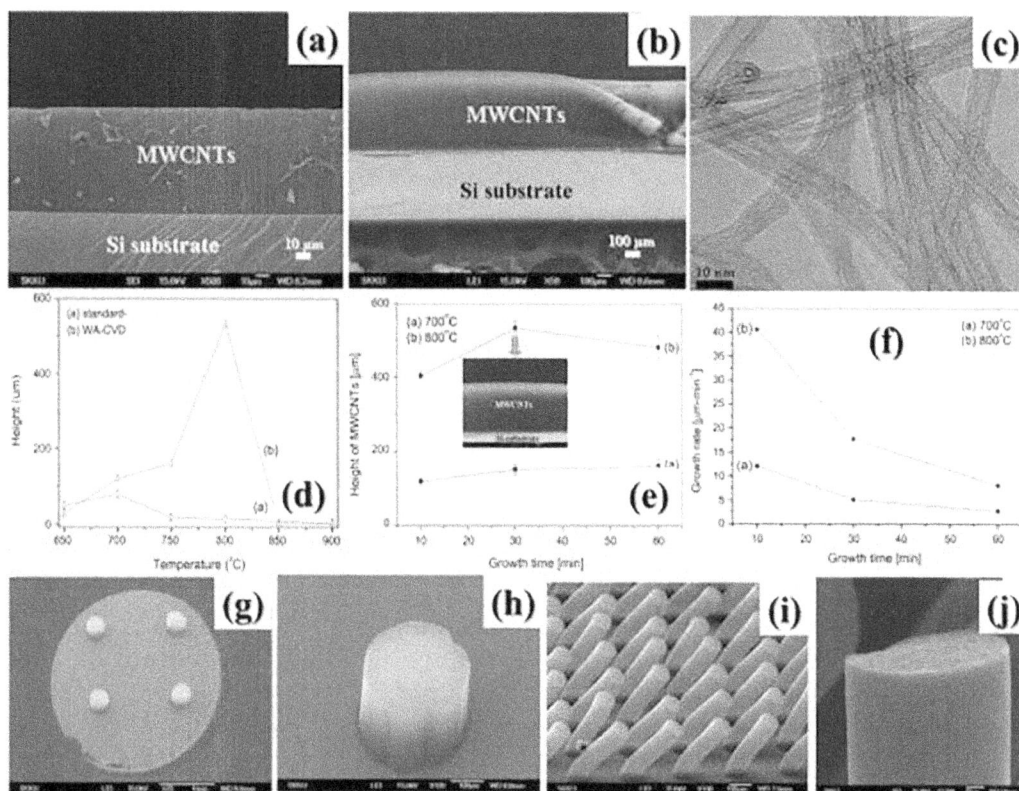

FIGURE 2.15 Recorded SEM micrographs for (a) standard and (b) WA-CVD-grown MWCNTs (scale bar: (a) 10 μm and (b) 100 μm). Recorded TEM micrographs for (c) WA-CVD-grown MWCNTs (growth temperature: 800°C), (d) variation in the height of MWCNTs as a function of temperature for (a) standard and (b) WA-CVD-grown MWCNTs, and (e) variations in the height of MWCNTs as a function of growth time at (a) 700°C and (b) 800°C growth temperatures. The MWCNTs have been grown by the WA–CVD technique. The inset shows the recorded SEM micrographs for maximum height of the MWCNT film grown at 800°C, (f) the estimated growth rate (μm/min) for MWCNTs grown at (a) 700°C and (b) 800°C, and (g)–(j) typical SEM micrographs for MWCNTs patterned on Si substrate. Photograph (a) 300 μm pillar diameter and (b) 100 μm pillar diameter.

poly(methyl methacrylate) (PMMA) or polydimethylsiloxane (PDMS) stamp, CO_2 snow-jet etching, and O_2-plasma etching.[4] However, these methods have limitations in commercial applications, such as lack of scalability, low resolution, and low reliability. This study reports that noble metals, such as Au, Pt, and Ag, can promote the oxidation of CNTs at a relatively low temperature (350.8°C) because of the reduction potential of CNTs (in this study, oxidation means decomposition to CO_2).

Based on these phenomena, a nanometre-sized, patterned, random network of CNTs is fabricated. This study also examines the difference in the reduction potentials of single-walled and multiwalled CNTs (SWCNTs and MWCNTs, respectively). Figure 2.16 summarizes the experimental procedure used to determine the reduction potential of the CNTs and to obtain the novel nanometre-sized patterns of SWCNTs. In order to make a CNT–metal junction, very thin metal films were deposited on a CNT film using electron beam evaporation. The samples were then annealed in a box furnace at 350°C in ambient air. In a reactive environment, a material system can be considered a "galvanic cell" if there is an electrical contact between two materials with different reduction potentials and they are in the same electrolyte. Under these conditions, the corrosion rate of the material with the lower reduction potential can be faster than that of the material with the higher reduction potential [31,32].

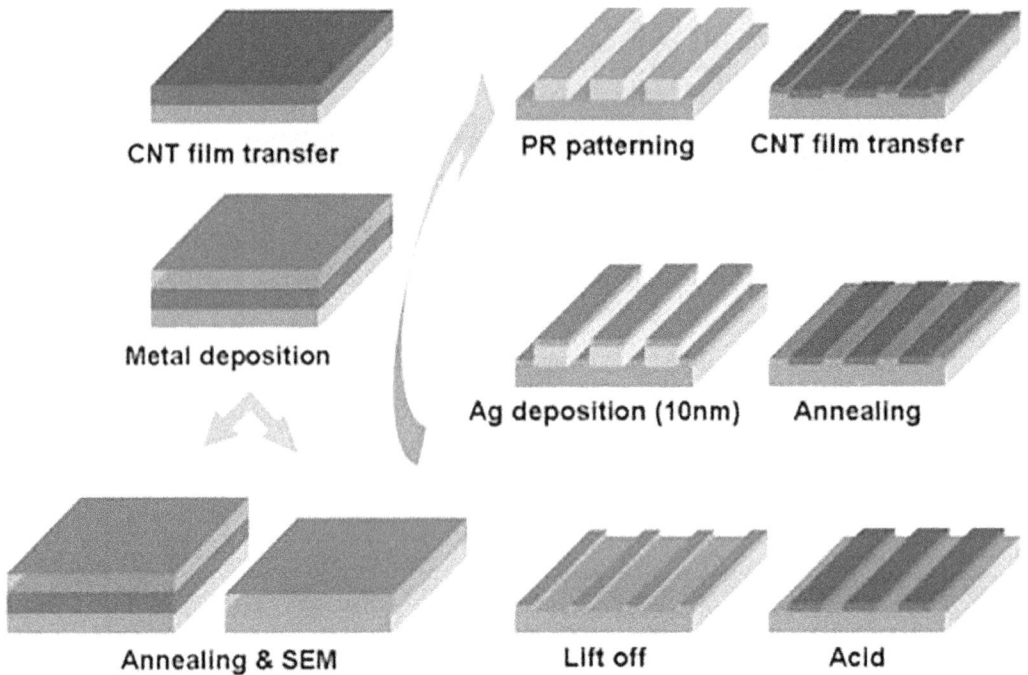

FIGURE 2.16 A flow diagram of the method used to estimate the CNT reduction potentials and to pattern the CNT films. PR: photoresist.

The novel method for patterning CNT films was developed by applying this phenomenon. In order to confine the CNTs to a selected area, lines and numbers were drawn with Ag (15 nm thick) using photolithography, before depositing the CNT film on the substrate. Figure 2.17a and b shows images of the patterned CNT films after annealing. The CNTs are confined to the areas without Ag. Only the CNT patterns remained after an aqua regia treatment. The widths of the lines patterned by the CNTs were 560 and 750 nm. Photolithography and metal deposition are well-developed techniques with which CNT patterns of several tens of nanometres in size can be obtained. In summary, a nanometre-sized pattern of a CNT film was successfully fabricated by employing the difference in the reduction potential between the CNTs and different metals. Moreover, this study examined the reduction potentials of CNTs using a very simple method. Noble metals (such as Au, Pt, and Ag) can induce the oxidation of CNTs at a relatively low temperature in air. The reduction potential of MWCNTs is located between those of W and Ni, while that of SWCNTs are located between those of Ni and In. Overall, these findings can be used as guidelines for designing CNT-based devices [32].

2.2.3.4 Selective Deposition

In this section, a potential method has been presented for depositing the catalyst particles required for the growth of CNTs, without resorting to the traditional PVD (physical vapour deposition) method and lift-off process, especially in a CNT interconnect system. Due to their unique electrical properties, such as high current density exceeding 10^9 A/cm^2 [33] and ultrahigh thermal conductivity as high as that of diamond [34], CNTs are of particular interest for electronic devices such as CNT-FEDs (field emission devices) and CNT interconnects. Recently, many attempts have been made to synthesize CNTs using nanoparticle catalysts; see [33,34] and references cited therein. Among the numerous potential applications of CNTs, intensive research has been conducted on their use as interconnects [33] (Figure 2.18).

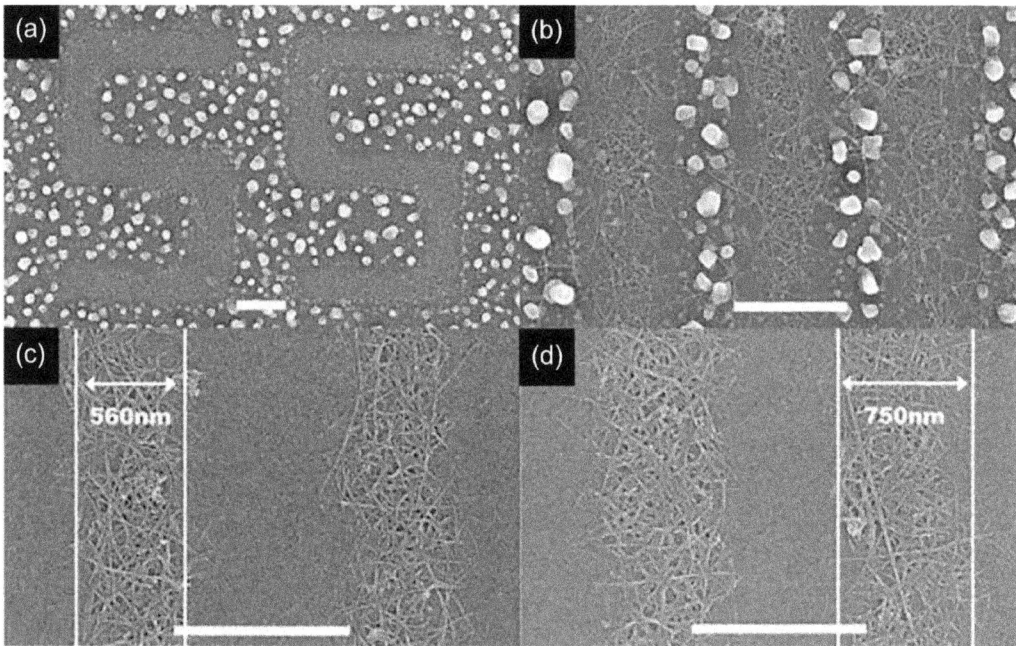

FIGURE 2.17 (a), (b) Patterned CNT films obtained after annealing with Ag. (c), (d) The films after acid treatment to remove the Ag (all scale bars are 1 mm).

The interconnect in an integrated circuit (IC) distributes the clock and other signals, as well as providing power or ground to the various circuits on a chip. The International Technology Roadmap for Semiconductors (ITRS) [33] emphasizes the high-speed transmission needs of integrated circuits as the motivation for future interconnects development. The size of interconnects in ultra-large-scale integrated circuits (ULSIs) is gradually decreasing. This scale reduction in the copper material, which is now being used as an interwiring conductor, in ULSIs results in an increase in the resistance, because the electrical resistivity of Cu increases with decreasing dimensions of the interconnect, due to the effects of electron surface scattering and grain boundary scattering [33,34]. As stated above, the shrinkage of the feature size will inevitably result in an increase in the copper resistivity, due to surface and grain boundary scattering, as well as aggravating the surface roughness. In contrast, CNTs exhibit the ballistic flow of electrons with electron mean free paths of several microns and are capable of conducting very large current densities. Therefore, CNTs have been proposed as potential candidates for power and signal interconnections. The extraordinary electrical, mechanical, and thermal properties of CNTs may provide near-term solutions for problems in interconnects in silicon IC technology. For example, a study showed that the current-carrying capacity of multiwalled CNTs (MWCNTs) did not degrade after 350 hours at current densities of 10^{10} A/cm^2 at 250°C. The thermal conductivity of CNTs [33] is about 1700–3000 W/m K. The mechanical properties of CNTs are also superior to those of the traditional materials used in the IC industry [33,34].

A simple methodology has been adopted in which iron acetate solution as a catalyst was used by dissolving iron acetate in ethanol and dissolving it again in ethylene glycol. After spin coating, the samples having several vias with iron acetate solution and the catalyst layer was pulled down into the bottom of the holes by gravity. It was found that the catalyst layer in the experiment was selectively dissolved by the TMAH solution. In this way, the catalyst layers coated over undesirable areas have been removed. Using this method, a demonstration has been made for the selective synthesis of CNTs for use in the CNT interconnect systems for ULSI technology [33,34]. There are a number

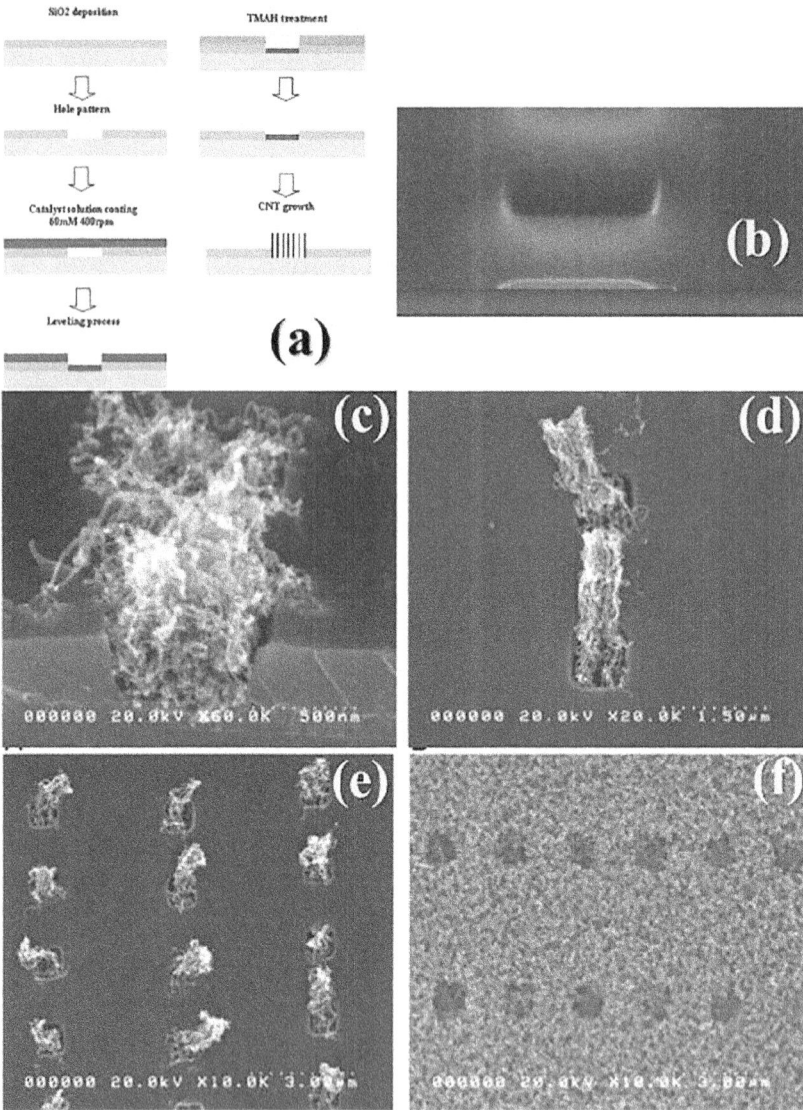

FIGURE 2.18 Schematic images of (a) production steps involved in the preparation of the sample, (b) SEM images of a typical substrate in which catalyst removed surface after TMAH solution treatment, (c) SEM images of the as-grown carbon nanotubes from the several vias; plane-view SEM images of CNTs grown from the iron acetate films on the Si substrate (e) with tetramethylammonium hydroxide (TMAH) treatment and (f) without TMAH treatment.

of other ways by which CNTs could be realized for widespread applications by transforming into a paste format [35–37], nano-reinforced composites [38–41], and material coating, especially by transition metal decoration [30,32].

2.3 2D GRAPHENE

Graphene, a one-atom-thick planar honeycomb crystal lattice, has excellent physical, chemical, and mechanical properties [42]. A considerable amount of work has been carried out on the production of graphene by mechanical/micromechanical exfoliation of graphite [43], epitaxial growth,

FIGURE 2.19 (a) Photograph of the expanded graphene-based paste; (b) a typical image of SEM of graphene cathode layers.

chemical vapour deposition, solvothermal synthesis, liquid phase, microwave exfoliation, oxidative techniques, and other techniques, of which references are cited in [44]. Graphene synthesized from the colloidal suspension [43] is both scalable and affordable. In this method, graphene is derived either by the reduction process of graphene oxide or from the intercalation of graphitic compound such as expanded graphite. The graphene produced from the reduction of graphene oxide has a significant amount of oxygen and defects [44], whereas the graphene produced from the expanded graphite consists of low density of defects and less production yield [44]. And hence, the graphene produced from these two approaches have distinctly different physical as well as chemical properties. However, most of the techniques mentioned above are optimized for the production of graphene with large amount of homogeneous sp^2 phase.

2.3.1 Expanded Graphite

The readily available expanded graphite (EG) ~0.5 mg was mixed with about 9.5 gm α–terpineol ($C_{10}H_{18}O$, (R)-2-(4-methyl-3-cyclohexenyl)-2-propenol) and subjected to the ultrasonication process for 12 hours. Further, the solution was mixed with ethyl cellulose and subjected to premixing for a period of 30 minutes. In addition to sonication, adequate shear stresses have been applied so that the graphene sheets untangle and disperse uniformly in the ethyl cellulose matrix. The mixture was employed to the calendering process, for a residence time of ~15 minutes, by utilizing a commercially available laboratory-scale three-roll mill. The intense shear mixing occurred between the narrow gap of the rolls as well as the mismatch in angular velocity ($\omega_3 = 3\omega_2 = 9\omega_1$) of the adjacent rolls due to compressive impact as well as shear stress. The mill setting has been controlled electronically, and the viscosity of the material was monitored during the calendering process. The details are provided in Section 9.1.1.5.1 [35,36] (Figure 2.19).

2.3.2 Disordered Graphene Like Nano-Carbons (GNCs)

Studies on disordered solids are important from the point of view of fundamental sciences and technological applications. In graphene, disorder could be in the form of extrinsic and intrinsic defects, and/or mixed phase sp^2–sp^3 carbon network [44,45]. The presence of disorder can induce modification in physico-chemical and mechanical properties [44] of the graphene. This is due to the change in the short-range interactions of honeycomb lattice [45]. Such defect-induced graphene can be used for wide range of applications [46–48]. Moreover, the reports on the presence of graphene and amorphous carbon (a–C) together showed interesting properties of the mixed structure. Zhang et al. mixed the two entities, graphene and a–C, to improve the hardness and the elastic recovery [49]. Diamond-like carbon (DLC) prepared by incorporating graphene oxide into a–C matrix showed significant improvement in Young's modulus, hardness, elastic recovery, and electrical conductivity of the synthesized DLC [49]. However, such mix mode structures are either difficult to synthesize or not scalable.

In this section, we have presented the synthesis of a mixed phase, sp^2–sp^3 bonded, disordered, few-layer graphene-like nanocarbon (GNC) network obtained from the combustion of softwood charcoal (C) precursor. The obtained, as-synthesized samples were intercalated and annealed. These samples were studied using Raman spectroscopy, SEM, HRTEM/SAED, STM/STS, whereas the composition was investigated using FTIR and XPS. The analysis revealed that sp^2 chains and polycyclic carbon rings (PCR) were formed during intercalation and annealing in the host a–C matrix generating a mixed phase, sp^2–sp^3 bonded, few-layer GNC network with local disorder [44,45].

GNC sheets were synthesized from the precursor of softwood charcoal (C) powder mixed with potassium nitrate (KNO_3) and sulphur (S) (stoichiometric ratio ~85:10:05, C:KNO3:S). Initially, the precursor along with KNO3 and S was admixed in the powder form and a large number of pellets of dimension ~15 mm (diameter) and ~5 mm (thickness) were prepared. The pellets were prepared by loading 1 g mixture into a mechanical press machine and applying a load of ~10 kPa. The prepared pellets were dried in a vacuum oven at ~80°C for 6 hours to remove residual moisture. These pellets were then stored in the dry atmosphere for further processing. One pellet at a time was taken in a glass plate, detonated under atmospheric conditions, and allowed for the flame-assisted combustion. The complete combustion of the detonated pellet took place in about 20–30 seconds, leading to the formation of a carbon-like ash column along with smoke. The column was crushed, and the weight of the obtained powder was measured. The weight of the powder was found to be decreased down to ~35 mg with respect to its original weight value ~1 g.

The obtained powder was immersed in a mixture of deionized water/acetone (8:2, v/v) and sonicated for a period of 1 hour, followed by the vacuum filtration using PTFE filter (pore size ~1.2 μm). The process was repeated for about three to four times in order to remove impurities such as potassium, sulphur, and other oxide elements. Thus, the powder obtained by such a method was termed as the as-synthesized samples. The as-synthesized samples were then subjected to intercalation followed by annealing process. The intercalation of the as-synthesized samples was adopted for etching the a–C and separating the conjugated GNC layers. The intercalation process was carried out at room temperature using a mixture of H_2SO_4 (98%):HNO_3 (60%) (4:1, v/v) for 48 hours with continuous stirring. It is previously reported that, during intercalation, nitronium ion (NO_2^+) is formed due to the dissociation of HNO_3, which could act as a weak etchant for a–C as well as an oxidizing agent [44].

This process leads to the evolution of CO_2 gas along with the liberation of N_2. Although NO_2^+ selectively attacks sp^3 sites, the presence of H_2SO_4 is necessary for the intercalation to form transient species such as $C–S–O_n$ and $C–N–O_n$. And this process leads to the formation of sp^2 chains and the enhancement of overall sp^2 content in the obtained samples. The samples processed in this way were then continuously washed using 5 mM NaOH solution. These samples were later washed in deionized water until the pH of the wash water reached ~7. The obtained samples were dried in a vacuum oven for 12 hours at 100°C. The samples obtained by such a chemical treatment were designated as intercalated samples. The intercalated samples were subjected to the annealing process at 1000°C for a period of 60 seconds. The annealing was carried out in a force convection oven, under inert atmosphere conditions. These samples were termed as annealed samples. During annealing, the reduction of $C–S–O_n$ and $C–N–O_n$ could occur with the evolution of species such as NO_n and SO_n. As a result, the state of hybridization of the reduced carbon atoms could get changed from sp^3 to sp^2 [44,45]. Moreover, the degree of disorder in sp^2 chains could be changed and these chains may get coupled into the polycyclic carbon ring (PCR) structure. Annealing, onset, may offer strength to intra-PCR coupling with the ordered π-states. A schematic representation of the formation of GNC sheets is shown in Figure 2.20.

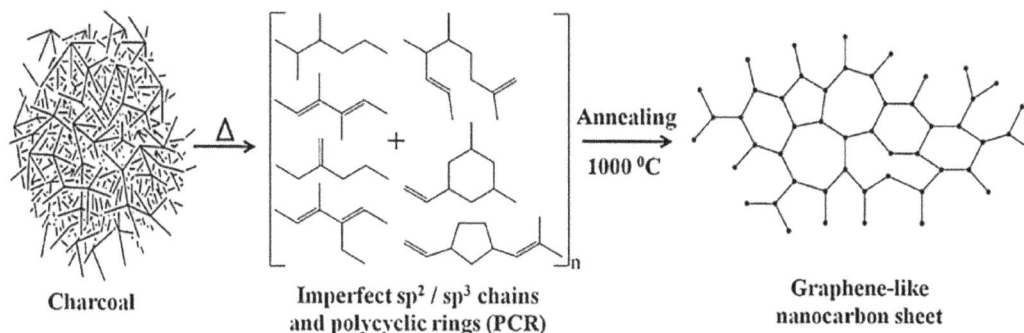

FIGURE 2.20 Schematic representation of the process of formation of GNC sheets from softwood charcoal. The scheme shows transformation of charcoal to imperfect sp^2–sp^3 chains and polycyclic rings (PCR) [44].

Figure 2.11 shows the recorded Raman spectrum for (a) the as-synthesized, (b) intercalated, and (c) annealed samples. The inset in Figure 2.21 shows the spectrum recorded for the precursor of GNC sheets, i.e. softwood charcoal. Raman analysis indicated that the charcoal after combustion could be converted into the mixed phase carbon network containing both sp^2 as well as sp^3 components. The presence and emergence of G and 2D peaks with the subsequent processing signifies the evolution of a phase of carbon which is a combination of amorphous and graphene-like carbon with 2D planar structure. Thus, Raman analysis revealed that the material obtained in this process is GNC sheets and not graphene, which contain mixed sp^2–sp^3 phase rather than pure sp^2 bonded graphene network along with local disorder [45].

Figure 2.21a shows the atomic-level resolution image of 1×1 nm area of GNC deposited on a highly oriented pyrolytic graphite (HOPG) substrate with set current 2.5 nA and bias voltage 51 mV. A slightly distorted honeycomb lattice could be observed. The corresponding current versus voltage (I–V) characteristics curve along with conductance plot (inset) is shown in Figure 2.21b. It is well proven that graphene LDOS on a single layer is linear in energy and vanishes at Dirac point. Therefore, a linear LDOS that vanishes at Dirac point can be used to identify the signature of graphene decoupled from the substrate [50]. The inset in Figure 2.2b does not vanish at Dirac point. This suggests that the synthesized material is not a pure form of graphene and the layers of this material are more coupled to the substrate. Figure 2.21c shows another atomic resolution image of 5×5 nm area of GNC deposited on HOPG under the same current and bias conditions. A magnified image of the marked region shown in Figure 2.22d clearly depicts the presence of only three atoms out of six in the honeycomb, showing triangular lattice. Such a triangular lattice is observed in the case of graphite [50]. Such a kind of effect is referred to as the carbon site asymmetry and is explained by the electronic origin [45–50]. It is well known that the surface of graphite consists of two types of non-equivalent carbon atom sites, namely A and B. A-site carbon atom sits directly above a carbon atom of the underlying layer, whereas a B-site carbon atom is located above the centre of a carbon hexagon of the layer underneath. The result of the interlayer interactions is that the p electronic levels around the Fermi energy, detected by STM, have a higher density on the B-site carbon atoms over the A-site atoms, so B-site atoms have more contribution to the tunnelling current than A-site atoms and hence the carbon site symmetry [50,51].

2.3.3 GRAPHENE NANO-RIBBON

The GNR was synthesized from chemical vapour deposition (CVD) processing of precursor of the cotton [52–54]. Initially, the cotton was dried in a vacuum oven at ~100°C for about 3 hours to remove the residual moisture and stored in dry atmosphere for further processing. Prior to CVD treatment, the substance was covered with a thin copper foil (thickness ~25 μm). The wrapping of

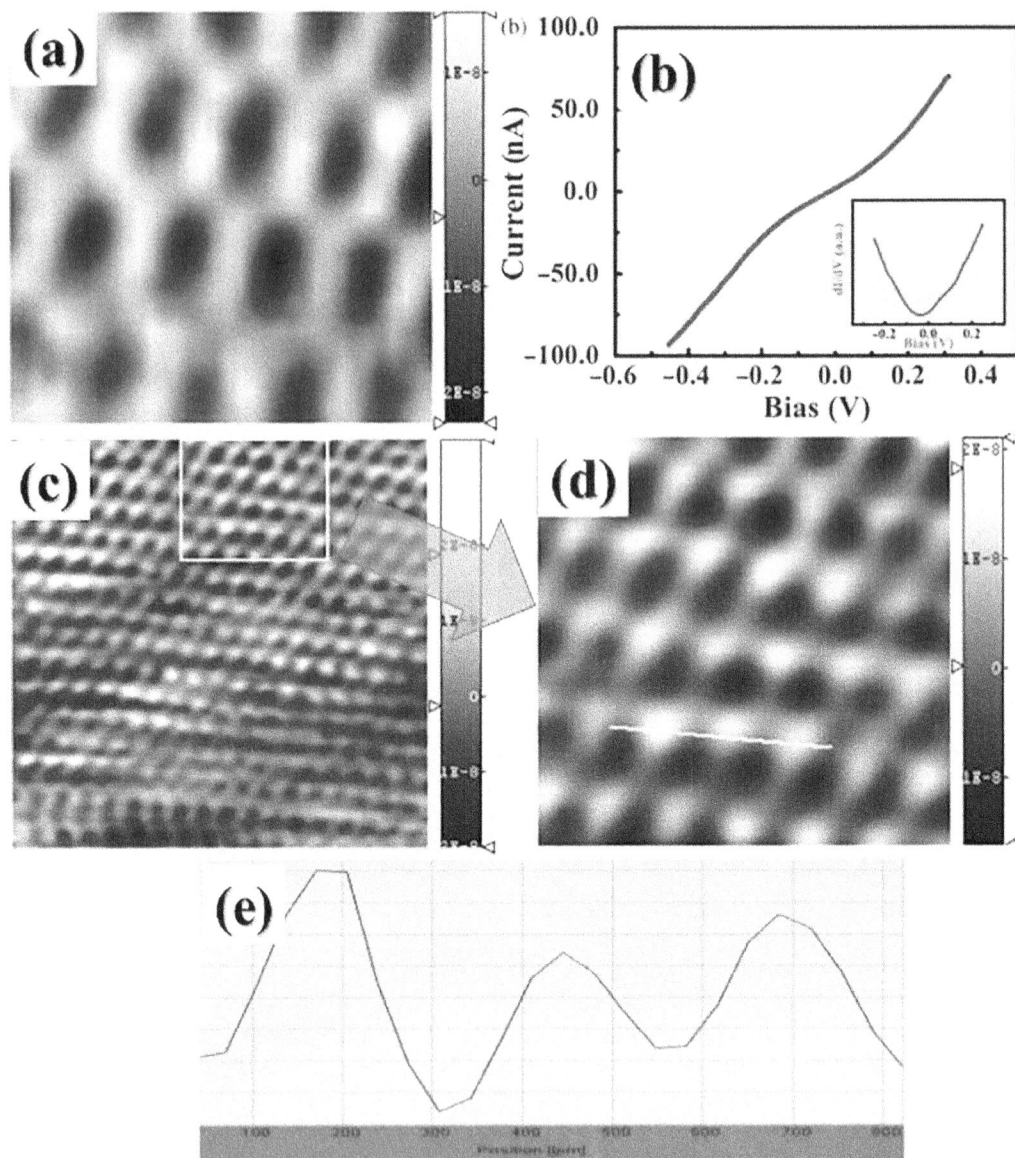

FIGURE 2.21 (a) Small area (1×1 nm) STM image of GNC with slightly distorted honeycomb lattice, (b) the corresponding *I–V* characteristics (inset shows averaged, d*I*/d*V*, tunnelling spectra as a function of bias voltage, V), (c) recorded STM images of disordered GNC of 5×5 nm area on HOPG (with set current 2.5 nA and voltage 51 mV) showing triangular lattice structure, (d) zoomed in area of the above image with clear evidence of triangular lattice, and (e) line scan performed on the zoomed in region.

foil was carried out in a metallic enclosure to ease out the handling and to shield precursor substance with the copper foil. Several such samples were prepared and one at a time was subjected to CVD processing, and the base pressure of the reactor was maintained ~2 mTorr. Argon gas was used as the carrier gas with a flow rate ~25 sccm (standard cubic centimetre), and the temperature of the reactor was ramped up to 200°C for about 2 minutes. The temperature ramping was carried out by ramping temperature by 200°C after every 2 minutes up to ~800°C [53]. The holding time was about 5 minutes at 800°C. Following this, the reactor temperature was brought down to room

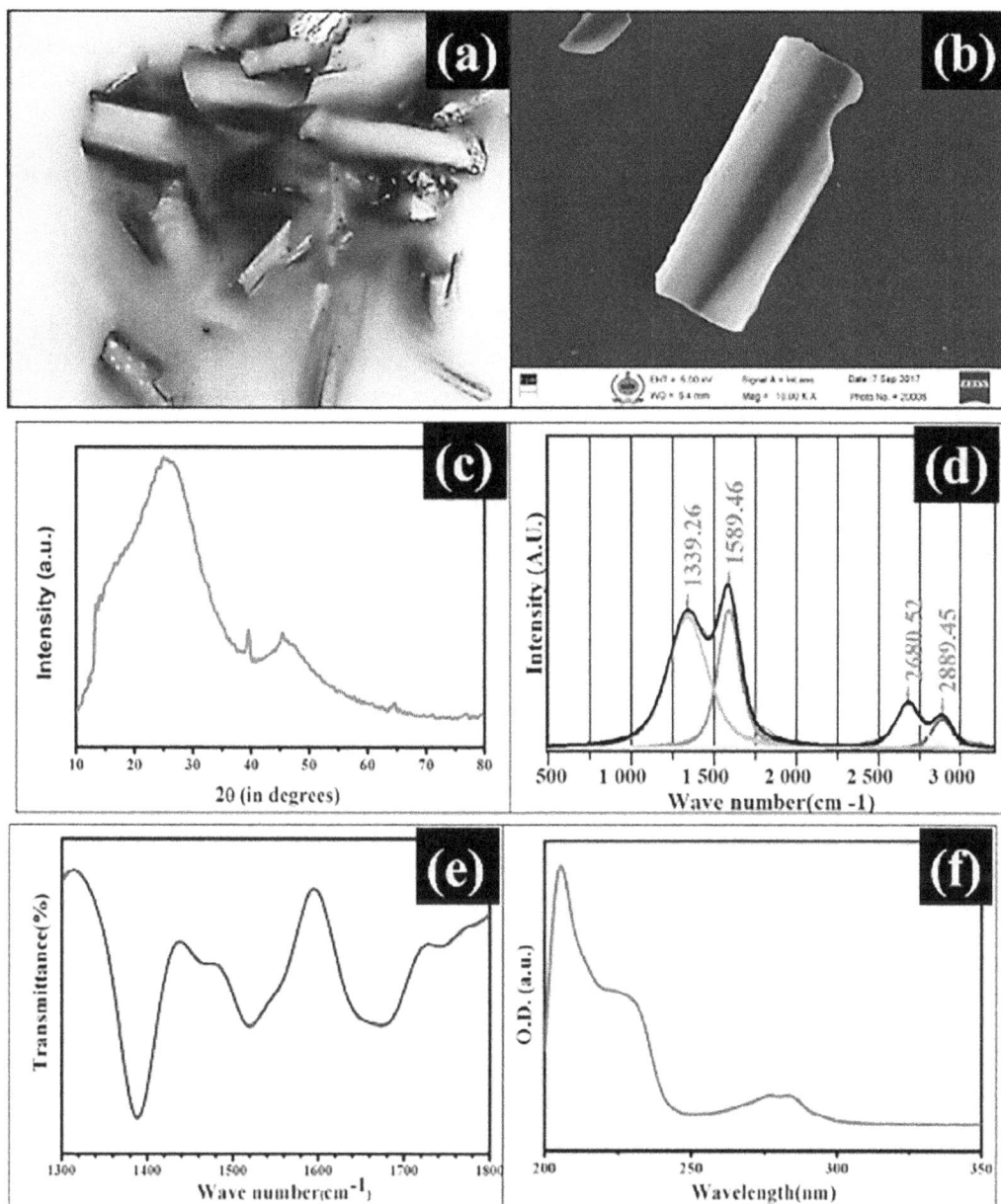

FIGURE 2.22   (a) A typical optical microscopy, (b) FESEM, (c) XRD, (d) Raman@632 nm, (e) FTIR over 1330–1800 cm^{-1}, and (f) UV–visible spectra recorded for GNR.

temperature by switching off the reactor. The cooling was carried out naturally till reactor reaches room temperature. The flow of argon was kept constant during processing and cooling [54]. The carbon powder was extracted from the enclosures and processed identically as described in Section 2.3.2, i.e. processing of GNCs [45,54].

In summary, the synthesis of GNR was carried out from cotton by CVD treatment followed by sonication, vacuum filtration, and intercalation and annealing for purification. The GNR was characterized by several techniques to investigate its morphology and structural properties. Optical imaging and SEM showed that the morphology of GNR is rectangular ribbon-like with sharp edges, dislocations, wrinkles, warps, consisting of few layers with a smaller degree of crystallinity. The Raman analysis and estimation of dynamic force constants kq, crystalline length L_a, disordered length L_D, electron–phonon coupling (EPC), and Fermi velocity (V_F) leads to the conclusion that the GNR consists of few layers stacked with disorder, dislocations, and defects. The analysis revealed the morphology and structural properties of the GNR. The dynamic deformation and high strain rate studies were performed using split Hopkinson pressure bar (sHPB) on GNR, and its response was studied (not shown in this section) [53,54].

2.3.4   REDUCED GRAPHENE OXIDE (rGO)

The GO synthesis was carried out using the modified Hummers method [55–57]. In brief, 2 g of graphite flakes and 2 g of $NaNO_3$ were added to 90 mL of concentrated H_2SO_4 under stirring. Twelve grams of $KMnO_4$ was added very slowly to the above mixture while the temperature was kept below 20°C. A further 184 mL of DI water was added very slowly at 35°C. H_2O_2 was added to the above solution mixture followed by 15–20 minutes of reflux at 95°C, which gave a solution with a brilliant yellow colour. The reaction mixture was washed using 10% HCl solution to remove residual sulphate ions. Further washing was done repeatedly with DI water until a pH of about 7 was obtained. The final centrifuged product was dried at 80°C [57] (Figure 2.23).

2.3.5   CVD GRAPHENE

Twenty-five micrometre thick copper foil of purity 99.8% (Alfa Aesar, CAS: 7440-50-8) had been used to catalytically deposit graphene [58]. Before placing the sample into the chemical reactor, it was washed with IPA, and after drying in N_2 atmosphere, samples were kept for further processing. Initially, the reactor was heated up to 1000°C for a period of 15 minutes or so in an Ar environment (300 sccm) and then cooled down to room temperature for the removal of moisture and other contaminants in the chemical reactor. This process is termed as preheating, and a set protocol for every synthesis is run. One sample at a time was loaded in the chemical reactor, and the evacuation door was locked to evacuate the chamber. The base pressure was reduced to 0.2 mTorr in a period of 5 minutes or so. After evacuation, Ar was introduced to purge reactor at a flow rate of 300 sccm at room temperature in order to remove the contaminants. This entire process was termed as sample conditioning process, as indicated in Figure 2.24 as zone 1. Following this, the temperature was ramped up to 500°C (ramp rate 20°C/s) and held for 5 minutes for pre-baking of the substrate, designated as zone 2. Subsequently, in zone 3, the temperature was ramped up to 800°C for annealing of the Cu substrate for a period of 10 minutes. This process was carried out for the surface preparation of Cu. Thereafter, in zone 4, the temperature was ramped up to 850°C, keeping ramp rate fixed, and the feedstock (C_2H_2) was released. The flow rate of C_2H_2 feedstock was kept at 20 sccm for 10 minutes. This was the zone in which growth was carried out. At the end of this zone, the flow of feedstock and temperature supply were terminated, abruptly. And the reactor was allowed to cool down (in zone 5) to room temperature in an Ar environment at the same flow rate. During the evacuation of the chamber, the Ar supply was discontinued along with the shutdown of the vacuum system, providing leak to the reactor. The sample was unloaded. The six cascading stages of the growth process established are shown in Figure 2.24. Since the growth processing was not optimized, samples were also subjected to catalyse carbon at various temperatures at an interval of 100°C. A similar procedure was adopted to carry out the growth as described in the paragraph above. In this fashion, a few number of samples each of dimension 1×3 inch were prepared and kept ready for preliminary characterizations. In addition to these samples, pre-baked and annealed Cu samples were subjected to structural and morphological characterization. The study of these samples may shed light on the improvement of growth and related mechanism [58]. In addition to this, there are graphene-type phases realized in the form of composite or hybrid format; however, their details are provided in [59–63].

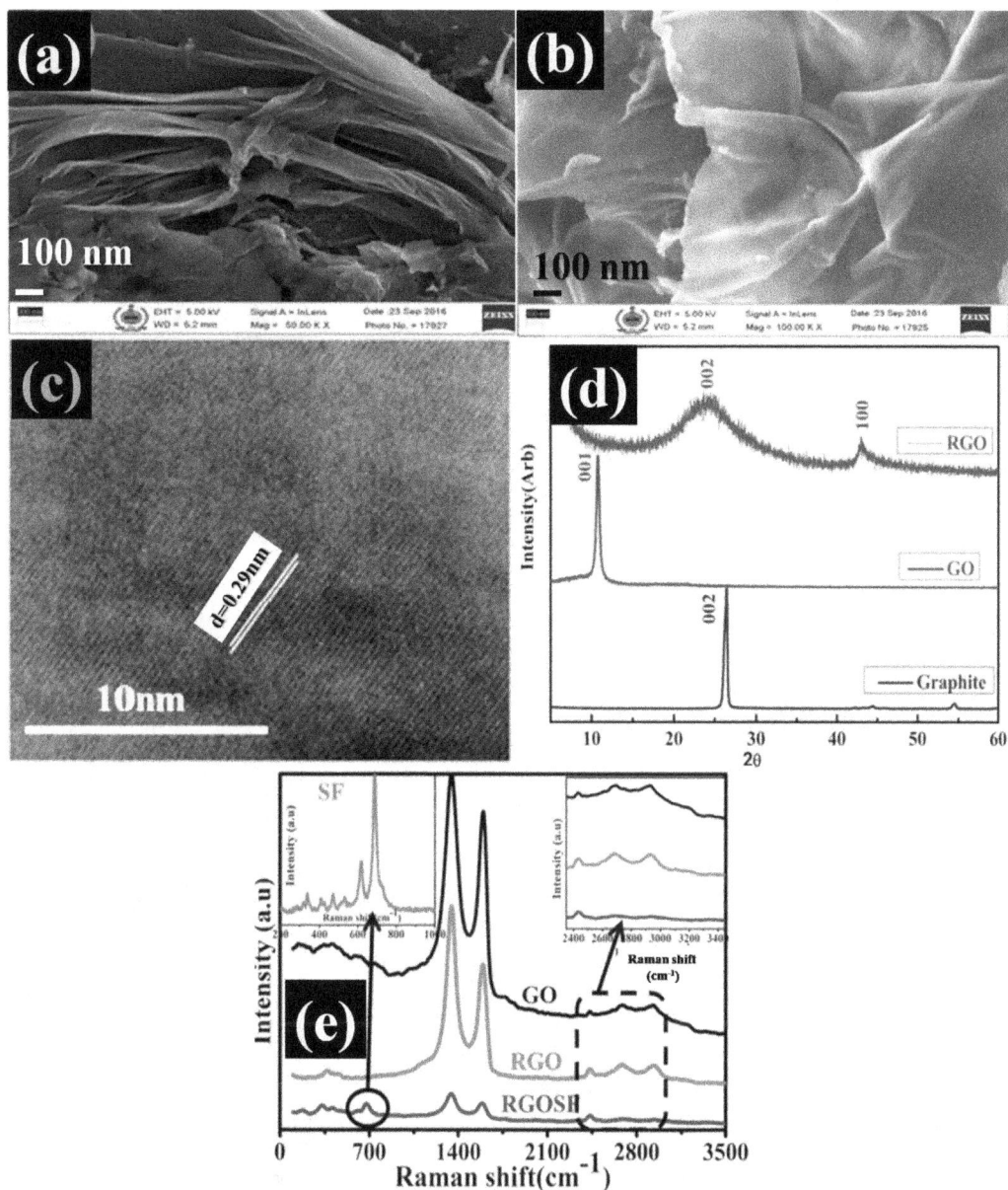

FIGURE 2.23 (a) Recorded FESEM image for GO and (b) the corresponding rGO, (c) fringing effect and crystallinity, (d) comparison of XRD patterns for GO, rGO, and graphite, and (e) Raman spectra recorded for GO, rGO, and strontium fluoride.

2.4 FOUNDRY-PROCESSED 3D GRAPHITE: VARIABLE DENSITY EFFECT

The VDG was fabricated by hot isostatic pressing (HIP). HIP is an industrial process in which materials of appropriate density could be obtained by compressing them isotropically under high pressure and temperature in an inert atmosphere [64,65].

The starting material in the form of powder was obtained from M/s Graphite India Pvt. Ltd. During processing, graphite powder was loaded into the basket and inserted into the hot zone. The hot zone was pressurized, simultaneously, by injecting hot argon gas from the top. The loaded

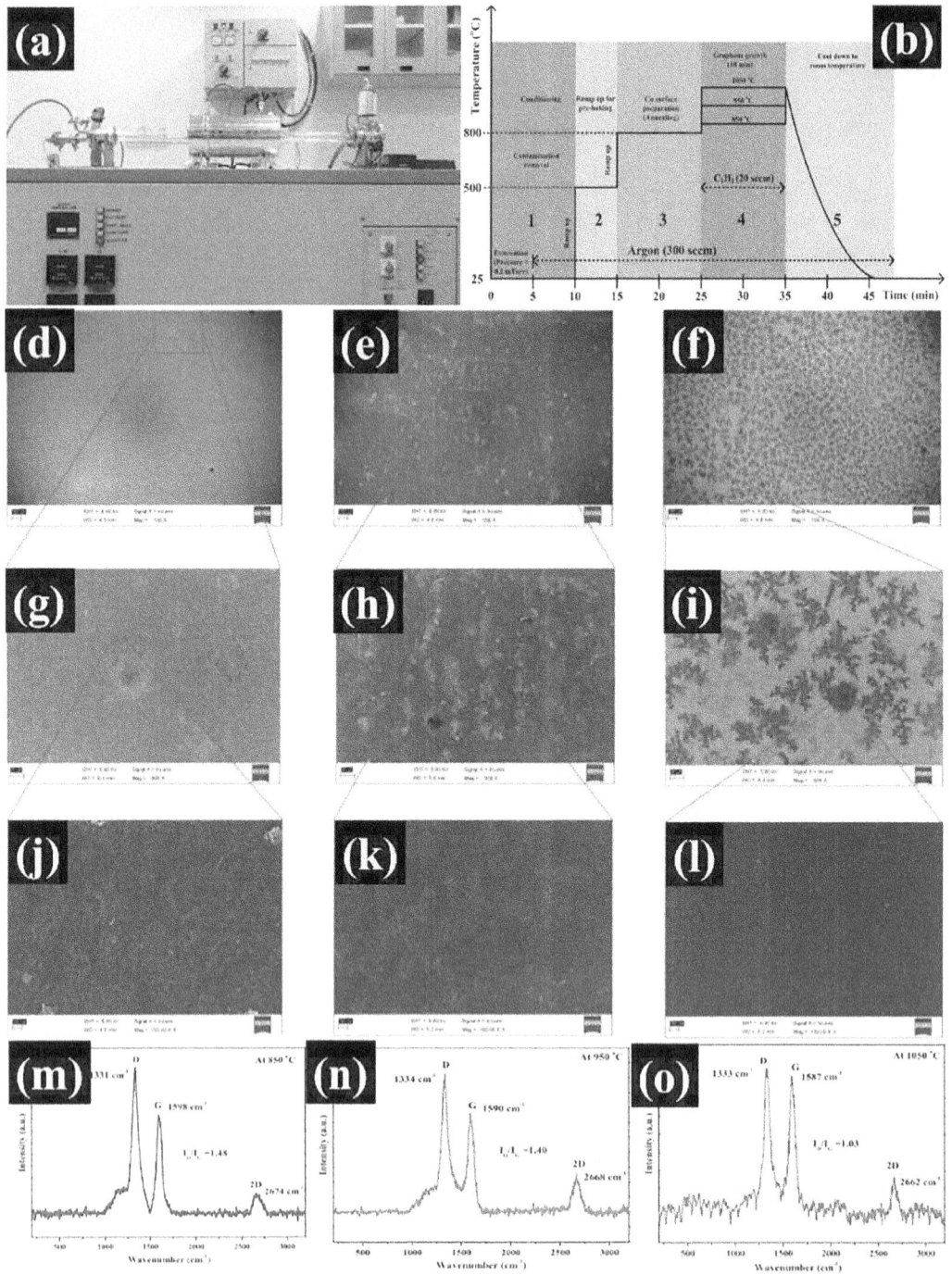

FIGURE 2.24 (a) CVD system @DIAT, Pune, India, (b) the growth process established for the synthesis of graphene is shown in the time–temperature (*T–t*) diagram. Five stages have been realized: (1) evacuation and contamination removal, (2) ramp-up for pre-baking, (3) catalyst surface preparation (annealing), (4) growth, and (5) cooldown. Pre-baking and catalyst surface preparation was carried out, respectively, at 500°C and 800°C; the recorded FESEM micrographs for graphene precludes on Cu substrate. Panel (d–f) for 850, (g–i) for 950, and (j–l) for 1050°C processing temperatures; typical Raman spectra recorded for CVD-grown graphene precludes, plot (m) for 850, (n) for 950, and (o) for 1050°C. The excitation wavelength was ~633 nm, and the beam power was 2 mW. All spectra were collected at 100× magnification. FESEM Scale bar: 100 micrometer.

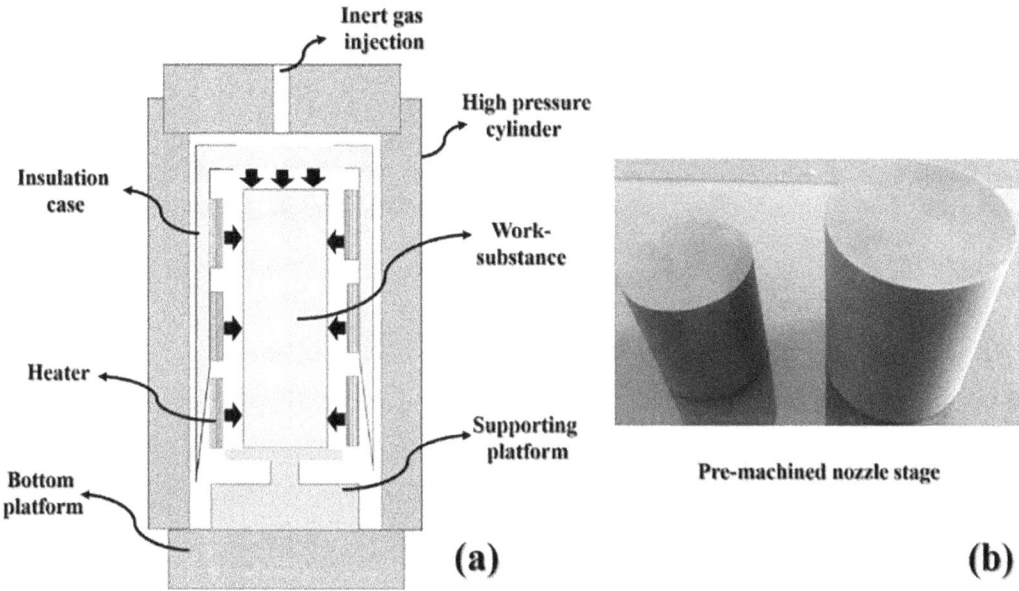

FIGURE 2.25 (a) Foundry layout of the HIP zone indicating various stages; (b) a typical HIP-fabricated VDG block.

powder was mounted onto a firm supporting platform and surrounded by heaters encased in a thermal insulator. A high-pressure cylinder was used to press the powder to achieve the desired density graphite in a few hours of waiting time. The VDG of densities 1611, 1744, 1811.7, and 1910.4 kg/m^3 was fabricated. During operation, the pressure was kept in the range of 250–300 MPa with a temperature of ~2000 K. Figure 2.25a shows the scheme of the HIP foundry zone, and Figure 2.25b, a typical fabricated VDG block [64,65].

2.5 4D ORTHOGONAL CARBON FABRIC

Fibre-reinforced polymer (FRP) composites or commonly known as composites have become an essential component of aircraft, ships, offshore rigs, locomotives, armoured vehicles, body armours, bridges, and automobiles. Their advantage of high strength-to-weight ratio has compelled all sectors to adapt the composites in place of metals [66].

The fibre-reinforced polymer (FRP) composites or commonly known as composites are made of two components: (a) textile fibre reinforcement – this is responsible for giving tensile strength to the composite component; (b) polymer plastic matrix – this is responsible for shape retention and gives rigidity to the component.

We will briefly discuss the textile fibre reinforcement: Carbon fibre or high-end glass fibre or polyaramide (Kevlar) or ultrahigh molecular weight polyethylene (UHMWPE, Dyneema) or other high-end fibres or combinations of these are mainly used to reinforce composites. The present-day process of composite manufacture, particularly, in India is by lamination of the carbon fabric (or other fibres), pre-impregnated with the resin monomer (this pre-impregnated carbon fabric is called pre-preg). The wet fabric is cut into shapes as per the component shape and the orientation required. Then they are placed in the die in a predefined sequence. Then the form is isolated and additional resin is introduced. Then the resin is cured in an autoclave. After curing, the component is machined to bring it to the final shape required.

Such a process has many drawbacks. With the introduction of 3D orthogonal woven preforms nearest to the shape of the final component, most of the drawbacks are eliminated or reduced.

The major performance-related and process-related drawbacks of laminating that are eliminated by 3D orthogonal preforms are given below:

1. The layup of laminates is a highly skilled job and is time-consuming.
2. The wastage of the costly and mostly imported pre-preg is high to the tune of 30%–40%, and this is saved in 3D preforms. This reduces costing as well as carbon footprint.
3. The laminated one requires extensive after-machining. The 3D process requires less machining.
4. The three-dimensional woven system has repeated accuracy of components. The production rate also is much higher.
5. Human-introduced defects are eliminated in 3D woven preforms.
6. The cost of the process is reduced drastically.

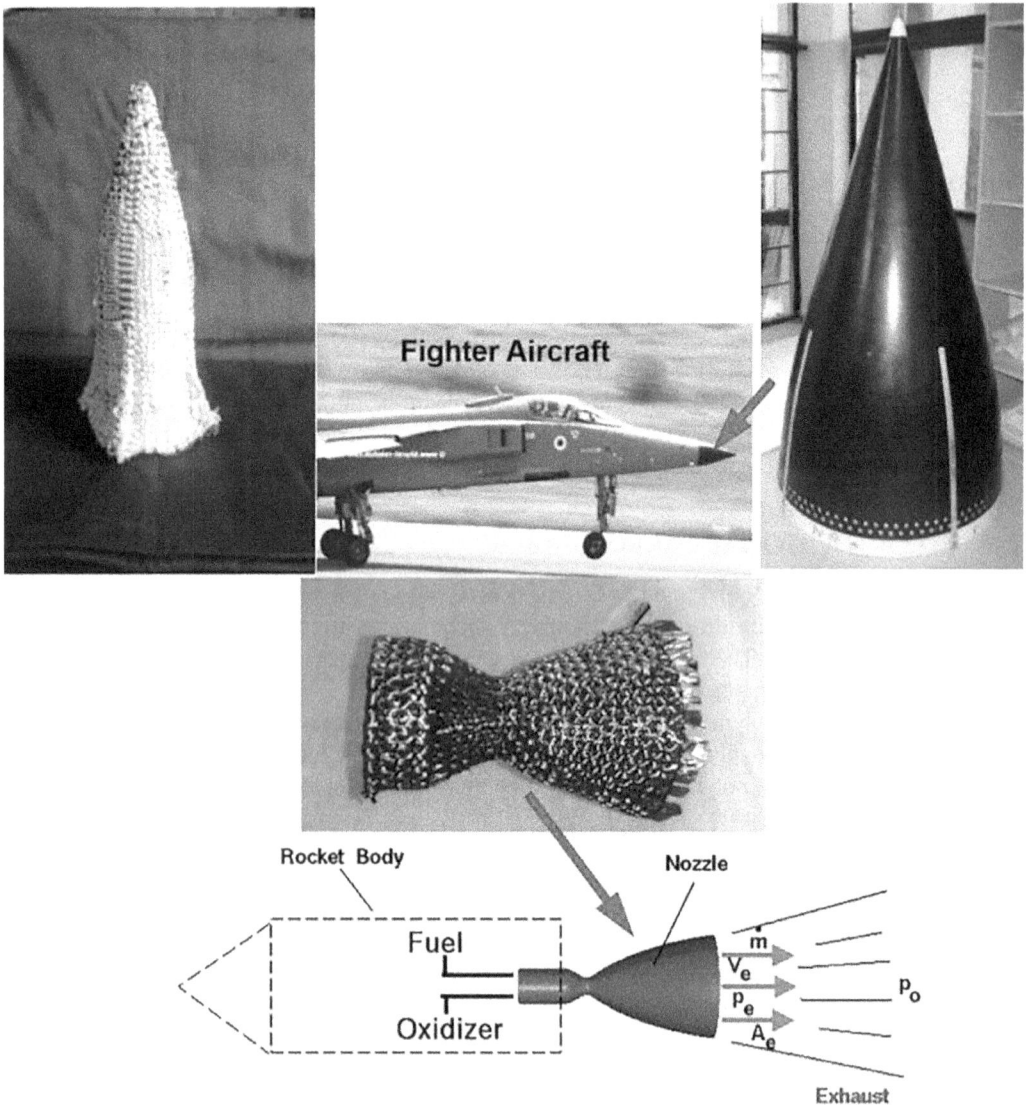

FIGURE 2.26 Typical applications of 4D orthogonal fibres.

Some of the applications of this technology are as below:

1. Aircraft – all the structural elements, skin, and stiffeners can be made out of the 3D woven composites. This will give a great advantage in manoeuvrability, speed, and efficiency. For example, the disc brakes in the landing gear are consumables and are required to be replaced from time to time. This causes downtime of the craft. These discs and brake shoes are imported. Our special polar 3D weaving machine can produce these discs here in India.
2. Jet engines – all the jet engine manufacturers have adapted 3D woven composites for most of the components, including fan blades, outer and inner casings, vanes, and flaps. This reduces the weight of the jet engine to get a better thrust-to-weight ratio. The vane units are produced by machining from a block of carbon fibre laminate impregnated with graphite. Given chance, we can develop the nearest to the shape avoiding excessive machining causing pulling of threads.
3. High temperature shields – after the Columbia spacecraft accident, the USA adapted 3D woven carbon fibres with graphite or ceramic matrix for heat shields. We can manufacture 3D woven preforms for the shield of required shape. This will help in re-entry vehicles, ICBMs, supersonic and hypersonic missile front tips, etc. The missile exhaust heat shields or ablatives case can be backed by 3D woven preforms.
4. Ships with a better weight-to-strength ratio can be produced using 3D woven composites and can be very useful.
5. Body armour and armoured vehicles will be better off with 3D woven composites. In the presence of through-thickness threads, the dissipation of energy will be on a larger area, rendering better energy absorption and hence better safety [66] (Figure 2.26).

REFERENCES

[1] Pierson, Hugh O. *Handbook of carbon, graphite, diamonds and fullerenes: processing, properties and applications.* New Maxico: Noyes, 2012.
[2] Harris, Peter J. F. "Carbon nanotubes and related structures: new materials for the twenty-first century." *American Journal of Physics* 72 (2004): 415–415.
[3] Suh, Jung Sang, and Jin Seung Lee. "Highly ordered two-dimensional carbon nanotube arrays." *Applied Physics Letters* 75, no. 14 (1999): 2047–2049.
[4] Barata-Rodrigues, P. M., Timothy J. Mays, and Geoffrey D. Moggridge. "Structured carbon adsorbents from clay, zeolite and mesoporous aluminosilicate templates." *Carbon* 41, no. 12 (2003): 2231–2246.
[5] Fonseca, Antonio, Klara Hernadi, Patricia Piedigrosso, Jean François Colomer, Kingsuk Mukhopadhyay, Roger Doome, Sorin Lazarescu et al. "Synthesis of single-and multi-wall carbon nanotubes over supported catalysts." *Applied Physics A* 67, no. 1 (1998): 11–22.
[6] Shi, Keying, Yujuan Chi, Haitao Yu, Baifu Xin, and Honggang Fu. "Controlled growth of mesostructured crystalline iron oxide nanowires and Fe-filled carbon nanotube arrays templated by mesoporous silica SBA-16 film." *The Journal of Physical Chemistry B* 109, no. 7 (2005): 2546–2551.
[7] Lan, Aidong, Zafar Iqbal, Abdelaziz Aitouchen, Mattew Libera, and Haim Grebel. "Growth of single-wall carbon nanotubes within an ordered array of nanosize silica spheres. " *Applied Physics Letters* 81, no. 3 (2002): 433–435.
[8] Lan, Aidong, Yan Zhang, Xueyan Zhang, Zafar Iqbal, and Haim Grebel. "Is molybdenum necessary for the growth of single-wall carbon nanotubes from CO?" *Chemical Physics Letters* 379, no. 5–6 (2003): 395–400.
[9] Huang, Limin, Shalom J. Wind, and Stephen P. O'Brien. "Controlled growth of single-walled carbon nanotubes from an ordered mesoporous silica template." *Nano Letters* 3, no. 3 (2003): 299–303.
[10] Li, Nan, Xiaotian Li, Wangchang Geng, Lan Zhao, Guangshan Zhu, Runwei Wang, and Shilun Qiu. "Template synthesis of boron nitride nanotubes in mesoporous silica SBA-15." *Materials Letters* 59, no. 8–9 (2005): 925–928.
[11] Wang, Weihsiang, K. M. Chao, and Chengtzuo Kuo. "Process and characteristics of the large area well-aligned CNTs with open ends by electron cyclotron resonance chemical vapor deposition." *Diamond and Related Materials* 14, no. 3–7 (2005): 753–757.

[12] Boccaccini, Aldo R. D. R. Acevedo, Giovanna Brusatin, and Paolo Colombo. "Borosilicate glass matrix composites containing multi-wall carbon nanotubes." *Journal of the European Ceramic Society* 25, no. 9 (2005): 1515–1523.

[13] Alegaonkar, Prashant S., H. C. Lee, S. H. Lee, A. F. Moses, D. Fink, and J. B. Yoo. "Carbon nanoparticles grown in the subsurface-region of porous SiO_2." *Journal of Physics D: Applied Physics* 40, no. 11 (2007): 3423.

[14] Shingubara, Shoso "Fabrication of nanomaterials using porous alumina templates" *Journal of Nanoparticle Research* 5, no. 1 (2003): 17–30.

[15] Alegaonkar, Prashant S., H. C. Lee, S. H. Lee, A. F. Moses, D. Fink, and J. B. Yoo. "Carbon nanoparticles grown in the subsurface-region of porous SiO_2." *Journal of Physics D: Applied Physics* 40, no. 11 (2007): 3423.

[16] Ugale, Ashok D., Resham V. Jagtap, Dnyandeo Pawar, Suwarna Datar, Sangeeta N. Kale, and Prashant S. Alegaonkar. "Nano-carbon: preparation, assessment, and applications for NH_3 gas sensor and electromagnetic interference shielding." *RSC Advances* 6, no. 99 (2016): 97266–97275.

[17] Tripathi Krishna Chand, Sayaad M. Abbas, Rishi Babu Sharma, and Prashant S. Alegaonkar. "Preparation and evaluation of carbon black-MWCNT nano-composites for microwave absorption." *International Journal of Science and Research (IJSR)* 3, no. 11 (2014): 2398–2402.

[18] Kumar, Arvind, and Prashant S. Alegaonkar. "Impressive transmission mode electromagnetic interference shielding parameters of graphene-like nanocarbon/polyurethane nanocomposites for short range tracking countermeasures." *ACS Applied Materials & Interfaces* 7, no. 27 (2015): 14833–14842.

[19] Haladkar, Sushant, and Prashant Alegaonkar. "Preparation and performance evaluation of carbon-nanosphere for electrode double layer capacitor." *Applied Surface Science* 449 (2018): 500–506.

[20] Jagtap, Resham V., Ashok D. Ugale, and Prashant S. Alegaonkar. "Ferro-nano-carbon split ring resonators a bianisotropic metamaterial in X-band: constitutive parameters analysis." *Materials Chemistry and Physics* 205 (2018): 366–375.

[21] Haladkar, Sushant A., Mangesh A. Desai, Shrikrishna D. Sartale, and Prashant S. Alegaonkar. "Assessment of ecologically prepared carbon-nano-spheres for fabrication of flexible and durable supercell devices." *Journal of Materials Chemistry A* 6, no. 16 (2018): 7246–7256.

[22] Tripathi, Krishna C., Sayaad M. Abbas, Rishi B. Sharma, Prashant S. Alegaonkar, and Manish Verma. "Microwave absorption properties of carbon black nano-filler in PU based nano-composites." *International Journal of Advanced Research in Science, Engineering and Technology* 3, no. 2 (2016).

[23] Lee, Su Hong, Tae Yong Lee, Jae Hee Han, Prashant S. Alegaonkar, Alexander S. Berdinsky, Drimtiv G. Khushinov, Ji Beom Yoo, and Chun Yun Park. "The growth of carbon nanotubes at the channel ends of the $SAPO_4$-5 zeolite structures." *Diamond and Related Materials* 14, no. 11–12 (2005): 1876–1881.

[24] Lee, Su Hong, Prashant S. Alegaonkar, Jae Hee Han, Alexander S. Berdinsky, D. Fink, Y-U. Kwon, Ji Beom Yoo, and Chun Yun Park. "Carbon nanotubes growth in $AlPO_4$–5 zeolites: evidence for density dependent field emission characteristics." *Diamond and Related Materials* 15, no. 10 (2006): 1759–1764.

[25] Nam, Jin Woo, Prashant S. Alegaonkar, Jae Hong Park, Ji Beom Yoo, Dhin Ho Choe, Jin Man Kim, and Wo Su Kim. "Field emission properties of plasma treated multiwalled carbon nanotube cathode layers." *Journal of Vacuum Science & Technology B: Microelectronics and Nanometer Structures Processing, Measurement, and Phenomena* 25, no. 2 (2007): 306–311.

[26] Lee, H. C., Prashant S. Alegaonkar, D. Y. Kim, J. H. Lee, T. Y. Lee, S. Y. Jeon, and J. B. Yoo. "Multibarrier layer-mediated growth of carbon nanotubes." *Thin Solid Films* 516, no. 11 (2008): 3646–3650.

[27] Lee, H. C., Prashant S. Alegaonkar, D. Y. Kim, J. H. Lee, and J. B. Yoo. "Growth of carbon nanotubes: effect of Fe diffusion and oxidation." *Philosophical Magazine Letters* 87, no. 10 (2007): 767–780.

[28] Patole, Shashikant P., Prashant S. Alegaonkar, Hyun-Chul Lee, and Ji-Beom Yoo. "Optimization of water assisted chemical vapor deposition parameters for super growth of carbon nanotubes." *Carbon* 46, no. 14 (2008): 1987–1993.

[29] Patole, Shashikant P., Prashant S. Alegaonkar, Jong Hak Lee, and Ji Beom Yoo. "Water-assisted synthesis of carbon nanotubes: acetylene partial pressure and height control." *EPL (Europhysics Letters)* 81, no. 3 (2008): 38002.

[30] Lee, Hyun-Chul, Prashant S. Alegaonkar, Do-Yoon Kim, Jong-Hak Lee, Shashikant P. Patole, and Ji-Beom Yoo. "Water-assisted synthesis of long, densely packed and patterned carbon nanotubes." *Electronic Materials Letters* 3 (2007): 47–52.

[31] Lee, Jong Hak, Jun Ho Shin, Yu Hee Kim, Sung Min Park, Prashant S. Alegaonkar, and Ji-Beom Yoo. "A new method of carbon-nanotube patterning using reduction potentials." *Advanced Materials* 21, no. 12 (2009): 1257–1260.

[32] Patole, Shashikant P., Prashant S. Alegaonkar, Hyun-Chang Shin, and Ji-Beom Yoo. "Alignment and wall control of ultra long carbon nanotubes in water assisted chemical vapour deposition." *Journal of Physics D: Applied Physics* 41, no. 15 (2008): 155311.

[33] Kim, Do-Yoon, Hyun-Chul Lee, Jong-Hak Lee, Jae-Hong Park, Prashant S. Alegaonkar, Ji-Beom Yoo, In-Taek Han et al. "Selective deposition of catalyst nanoparticles using the gravitational force for carbon nanotubes interconnect." *Thin Solid Films* 516, no. 11 (2008): 3534–3537.

[34] Berdinsky, A. S., Prashant S. Alegaonkar, H. C. Lee, J. S. Jung, J. H. Han, J. B. Yoo, D. Fink, and L. T. Chadderton. "Growth of carbon nanotubes in etched ion tracks in silicon oxide on silicon." *Nano* 2, no. 1 (2007): 59–67.

[35] Park, J. H., Prashant S. Alegaonkar, D. Y. Kim, and J. B. Yoo. "Electrical ageing of carbon nanotube composite cathode layers." *Diamond and Related Materials* 17, no. 6 (2008): 980–985.

[36] Park, J. H., Prashant S. Alegaonkar, S. Y. Jeon, and J. B. Yoo. "Carbon nanotube composite: dispersion routes and field emission parameters." *Composites Science and Technology* 68, no. 3–4 (2008): 753–759.

[37] Berdinsky, A. S., A. V. Shaporin, J-B. Yoo, J-H. Park, Prashant S. Alegaonkar, J-H. Han, and G-H. Son. "Field enhancement factor for an array of MWNTs in CNT paste." *Applied Physics A* 83, no. 3 (2006): 377–383.

[38] Jeong, J. S., S. Y. Jeon, T. Y. Lee, J. H. Park, J. H. Shin, Prashant S. Alegaonkar, A. S. Berdinsky, and J. B. Yoo. "Fabrication of MWNTs/nylon conductive composite nanofibers by electrospinning." *Diamond and Related Materials* 15, no. 11–12 (2006): 1839–1843.

[39] Jeong, Jin Su, Sung Joon Park, Yun Hee Shin, Yong Jun Jung, Prashant Sudhir Alegaonkar, and Ji Beom Yoo. "Fabrication of carbon nanotube embedded nylon nanofiber bundles by electrospinning." In *Solid state phenomena*, vol. 124, pp. 1125–1128. Trans Tech Publications Ltd, Korea 2007.

[40] Jeong, Jin Su, Jin-San Moon, Sung Yun Jeon, Jae Hong Park, Prashant Sudhir Alegaonkar, and Ji Beom Yoo. "Mechanical properties of electrospun PVA/MWNTs composite nanofibers." *Thin Solid Films* 515, no. 12 (2007): 5136–5141.

[41] Tripathi, Krishna Chandra, S. M. Abbas, R. B. Sharma, and Prashant S. Alegaonkar. "Microwave absorbing properties of MWCNT/Carbon black-PU Nano-composites." In *2017 IEEE International Conference on Power, Control, Signals and Instrumentation Engineering (ICPCSI)*, pp. 489–496. IEEE, 2017.

[42] Alegaonkar, Prashant S., Jaeo Hong Park, Sun Yung Jeon, Jin Ho Shin, Alexander S. Berdinsky, and Ji Beom Yoo. "Field emission properties of expanded graphite composite." In 한국정보디스플레이학회: 학술대회논문집, pp. 775–777, 2007.

[43] Lee, Jong Hak, Seong Man Yoo, Dong Wook Shin, Ji Beom Yoo, Jaon Hon Park, Yu Hee Kim, Prashant S. Alegaonkar et al. "Graphene composite using easy soluble expanded graphite: synthesis and emission parameters." In *2009 22nd International Vacuum Nanoelectronics Conference*, pp. 19–20. IEEE, 2009.

[44] Kumar, Arvind, Sumati Patil, Anupama Joshi, Vasant Bhoraskar, Suwarna Datar, and Prashant Alegaonkar. "Mixed phase, sp^2–sp^3 bonded, and disordered few layer graphene-like nanocarbon: synthesis and characterizations." *Applied Surface Science* 271 (2013): 86–92.

[45] Patil, Sumati, Sadhu Kolekar, Arvind Kumar, Prashant Alegaonkar, Suwarna Datar, and C. V. Dharmadhikari. "Investigation of disorder in mixed phase, sp^2–sp^3 bonded graphene-like nanocarbon." *Journal of Nanoscience and Nanotechnology* 18, no. 4 (2018): 2504–2512.

[46] Alegaonkar, Ashwini P., Arvind Kumar, Sagar H. Patil, Kashinath R. Patil, Satish K. Pardeshi, and Prashant S. Alegaonkar. "Spin transport and magnetic correlation parameters for graphene-like nanocarbon sheets doped with nitrogen." *The Journal of Physical Chemistry C* 117, no. 51 (2013): 27105–27113.

[47] Kumar, Arvind, and Prashant S. Alegaonkar. "Properties of spin bath of graphene–like nanocarbon." *International Journal of Innovative Research in Science, Engineering and Technology* 3 (2014): 14049–14055.

[48] Zhang, Yong-Hui, Ya-Bin Chen, Kai-Ge Zhou, Cai-Hong Liu, Jing Zeng, Hao-Li Zhang, and Yong Peng. "Improving gas sensing properties of graphene by introducing dopants and defects: a first-principles study." *Nanotechnology* 20, no. 18 (2009): 185504.

[49] Zhang, Junyan, Yuanlie Yu, and Deming Huang. "Good electrical and mechanical properties induced by the multilayer graphene oxide sheets incorporated to amorphous carbon films." *Solid State Sciences* 12, no. 7 (2010): 1183–1187.

[50] Klusek, Z., P. Dabrowski, P. Kowalczyk, W. Kozlowski, W. Olejniczak, P. Blake, M. Szybowicz, and T. Runka. "Graphene on gold: electron density of states studies by scanning tunneling spectroscopy." *Applied Physics Letters* 95, no. 11 (2009): 113114.

[51] Chinke, Shamal, Rohini Gawade, and Prashant Alegaonkar. "Synthesis and characterization of graphene like nano flakes (GNF) using chemical vapor deposition." In *AIP Conference Proceedings*, vol. 2142, no. 1, p. 060004. AIP Publishing LLC, 2019.

[52] Joshi, Anupama, Anil Bajaj, Rajvinder Singh, Prashant S. Alegaonkar, Kandaswami Balasubramanian, and Suwarna Datar. "Graphene nanoribbon–PVA composite as EMI shielding material in the X band." *Nanotechnology* 24, no. 45 (2013): 455705.

[53] Chinke, Shamal L., Inderpal Singh Sandhu, Tejashree M. Bhave, and Prashant S. Alegaonkar. "Surface interactions of transonic shock waves with graphene-like nanoribbons." *Surfaces* 3, no. 3 (2020): 505–515.

[54] Chinke, Shamal, Rohini Gawade, and Prashant Alegaonkar. "Synthesis and characterization of graphene-like nano ribbons (GNR) using chemical vapor deposition for shock absorbent application." In *AIP Conference Proceedings*, vol. 2142, no. 1, p. 060005. AIP Publishing LLC, 2019.

[55] Acharya, Sanghamitra, Jay Ray, Tutiki Umasankar Patro, Prashant Alegaonkar, and Suwarna Datar. "Microwave absorption properties of reduced graphene oxide strontium hexaferrite/poly (methyl methacrylate) composites." *Nanotechnology* 29, no. 11 (2018): 115605.

[56] Acharya, Sanghamitra, Chinnakonda S. Gopinath, Prashant Alegaonkar, and Suwarna Datar. "Enhanced microwave absorption property of reduced graphene oxide (RGO)–strontium hexaferrite (SF)/poly (vinylidene) fluoride (PVDF)." *Diamond and Related Materials* 89 (2018): 28–34.

[57] Chinke, Shamal L., Inperpal S. Sandhu, Tejashree M. Bhave, and Prashant S. Alegaonkar. "Shock wave hydrodynamics of nano-carbons." *Materials Chemistry and Physics* 263 (2021): 124337.

[58] Kumar, Arvind. "Synthesis of graphene-like nanocarbon for structural, tactical shielding, and spin based applications and indigenous development of chemical vapor deposition set up." PhD diss., DIAT deemed to be University, (2009), 2015.

[59] Kumar, Arvind, Devesh Kumar Chouhan, Prashant S. Alegaonkar, and Tutiki Umasankar Patro. "Graphene-like nanocarbon: an effective nanofiller for improving the mechanical and thermal properties of polymer at low weight fractions." *Composites Science and Technology* 127 (2016): 79–87.

[60] Joshi, Anupama, Arvind Kumar, Prashant S. Alegaonkar, and Suwarna Datar. "Graphene-like-nanocarbon—polyaniline composite as supercapacitor." *Energy and Environment Focus* 2, no. 3 (2013): 176–180.

[61] Joshi, Anupama, Anil Bajaj, Rajvinder Singh, Anoop Anand, Prashant S. Alegaonkar, and Suwarna Datar. "Processing of graphene nanoribbon based hybrid composite for electromagnetic shielding." *Composites Part B: Engineering* 69 (2015): 472–477.

[62] Arora, Rajat, Nitesh Singh, K. Balasubramanian, and Prashant Alegaonkar. "Electroless nickel coated nano-clay for electrolytic removal of Hg (ii) ions." *RSC Advances* 4, no. 92 (2014): 50614–50623.

[63] Kumar, Arvind, and Prashant S. Alegaonkar. "Impressive transmission mode electromagnetic interference shielding parameters of graphene-like nanocarbon/polyurethane nanocomposites for short range tracking countermeasures." *ACS Applied Materials & interfaces* 7, no. 27 (2015): 14833–14842.

[64] Kalal, Rakesh Kumar, Balesh Ropia, Prashant S. Alegaonkar. "High temperature thermodynamics in rocket motor nozzles" *International Journal of Nanomaterial Molecular Nanotechnology* 4, no. 1, (2022): 123–130.

[65] Kalal, Rakesh Kumar, Balesh Ropia, Himanshu Shekher and Prashant S. Alegaonkar. "Thermophysical property assessment of variable density graphite: improving jet propulsion nozzle designs for aerospace applications" *International Journal of Nanomaterial Molecular Nanotechnology* 4, no. 1, (2022): 136–147.

[66] Information brochure on 3-D woven composite for defence under make in India, (2022).

3 Hydrodynamics and Shock Absorption Properties of Nanocarbons

3.1 EXPLOSION: BACKGROUND

The detonation of a high explosive generates an intense pressure wave (a blast wave) in the adjoining air medium. Exposure to blast waves could cause severe injuries, casualties, and catastrophic damage to the military or the civilian structures. Therefore, blast waves are considered to be a major and, potentially, dangerous threat for such structures, armed forces, and their systems/vehicles. The armed forces require absolute protection or minimal collateral damage from the effects of detonation of explosive devices. Therefore, the effects of blast waves on structures, vehicles, and soldiers are of major interest to the blast protection research industry. In recent years, the conflicts have increasingly been extending to urban warfare wherein a high explosive system like improvised explosive device (IED) has become a preferred weaponry option to damage and terrorize the community in majority of the attacks, for example Murrah Federal Building bombing, United States attack, and 26/11 Mumbai attack [1].

3.1.1 ORIGIN: THE INTRODUCTION

Broadly, a blast (or an explosion) is an erratic phenomenon that involves flow of energy carried by the constituents of the medium in the form of compression and rarefaction of the surrounding medium. A blast is an intense acoustic (sound) wave transformed into a discontinuous and stepper wave front of impulse pressure that has devastating effects on the surrounding medium. A blast needs explosive materials, projectiles forming the fragmenting shells, and munitions. For example, an explosion near a building in addition to an intense blast wave generates numerous fragments perforated due to the failure of concrete and structural components such as bricks, ceiling, window frames, glasses, and interior items. Typically, an IED detonation could cause severe injuries, deaths, and structural damage to buildings, including vehicles and their occupants. The local effect involves shock loading and fragment hitting, whereas shock momentum can completely get transferred to the vehicle structure as a global effect, leading to dynamic effects such as overturning or displacement of the vehicle. An increase in casualties in the UN forces' vehicles has been observed during the Iraq and Afghanistan operations due to a rise in the use of the IEDs. Therefore, for military vehicles, it became an important issue to effectively resist the blast loads and provide safety to the occupants. In order to counter the terrorist activities, the most primitive, however, effective countermeasure is to gather sufficient intelligence and arrest the attack, but in its absence, the mitigation of the blast waves has manifested as a field of great attention to the scientific community due to the intense increase in the man-made blast hazards as well as accidental industrial explosions [2].

3.1.2 OCCURRENCE: AN ACOUSTIC-MECHANICAL ANALOGUE

Due to a solid and liquid explosive class material, a rapid release of large amount of energy, in the form of light, heat, sound, and high-pressure gases, occurs during the detonation of an explosive. This leads to the generation of an impulse pressure (kPa–GPa) with a transient temperature

DOI: 10.1201/9781003317258-3

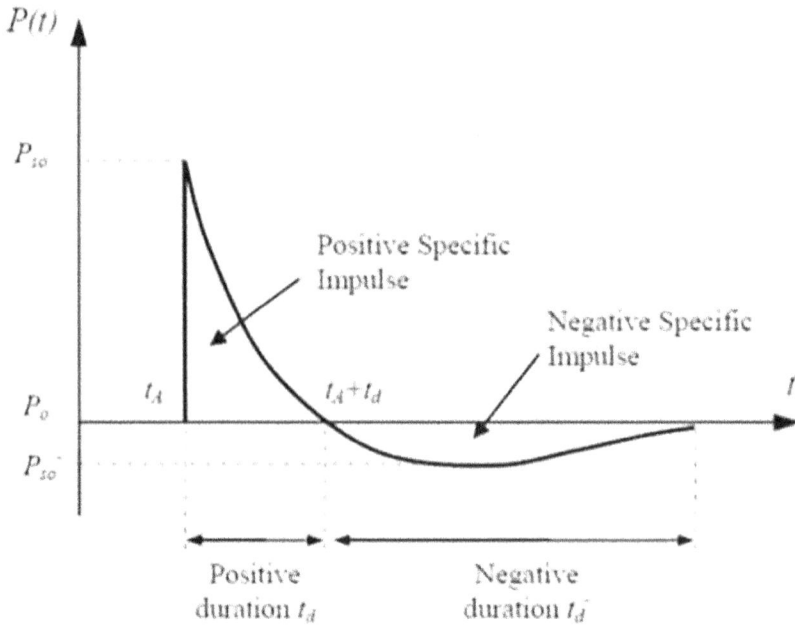

FIGURE 3.1 A typical profile of a blast wave.

(2000–4000 K) rise in the surroundings. Such a temperature and pressure change is a result of a swift chemical reaction that takes place inside the explosive. The chemical reaction converts the explosive into gases in a very short period of time. These gases expand violently by forcing out the surrounding air and forming a compressed air layer in front of the gases known as a blast wave [2]. The blast wave so formed during explosion is mainly highly compressed air travelling radially away at supersonic velocities from the point of explosion. As the blast wave travel in air, its peak overpressure decreases rapidly and it is reflected and amplified by a factor of up to eight [2,3] upon interaction with a rigid surface. The peak pressure also decays exponentially with time. The blast waves generated by conventional explosives have a very brief span of existence, measured typically in one-thousandth of a second [3,4].

The typical profile of a blast wave is shown in Figure 3.1 and is comprised of:

   i. **Peak overpressure (P_{so}):** An abrupt pressure jump above the atmospheric pressure.
  ii. **Positive duration (t_d):** The time duration between the arrival of blast wave to the decay of pressure below the atmospheric pressure.
 iii. **Positive impulse ($I+$):** A measure of the area under the pressure–time profile for the positive duration.
  iv. **Negative pressure (P_{so-}):** A measure of maximum pressure below the atmospheric pressure that occurs due to the overexpansion of the medium by the blast wave.
   v. **Negative duration (t_d^-):** The time duration for which the pressure remained below the atmospheric pressure.
  vi. **Negative impulse (I_-):** The area enclosed by the negative pressure for the negative phase.

The positive phase of the incident blast loading at any time t on a target can be described by a modified Friedlander equation [2]: $p(t) = p_{so}\left\{1 - \dfrac{t}{t_d}\right\}e^{-bt/t_d}$, where b is known as the decay or waveform decay parameter and gives the decay rate of peak pressure.

3.1.3 SHOCK CHARACTERISTICS

The outcome of an explosion is a shock wave, which is, typically, an impulse pressure of microsecond duration with a peak energy in the GPa range. It disperses at a cubical power with respect to the distance. Therefore, the shock wave can be treated as a dynamic strain (ε) propagating through a medium and, thus, be simulated at laboratory level. The experimental set-ups such as shock wave tube, plate impact, pendulum, Kolsky bar, as well as field testing of the actual explosives are the avenues available to study the blast and its multi-fold effects [5].

There are several peculiar features of a shock wave. Particularly, it's a manifestation of the intense pressure disturbance propagating in a medium in which the blast energy propagates away from the source of explosion. On its travel, with distance and time, gradual dissipation of energy occurs, transforming shock progressively into an acoustic (sound) wave form. The consequence of any blast depends on the strength of the explosion and properties of the medium, such as elasticity (elastic wave), elastic yield (plastic), compressibility (flexural), porosity (distortional or shear), free surfaces (Rayleigh), layers (Love), interfaces (Stoneley), flow direction (longitudinal), and the density of the medium. Shock waves are lethal and may cause shear instability, fracture, disintegration, perforation, and spall of the structure instantaneously [6].

While an ideal shock wave profile would show a discontinuity at the front, a plateau at the top, and a gradual return to zero pressure, however, the real shock waves exhibit a number of peculiarities that are material and pressure dependent. A few of them are discussed below. A generic profile of the interface velocity (which represents, after appropriate conversion of units, the shock wave pressure) is shown in Figure 3.2a [7]. The experimental techniques used to obtain such a pulse profile are discussed in subsequent sections. The specific profile is obtained by the VISAR technique. The pressure–volume curve of a real material is not identical to the Hugoniot (hydrostatic) curve because of the deviatoric component of stress. This is shown in Figure 3.2b. The rate of rise in stress with volume is much higher in the elastic range. When the elastic limit under the imposed stress and strain rate conditions is reached [this is called the Hugoniot elastic limit (HEL)], the pressure–volume curve shows a change in slope [8–10].

In metals, the flow stress (HEL) is fairly low, and these effects are reasonably unimportant. However, in ceramics, it is not the case. For instance, the HEL of sapphire is close to 20 GPa, whereas that of alumina is around 6–8 GPa. In Figure 3.2b, the elastic portion of the wave is separated from the plastic portion. This elastic portion, below the HEL, travels at a velocity higher than the plastic wave. Thus, the effects observed in Figure 3.2 can be explained. After an initial steep rise in pressure (or particle velocity, measured by VISAR), the HEL is reached. Beyond the HEL, the pressure rises continuously to the top (actually, we do not have a discontinuity). The rate of rise

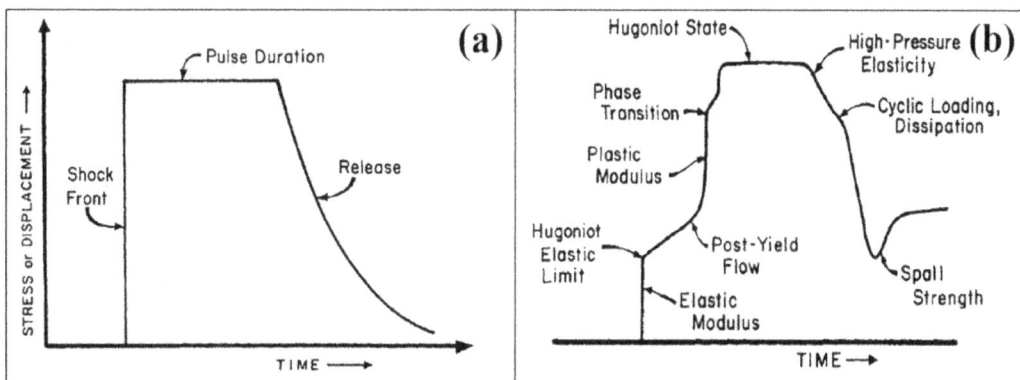

FIGURE 3.2 A shock wave profile: (a) ideal and (b) realistic showing various features.

in this pressure is dictated by the constitutive behaviour of the material. If there is a phase transition (transformation), there can be a clear signal in the wave profile. The wave may separate into two waves. At the top of the plot in Figure 3.2b, we have the pulse duration plateau. When unloading starts, this occurs initially elastically and then plastically. This elastoplastic transition in unloading leaves a signal, in an analogous manner to the HEL on loading. Since the free surface velocity is being measured by VISAR, the wave reflects and can fracture the material [11–13].

3.2 LABORATORY SYNTHESIS: THE SURVEY

In recent years, there have been few reports on the determination of mechanical properties of aluminium alloys subjected to HSR at room temperature [14], as well as at high temperatures [15]. Particularly, the energy absorption capability of individual single-walled [14], multiwalled [16], and polymer-dispersed carbon nanotubes [16–19] has been studied by experimental tensile force–displacement curves, structural mechanics compressive data, sHPB compressive technique, and MD (tension) simulations. These studies are based on quasi-static measurements rather than HSR and mainly focused on the amount of enhancement in the reinforcement properties of the composites [20,21]. There are few reports on mechanical deformation in graphene [22–24].

3.2.1 BLAST MITIGATION STUDIES

As a blast wave is a mechanical wave similar to sound waves, it mitigates with distance. Therefore, the easiest and effective means to reduce the blast wave effects is to increase the standoff distance between the explosion point and probable target. Many other means are suggested in the literature to mitigate the deadly effects of a blast and to help in preventing structural damage and human causalities. The traditional method of mitigation of structural damage to the target from blast loads is increasing its strength and rigidity. However, if a structure is damaged due to blast loads, its repair and retrofitting is not only laborious and costly, but also time-consuming. The strengthening of a building generally reduces its resilience. Some other techniques such as the use of barricades and reflecting surfaces have also been reported to reduce the loading on the main structure and its occupants. The use of an add-on or sacrificial cladding is also very promising for blast mitigation. A substantial amount of energy is absorbed by the cladding during its plastic deformation, and the incident load on the protected structure is reduced. The damaged cladding is replaced with a new one after the blast to restore the protection to structure. These requirements have excited the researchers for rigorous studies on different materials such as cellular, porous, and layered composite materials. These studies include the determination of dynamic material properties, blast tests and simulations, and the design of lightweight structures for blast-resistant applications [1].

The extensive use of porous materials in blast mitigation and shock isolation applications is because of their lower weight and higher efficacy in mitigating the blast wave by undergoing compression. A porous material is basically a solid having a network of interconnected pores (voids). These pores may be filled with a liquid or gas. The porous material undergoes compaction under blast loading and efficiently lowers the amplitude of the blast wave and delays its propagation [1,19].

3.2.2 BLAST-MITIGATING MATERIALS AND SHIELD

A soldier in conflict situation may either be exposed to blast pressure due to a standoff explosion of ammunitions such as a shell, bomb, or missile, or to a high shock pressure due to a very near or in-contact explosion of ammunitions such as anti-personnel mines. Therefore, the protection of vehicles, soldiers, and structures to meet increased survivability requirements is the main concern in military applications. These days, in civilian world too new materials and systems are required for providing protection to a common man, vehicles, an existing building, etc., against the effects of IEDs due to the increase in terrorism.

There is a need to identify different blast mitigation materials and study their individual blast mitigation capabilities under different situations. As reported by Gelfand et al. [25] and Petel et al. [26], a combination of two or more such materials can mitigate blast pressure better than an individual material; therefore, the study of materials in combination with each other is also very important. At the same time, there is a requirement to mitigate the large acceleration that is experienced by the vehicle occupants in landmine blasts.

There are several materials/systems that exhibit the property of blast mitigation. But the same material cannot be deployed for protection against all types of threats as such; in other words, each material/system is unique for a particular situation. According to Yu et al. [16], the design of energy-absorbing structures and the selection of energy-absorbing materials should suit the particular purpose and circumstances under which they are required to work. Although the design and material selection can vary notably from one application to another, yet in all cases, it is aimed to dissipate energy in a controlled manner or at a predetermined rate. The important and desirable characteristics of a blast mitigation material are as follows:

a. The irreversible energy conversion by structures/materials. This means that the structures/materials should not store the energy and convert maximum input energy into inelastic energy by undergoing plastic deformation or by other dissipation processes.
b. The peak reaction force of an energy-absorbing material should be below a threshold. Ideally, during the large deformation process of energy absorption, the reaction force should remain constant.
c. The reactive force should be restricted and almost constant. But as the work done by a force is equal to its magnitude times the displacement, therefore, if the structure is to absorb a large amount of input energy, the displacement, i.e. the compression length, should be sufficiently long.
d. To cope with very uncertain and harsh working loads, the deformation mode and energy absorption capacity of the system should be stable and repeatable so as to ensure the reliability of the structure in its service.
e. The energy-absorbing component should be lightweight and possess high specific energy absorption capacity, which are of vital importance for many applications such as vehicles, aircraft, and personal safety devices.
f. The manufacture, installation, and maintenance of energy absorption devices should be easy and cost-effective.

3.3 EXPERIMENTAL SIMULATION OF BLAST: THE INSTRUMENTATION

3.3.1 CONTACT EXPLOSION

A test set-up for blast evaluation of materials under in-contact or near-contact explosions such as land mine blast was reported by Hamilton et al. [27]. It consists of a mechanical (stainless steel) rod deployed in vertical orientation and having a piezoelectric pressure sensor deployed at its lower end to measure the transmitted pressure through different materials for the evaluation of their blast mitigation efficiency. The rod can move freely in upward/downward direction through a guide so that protective materials of different thicknesses can be placed under. In this test set-up, provisions for future deployment of accelerometer and force sensor have also been made. There is also a provision to increase or decrease the weight of the mechanical rod. The schematic and photograph of the set-up is shown in Figure 3.3.

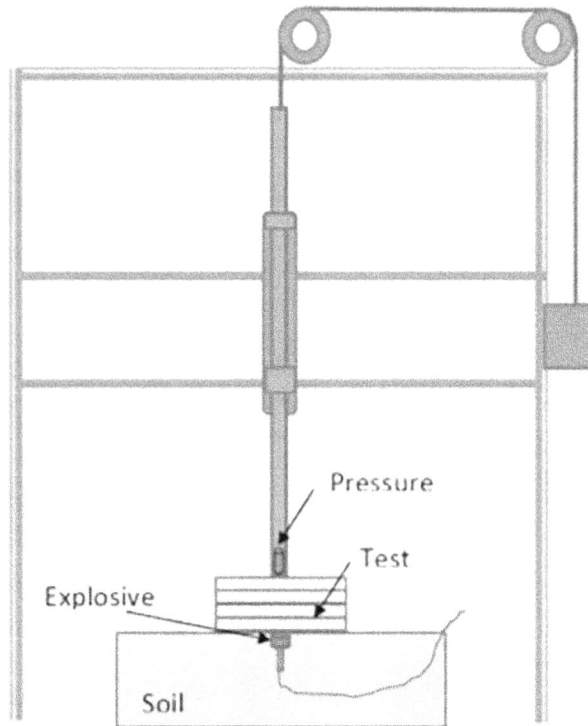

FIGURE 3.3 Schematic of the field set-up.

3.3.2 STANDOFF EXPLOSION TECHNIQUE

A set-up for the evaluation of different materials for their blast mitigation characteristics in case of standoff was reported by Petel et al. [26]. It consists of a rigid 1 m × 1 m size E-glass composite plate that is mounted on a mild steel (MS) square channel frame using M15 nut bolts. The thickness of the composite plate was 25 mm, and its density was 2.01 g/cc with a tensile strength of 350 MPa. The strength of the loading blast wave was kept below this value so that it did not damage during tests. The MS frame was rigidly attached to four concrete blocks of about 1000 kg each by welding. The test material is placed over the E-glass plate for evaluating its blast/shock mitigation efficiency. The schematic of the experimental set-up fabricated for performing blast wave mitigation study is shown in Figure 3.4.

3.3.3 SHOCK WAVE TUBE

A shock tube is mainly used to generate and study the effect of high-speed flows on materials and aerospace structures/applications [28]. It is basically a long tube having two sections, viz. driver and driven sections. These sections are separated by thin polymer/metal diaphragms. The driven section of the shock tube is generally at atmospheric pressure for blast applications, and a higher pressure is achieved in driver section using compressed gases. Two waves are formed inside the shock tube at the diaphragm rupture: a shock wave moving towards the driven section, and simultaneously, a rarefaction wave moving upstream towards the end wall of the driver section. The time after the arrival of shock wave and the reflected rarefaction wave is a useful measure for aerospace researchers. Researchers have also used the shock tube for evaluating different materials for their shock mitigation applications. Different researchers have used shock tubes of different sizes and shapes for blast mitigation studies in different materials [28] (Figure 3.5).

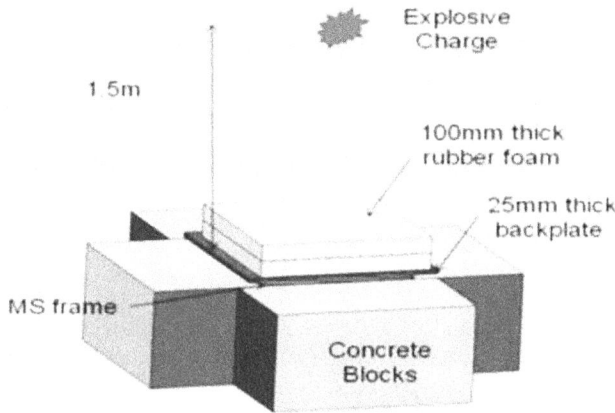

FIGURE 3.4 Schematic representation of a standoff explosion set-up.

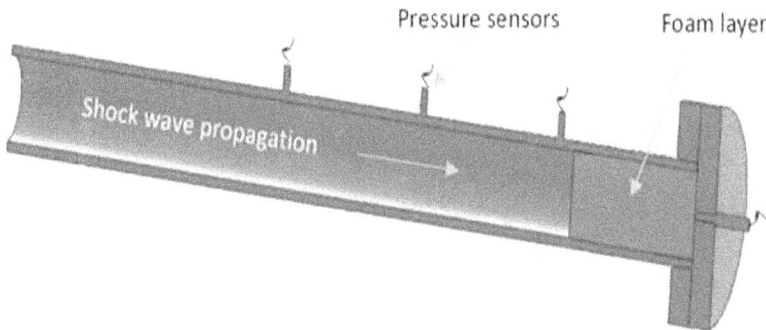

FIGURE 3.5 Experimental set-up of a shock tube.

3.3.4 SPLIT HOPKINSON PRESSURE BAR (SHPB)

One could realize a shock wave's impacts at laboratory level. The underlying parameter to simulate the shock loading condition is the temporal variations in the strain rate ($\dot{\varepsilon}$). In general, dynamic deformation occurs at a higher strain rate ($\dot{\varepsilon} = 10$–10^7/s) in which the propagating elastic and plastic waves could build-up a huge transient pressure causing shear instabilities in the specimen. There are numerous techniques; however, Kolsky bar has widely been used for the intermediate strain rate ($\dot{\varepsilon} = 10^2$–10^4/s) shock testing [29]. The high strain rate measurements were performed using the split Hopkinson pressure bar (sHPB), also known as the Kolsky bar technique. In this method, a striker bar impacts the incident bar and produces a pressure impulse of peak amplitude of the order of GPa. The pulse propagates through the incident bar in the form of an elastic wave at a transonic speed of 1–1.5 Mach number and reaches the sample specimen sandwiched between the incident and transmitted bars. Data for incident, transmitted, and reflected waves were recorded having amplitudes ε_I, ε_T, and ε_R, respectively. From the obtained data, one can derive the stress–strain relationship for the samples. Strain rates in the range of 10^2–10^4/s were achieved by means of the Kolsky technique [29]. The uncertainty in the measurements was 3%–5% and estimated as per the protocols in reference [30] (Figure 3.6).

3.4 GNF FOR SHOCK-ABSORBING APPLICATIONS

Thus far, not much attention has been paid to the modifications in morphology and microstructure of graphene-like nano-flakes (GNF) upon impact. The propagation and transmission of shock

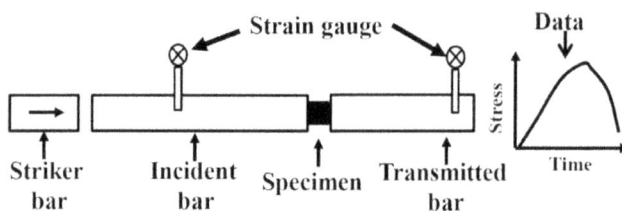

FIGURE 3.6 Schematic of the sHPB/Kolsky bar for dynamic strain measurements.

energy is responsible for the deformation of flakes, plane slip, mobilization of inherent defects, etc., to absorb the shock. Shock wave attenuation by GNF is very important from the application point of view in defence.

The synthesis methodology of GNF is presented in Section 2.3.2 dedicated to the synthesis and characterization of GNF [31].

In the subsequent section, we present the results on high strain rate (HSR) measurements on the GNF to obtain stress, strain, and strain rate parameters for the analysis of deformation mechanics. HSR GNF samples were also investigated using a number of characterization techniques. The obtained data provided insights into the processes that emerge in GNF due to shock, such as glissile dynamics of disorder, flow hardening, phase transformation, physico-chemical modifications, shock damping, and phonon drag.

3.4.1 Dynamic Deformation at Nanoscale Level: Methodology

For HSR measurements, the preparation sequence is displayed in Figure 3.7. The requirement was an Al base (dimensions: 10 (dia.) mm×5 (h) mm) on which samples can be mounted as shown in image (b). The base was fabricated using the machining and surface grinding technique (flatness ~10 μm). To mount a sample (thickness ~5 mm) on the base, a die (50 mm (l)×15 mm (w)×15 mm (h)) was made, as shown in image (c). The die consisted of five cavities in which the Al base can be loaded to cast five samples at a time. Image (d) shows a typical batch of epoxy and GNF/epoxy. For epoxy samples, epoxy resin was added in 1:1 w/w ratio and mixed manually for about 15 minutes. In parallel, each die cavity was lubricated with silicon grease and the Al base was loaded in. The lubrication was done to extract sample pellets after polymer composite solidification. The base-loaded cavity was filled with epoxy resin paste and allowed to solidify for about 30 minutes. In a similar fashion, GNF:epoxy resin (samples named as HSR GNF) was prepared in a ratio of 1:10 w/w %. The epoxy or GNF/epoxy samples were separated from the cavity/die by rear impact extraction technique. Due to high viscosity, the samples were having high surface tension while solidifying. This resulted in the positive curvature of the sample surface (k).

For HSR test, the critical requirement is flat surface to have uniform compression. For surface flattening, the specimens were subjected to polishing process, as shown in (e). The side and top views of the samples ready for test are shown as photographic images in (f) and (g), respectively.

3.4.1.1 Split Hopkinson Pressure Bar (sHPB): HSR Measurement Details

As there is no standard design for sHPB equipment, the geometry of the bars is designed in such a manner that conditions for uniform elastic deformation within the samples are met. In general, it is difficult to achieve ideal elastic deformation without considering plastic wave propagation, friction, and inertia effects in samples. The configuration of the sHPB is schematically shown in Figure 3.8, which consists of two elastic pressure bars each of diameter ~15.9 mm and length 1.4 m. They are cast from maraging 300 steel, with a yield stress exceeding 2500 MPa. In order to satisfy the assumption of one-dimensional wave propagation, the ratio of pressure bar length to diameter is >20. Further, the longitudinal and radial inertia due to the rapid particle acceleration can influence

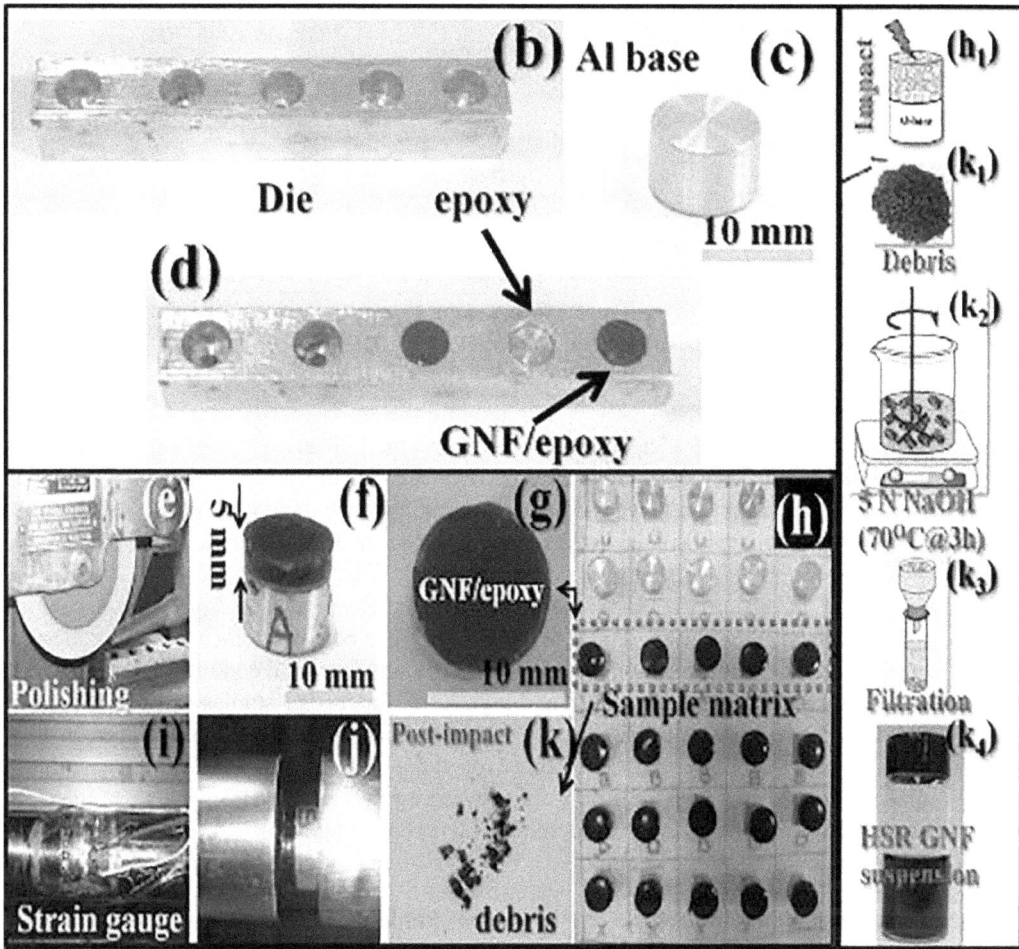

FIGURE 3.7 Image gallery: (b) fabricated die, (c) Al-base, (d) sample-filled die, (e) polishing process, (f) HSR GNF/epoxy, (g) top view of (f), (h) variety of nanocarbon samples (red square GNF) prepared on Al-base, (i) bonded rosette strain gauges to bars, (j) sandwiched position of a typical sample in sHPB, and (k) collected debris.

the stress–strain curve and enhance errors; hence, the ratio of sample diameter to thickness is ~ 1.5 and 2 (see Figure 3.7b, f, and g). In nanocomposites, failure occurs at low plastic strain within the first reverberation of the shock wave. In order to delay the sharp step, the rise time of the wave can be increased by a copper pad (labelled as pulse shaper), as shown in Figure 3.8. The pad deforms plastically upon impact and increases the wave rise time. The specimens are sandwiched between the bars as shown in Figure 3.7j. The projectile, which is a striker bar, is propelled towards the incident bar by the gas gun. The stresses generated are derived by measuring the strains produced by elastic wave at any point, as it propagates along the bars, using semiconductor strain gage rosettes bonded to both incident and transmitted bars (shown for one such bar in Figure 3.7h). The gages are electronically connected to a sensor instrument, data acquisition unit, and processing system. After configuring, an impulse of peak pressure 1500 MPa is imparted to the samples of Al, epoxy/Al, and HSR GNF. The debris was collected for epoxy/Al and HSR GNF. The velocity of the striker bar was determined using a laser velocimeter measurement system. For Al base, the thickness and diameter were measured after the impact. For epoxy and HSR GNF, the debris was collected in a plastic casing.

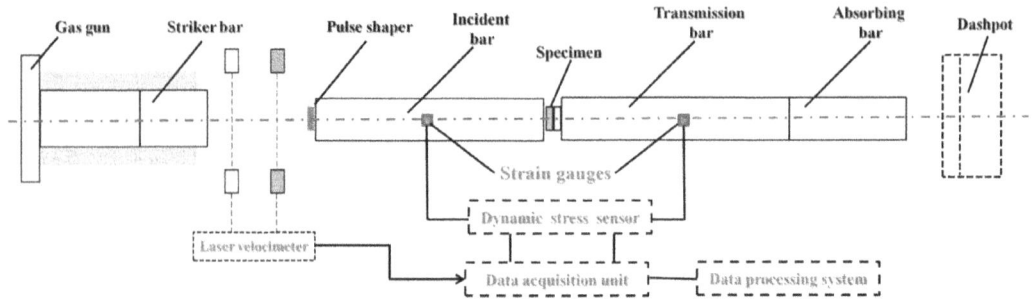

FIGURE 3.8 The details of sHPB experimental scheme.

After shock treatment, the GNF that was arrested into the epoxy matrix as a pellet onto the Al base has been recovered. Figure 3.7h$_1$ schematically shows the impact on the sample that turned it into debris (Figure 3.7k$_1$). The debris was small particles that contain HSR GNF arrested/trapped into the epoxy polymer. After gentle washing and filtration, they were immersed in 5N NaOH solution and the bath was constantly stirred and kept at 70°C (Figure 3.7k$_2$). After about 3–5 hours, the solution of NaOH started becoming turbid, which gradually increased indicating release of more and more HSR GNF flakes into the NaOH medium. The heating was switched off, and the solution was allowed to cool and filtered using PTFE (Figure 3.7k$_3$). The powder on the filter was collected, washed with DI water/isopropyl alcohol, and allowed to dry under normal conditions. From the collected powder, the suspension was prepared and kept ready for further studies (Figure 3.7j$_4$). The same procedure was adopted for all five samples that were shock-treated (rectangle, Figure 3.7h).

The analysis of mechanical properties is based on three assumptions: (a) The deformation in GNF is uniform; that is, stress equilibrium is maintained throughout the experiment, (b) there are no frictional or other dynamic effects considered, and (c) propagating volume strain (particle velocity) is linearly related to the shock velocity.

3.4.1.2 Hydrodynamic Parameters: The Lagrange–Rankine–Hugoniot Approach

Briefly, when the stress in a ductile matrix, such as GNF, exceeds the elastic limit, then plastic deformation sets in, both in quasi-static and in dynamic range. When the transmitted pulse *peaked up* with amplitude beyond the elastic limit, it decomposes into an elastic and a plastic wave. While transforming into shock waves, they have a steep wave front with a state of uniaxial strain that allows the build-up of the hydrostatic stress component exceeding the dynamic flow stress (i.e. flow stress at strain rate is established at the front) by several factors. One can assume that the solid has no resistance to shear as a first approximation. The calculation of the shock wave parameters presented, for GNF, is based, in its simplest form, on *Lagrange–Rankine–Hugoniot* formulism [2,3].

In this, the propagation of stress disturbance is approximated by: $\dfrac{\partial^2 u}{\partial t^2} = \dfrac{d\sigma/d\epsilon}{\rho_0}\left(\dfrac{\partial^2 u}{\partial x^2}\right)$ at a constant strain, ε, given by: $V_o = \left(\dfrac{d\sigma/d\epsilon}{\rho_o}\right)^{1/2}$ and $\dfrac{d\sigma}{d\epsilon} = E$ (within elastic limit), where u is the speed of propagating volume strain, ρ_o, the density ~0.0213 kg/m^3 (for GNF), and E, the elastic energy (in GPa). The boundary conditions are as follows: $u = V_o$ t@ $x_i = 0$ and any $t > 0$; $u = 0$ @ $x_f = \infty$. The spatial, x, and temporal, t, variations are approximated by: $\dfrac{x}{t} = \dfrac{\dot{\sigma}}{\rho_0}$. The variable, stress rate, $\dot{\sigma}$, is explained in detail in analysis.

A qualitative understanding can be developed from the analysis of the stress (σ)–strain (ε) curve. The structural strength of GNF is dependent on strain, ε, and strain rate, $\dot{\varepsilon}$ (i.e. $\dfrac{d\varepsilon}{dt}$), as seen in Figure 3.9. For GNF, the σ–ε curve is represented by a bilinear function, where the first stage is elastic,

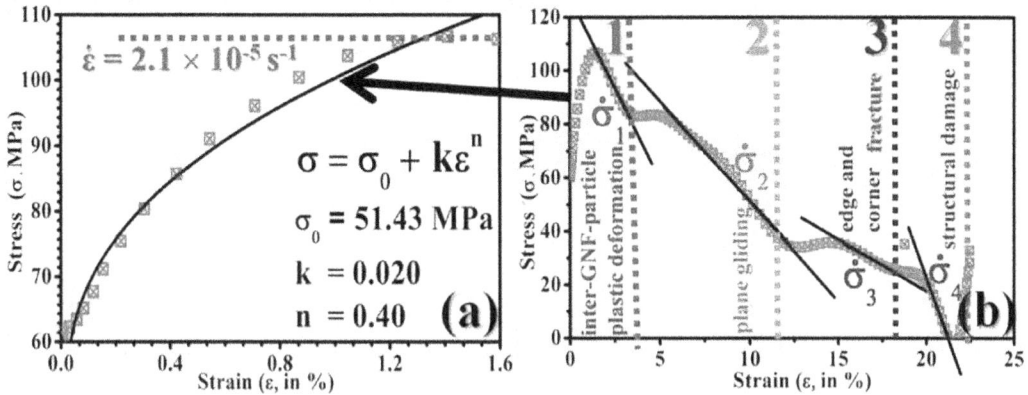

FIGURE 3.9 (a) Magnified region of stress, σ, (MPa) vs % strain, ε, curve (arrow: elastic region); the tangent shows the estimated strain hardening time parameter; (b) the corresponding plastic regime showing variations in stress rate, $\dot{\sigma}_i$, in four stages (separated by vertical lines: 1 – plastic deformation, 2 – plane gliding, 3 – fracture, and 4 – structural damage).

plot (a), and the second is plastic, plot (b). In (a), the σ–ε curve for GNF can be expressed nearly as a power function of the type: $\sigma = \sigma_0 + k\,\varepsilon^n$, where k is the pre-exponential factor (0.020) and the work hardening coefficient $n=0.40$. The velocity, V_S, with which the shock wave is incident is $\sim 5.0 \times 10^3$ m/s, given as above. The GNF departs from the elastic limit with the large spread out in V_o. The characteristic time parameter t_c involved in the phase transformation, from elastic to plastic, is given by: $t_c = \dfrac{\partial\sigma/\partial\dot{\varepsilon}}{\partial\sigma/\partial\varepsilon}$ and is estimated to be 2.1×10^{-5}/s as given in (a). The value suggests that the time scale over which the plastic flow inter-transforms the sp² and sp³ content in GNF is of the order of a few microseconds, close to metal phase transformation like titanium. Moreover, GNF have a strain rate sensitivity and strain hardening similar to a metal matrix. Although the change is *metal-like*, the flow stress due to the change in GNF has its strain rate and strain dependent on ε and $\dot{\varepsilon}$. In (b), the effect of stress rate change, $\dot{\sigma}$ (i.e. $\dfrac{d\sigma}{d\epsilon}$), on the strain is seen. Broadly, GNF is departed, significantly, from the ductile behaviour, in plastic regime. It shows a change in $\dot{\sigma}$ at several (typically four) stages over strain, ε, range that shows how GNF responds to the shock waves.

The change in $\dot{\sigma}$ is indicative of the variations in flow strain. Moreover, the gradual decrease in σ is indicative of stress propagation in GNF matrix has experienced the strain flow hardening. The change that occurred at the initial stage of $\dot{\sigma}_1$ is due to the impediment offered by GNF inter-particle plastic deformation. In this, there could be a possibility of separation of two nearest neighbour clusters. Following this, in the next stage, the stress rate, $\dot{\sigma}_2$, introduces plane gliding to the conjugated graphene stacks (also evident in the subsequent Raman and microscopy analyses). In stage three, the impact is sequentially percolated down to the individual layer in which the edge and the corner fractures, i.e. $\dot{\sigma}_3$, finally result in the structural damage of the flake at the terminal stage. The effect is cascading. From the slope estimation, the temporal variation at stage one is three times high compared to the third stage, whereas almost the same time is required for the plastic deformation that sets in compared to the dynamic fracture of GNF towards the end stage. The V_s is reduced quite significantly with an average magnitude of $\sim 2.5\times10^3$ m/s in the plastic regime. The stage-wise estimation indicates that the inter-particle separation occurs at $\sim 3.5\times10^3$ m/s, stacking faults $< 2.5\times10^3$ m/s, edge/corner fracture $\sim 2.0\times10^3$ m/s, and structural spall below $\sim 1.5\times10^3$ m/s. For plastic deformation, V_s is computed using: $V_s = \left(\dfrac{k\,n}{\rho_0}\right)^{\frac{1}{2}} \times \left(\dfrac{n+1}{2}\right)^{-1} \times \varepsilon_L^{n+1/2}$, where ε_L is the elastic

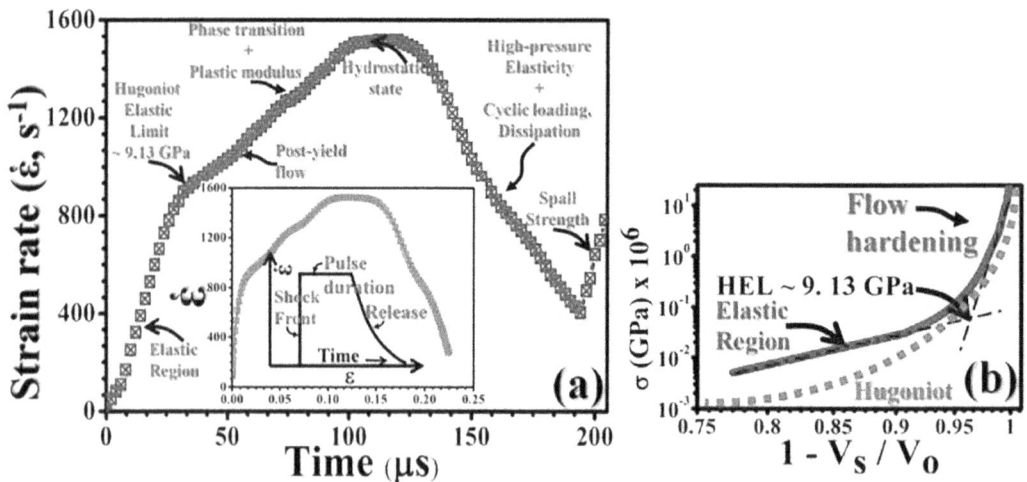

FIGURE 3.10 (a) Variations in strain rate, $\dot\varepsilon$, with time, imprinting a specific shock wave profile on 2D GNF carbon layers. The curve exhibits a number of stages that are dependent on the peculiar response of GNF to the pressure pulse. The inset shows $\dot\varepsilon$ vs ε indicating the hydrodynamic response of GNF and inside it an idealized shock profile. Plot (b) shows the shock response of GNF (blue frozen line) and Hugoniot solid (red dotted) indicating HEL point and flow strength.

strain limit ($\sim 1.5 \times 10^6$ for GNF). The decrease in velocity is indicative of the decrease in work hardening rate in GNF as it shatters [32,33].

An ideal shock wave profile exhibits a discontinuity at the front, a plateau at the top, and a gradual arrival to zero pressure, as schematically shown inside the inset in Figure 3.10. However, real shock waves imprint a number of peculiarities that are pressure and material dependent. For GNF, a few of them are discussed below.

For GNF, the rate of rise in volumetric stress is much higher, under elastic behaviour. It imposes stress and strain rate conditions leading to the Hugoniot elastic limit (HEL), which is ~9.13 GPa (GNF). Plot (b) indicates flow stress of GNF at the imposed strains. The Hugoniot behaviour is compared with GNF. GNF departs, drastically, from the elastic region. Some disagreement is seen, even at the plastic regime. In general, the flow strength of material either is independent (Hugoniot) or decreases (softening) or increases (hardening) with pressure. Broadly, in metals, due to rich free carriers the flow stress (HEL ~20 GPa, sapphire) is quite low, and these effects are somewhat unimportant. However, this is not the case for insulators (6–8 GPa, alumina) [33].

In Figure 3.10a, at HEL, the elastic fraction of the wave is separated from the plastic portion. This elastic fraction, below the HEL, travels at a velocity higher than that of the plastic wave. The effects observed in (a) can thus be explained. Beyond the HEL, the pressure rises continuously to the top due to the non-discontinuous nature of plastic wave front. The rate of increase in this pressure is dictated by the constitutive behaviour of GNF. The involved phase transition cannot be seen clearly in the wave profile. It seems that the wave may be separated into two wave fronts, marginally. At the top of the plot, the pulse duration plateau is seen. When unloading starts in GNF, the profiles suggest that it occurs initially elastically and then plastically. This elasto-plastic transition in unloading GNF leaves a shock signal, in an analogous manner to the HEL on loading. The wave on its reflection seems to have fractured the material as seen in spalling *tail* towards the end. A similar trend is seen for $\dot\varepsilon$ vs. ε. The other profiles related to $\dot\varepsilon$, force vs. time are provided in Appendix A (Figures A.3.1 and A.3.2). Figure 3.11 shows the schematic of the effect of shock impulse on GNF.

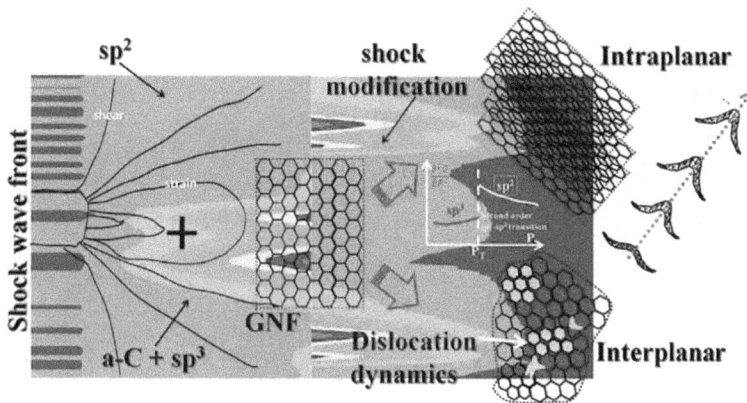

FIGURE 3.11 Scheme of effect of shock on GNF within layer and layer conjugation. Background colour contours show inherent and modified sp^2 and sp^3 zones, obtained by the algorithm of signal processing of Raman bands. The emergence of disorder and dislocation dynamics are shown.

3.4.1.3 Fractography Analysis: Electron Microscopy Imaging

After drying, under IR heating, the samples were extensively subjected to fractography analysis using optical as well as electron microscopy. The optical microscopy was carried out at 100× magnification. A large number of sites were surveyed (500+), and a few typical micrographs of the HSR GNF and GNF are provided in Appendix A (Figure A.3.3).

The effects of shock waves on local structural transitions such as dislocation/slip in GNF were investigated using dark field STEM imaging. The studies revealed the effect of impact on the distribution of transition regions. The observed transitions could be described in terms of strain solitons either with the help of a *Frenkel–Kontorova* model, or with more complex atomistic models that include the structural transition as well as the delamination of GNF layers [34,35]. The structural transition is also briefly discussed. Figure 3.12 is a typical dark field scanning (S) TEM micrograph recorded for GNF and HSR GNF.

In image (a), one can see the smooth wrinkles or contours on the GNF, prior to impact. The contour pattern is almost random in direction and seems to be intercepted at several places (shown by the dotted circles). These interceptions are the sources of in-plane topological disorder and may run parallel to the layer plane, on impact. They are seldom observed at the flake edge. The topology of wrinkles are somewhat protruded at some places. From this image, one cannot definitely comment on the basal plane that exists with dislocations. Moreover, image (a) seems to be a bit dark and non-transparent compared to image (b), recorded for HSR GNF. Nevertheless, for most of the images we have recorded a similar type of pattern for pristine GNF having an area ~25 μm² (5+ graphitic layers) as revealed by the optical and Raman analyses, presented appropriately. In (b), including inset, the scenario is somewhat different indicating a large amount of contours randomly overlapped, after impact. It shows that a large amount of dislocations are generated, glided/tilted in and out of the GNF flakes. At the dislocation sites, the effective shear force on GNF with respect to its disordered part plays a mutual role in the motion of dislocation, which may cause variation in basal plane stress. This results in the expansion or the contraction of the conjugated graphitic layers. Such dislocations have planar movements through the crystal on which they can glide easily. However, their effective dynamics depends on the type of dislocation, the alignment of its **b,** and the overall crystallographic structure of GNF.

Image (b) also shows that the thickness of HSR GNF is somewhat decreased compared to the thickness of GNF in (a). After impact, the wrinkle patterns are prominent at the edges. The two left arrows in (b) indicate the introduction of gliding slip due to the movement of GNF over-layers

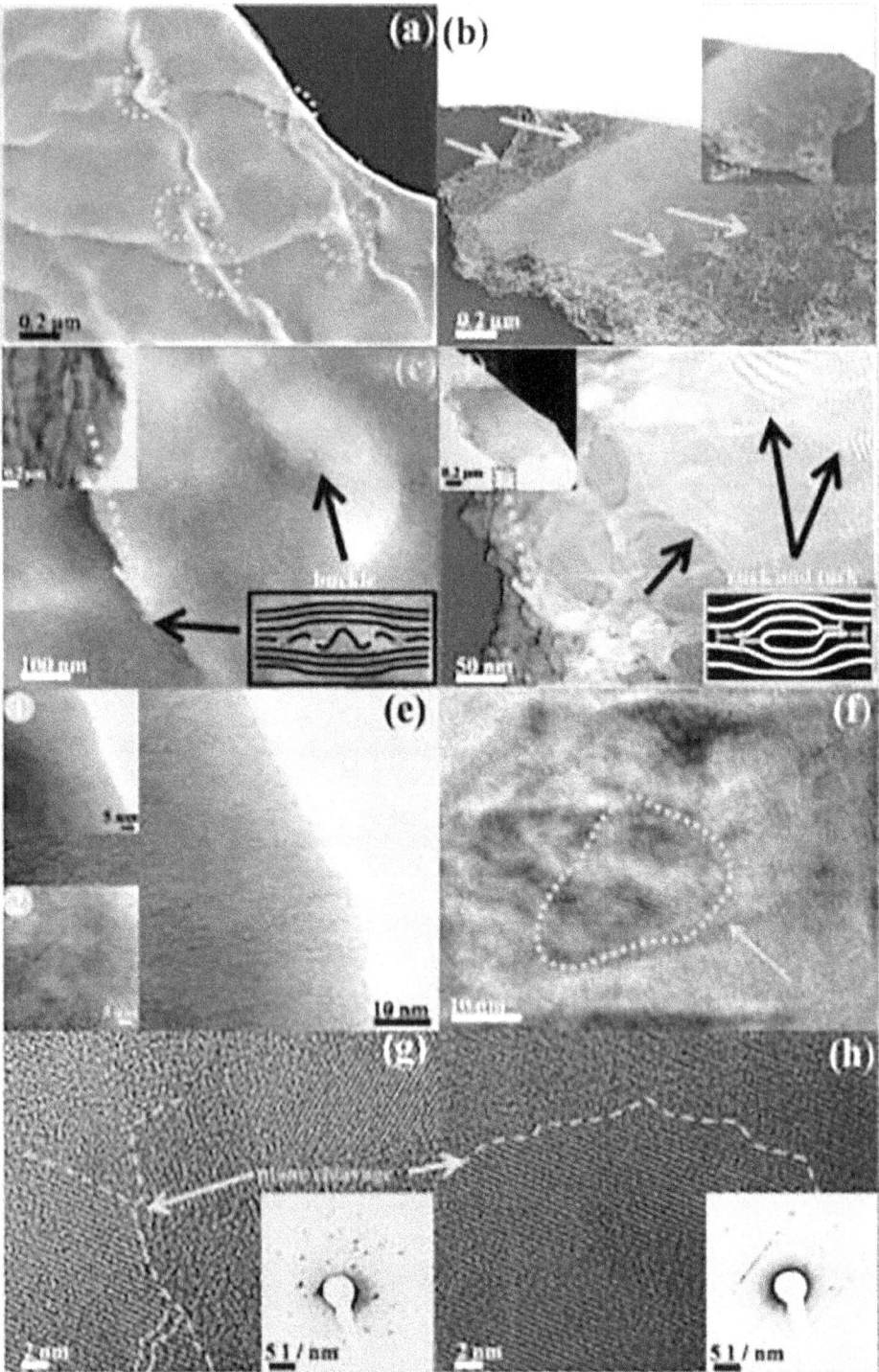

FIGURE 3.12 Scanning TEM image typically recorded for GNF (a, c, e) and HSR GNF (b, d, f, g, h). Image (a) dotted circles – intercepted wrinkle contours, (b) arrows – over-layer gliding of GNF, (c) zoomed in portion of inset revealing *buckled ripplocations*, (d) stacking faults, post-impact, generating *ruck and tuck* type of prismatic dislocation loops, (e) and (f) high-magnification images in which (f) shows a crystalline region (*line contour*), (g) and (h) *line contour* showing post-impact modification in local crystallinity of GNF (insets show SAED patterns).

(evident in (d), shown by a black arrow) imprinting buckling pattern on lower layers on either region of GNF (right-hand side arrows). The top portion of GNF is seen intact. Under the impact conditions, GNF experiences several dimensional changes. Such dimensional changes have also been observed previously in graphitic nano-flakes, giving rise to the contraction of the basal plane and exfoliation of the conjugated layers. As a result, the effective thickness of GNF seems to be decreased, pushing the conjugated planes apart and possibly causing expansion of planes in lateral direction. The area is further zoomed in, and more images were recorded at a higher magnification in order to investigate the transition regions of both the specimens.

In image (c), the magnified portion of warp in GNF is shown by an arrow connected to the corresponding inset from where a typical location is selected. There are two planes: upper and lower; both are having a homogeneous sp^2–sp^3 environment. For GNF, under no stress conditions, the periodic sp^2 lattice potential opposes the dislocation glide by lattice friction. The external stress required to overcome this potential and move the dislocation, known as *Peierls–Nabarro* stress, is effectively zero in GNF. It can glide the upper layer plane and pile up some residual stress as the lower layer plane repels this pile-up process. It results in the bending of the basal plane and delimitation of the adjacent plane to the dislocation glide plane that takes place in the form of *buckled planes* (schematically shown in (c)). These are purely van der Waals interactions that could be in the form of line defects compounded with surface ripples. The crystallographic dislocation character is known as *ripplocations*. After impact, there seems to be a basal movement of the GNF scaffolds as seen in image (d), as recorded for HSR GNF.

A large number of contours are seen with a larger interplanar displacement generating stacking faults (indicated by an upward black arrow). The stacking faults lead to the creation and migration of interstitial carbon atoms, which may form *ruck and tuck* type of prismatic dislocation loops (arrows in (d) with schematic representation) between the conjugated layers of GNF. This occurs due to the basal plane that freely moves along their glide planes (parallel to graphene sheets) without any resistance. Their motion can be described using the *Peierls–Nabarro* model [36–38].

At higher magnifications ((e) and (f)), the morphology of GNF and HSR GNF can be seen in greater detail. For GNF, no prominent crystalline region is observed, whereas for HSR GNF, i.e. (f), crystalline order is clearly visible (*dashed line contour* indicated by an arrow). It seems that the shock has introduced prominent ordering in GNF due to the *pinning* of dislocations at impurity (non-carbonaceous) sites or accumulation of a large amount of *Peierls–Nabarro* force at the impurity sites. The high level stress at such dislocations, away from prismatic dislocation loop, has different proportions of the edge character resulting in different amounts of line tensions. This is more prominently seen in images (g) and (h), as recorded for HSR GNF.

It seems that the impact has significantly modified the local homogeneity of sp^2/sp^3 zone, enriching the sp^2 content of GNF. The crystalline character and degree of crystallinity seem to be different along the plane. One can see various crystalline zones marked by the *dashed line contours*. The results also suggest that, on impact, multiple dislocations may be diffused out from the inner area of GNF and pile-up at the boundaries. Correspondingly, the recorded selected area electron diffraction (SAED) of crystalline zone is shown as an inset.

We have observed twinning in GNF as characterized by microscopy. However, we have not evaluated any structural parameters such as twist angle, stacking angle, and interlayer spacing. Figure 3.13 shown above is the recorded dark field TEM image for HSR GNF, including the SAED pattern. From (a), one can see long-range atomic ordering for GNF over-layers and it also confirms a kind of stacking that could be a Bernal or non-Bernal type. From (b), the twisted multilayer also shows long-range atomic ordering. The atomic ordering within GNF and on twisted surface is indicative of multiple strain-induced deformations in HSR GNF. The orientation (although not quantified) seems to be different as indicated by arrows in (a). The relative density of twin domains in (a) seems to be high compared to (b). The multiple zones seem to coexist with different stacking configurations and might be connected by discrete twin boundaries, which are not visible in both (a) and (b). The individual over-layer in twisted layer may experience continuous stress and

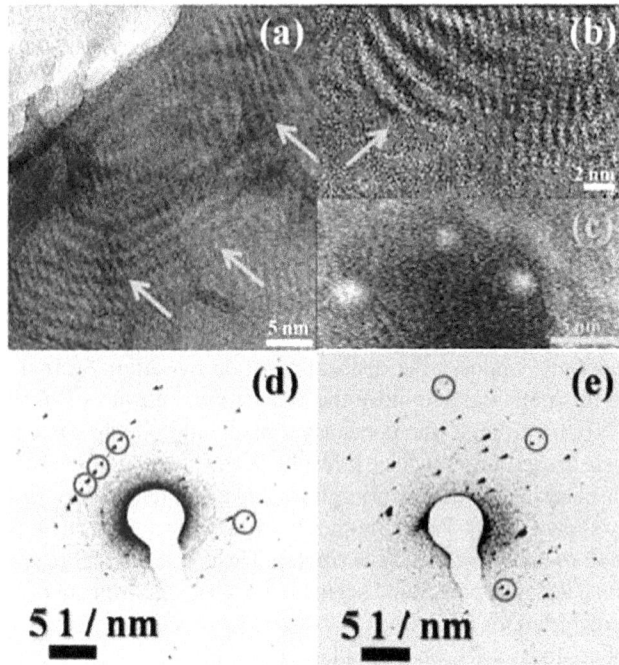

FIGURE 3.13 (a) and (b) Typical dark field TEM for HSR GNF. Arrows indicate twining and twist. (c) Dashed circles display point defects that emerged in GNF due to impact. (d) and (e) Corresponding SAED patterns for (a) and (b). Closed circles: coupled diffraction spots revealing twining.

shear independently. We have not estimated the twist angle distribution. Further, the SAED pattern recorded for these zones interestingly show coupled diffraction spots indicated by circles. In case of (a), the corresponding SAED shows a quite prominent twin-spot pattern, whereas in case of (b), twin spots were observed in few places (shown by the circles). In a metallic system such as GNF, the plastic deformation gets suppressed by the motion of dislocations when subjected to high strain rates. The twinning process while deformation is responsible for the hindrance of plastic deformation. In case of GNF, the observed dislocation slip and twinning might have promoted plastic deformation in a limited fashion, Moreover, (c) shows the existence of point defect zones in GNF as typically observed.

3.4.2 RAMAN SPECTROSCOPY: STATISTICAL MODEL

Raman spectroscopy is a simple, non-destructive, and standard characterization tool that provides maximum structural and electronic information on a large scale in a swift fashion. Mostly, carbon compounds are Raman active. However, the toll for its simplicity is paid when it comes to spectral interpretation [39,40]. In general, Raman spectra of carbon systems show only a few prominent features independent of the final nanocarbon structure, comprising of an intense band appearing between 1000 and 2000 cm^{-1} and few other second-order modulations. However, the shape, intensity, and position allow us to differentiate between amorphous carbon, metallic nanotube, few-layer graphene, and monolayer graphene [40].

Compounded with optical microscopy, a statistical analysis using Raman spectroscopy is presented. The optical spectroscopy was used for lateral area estimation of GNF and HSR GNF, whereas Raman spectroscopy determined a number of parameters such as crystalline length, L_a, disordered length, L_D, areal defect density (in cm^{-2}), the number of layers, and electron phonon coupling.

Figure 3.14 presents the Raman measurements carried out on GNF and HSR GNF samples at the excitation wavelength of ~532 nm. The Raman scanning was performed in a cascading fashion to optical imaging of the flakes (Figure 3.15).

In plot (a), the integrated I_D/I_G ratio as a function of the flake area is seen. The figures indicate an approximate lateral area (in μm^2). For GNF, the ratio is gradually increased by ~6 times. The large variation in the ratio is indicative of complex phases associated with graphitic carbon present in GNF in terms of bond disorder, heterogeneous clustering of the sp^3 phase, and the presence of sp^{2-} rings and chains. It also provides an idea that the amount of amorphization is high for larger flakes. In general, the amorphization trajectory consists of various stages such as graphite to nanocrystal-line graphite phase, nanocrystalline graphite to nucleation of low sp^3 amorphous carbon phases, and low sp^3 stage to high sp^3 amorphous carbon phase [40]. The trend is indicative of graphitized carbon rich in sp^3 amorphous carbon phase. For smaller GNF flakes, the sp^2 content is high as compared to the large-area flakes. After shock impact, the changes observed in the ratio are, dramatically, oppo-site. It shows a large amount of amorphization in smaller flakes, whereas larger flakes transform into crystalline graphitic carbon rich in sp^2 content. There could be several possibilities: The larger flakes might have shattered into the smaller flakes (a) by the inherent glissile nature of the disorder present in GNF, (b) by the large amount of interface modifications due to shock, and (c) by the com-plete detachment of one or more over-layer(s) of GNF from the parent flake. Currently, the origin of such behaviour is unclear and more experimentation is needed on the shock impact on GNF.

In plot (b), the estimated I_D/I_G ratio is connected to the crystalline length, L_a, by the *Tuinstra and Koening* (TK) relationship: $L_a (nm) = (2.4 \times 10^{-10}) \lambda_L^4 \dfrac{I_D}{I_G}$, where λ_L is the laser excitation. For GNF, the smaller flakes have a smaller magnitude of L_a (1.0±0.1 nm) due to the nearly crys-talline nature of the flakes, and as the area of the flake is increased, the estimated values of L_a (110±10 nm) are increased by more than 100 times. In the TK relation, it is assumed that the gra-phitic phase becomes uniformly nanocrystalline. Further, the TK relation is canonically connected to the ratio of phonon confinement (D peak) to the allowed phonon branching (G peak), which quan-tifies the forbidden phonon participating in the confinement process governed by the selection rule of the *amount of breaking* estimated in terms of an average inter-defect distance, L_D. It is given by: $\dfrac{I_D}{I_G} = \dfrac{C(\lambda_L)}{L_D^2}$, where $C(\lambda)$ is the phenomenological parameter, which is ~102 nm² (at 532 nm) and, L_D, in terms of defect density n_D, one can write: $n_D (cm^{-2}) = \dfrac{10^{-14}}{L_D^2}$. For HSR GNF, a drastic change in L_a, L_D, and n_D is consistent with the variations in I_D/I_G attributed to the emergence of mixed phase grain sizes, their impact modification resulting in the change in the volume fractions of sp^2/sp^3.

Plots (c) and (d), respectively, show variations in the FWHM of G peak, for GNF and HSR GNF to their extracted L_a. The G peak originates due to the doubly degenerated zone centre E_{2g} mode associated with the second-order zone–boundary phonons and is independent of 2D peak. For GNF, the FWHM is observed to be increased by eight times with a sequential increase in L_a, whereas the broad G peak for HSR GNF was found to be reduced by almost the same amount of order with an increase in L_a. On impact, the composition of GNF seems to be modified heavily in which the amount of amorphous C might be transformed into the smaller sp^2 fraction, topologically disordered sp^3 phases, and change into the ring configuration to form the chains. As a result, the narrowing of FWHM for higher L_a is observed.

Figure 3.16 shows the features of 2D band that emerged (for GNF: left panel plots (a)–(d) and for HSR GNF: right panel plots (e)–(h)). The band emerged between 2635 and 2710 cm⁻¹ and is associated with the D peak that in general emerges at ~1350 cm⁻¹ for both the systems. In plot (a), the 2D band for smaller flakes was found to be single and sharp at ~2644 cm⁻¹, indicating the presence of 1L of GNF. However, for a relatively large area (10–15 μm^2), the band splits into two components showing 2–3L of GNF (plot (b)). In (c), as the area increases further, the peak splits into several components, typically 4, indicating the increase in the number of layers (4–5L). For largest GNF (shown in (d)), a *graphite-like* band split is noted. For HSR GNF, the graphitic phase is observed for a smaller area (plot

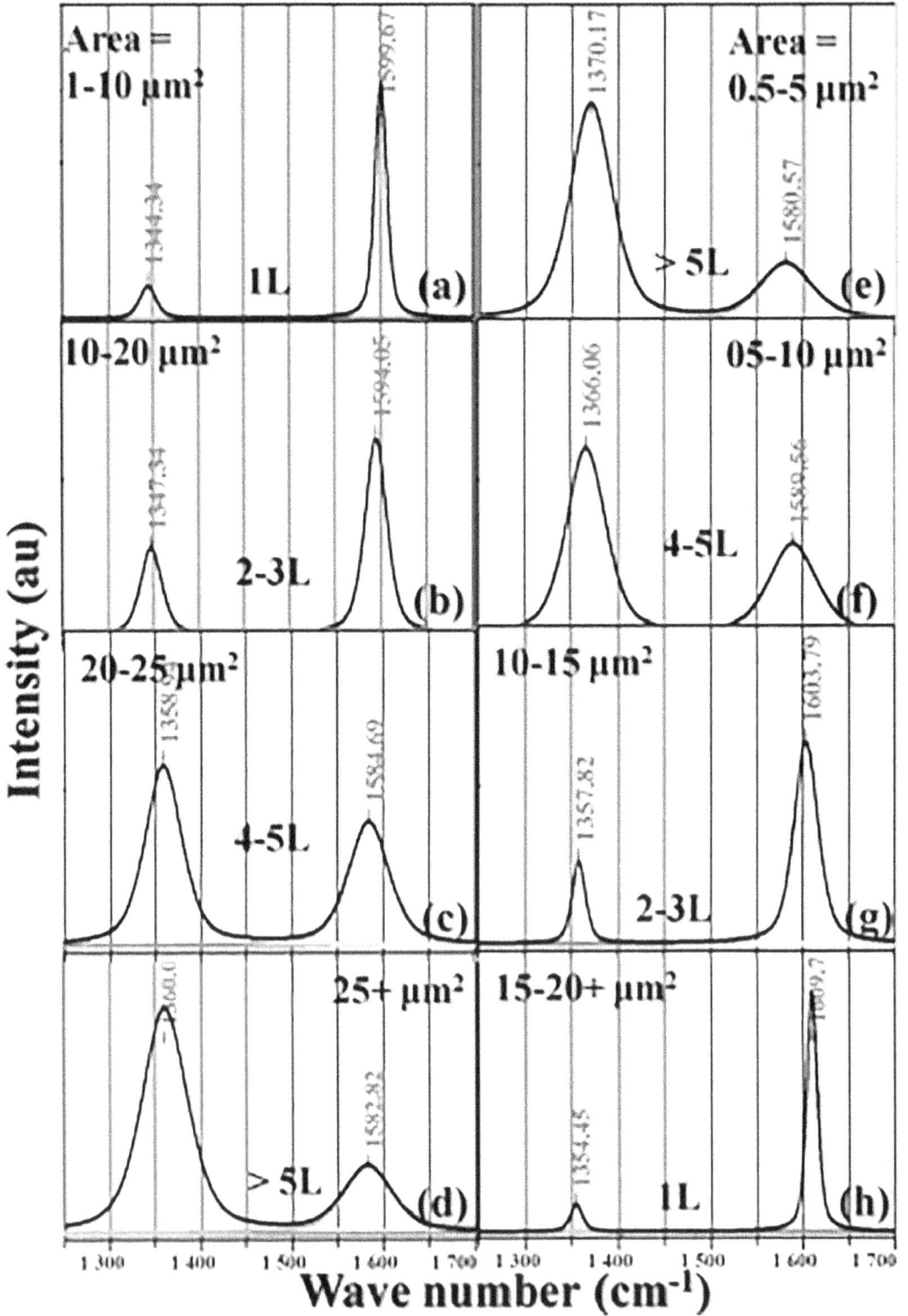

FIGURE 3.14 Recorded Raman spectra over 1000–2000 cm⁻¹ for GNF (a–d) and HSR GNF (e–h).

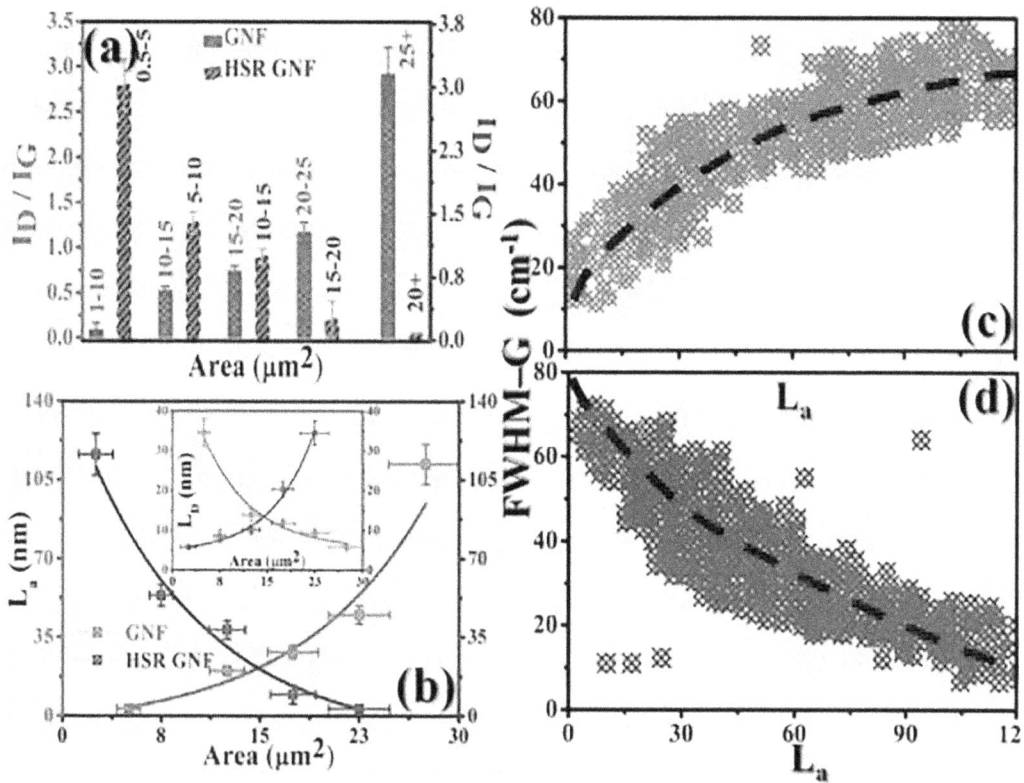

FIGURE 3.15 Raman spectroscopy measurements ($\lambda \sim 532\,nm$) for GNF and HSR GNF revealing various statistical parameters such as (a) I_D/I_G, (b) crystalline length, L_a ((inset) disordered length, L_D) vs approximate lateral flake area (μm^2). In plots (c) and (d), variations in FWHM-G (cm^{-1}) with the estimated L_a are plotted, roughly, for ~500 data points.

€). In contrast, for larger areas, a single sharp peak indicates the presence of 1L of HSR GNF (h). For turbostratic graphite, a single 2D peak is reported in the literature. However, it is found to be absent in our studies. From the estimated L_a and L_D, a number of other parameters such as the defect present and populated, i.e. n_D, electron–phonon coupling, γ_{EPC}, Fermi velocity, V_F, and the number of layers as a function of lateral flake area have been estimated and are provided in Table 3.1.

The EPC is computed from the G peak line width and correlated with the Fermi velocity, V_F, (cm/s) by the relation: $EPC = \dfrac{\sqrt{3}}{4} \dfrac{a_0^2}{\mu} \dfrac{EPC\Gamma^2}{v_F^2}$, where μ is the reduced mass, $EPC\Gamma^2$ is the theoreti-

cal Γ–phonon branching parameter (47 $(eV/Å)^2$), and a_o is 3.14 Å. However, the estimated L_a, L_D, and n_D are associated with γ (shear strain). For GNF, the movement of chiral dislocations will produce elastic shear strain, γ_E, where it is assumed that the dislocations do not interact, within elastic limit. The γ_E, n_D, and L_D correlation is given by: $\gamma_E = n_D \cdot b \cdot L_D$, yielding $\approx 2 \times 10^{-10}$, for GNF, whereas, in plastic regime, $\gamma_P \approx 8 \times 10^{-10}$. The disorder velocity, ν, given by: $\dot{\gamma} = n_D \cdot b \cdot \nu$, is estimated to be $\approx 2000 \pm 50$ cm/s.

In another study, the surface area, S_A, values for GNF and HSR GNF were determined by the BET – low-temperature nitrogen adsorption–desorption measurements. Prior to these studies, the samples were dried at 110°C under vacuum conditions for about 2 hours to remove moisture content. Figure 3.17 shows the pore size distribution for both the systems. The curves provide information about the amount of adsorbents condensed by the capillary action onto the surface under examination. The action depends on the relative pressure, the molecular dimensions of the adsorbents on the available surface area, and the nature of the adsorbed layers (mono-/multimolecular, condensed,

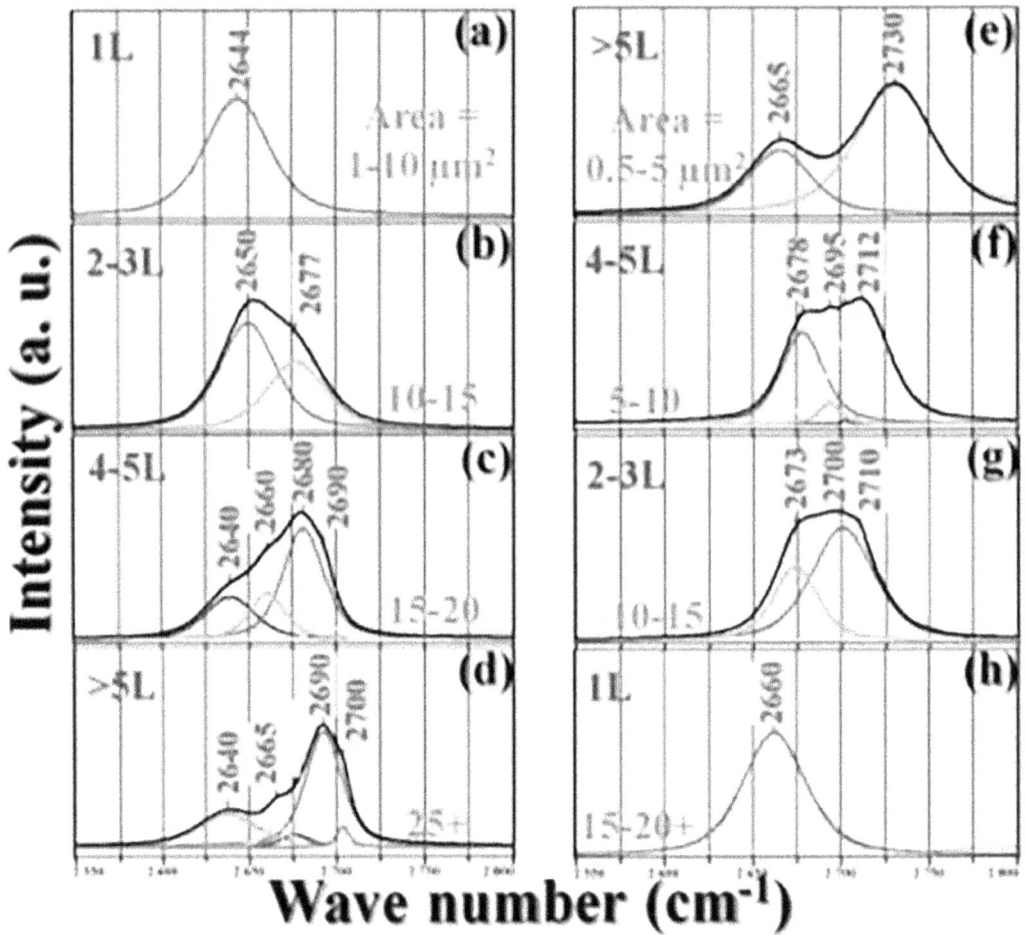

FIGURE 3.16 Evolution of the 2D Raman spectra at $\lambda \sim 532$ nm recorded for a representative (a–d) GNF and (e–h) HSR GNF indicating lateral flake area and the corresponding number of layers. The fit to various 2D components for both the systems is shown.

TABLE 3.1

Estimated Parameters for GNF and HSR GNF from Collected Raman Statistics

	GNF					HSR GNF			
Area (μm^2)	$n_D \times 10^{-16}$ (cm^{-2})	γ_{EPC} (cm^{-1})	$V_F \times 10^6$ (cm^2/s)	No. of Layers	Area (μm^2)	$n_D \times 10^{-16}$ (cm^{-2})	γ_{EPC} (cm^{-1})	$V_F \times 10^6$ (cm^2/s)	No. of Layers
1–10	0.1±0.03	15±04	~2.00	1	0.5–5	3.3±0.4	70±05	~8.73	10+
10–15	0.5±0.02	35±10	~2.45	2–3	05–10	2.5±0.6	53±09	~6.56	5–6
15–20	0.7±0.03	64±07	~3.17	4–5	10–15	1.7±0.9	37±04	~6.03	4–5
20–25	1.1±0.60	95±12	~4.20	5–7	15–20	1.0±0.20	25±12	~3.96	2–3
25+	2.9±0.10	115±10	~4.89	7+	20+	0.9±0.1	10±02	~2.50	1

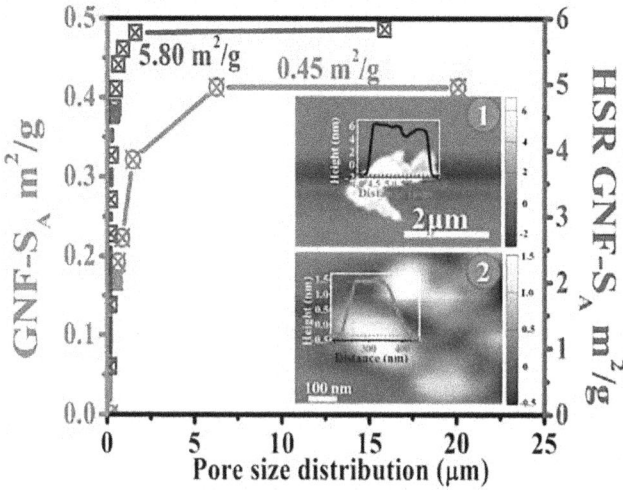

FIGURE 3.17 Measured BET–BJH curves for pore size distribution of GNF and HSR GNF. Inset: AFM images of (1) GNF and (2) HSR GNF (typical small flake) with line scan indicating the height of the flake in nm.

etc.). For GNF and HSR GNF, the average S_A was estimated to be ~0.45 and 5.8 m^2/g, respectively. The porosity of the HSR GNF was found to be quite high as compared to GNF. As per IUPAC categorization, the porosity of the medium can be classified into three channels: micropores (<2 nm), mesopores (2–50 nm), and macropores (>50 nm). The values indicate that GNF consisted of a widely distributed, heterogeneous laminar network of carbon which, on impact, is expanded, revealing relatively high porosity and S_A.

In Figure 3.17, we present Raman fingerprints for 1L to few layers of both GNF systems; however, AFM is another method to identify the number of layers, approximately. Due to chemical contrast between GNF and the substrate, one may find an apparent chemical thickness of 0.5–1 nm, which is comparably higher than the interlayer graphitic spacing. In practice, it is difficult to distinguish selectively mono-, bi-, tri-, etc., layers by AFM because GNF contains numerous topological disorders such as folds, warps, and wrinkles that change the local electronic band structure and lead to ambiguous thickness measurements. The AFM measurements were performed on GNF and HSR GNF and are shown in Figure 3.17 inset (1) and (2). The line scan of GNF shows ~6 nm of height, which may contain >5 L. For HSR GNF, the line scan shows ~1.5 nm height, showing the presence of 2–3 L of graphene. The energy of the shock wave, dissipated within GNF, as it travels through GNF can lead to a numerous physical and chemical changes, as briefly discussed below.

3.4.2.1 Shock Wave–GNF Interactions: Slip Mechanism and Peierls–Nabarro Stresses

The shock impulse is a transient *stress* acting as a δ-function kind of excitation perturbation [1]. It offers a negligibly small shear compared to the compressive hydrostatic component of the medium like GNF. Moreover, it induces a velocity of a disturbance, propagating isentropically, given, approximately, by: $\sqrt{\dfrac{d\sigma}{d\varepsilon}\Big/\rho}$ which is equivalent to $\sqrt{dP\big/dV}$ (where ρ is density, and P and V are pressure and volume, respectively). Such perturbation varies with time, rises in strength as it travels through, and activates disorder [40]. The disorder plays a pivotal role against the dynamic flow strength of the GNF. On propagation, there occurs a mismatch between the material flow in GNF and the speed of the shock wave front causing a deformation beyond the elastic limit setting in a plastic deformation that leads to spall/fracture. GNF is a covalent system that includes few graphene layers having interlayer spacing (*van der Waals distance, r_{vw}*)$\geq$0.335 nm with Bernal (ABAB...) and non-Bernal stacking order. GNF is also compounded with inherent disorders such as vacancy defects, edge dislocation, slip, prismatic loop, and twinning making them act like a *metal* matrix that undergoes strain deformation, which can be explained reasonably well under the elastic approximation. Strain deformation generates movement of disorder; however, each one possesses complex dynamics.

Further, in monolayer graphene, the dislocation motion is majorly limited because of the fact that it has only a single atomic layer in thickness, hence no slip action [38]. The slip system in GNF is complex and originates due to topological imperfections such as edge, extra/missing lattice plane, and screw dislocations appearing in an isolated or in hybrid form. The amount of slip could, indirectly, be quantified by the magnitude of Burgers (offset) vector (**b**). For an ideal orientation arrangement, the force on the dislocation per unit length is given by: $F = \tau \, \boldsymbol{b}$, where τ is the shear stress. Under the action of τ, the dislocation moves and assumes to produce a shear strain, γ.

It is difficult to apply a specific elastic theory that can describe the entire mechanical behaviour of GNF. Most of the formalisms have exceptions due to the wide number of deformation mechanisms. Using Raman analysis, the slip movement has been quantified as the agent for such deformation by treating the dynamical variable of force constant, k_q, at the *footy* of τ. The estimate for k_q is approximately given by the relation: $\omega_q = \sqrt{k_q/\mu}$, where μ is the reduced atomic mass of carbon. For GNF, from the positions of D and G peaks, the values estimated for k_q were 650$\pm$05 and 920$\pm$12 N/m, respectively. The corresponding bond length of –C–C– and –C=C– was 1.54$\pm$0.12 and 1.34$\pm$0.20 Å, respectively. After impact, they were reduced to 600$\pm$10 (D) and 880$\pm$17 (G) N/m with respective bond deformations of 2.80$\pm$0.10 and 3.65$\pm$0.17 Å. There are frictional forces resisting the movement of a slip; thus, a force is required to make it move. At sp^3 site, such a force movement, i.e. estimated F_D, is ~0.5$\pm$0.02 nN, whereas, at sp^2, F_G ~1.0$\pm$0.22 nN. Thus, the out-plane slip motion is almost 50% high compared to in-plane slip motion.

Further, the slip in GNF crystal is assumed to be parallel to the layer plane with a vectorially in-plane-oriented Burger component. This would build up residual internal stresses. It separates the crystallographic δ-neighbourhood of GNF into two components: one having an extra plane and the other missing one. Such a δ-neighbourhood above and below the gliding plane of GNF crystal can be modelled with classical *Peierls–Nabarro* stress. The interfacing gliding plane separating these two regions experiences two net resultant forces: one compression from the zone having an extra plane to extend the dislocation and, second, missing plane zone to neutralize the force by layer alignment of missing plane to shorten the dislocation. They balance each other at the equilibrium; moreover, the extension of dislocation depends on the Poisson's ratio, σ. Under the influence of external shear stress, the slip dislocation moves and exceeds the critical stress limit that strongly depends on the ratio of in-plane to out-planes stress components and anisotropy of the crystal. The slip ratio is given by: $\dfrac{\sigma_\parallel}{\sigma_\perp} = \dfrac{4\pi}{(1-\sigma)} \times \left\{5.8 - \log(1-\sigma)\right\} \times e^{\frac{-4\pi}{(1-\sigma)}}$, where $\sigma_\parallel$ and $\sigma_\perp$ are in-plane and out-plane stress components. From the estimated values of σ, for GNF (0.13) and HSR GNF (0.45), the value of slip ratio $\sigma_\parallel/\sigma_\perp$ is ~7.1$\times$10^{-5} for GNF, which is lowered to ~2.8$\times$10^{-8} after impact.

For monolayer graphene, although the *Peierls–Nabarro* stress is substantially high due to the homogeneous, impurity-free, and strong covalent environment, one would expect the release of such stress only at the edge sites and not within the graphene layer. In case of bilayer graphene, the dislocation cannot escape as it is confined between two monolayers . As a result, for such a bilayer system, there is no stacking fault, which leads to a characteristic dislocation pattern that corresponds to the alternation of AB to AC change of the stacking order. The strain stored in the bilayer graphene results in the buckling of the bilayer graphene. For multilayered bulk graphite, the existence of *Peierls–Nabarro* stresses is restricted to the Basal plane dislocation movement. In case of GNF, which is having a finite size as well as the number of layer distribution, they may experience extensive image forces, which may attract the movement of dislocation to pop up onto the surface to

TABLE 3.2

Estimated Dynamic Mechanical Parameters for GNF from *Lagrange–Rankine–Hugoniot* Model under *Peierls–Nabarro* Stress Matching Condition

S. No.	Parameter (SI Unit)	GNF	HSR GNF
1	Young's modulus	>50 MPa	-
2	Incident shock wave velocity, V_o (m/s)	5.0×10^3	2.5×10^3 (Avg.)
3	Inter-particle separation (m/s)	-	3.5×10^3
4	Stacking faults (m/s)	-	$< 2.5 \times 10^3$
5	Edge/corner fracture	-	2.0×10^3
6	Elasto-plastic time constant, t_c	2.1×10^{-5}/s (*metal-like*)	-
7	Hugoniot elastic limit (GPa)	-	9.13 (*ceramic-like*)
8	Flake size reduction	-	50%
9	Force constant (–C–C–) (N/m)	650±35	600±15
10	Force constant (–C=C–) (N/m)	920±70	850±40
11	Bond length (Å)	1.54	1.34
12	Bond deformation (Å)	2.80 (80%)	3.65 (>150%)
13	Dynamic force (nN)	-	$0.5@sp^3$ and $1.0@sp^2$
14	Crystalline length (nm)	30	10
15	Inter-defect distance (nm)	05	20
16	Dislocation density (cm^{-2})	6.6×10^{-13}	4.0×10^{-12}
17	Shear strain	$\approx 2 \times 10^{-10}$ (elastic)	$\approx 8 \times 10^{-10}$ (plastic)
18	Disorder velocity (cm/s)	-	2000±50
19	Electron–phonon coupling (N/m)@300 K	10	25
20	Fermi velocity (m/s)	2×10^6	8×10^6
21	Eff. sonic impedance $\left(\dfrac{kg}{m^2-s}\right)$	1100	550
22	Shock impedance $\left(\dfrac{kg}{m^2-s}\right)$	30@(–C–C–); 25@(–C=C–)	60@(–C–C–); 80@(–C=C–)
23	Compressive strength $\left(\dfrac{m^2}{N-s}\right)$	4×10^{-5}@(–C–C–) 5×10^{-5}@(–C=C–)	-
24	Gibbs free energy (eV)	-	~50 meV
25	Damping coefficient $\left(\dfrac{N-s}{m^2}\right)$	3.7×10^{-5} (elastic)	9.4×10^{-5} (plastic)
26	Phonon viscosity $\left(\dfrac{N-s}{m^2}\right)$	3×10^9@(–C–C–) 5×10^9@(–C=C–)	6×10^9@(–C–C–) 2×10^9@(–C=C–)

release strain energy. Due to inherent disorders present in GNF, under applied stresses, the dislocation starts gliding and results in the lowering of *Peierls–Nabarro* stress, as estimated above. This is how the slip system in GNF can be quantified [36]. Table 3.2 shows the summary of dynamic parameters estimated for GNF.

By and large, the incident transonic impulse (>1.1 Mach number, ratio of sound (340.3 m/s) to incident wave speed) damps its ~65% of energy within GNF at 250 m/s. Only a small fraction of wave, ~15%, was transmitted.

Sometimes, it is very important to enhance energy dissipation rather than its storage to damp shock and acceleration, effectively, useful in building armours, blocks, bunkers, etc. Our study provides an insight into how shock waves are propagated and damped within nanocarbon materials such as GNF, which is important from the point of view of armour applications in the strategic sector.

3.5 SURFACE INTERACTIONS OF TRANSONIC SHOCK WAVE WITH GLNR

In this section, we present the shock mitigation properties of graphene-like nano-ribbons (GLNR). Particularly, we address the fundamental questions such as how shock waves interact with GLNR, what happens to its morphology and microstructure with dynamic pressure, how inherent disorders and defects play a role in this resulting various microscopic failures such as slip and twinning. The samples were further investigated using optical, field emission scanning electron microscopy (FESEM), Raman, Fourier transform infrared (FTIR), and UV–visible spectroscopies. The transmitted and reflected shock wave signals were *signal processed* for damage feature extraction in GLNR. Broadly, the shock wave–GLNR interaction process involves microstructure modifications related to glissile dynamics, flow hardening, transient phase transformation, physico-chemical modifications, phonon drag, and importantly, shock damping. Details are presented.

For this study, GLNR were obtained by the synthesis protocols presented in Section 2.3.3, whereas sHPB measurements were performed as per the procedures mentioned in Section 3.6. GLNR samples subjected to sHPB were termed as HSR GLNR samples [41,42].

For analysing mechanical properties, we assumed that the deformation in GLNR is uniform maintaining stress equilibrium, while on impact, the frictional and other dynamic effects were neglected, including linear variations in particle velocity (volume strain) with shock velocity.

3.5.1 MECHANICAL BEHAVIOUR

Figure 3.18a shows the σ–ε curve for GLNR fitted by a bilinear function, in the first stage, which is elastic in nature. The power function is fitted of the type: $\sigma = \sigma_0 + k\,\epsilon^n$, where k is the pre-exponential factor (0.13) and the work hardening coefficient $n = 0.17$. The velocity, V_s, with which shock wave is incident is ~3.5×10^3 m/s, given as above. The GLNR departs from its elastic limit with large dispersion in V_o. The characteristic time parameter, t_c, involved in the phase transformation, from elastic to plastic region, is given by: $t_c = \dfrac{\partial\sigma/\partial\dot{\varepsilon}}{\partial\sigma/\partial\varepsilon}$ and is estimated to be ~10^{-5}/s, as given in plot (b).

The gradual decrease in σ is indicative of ε propagation in GLNR matrix has experienced the strain flow hardening. The change has occurred in three stages. The $\dot{\sigma}_1$ is attributed to the strain impediment offered by GLNR in the form of inter-particle plastic deformation. The $\dot{\sigma}_2$ for plane gliding of conjugated ribbon stacks (evident prominently in, subsequent, Raman and microscopy analyses). The $\dot{\sigma}_3$, finally, results in the structural damage of the ribbon, at the terminal stage. The effect seems to be cascading. For an ideal shock wave, a discontinuous shear wave front follows a plateau at the top followed by the gradual decay of impulse, as shown in the inset in Figure 3.18. In

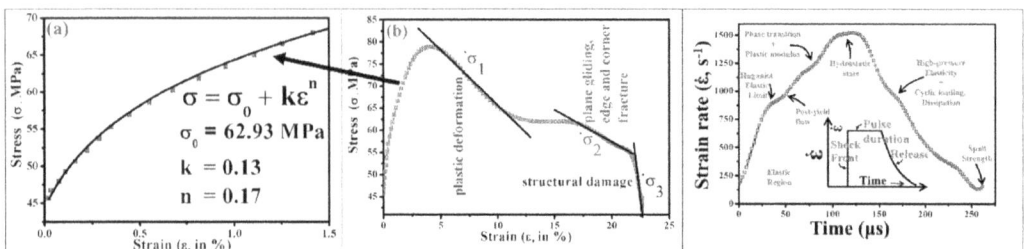

FIGURE 3.18 (a) Magnified region of the stress, (σ, MPa) vs % strain, ε, curve (arrow: elastic region). The tangent shows the estimated strain hardening time parameter, (b) corresponding plastic regime showing variations in stress rate, σ_i, in three stages (shown by slope lines) of dynamic deformation in GLNR, and (c) variations in strain rate, $\dot{\varepsilon}$, with time, imprinting a specific shock wave profile on 2D GLNR exhibiting number of stages.

contrast, realistic shock waves imprint a number of peculiar features that are pressure and material dependent. For GLNR, the rate of rise in volumetric stress is much higher, under elastic conditions, which leads to Hugoniot elastic limit (HEL) at about 8 GPa. The flow strength of any medium depends on the pressure employed. It can be independent (Hugoniot) of pressure or change causing softening or hardening of material. In plot (c), at HEL, elastic and plastic wave components get separated. The elastic fraction, within HEL, propagates with a higher velocity than plastic wave. After crossing HEL, the rise in pressure is steady till the approach of top plateau region due to continuous plastic wave front nature, and mainly, such an increase in pressure is dictated by constitutive parameters of GLNR. The resultant phase transition is not prominent in the obtained wave profile due to a significant amount of elastic and plastic wave front separation. The plateau zone is observed at the top of the recorded profile. The pressure unloading region in GLNR suggests elastic deformation followed by plastic one and identical to shock loading subsequent to HEL. The short tail shows the fracture of GLNR [42].

3.5.1.1 Fractographic Analysis

Both GLNR and HSR GLNR were extensively studied using optical as well as electron microscopy. The optical microscopy was carried out at 100× magnification, surveying a large number of sites, and the typical micrographs are shown in Figure 3.19.

Figure 3.19a and (b) shows GLNR and HSR GLNR, respectively. The images were processed for their contrast with optimum brightness and sharpness to reveal the features of the ribbons. The GLNR were mostly seen with sharp edges and a prominent rectangular shape. Even smaller ribbons carried similar structural features; however, they are not provided here. The ribbons were thick and may have several conjugated graphene layers within them. Image (b) is a typical HSR GLNR, which is observed to be different in its morphology. It seems that, after impact, the size and shape have changed with a bit of elongation and loss of conjugated layers. Images (c) and (d) are representative FESEM images of GLNR and fractured HSR GLNR. The resemblance of optical images can be seen in FESEM imaging, in terms of shape, sharp edges, and rectangular shape of the ribbon geometry. Further, the fractography analysis provides important clue for the underlying mechanism of deformation, fracture, and spall due to the shock experienced by GLNR. Images (e) and (f) are HRTEM images recorded for HSR GLNR. In FESEM, it is quite evident that the individual surface of the ribbon is acting as a stress concentration zone. The *warps* that are visible on the surface of GLNR, in a moderate fashion, are indicative of the flow of stress wave front within GLNR, which may have undergone breakage of the entire ribbon structure. The effect of shock on local structure modifications such as dislocations, slips, and disorders can be revealed in depth in TEM investigations. By and large, the images showed a detail change in morphology of ribbon and effect of impact. The HSR GLNR has displayed typical two/three *step-like* features in image (e) inset. The arrow connected to the inset shows the existence of *micro-cracking* originated in one of the steps. The *crack tip* is located around the junction of two nearly orthogonal cut sides of the step. The track of the crack is almost straight and narrowing towards the termination. The track length is ~3–3.5 μm with width roughly 150–200 nm. It seems that the crack ends with generating stacking faults in the over-layer graphene ribbon. Image (f) displays more features such as a circular bend at the end of the track of the crack followed by warp-like exfoliation. Startlingly, in another step crack is not seen. The blur rings recorded in the SAED pattern is indicative of less degree of crystallinity in GLNR. Broadly, the degree of crystallinity is not changed even after impact [42].

3.5.1.2 Raman Spectroscopy

Figure 3.20 shows the recorded Raman spectra for (a) GLNR and (b) HSR GLNR at 532 nm photoexcitation; the corresponding confocal Raman imaging is shown in (c) and (d). For GLNR, D and G peaks appear, respectively, at ~1333 and 1593 cm^{-1}. The estimated dynamic variable of force constant, k_q, was ~460 and 650 N/m, respectively, at D and G sites. The corresponding bond length of –C–C– and –C=C– was ~1.60 and 1.20 Å, respectively. After impact, they were reduced to ~410

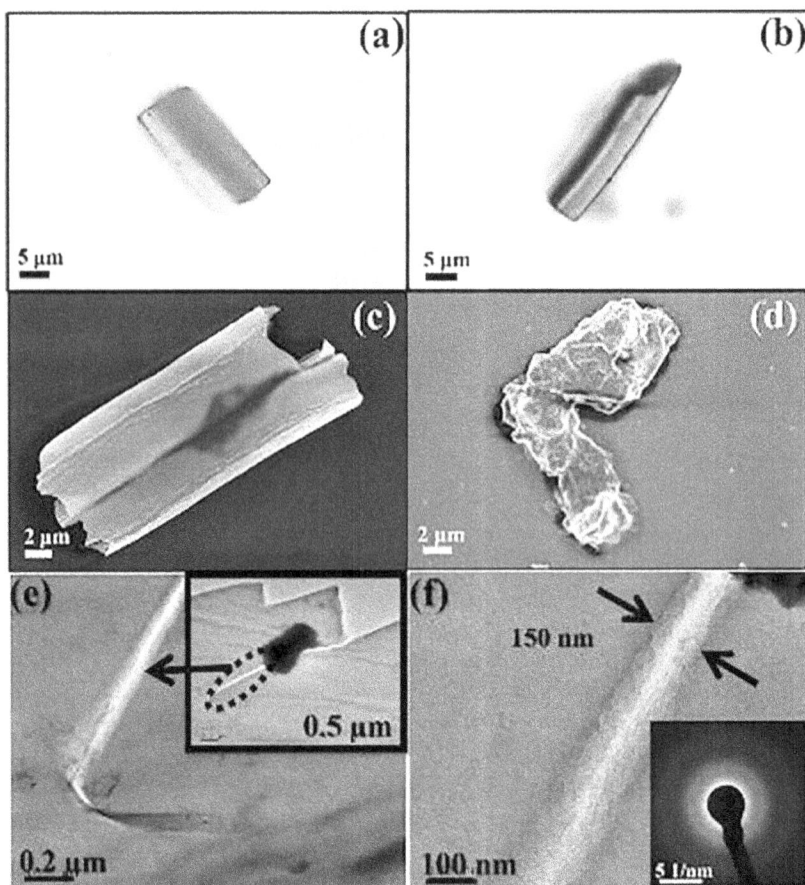

FIGURE 3.19 Typical optical micrographs at 100× for (a) GLNR and (b) HSR GLNR, respectively, (c), (d) FESEM images, and (e), (f) HRTEM images of fractured HSR GLNR. Optical micrography was used to obtain more insights into the nature shape and size of pre- and post-impacted GLNR. For better imaging, the suspended GLNR (HSR GLNR) solution was drop-cast onto silicon wafer and dried. The optical imaging was carried out at several sites in a *raster scan line* fashion to collect statistics on ribbons.

(D) and 600 (G) N/m with respective bond deformation of 2.85 and 3.50 Å. The plastic deformation in GLNR cannot be explained by applying a specific theory. The challenges come due to the for-mulism having numerous deformation mechanisms such as dislocation glide, mechanical twinning, and phase transformations. There are frictional forces resisting the movement of a dislocation; thus, a mechanical force is required to make it move. At sp^3 site, such a force movement, i.e. estimated F_D, is ~0.3 nN, whereas, at sp^2, F_G is ~0.6 nN. The in-plane shear stress is 33% high compared to out plane.

There were a number of Raman features noted for GLNR, similar to GNF, as presented in Section 3.5. For GLNR, D peak deconvolutions were composed of two prominent components: one at ~1200 cm^{-1} attributed to amorphous (a–C–O) carboxyl phase and another broad one at 616 cm^{-1} attributed to quasi-crystalline carbon (qc-C). There was a significant upshift in vibration modes, on impact. The D peak was shifted to ~1326 cm^{-1}, whereas the G peak was shifted to 1583 cm^{-1} for HSR GLNR. The a-C is also blueshifted by ~90 cm^{-1} with total disappearance of the qc-C com-ponent. The upshifted D and G show strain and frustration at molecular levels of HSR GLNR. At 532 nm excitation, the value of I_D/I_G for GLNR was 1.55, which was reduced to 0.44 after impact. Correspondingly, the crystalline length, L_a, was found to be ~17 nm for GLNR and 7 nm for HSR

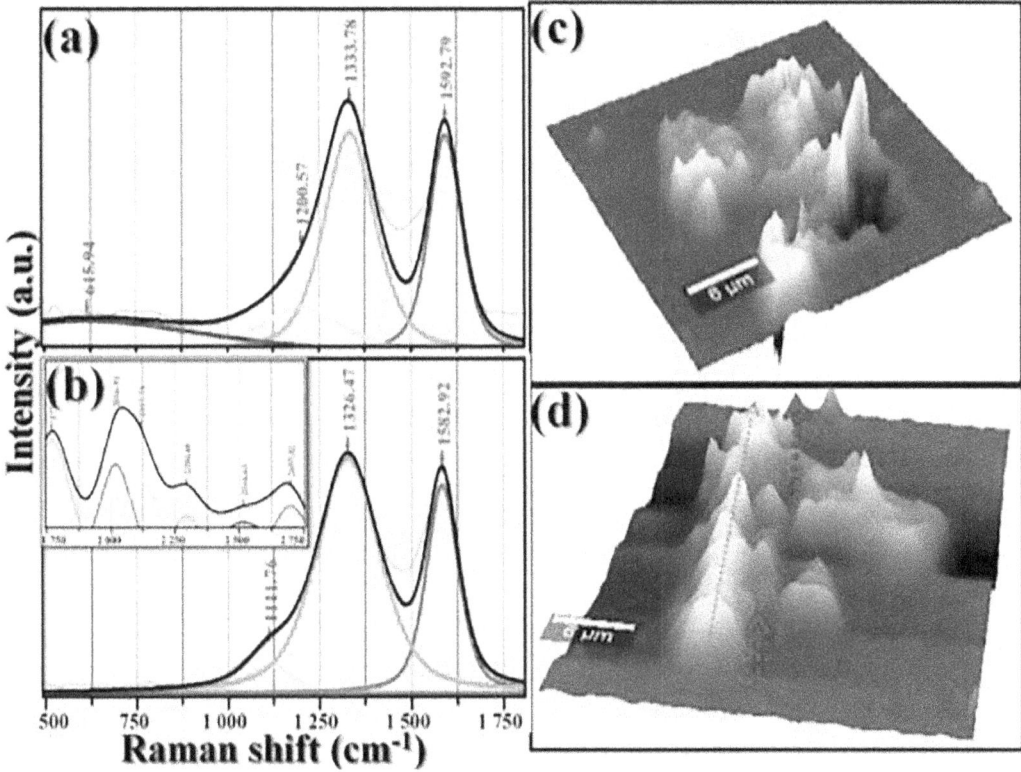

FIGURE 3.20 Raman spectrographs recorded for (a) GLNR and (b) HSR GLNR (inset: 2D peaks), corresponding confocal Raman imaging (c) and (d) (scan area 40 × 40 μm²). Arrows indicate shock wave-induced redistribution of *D* and *G* zones.

GLNR. The change was attributed to the decrease in sp^2 content and the increase in sp^3 zone by impact. In addition, the I_D/I_G is correlated to inter-defect distance L_D and defect density n_D. For GLNR, L_D and n_D were, respectively, ~ 15 nm and ~10^{-13}/cm², whereas, for HSR GLNR, they were ~10 nm and ~10^{-12}/cm², respectively. The derived values of L_a, L_D, and n_D are, particularly, connected to shear strain, γ. The movement of chiral dislocations in GLNR may produce elastic shear strain, γ_E, with the assumption that dislocations are interacting within the elastic limit. The relation between n_D, γ_E, and L_D is given by: $\gamma_E = n_D \times b \times L_D$, which yields a value >$10^{-9}$, whereas in plastic regime, γ_P is >10^{-9}, for GLNR. The disorder speed, ν, is given by: $\dot{\gamma} = n_D \times b \times v$, and ν is found to be more than 1000 cm/s.

Further, the width of D peak in 2D nanocarbon system is connected with EPC having acoustic longitudinal and transverse branches. The FWHM for GLNR was 165 and 205 cm⁻¹ for HSR GLNR, and the corresponding estimated EPC was 5 and 35 Nm⁻¹. The EPC has a relation with Fermi velocity, V_F, which was calculated to be ~1×10^6 and 5×10^5 ms⁻¹ for GLNR and HSR GLNR, respectively. Interestingly, there was no 2D peak recorded for GLNR. However, for HSR GLNR, a prominent 2D band was observed, extending to 2000–2750 cm⁻¹. This shows the longitudinal effect of the impulse, which has exfoliated the conjugated layers of GLNR, introducing stacking faults.

Further, spectral confocal mapping is a promising technique successfully used, previously, to evaluate the nanofiller dispersion in polymer matrix [42]. We adopted a similar technique to assess our post-impacted GLNR. The features of Raman spectra collected from each specific spatial coordinate within the selected scan area are used as pixels for final imaging of the specific zone. The typical Raman spectra were recorded in the range of 1000–1800 cm⁻¹. However, the spectral variation over the range 1200–1600 cm⁻¹ (range covering D and G peaks of GLNR) was used to map the modifications in the post-impacted nanocomposites. The obtained images consisted of topology of sp^3 and sp^2 zones, which are shown in Figure 3.18c and d. The colour contrast between sp^3 and sp^2 zones is seen in both the images. The spectral image was taken in large areas (50×50 µm^2). The sp^3 phase, in image (c), was distributed homogeneously with a uniform spread within the planar sp^2 network. At some places, the degree of disorder in sp^3 zone was seen to be higher and appeared as a distinct dark region. In planar region, no such peculiarities have been observed. They were mostly smooth and flat. The features seen in image (d) are significantly different. There are two important observations: First, the sp^3 zone is redistributed in a systematic pattern, and second, it shows small warps in the sp^2 zone. Perhaps, it seems to be the imprint of the propagated shock impulse, which sequentially stressed the sp^3 zone (indicated by a horizontal arrow). The vertical arrows show its relative strength, which is enhanced at the boundary layers of GLNR. After impact, the trail of the damaged zone is seen [42].

3.5.1.3 Signal Processing Studies: Pressure Impulse Interaction with GLNR

The reflected and transmitted strain signals were processed further to extract information about how the strain energy is propagated in GLNR. The signals received are shown in Figure 3.21a. For this, signal processing studies were performed using short-time Fourier transform (STFT) technique to investigate the flow of strain energy in time and frequency domains. Images (b) and (c), respectively, show the reflected and transmitted strains in GLNR.

The STFT function is given by: $x\left([n], \omega\right) = X\left(m, \omega\right) = \sum_{m=0}^{m=\infty} x[n] \bullet \omega[n-m] e^{-j\omega n}$. The operation was performed on a time sequence, where $x[n]$ is the data array variable of time, ω, continuous signal frequency, and m, discrete time locality. The equation shows the DTFT of the signal with a window function described below. The STFT was computed by adopting the following procedure: (a) A finite number of signal data points (N) were taken from the input shock signal, where these N points were equal to the window size, (b) the window of the chosen type was used to multiply the extracted data, point by point, and (c) zeros were padded on both sides of the window. The provision was made if in case the window width becomes smaller than the size of the FFT, and (d) FFT was computed. Further, the following steps were performed to compute STFT by setting: (a) sampling interval as 1, (b) FFT length as 256, (c) window length as 256, (d) overlap as 128, and (e) the type of window as rectangle. The window function is given by: $\omega[n] = 1$, if $0 \leq n \leq N-1$; $= 0$, otherwise. The output results are complex valued data matrix (N×M) with the conditions: $N = \dfrac{F_w}{2} + 1$ if input signal data real; else, $= F_w$ complex and

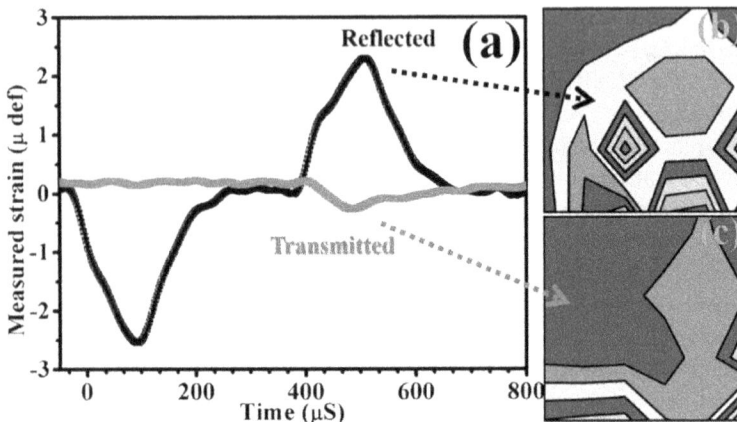

FIGURE 3.21 (a) Recorded shock wave signals by strain gauge sensors for the incident, reflected, and transmitted impacts within GLNR; (b) and (c) STFT signal processing of the collected signals and corresponding strain contours showing dispersion of shock energy, respectively, for reflected and transmitted waves.

$M = f\left(\dfrac{D_s - W_s}{W_s - O_s}\right) + 1$. Here, F_w is the transform width, D_s, the size of data, W_s, the width of win-

dow, O_s, the overlap function, and f, the floor function. In addition, one can analyse the frequency content of the signal as well as variations in frequency with time. The segments on the signal apply the discrete Fourier transform to slide the window onto the entire signal to get the coefficients of the transform. The spectral content of the signal can be obtained using $|X(m,)|^2$, i.e. by squaring the coefficients and plotting the two-dimensional contour plots. The spectrograph shows temporal variations in the shock rate of recurrence. The blocks represent how the power is dispersed within GLNR. Each colour area covered represents the corresponding energy dissipation, and the size of contour represents the amount of energy loss. The energy of the shock wave, dissipated within GLNR, as it travels through GLNR can lead to a numerous physical and chemical changes, as briefly discussed below (Figure 3.22).

The impulse pressure is a type of δ-function exerting compressive hydrostatic pressure with a negligibly small shear that induces disturbance, moving in an isentropic fashion. Such perturbation varies with time, rises in its strength as it travels through, and activates disorder . The disorder plays a pivotal role against the dynamic flow strength of the ribbon. On its propagation, there occurs a mismatch between the material flow and the speed of the shock wave front causing a deformation beyond the elastic limit setting in a plastic deformation that led to spall/fracture. The plastic deformation built up residual internal stresses, leading to the separation of conjugated layers. The gliding ribbon layers could be modelled by the Peierls–Nabarro stress mechanism. At the gliding interface, conjugated layers experience two resultant forces: first, compression due to the extra plane present, leading to edge dislocation and second, missing plane to neutralize the compressive force due to a shortfall in the dislocated edge. It's a balancing phenomenon, and the dislocation extension depends on the Poisson's ratio. The slip dislocation starts moving in response to the applied shear stress, and the ratio of in-plane to out-of-plane stress sets the critical stress limit [42]. The slip ratio, $\sigma_\parallel/\sigma_\perp$, where $\sigma_\parallel$ and $\sigma_\perp$ are the in-plane and out-plane stress components, was estimated to be 0.07 for GLNR and ~0.1 for HSR GLNR. The value of slip ratio $\sigma_\parallel/\sigma_\perp$ is ~10^{-4} for GLNR, which is lowered to >10^{-6} after the impact. Thus, the incident transonic impulse (>1.1 Mach number, the ratio of sound (340.3 m/s) to incident wave speed) lost more than 80% of its energy within GLNR. Only a small fraction of energy, ~5%, was transmitted, as schematically shown in Figure 3.20.

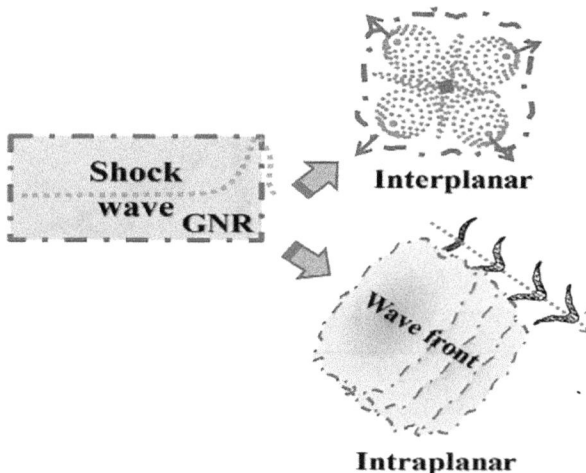

FIGURE 3.22 Scheme of effect of impulse pressure on GLNR within layer and layer conjugation.

3.6 HYDRODYNAMICS RESPONSE OF NANOCARBONS: CNS VS GNF

The purpose of this section is to extract a number of hydrodynamic parameters from the measured high strain rate data on nanocarbons such as GNF and CNS, to investigate what physics goes in and what shock and particle velocities are, to develop a construct for the exotic equation of state, and to analyse the transient pressure distribution within the nanocarbons with the details as presented below.

Two types of nanocarbons are investigated in this study: One is planar, two-dimensional (2D) graphitic flakes, the GNF, and the other is geometrically 0D CNS [43]. The synthesis methodology, preparation protocols and conditions, and the assessment of the structure–property relationship are presented in Section 2.1.2 for CNS and Section 2.3.3 for GNF.

3.6.1 STRESS (σ)–STRAIN (ε) BEHAVIOUR

Figure 3.23a shows the recorded behaviour of the system in the elastic region. The elastic yield strength of the CNS was noted to be remarkably large as compared to GNF. The analysis of the fitting parameters (displayed in (a)) showed different deformation strengths offered by the two nano-systems. From the obtained value of the exponent, n was found to be small by one order of magnitude for GNF (0.05) when compared to CNS (0.5). It showed the better stress accumulation ability of CNS over GNF. Perhaps the estimated pre-exponential factor (k) showed that the stress was getting built up swiftly in GNF (~300), approximately five times higher than in CNS (~60). Such a material response has a significant impact on the static strength of nanocarbons. Moreover, for CNS, the value of σ_0 was estimated to be ~50+ MPa (representing the static strength of the material) higher than the estimated σ_0 of ~200 MPa for GNF. It seems that in the static region morphology, geometry and areal dimensionality may have played a significant role in GNF compared to zero-dimensional CNS. The arrow shown in Figure 3.23 connects the elastic region (a) of the plot to the plastic deformation profiles recorded for the nanocarbons, as shown in (b). In general, the plastic deformation behaviour for both the systems is far more complex to analyse over the elastic region. Typically, four regions of deformation have been identified, namely (a) inter-particle plastic separation, (b) gliding of microstructures, (c) fracture, followed by (d) damage and spall, and the analysis is done in a similar fashion as presented in the previous section. From the slope, $d\sigma/d\varepsilon$, of σ–ε curves in

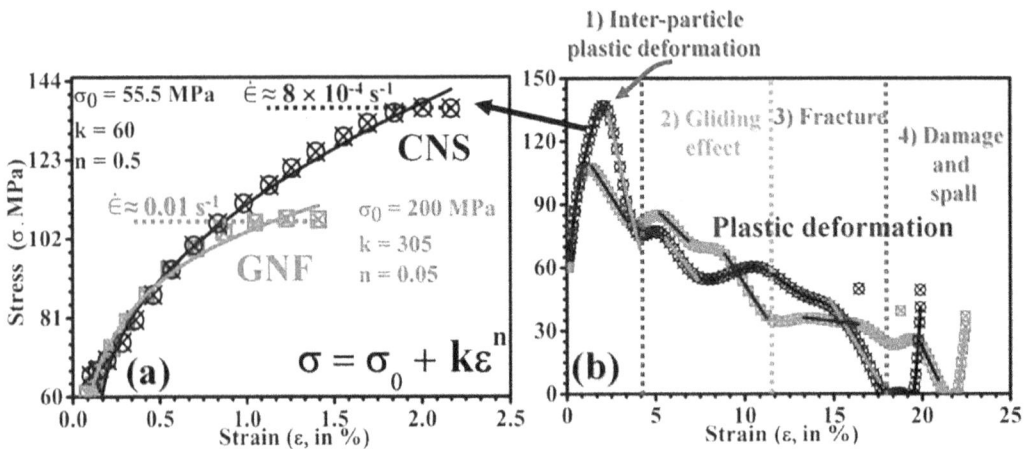

FIGURE 3.23 The stress (σ) vs strain (ε) curve recorded for the nanocarbons. (a) The elastic region showing the power law-based uniaxial extension, and (b) the plastic deformation region. It shows various stages of plastic deformation. The linear fitting in (b) is made at various regions for both the curves to extract the slope $\dfrac{d\sigma}{d\varepsilon}$ for the evaluation of hydrodynamic variables as discussed [43].

(b), the inputs were obtained to extract the number of hydrodynamic variables operational for the two systems. This enabled to examine the hydrodynamic interplay, using the Rankine–Hugoniot model, with the assumptions of matrix homogeneity and linear volume expansion, and neglecting the transient temperature rise. It provided information about several variables such as pressure (P), deforming volume (V_o), particle velocity (U_p), shock velocity (U_s), specific volume (V/V_o), density (ρ), and energy (E) behind and ahead of the shock front [43].

3.6.1.1 Equation of State

The governing Rankine–Hugoniot equation of state for the nanocarbons is approximated as:

$$P = P_0 \frac{\left[\frac{m}{m_0} \left(\frac{\gamma+1}{\gamma-1} \right) \left(1 - \frac{V}{V_0} \right) \right]}{\frac{V}{V_0}}, \text{ where } P, V, \text{ and } m \text{ are, respectively, pressure, volume, and mass}$$

ahead of the shock front and P_0, V_0, and m_0 are behind the shock front, and γ is the Gruneisen constant (1.5). Correspondingly, the calculated P–V/V_0 curves are displayed in Figure 3.24a. The blue curve shows an ideal shock Hugoniot. The constructed straight line, known as the Rayleigh line, joins the initial (P_0, V_0) and the final (P_1, V_1) hydrodynamic states of the medium and refers to the effective shock state with respect to P_1. From (a), one can see that the loci of the shock states were observed to be varied dramatically in GNF and CNS with reference to the ideal Hugoniot medium. It essentially underlined the fact that both systems mechanically responded distinctly different to the shock. Importantly, with an increase in the pressure in a shock front, it departed from the linear change in the P–V/V_0 path. Rather, for both the nanocarbon systems, there were changes observed discontinuously from its initial value P_0 to its value P_1. Moreover, in CNS, due to the reduction in the flow strength with the rise in pressure, matrix softening might had taken place. In contrast, in GNF, with a sequential increase in pressure, the flow strength was observed to be increased, which could be attributed to the matrix hardening. The projected shock pulse peak amplitude was found to be attained ≈100 GPa for GNF and ≈150 GPa for CNS. Notably, in both the nano-systems, the peak amplitude did not reach the final hydrodynamic state by following the linear Rayleigh relationship. Such a behaviour suggested discontinuity in the pressure change with variations in the local density. Both systems displayed non-isentropic character to the shock, indicating swift dissipation of the impulse pressure [43].

3.6.1.2 Hydrodynamic State Variables: The Interplay

It can be, comparatively, analysed by examining the slope of the Rayleigh line, which is proportional to U_s^2 of the shock wave. The exact relation is given by: $\frac{P_1 - P_0}{V_1 - V_0} = - \rho_{NC}^2 \, U_s^2$, where the density ($\rho_{NC}$) of the nanocarbon was assumed to be ≈10^{-3} kg/m^3. The estimated value of U_s for GNF was

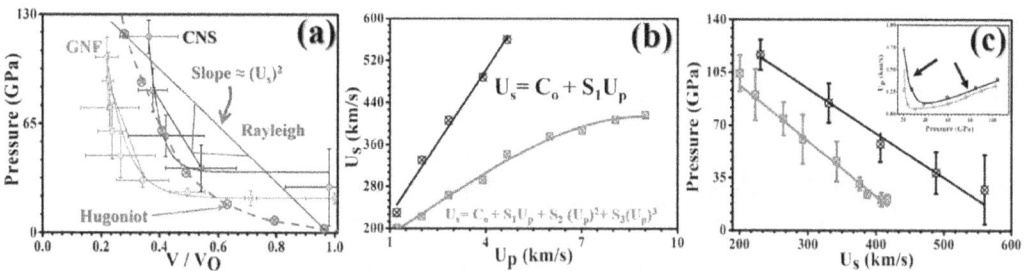

FIGURE 3.24 (a) The estimated pressure (P)–specific volume (V/V_0) curves for GNF and CNS, (b) transient shock velocity (U_s) and particle velocity (U_p) profiles revealing equation of state (EOS) for the nanocarbons, and (c) U_S–P and P–U_p (inset) relations. The black plot corresponds to CNS, and the red one is for GNF.

found to be ~100 km/h and was a bit higher, i.e. ~150 km/h, for CNS. Further, from the experimentally determined data, we established the relationship between U_S and U_P as seen in the plot (b) of Figure 3.24. Conceptually, U_S is the shock velocity followed by the particle flow U_P. Their interplay displayed another form of EOS. Both the plots were fitted for a polynomial equation with parameters C_O, S_1, S_2, and S_3, where C_O is the sound velocity and S_i's are the empirical structure factors of nanocarbons. For CNS, the linear relationship between U_s and U_p displayed fairly well-received shock response by the material that have not undergone any phase transition.

However, for GNF the departure from non-linearity is indicative of the involved phase transition attributed to its inherent structure deformations such as the gliding planes, the change in the topology of prismatic loops, and modulations in the porosity. Plot (c) in Figure 3.22 shows variations in P with U_s, which revealed a reduction in the shock velocity with pressure. The overall pressure change in CNS was observed to be very small compared to GNF. By and large, the prominent impedance due to interfacial effects had been experienced by the shock in progressing through GNF. The inset in (c) shows the change in U_P with pressure. The nature of both the curves exhibited the propagation of the particle flow (U_P) that has almost cubic dependence on the pressure experienced by these nanocarbons. The governing equation of state for the flow is approximated by: $\left(U_P + \dfrac{\gamma_1(\rho)}{P^2} \right) \approx \dfrac{\gamma_2(T)}{(P - \gamma_3)}$, where $\gamma_i s$ are the structure factors that depend chiefly on the matrix density (ρ) and the transient temperature (T) rise during pulse progression. The fitted equation, though not in the scale, showed a breakdown of the molecular attractions between the carbon atoms connected by the covalent bond [43].

Mild oxidation is predicted. Specifically, carbon atoms have the tendency to form unsaturated covalent bonds with oxygen anions and to increase the local charge state. As a result, the electron density gets shifted from the carbon cations to more electronegative oxygen anions. This in turn will affect the electronic structure and lead to an increase in the structure of the band gap. Hence, the shock may result in the change in electronic properties of nanocarbon systems. It would be of utmost interest to investigate the post-impacted nanocarbons by the Raman spectroscopy. Such investigations would confirm the percentage deformation in the dynamic force constant, k, of both the nanocarbons and confirm the band gap modifications. The two arrows in the inset show two extreme regions: (a) steep fall in the particle velocity and (b) slow build-up of multi-fold magnitudes, intimating non-homogeneous isothermal compressibility of the material that underwent compression.

3.6.1.3 Shock Contour: Realistic vs Theoretical

The shock is, truly speaking, an intense acoustic wave that has discontinuous wave front propagating with U_S. It has typical pressure (P), density (ρ), and temperature (T) parameters of the medium behind the shock correlated with P_o, ρ_o, and T_o ahead of the shock wave. This is a model scheme. It is, merely, the oversimplification of the problem statement in which the medium is assumed to be a homogenous, single component, isotropically interacting with an ideal impulse pressure of the δ-function type. In realistic situations, the hydrodynamic (P, ρ, T) variables may not be simply connected to P_o, ρ_o, T_o. Figure 3.25a shows such an ideal shock wave profile predicting a discontinuity at the front, a plateau on the top, and a gradual pressure drop while unloading.

There are numerous peculiarities/features seen before and after the shock loading, dictated uniquely by the nature of the material and the generated pressure impulse, as displayed in Figure 3.25b. One can see that even the overall shock profile distinctly differs from the derived Hugoniot curve. Particularly, in the elastic limit, this occurs, mainly, because of the non-linear change in the volumetric stress rate until the strain rate attains the limiting elastic value, termed as the Hugoniot elastic limit (HEL). From this region, one can determine the elastic properties of the material (for nanocarbons, shown in Figure 3.25a). The generated elastic compression gets separated from the plastic portion beyond the HEL, and thereafter, the plastic wave starts dominating and further leads to the plastic deformation of the matrix. Since the portion of the elastic wave gets separated,

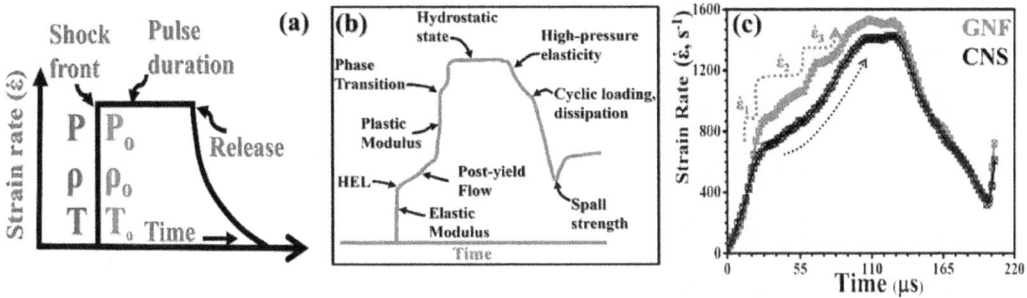

FIGURE 3.25 (a) Ideal shock profile scheme, (b) realistic shock contour exhibiting numerous peculiarities, (c) temporal variations in strain rate ($\dot{\varepsilon}$) for the two nanocarbons, obtained experimentally (dotted curves represent variations in the strain rate ($\dot{\varepsilon}$) while shock loading).

beyond HEL, from the plastic one, the carried particle flow, after the elastic yield, gives rise to a steep increase in the strain rate. In this region, the matrix experiences one or several phase transformations dictated by the constitutive behaviour of the material. This follows the pulse duration plateau, a typical hydrostatic state of the material. The pressure unloading initially occurs elastically (labelled as the high-pressure elasticity) and then plastically (the cyclic dissipation stage) as seen in Figure 3.25b. This takes place in an analogues way; the HEL is attained. The transported shock wave reflects and constructively interferes with the incident shock wave amplitude, resulting in the fracture of the material, labelled as the spall.

The actual shock wave profiles that were measured for the nanocarbons (red: GNF; black: CNS) are shown in Figure 3.25c. The estimated HEL attained by GNF was ~12 GPa, whereas it was ~06 GPa for CNS. In the plastic regime, the separation of elastic and plastic portions was quite evident and consistent with the theoretical profile. The post-yield material flow was observed to be, prominently, jagged in the GNF and shown by the red dotted contour suggested the phase transformation. Notably, the three distinct strain rates ($\dot{\varepsilon}_1$, $\dot{\varepsilon}_2$, and $\dot{\varepsilon}_3$) indicated that the impulse pressure met various structural features that inherently existed in the form of conjugated planes, edges, glide plane defects, etc. However, one cannot differentiate and assign them clearly and legitimately, at this juncture. For CNS, no such prominent feature was observed that proposed the softening of plastically flowing material, as discussed above. The hydrostatic holding of the impulse was somewhat dilated within the GNF over the time dilation in the CNS, showing that the degree of interaction was more preventive in the GNF compared to the CNS. Unloading and spall had almost taken place identically in both the nanocarbons.

3.6.2 Shock Imprinting

Figure 3.26 shows that the areal pressure was distributed in a non-uniform fashion for both the nanocarbons. The distribution was obtained from the force–time curve. In this, the force was appropriately transformed and scaled into the $F/A \leftarrow P$, where A is the estimated cross-sectional area, which was ~100 nm^2 for CNS and ~10^6 nm^2 for GNF. The temporal variations were obtained through the measured velocity estimations. For CNS, the moderate pressure areas were quite prominently distributed in a discrete fashion with the embedment of both the high- and the low-pressure areas within. At some portions of CNS, high-pressure regions were seen followed by the existence of low-pressure zone surroundings. For GNF, the scenario was distinctly different. The presence of the high-pressure zone conjugated with the moderate pressure zones was quite dominant. Nonetheless, the overall pressure distribution was inhomogeneous. The strain volume non-uniformity was a major common feature seen in both the nanocarbons, due to the compression wave propagation in an inhomogeneous fashion. Our calculation revealed that almost 65% shock energy damped within GNF and ~89% in CNS [43].

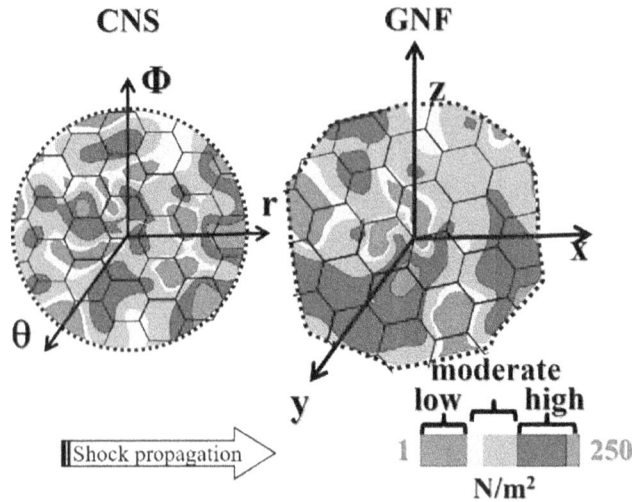

FIGURE 3.26 Scheme of the areal cross section of the volumetric pressure distribution of the nanocarbons compounded with the colour scale bar showing low-, moderate-, and high-pressure regions.

Notably, temperature (T) is another variable that is important in the deformation process. The elastic deformation did not give rise to any heating effect, since in such a process, the heat is carried away and dissipated due to the linear deformation of the matrix. Further, plastic deformation at a low strain rate in the range of 10^{-4} to 10^{-3}/s is also an isothermal process; however, high-strain-rate deformation is adiabatic and deformation work is transformed into heat with a consequent rise in the temperature of the specimen. In general, the thermal diffusion length (λ_L) is given by $2\sqrt{\alpha \cdot t}$, where α is the thermal diffusivity measured in cm²/s and t is the time of diffusion. To estimate temperature rise, one needs to calculate these variables in the above-mentioned equation. At this juncture, we are not in a position to provide any measurements or data on this. However, one can use the constitutive equation: $\sigma = (\sigma_o + k\,\varepsilon^n) \cdot (1 + j \log \varepsilon^*) \cdot \left(1 - T^{*m}\right)$, where $j = \dfrac{1}{\rho C_P}$ and m is 0.6–0.9 for a carbonaceous medium of density (ρ) ≈ 10^{-3} kg/m³ with a specific heat capacity (C_P at 300°K) ≈ 0.5 J/kg/°K. For GNF, the temperature rise was estimated to be ~237°K, and for CNS, it was ~189°K with respect to 300°K and have a negligible impact on deformation as compared to impulse pressure variations [43].

By and large, our study reveals that what hydrodynamics goes in when such a pressure impulse travels through the nanocarbons. The measured strain, strain rate, and stress with time lead to deriving a number of intricate parameters for CNS and GNF. This sheds light on the hydrodynamic state of such nanomaterials. More study is still required.

3.7 BLAST MITIGATION PARAMETERS FOR PNCs

In this section, we present the blast mitigation characteristics of porous nanocarbons (PNCs), experimentally, using the Kolsky bar (sHPB) technique. The effect of impulse pressure on the shock absorption capability of PNCs is discussed. The details of PNC pyrolysis are presented in Section 2.3.8 [44]. After moulding into the disc-shaped samples, the PNC was subjected to high strain rate (10^{-4}/s) measurements to obtain the data of stress (σ), strain (ε), and strain rate ($\dot{\varepsilon}$) with time. The hydrophysical parameters inferred from the spectro-microscopic techniques are comparatively studied for the PNC as well as for the other forms of nanocarbons such as GNF, GLNR, and CNS. By and large, 70% impact energy damped in the PNC. The details are presented below.

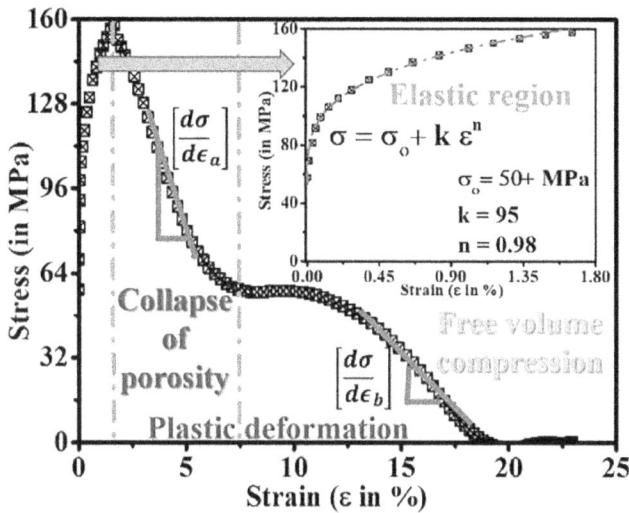

FIGURE 3.27 Behaviour of constituent variables σ–ε for PNCs in the plastic and elastic (inset: shown by a green arrow) deformation regions.

3.7.1 Constitutive Analysis

Actually, when the shock wave enters the specimen, there are reverberations, and after ~03–04 reverberations, an equilibrium is set up in the sample.

The main panel in Figure 3.27 shows the plasticity exhibited by the PNC, and the inset shows the corresponding elastic region. The constitutive variables, the σ–ε curve obeys the exponent law in which the magnitude of n is found to be 0.98, which is closer to one. This showed that the stress generated in PNCs is almost independent of the applied strain and indicative of the moderate stress accumulation character of the PNC. The obtained pre-exponential factor, k, is associated with the stress build-up tendency. The more the k, the higher the build-up stress. When compared to the previously reported systems such as GNF and CNS, the build-up stresses estimated for PNCs are found to be moderate or rather intermediate and followed the order GNF < PNC < CNS. The value of static strength σ_o was recorded to be ~50 MPa for PNC, which was the least compared to GNF (200) and CNS (60 MPa) [40,43]. Porosity, free volume fraction, and interconnects of voids in PNCs seem to have played a significant role in the microstructure that existed in GNF and CNS for the recorded deformation region.

3.7.1.1 Shock Hugoniot: P–V, U_S–U_P, and P–U_S Variations

A shock Hugoniot is a mathematical construct set-up to understand the hydrodynamic behaviour of the PNC. The presented calculation for the shock parameters is based on the Rankine–Hugoniot formulations and in line with the assumptions presented before for other NC systems. Typically, a Hugoniot is a pressure (P)–volume (V), pressure–particle velocity (U_P), and shock (U_S)–particle velocity (U_P) relationship dictated by the dynamic pressure. It largely departs from an ideal Rayleigh characteristic. The shock Hugoniot describes the locus of the shock (P–V) states the PNC exhibits and, essentially, labels its mechanical response. Notably, the PNC has porosity and it undergoes phase transformation, as observed in the obtained shock profile data. Hence, the linear behaviour of the equation of state is no longer applicable. The shock data plots and the corresponding Hugoniot calculations, including the extraction of impact parameters with reference to the standard material, have been done as per the reference and references cited therein. Broadly, two shock parameters are sufficient to determine the remaining others and the constants in the equation of states are derived.

In general, there are five shock parameters with which one can analyse the obtained data. They are P, E, ρ (or V/V_0), U_S, and U_P. With the use of EOS, these relationships can be separated into ten pairs. These ten pairs provide, in turn, 20 equations. However, we have presented the most relevant data which are important from the perspective of blast mitigation and control. From the measurements of the slope of the $\sigma-\varepsilon$ plot at various regions (shown in Figure 3.27), a hydrodynamic relationship was established and is displayed in Figure 3.27.

By and large, to determine the EOS for a porous/powdery medium, the Kolsky bar is a well-suited instrument for such kind of studies. For analysing and setting up the EOS for a porous medium, there are two approaches: One is the Mie–Gruneisen route in which the Gruneisen parameter, γ, is a function of both the shock dampening energy (dependence comes through pressure) and volume. However, this approach neglects the strength of porous material. In the second approach, γ is a function of energy alone and incorporates the material strength of the porous medium. We have used the second approach. Figure 3.28 shows the U_S–P relationship dictated by the shock in the PNC (Section 3.8.3). In this, the volume of PNC comprised random interconnects of solid carbon volume and micro-void gaps. Upon impact, the solid carbon gets disintegrated and accelerates in random orientations into the gaps/voids until it imparts the impact to the next neighbouring zones. A succession of impacts could be represented by the shock wave traveling through the PNC with a velocity that is the ratio of the product of shock, U_S, and particle, U_P, velocities to the sum of their of velocity components via mass fractions $\zeta\,(m/m_o)$. The $\zeta\,(m/m_o)$ played a crucial role, which depends upon the initial porous and effective densities of the PNC. As a result, behind the shock linear state equation was seen to be transformed into a non-linear equation, given by Mie–Gruneisen, and expressed by the relation: $\dfrac{P}{P_o} = n\,k_B\,T\left\{\left(\dfrac{V}{V_o}\right)^{-\gamma} + \zeta\left(\dfrac{V}{V_o}\right)^{-\gamma/2}\right\}$, where n is the molar fraction, k_B, Boltzmann's constant, T, the temperature, γ, the Gruneisen compression parameter, and ζ the mass fraction expressed as $\zeta\,(m/m_o)$. For a metallic solid, the value of γ was reported to be 1.0–1.5, whereas for PNC, γ was estimated to be 0.92 ± 0.25. The observed reduction was attributed to the inherent porosity in PNC. The value of $\zeta\,(m/m_o)$ was projected to be 0.79 ± 0.15 and implicative of 80% collapse in mass fraction due to impact pressure. The straight lines constructed onto Rayleigh

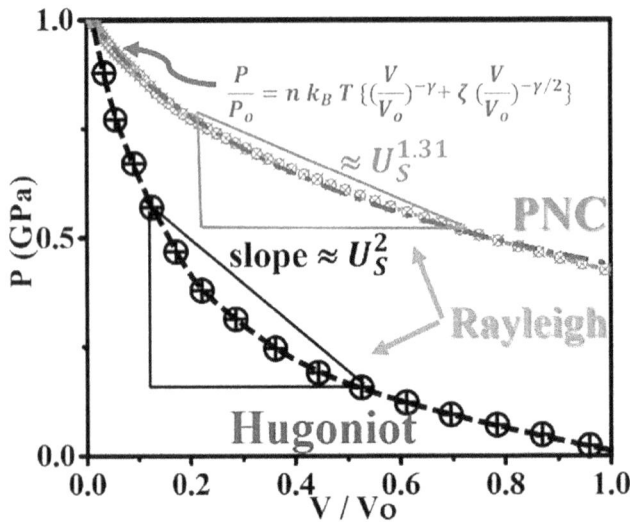

FIGURE 3.28 Calculated pressure (P)–specific volume ($\dfrac{V}{V_o}$) Hugoniot displaying hydrodynamic state of the PNC in comparison with a theoretical Rayleigh construct. The equation of state is displayed, and the value of slope was an exponent μ-function of shock velocity (U_S^μ).

and PNC curve referred to a particular shock state and was given by the Rankine–Hugoniot relationship. It had been realized that when the pressure was changed transiently, the relationship does not hold the linearity. The measure of discontinuity could be quantified by analysing the slope of Rayleigh line that is in propionate to the square of the shock velocity (U_S). However, for PNC, the modified state equation is given by: $\left(\dfrac{P}{P_o}\right) \Big/ \left\{\left(\dfrac{V}{V_o}\right)^{-\gamma} + \zeta \left(\dfrac{V}{V_o}\right)^{-\gamma/2}\right\} \approx -\rho_o^2 U_S$. In PNC, for discretely changing shock states, the degree was found to be somewhat less ($\mu = 1.31$) than the theoretical construct. To describe a hydrodynamic response, the knowledge of both U_S and U_P is crucial because the shock velocity is accompanied by the particle flow [44].

3.7.1.2 Shock and Particle Velocity Considerations

For a Rayleigh medium, the P–U_S state equation is given by: $P = \dfrac{\rho}{S_o}\left(U_S^2 - C_o U_S\right)$, where ρ is the density (in ≈ 1 kg/m^3), S_o is the structure factor (≈ 1.45), and C_o is 0.34 km/s (velocity of sound). For PNC, the modified state equation is: $P = \dfrac{\rho}{S_o}\left(U_S^{1.31} - C_o U_S^{0.5}\right)$, where the value of C_o was obtained to be 260 m/s, with $S_o \sim 2.49$ and $\rho \approx 10^{-3}$ kg/m^3. The value of C_o was reduced almost by ~25%. Moreover, the reduction in the value of exponent to 1.31 intricately suggests that the slope value was in proportionate to $U_S^{1.31}$ rather than U_S^2. The P–U_P correlation was established to yield an equation: $U_P = \dfrac{C_o}{2\,S_o}\left(\sqrt{1 + \dfrac{4 S_o}{\rho_o C_o^2} P}\; -1\right)^{0.97}$. From P–U_P, one can see that, behind the shock wave front, the velocity of matrix mass reached a value ≈ 100 km/s as the pressure was reduced to lower values. The shock pulse peak amplitude attained was ≈ 130 GPa; however, the amplitude did not attain the final hydrodynamic state as dictated by the Rayleigh relationship. This led to a conclusion that the pressure change is discontinuous compounded with the fall of local void and free volume and changing the density of the PNC matrix. The system displayed non-isentropic behaviour and seems to be dissipating the shock swiftly. It seems that, during the initial stage of shock interaction, there might be instantaneous disintegration of solid carbon to collapse within the porous region, resulting in an abrupt increase in pressure at smaller magnitudes of U_S. It seems that, thereafter, the pressure behind the shock front was seen to be almost constant with an increase in the velocity of the shock wave. The values of U_P were found to be high during the unloading of shock even at such a low amplitude value of pressure ~1 GPa. At lower pressures, the constitutive model seems to be describing further rarefication of the PNC matrix with a high flow stress [44] (Figure 3.29).

3.7.1.3 Loading–Unloading Characteristics

Further, a blast being a pressure impulse, in principle, was an intense acoustic wave. It has a discontinuous wave front propagating with the energy in square proportion to the shock velocity (U_S). Figure 3.30a, typically, depicts a shock profile comprised of a shock front, a hold (pulse duration) state, and a release. Moreover, it displays the variables pressure (P), density (ρ), and temperature (T) behind the shock that could be correlated to the P_o, ρ_o, and T_o ahead of the shock wave front. However, this is just a schematic in which the medium was assumed to be homogenous, mono-element, and isotropous and interacting with the δ-function-type pressure. In reality, circumstances are complex; the hydrodynamic variables behind and ahead of the shock may not be simply connected functions. Figure 3.30b shows the intricate stages a material may experience, such as a Hugoniot elastic limit (HEL), post-yield flow, phase transformation, hydrostatic state, high-pressure elastic unloading, and dissipation followed by a spall stage.

Figure 3.30c shows the actual shock profile recorded for the PNC. The profile is labelled with a loading region comprised of elastic and plastic zones; on top, a typical shock holding (hydrostatic) state could be seen followed by an unloading zone accompanied, with a hierarchy of elasto-plastic sequence. The estimated value of HEL was found to be ~8.12 GPa, as labelled in the plot. We have

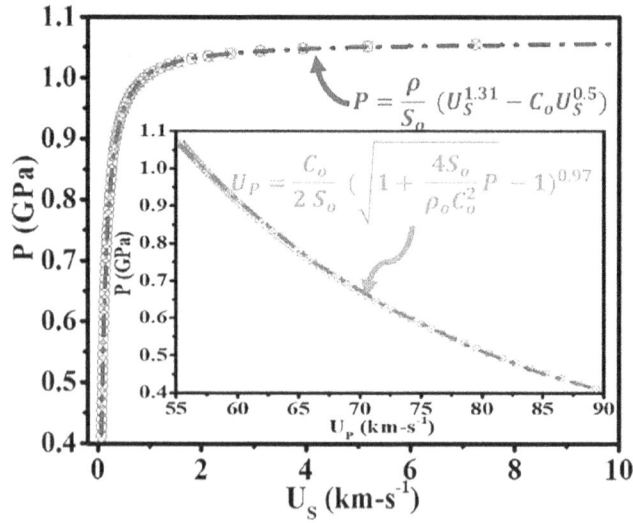

FIGURE 3.29 Calculated U_S–P connection with the corresponding equation that showed departure from the square dependence. At higher pressures, the slope becomes almost parallel to x-axis and displays an unprecedented rise in shock velocity. Inset shows the corresponding P–U_P plot along with state equation.

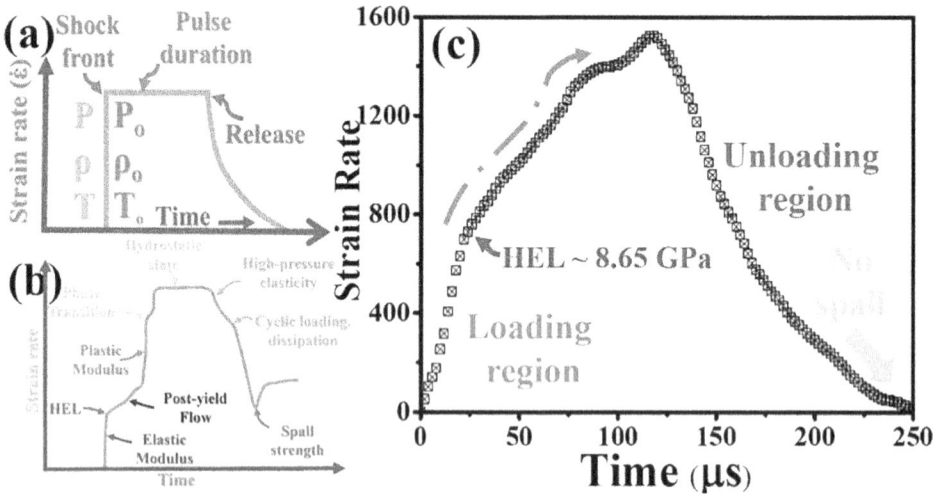

FIGURE 3.30 (a) Theoretical shock profile displaying the evolution of strain rate ($\dot{\varepsilon}$) with time. It shows prominent features such as a shock front, a hydrostatic (hold) state, and a release, (b) intricate features on a theoretical shock contour, (c) actual profile that indicates loading and unloading, and phase transformation (enveloped by a red arrow) with no evidence of spall feature.

evidently observed the separation of elastic and plastic regions, at HEL, and is consistent with the feature that exists in a theoretical contour. It seems that, after crossing the elastic yield, the PNC matrix atoms got set in the motion and followed the phase velocity of impulse wave front. As a result, the jagging in the plastic curve is observed as shown by a red envelope. The observation of such a feature, conclusively, suggested the phase transformation. The strain rates, in this region, are seen to be dramatically changed. It revealed that the incident pressure impulse met various structural features that inherently existed in the form of porous topographies such as voids, free volume, and interconnected bulk region. Interestingly, the hydrostatic holding state of the impulse is somewhat found to be degenerated over the total time expended within the PNC. Unloading had almost taken place in an identical fashion as that of loading with no record of spall feature. After investigating hydrodynamic variables, it was of our interest to study the structure–property correlations in PNC before and after the shock treatment [44].

3.7.1.4 Electron and Force Microscopy Studies

Figure 3.31 shows a couple of representative field emission scanning electron micrographs recorded for PNC before and after the shock treatment. To obtain the images, pristine and shock-treated powder specimens were dispersed in isopropanol for a period of 1 hour or so. As the solution became uniformly turbid, a few drops were collected in a micropipette and released on a cleaned silicon–silicon dioxide wafer surface. These samples were allowed to dry under atmospheric conditions in a closed petri dish. Before subjecting to electron microscopy, the samples were coated with a conducting carbon spray. A large number of micrographs were recorded at a fixed working distance of ~6 mm with magnification varying from 2.5 to 25 k. Figure 3.31a displays the recorded surface morphology of the PNC in which the top surface was found to be contaminated by a large amount of amorphous carbon debris. It was assumed that such debris might have fallen onto the PNC during the sample preparation process.

The channel interconnects are clearly visible on the pristine PNC. These interconnects were found to be in the range of diameter between 1 and 2 μm. At some portions, an opening end of the channels is seen. It seems that these interconnects offered a peculiar resistance and stability against the shock imparted on them. In Figure 3.31b, recorded under the similar micrographic conditions, we observe a complete collapse of the structure. No traces, tracks, and signs were observed that provided a clue about the intact condition of these channels. The entire PNC structure seems to be constricted. Our analysis revealed that there is ~20%–30% contraction in the volume of PNC. However, we were not able to quantify how much impact the pores had taken when compared to the interconnecting channels [44].

These microstructures were also examined by the high-resolution transmission electron microscopy. For this purpose, from the prepared isopropanol suspension, the samples were drop-cast onto a microscopy-grade sample grid. After drop-casting, these grids were dried and used for HRTEM studies.

A few representative images are shown in Figure 3.32, in which (a) pristine and (b) shock-treated PNCs are shown. Insets in the respective micrographs show the recorded selected area electron diffraction (SAED) pattern. Both the images are having different scale bars. In Figure 3.30a, the arrows indicate the inherent disorder. Truly speaking, these features are associated with the nanopores that the PNC possessed. The recorded SAED pattern showed blur rings with no evidence of

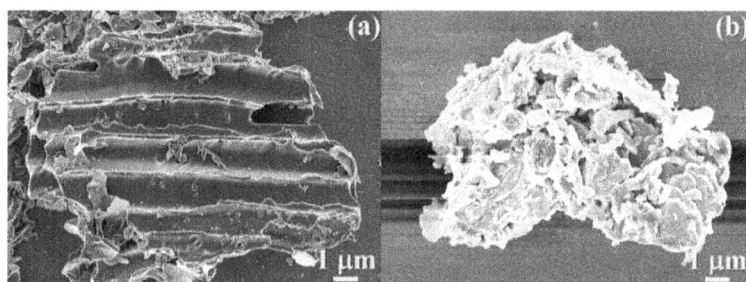

FIGURE 3.31 (a) Pristine PNC displaying micro-channels and interconnects; (b) shock-treated PNC showing rupture of the entire structure including effective shrinkage. Both images are representative and recorded at 2.5 k magnification at a working distance of 6 mm.

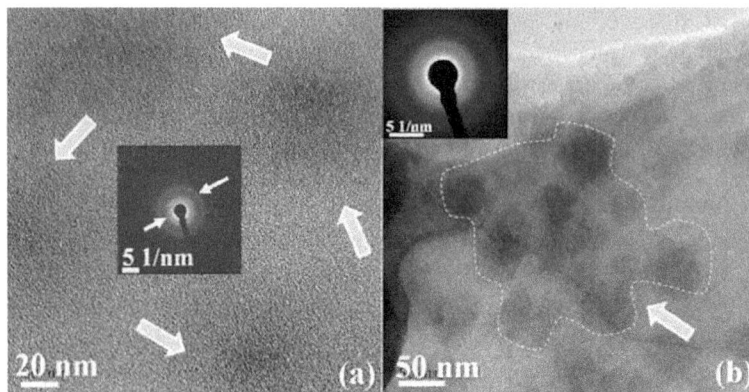

FIGURE 3.32 (a) Pristine and (b) shock-processed PNCs. Inset displays a selected area electron diffraction pattern.

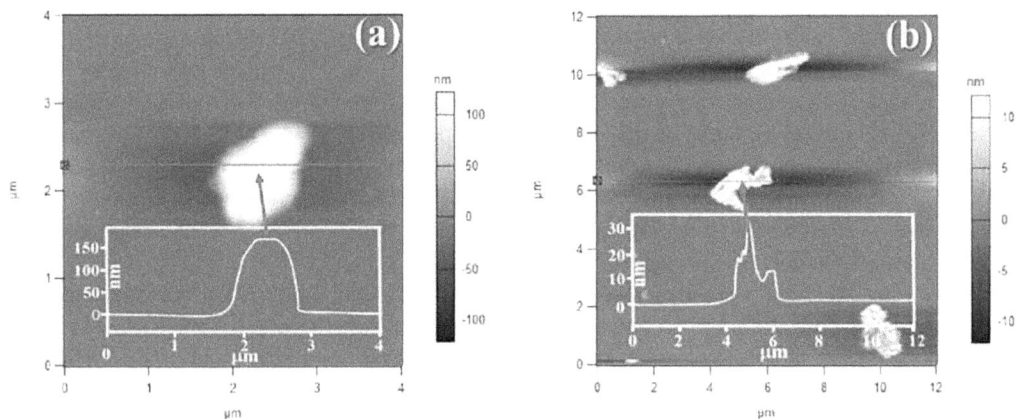

FIGURE 3.33 (a) Pristine and (b) shock-treated PNCs. Inset in respective plot displays line scan and thickness.

Laue points. This indicated, nearly, short-range ordering of the carbon lattices. Interestingly, after the impulse interaction, a dramatic modification is evident in size, shape, and densification of the material associated with pores, revealing collapse state. Figure 3.32b shows a typical micrograph displaying an increase in the collapse pore-densified zone. The SAED pattern recorded for the shock-treated PNC showed no peculiar features such as crystallization of carbon networks. The degree of disorder is thought to be of the same order of magnitude at the atomic level.

In another study, atomic force microscopic (AFM) imaging was carried out on both kinds of PNCs. For this purpose, the suspension was drop-cast onto highly oriented pyrolytic graphite. Figure 3.33 illustrates the effect of impact on PNCs vis-a-vis compared.

The morphological analysis carried out by the scanning and transmission electron microscopy was found to be consistent with the observations picked up by the atomic force microscope. Figure 3.33a shows the image of pristine PNC. It, characteristically, has a random shape and size. The thickness of pristine foam was found to be ~150 nm, or so, as measured by the line scan. On impact, a collapsed structure seemed to be compacted, however, not disintegrated. The thickness was observed to be reduced at least by a factor of six.

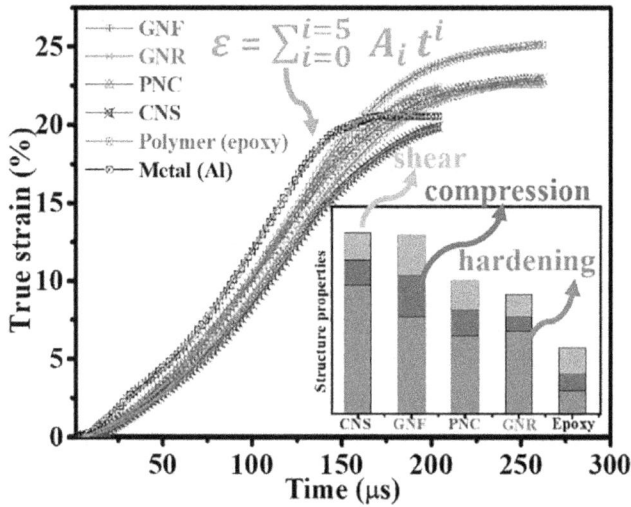

FIGURE 3.34 Variations in strain with time recorded for various nanocarbons including metal and polymer as a reference with known deformation mechanics. The curve obeys a fifth-degree polynomial. Inset shows a comparison of typical dynamic coefficients obtained from the fitting, such as shear, compression, and hardening.

3.7.2 STRAIN–IMPULSE INVESTIGATIONS: SHEAR, COMPRESSION, AND HARDENING

Notably, in our previous sections, we presented dynamic deformation in GNF, CNS, and GLNR. Figure 3.32 shows a comparison of the strain exerted, with time, on such nanocarbons, including metal (aluminium) and polymeric (epoxy) systems. Figure 3.34 represents the effective merits and demerits of these nanocarbons over each other with reference to a metal and a polymer base that exhibited a known mechanical performance. The strain and time parameters are found to mathematically obey an equation: $\varepsilon = A_o + A_1 \cdot t^1 + A_2 \cdot t^2 + A_3 \cdot t^3 + A_4 \cdot t^4 + A_5 \cdot t^5$, where $A_i s$ are the coefficients of polynomial associated with the dynamic structure factors of the matrix undergoing the deformation. One can see that each system has exhibited a different strain response with time. Further, some dynamic structure factors such as shear (A_1), compression (A_3), and work hardening (A_5) are correlated to judge the overall response and quantify the strand of merit of the nanocarbon systems.

Analyses revealed that GLNR exhibited a high degree of work hardening (~35%); however, they possessed the least compressive strength (~17%) compared to others. The compressive strength of GNF was found to be the highest among the nanocarbons, but the hardening capability was less than that of CNS and comparable to PNC and GLNR. In contrast, the work hardening of CNS was found to be far superior among the studied nano-systems and this is attributed to the structure and spherically symmetric geometry, which seemed to be advantageous for CNS over the 2D flakes, ribbons, and porous carbons. CNS exhibited no structural phase transformation, and the variations in the strain rates were observed to be very smooth. For GNF as well as for ribbon, due to their 2D and laminated nature including disorder-enriched phase, these nanocarbons experienced a larger shear and shear-assisted deformation. The PNC stands as the intermediate compound. Notably, the dynamic deformation in the epoxy resin was found to be the least among the nanocarbons in the quantified shear, compression, and hardening characteristics.

Moreover, these nanocarbons were also subjected to Raman spectroscopy to investigate the impact of impulse pressure at molecular, atomic, and electronic levels. From these techniques, intricate parameters were extracted together with the blast variables. Table 3.3 shows the compilation for the nanocarbons studied [44].

TABLE 3.3

Physical Parameters Extracted for the Nanocarbon Systems after the Shock Treatment

		Nanocarbon Systems			
S. No.	Physical Parameters (SI unit, in %, in×Fold Factor)	PNC	GNF	CNS	GNR
1	Young's modulus (MPa)	75	>50	~90	60
2	Shock velocity dispersion (in %)	88	50	73	45
3	Inter-particle separation speed (m/s)$\times 10^3$	-	3.5	1.3	4.3
4	Stacking fault separation speed (m/s)$\times 10^3$	>1.3	<2.5	-	<2
5	Elasto-plastic time constant (s^{-1})	0.1	0.01	8×10^{-4}	~10^{-4}
6	Hugoniot elastic limit (HEL in GPa)	8.93	12	06	08
7	Structural reduction (in %)	25–30	50	10	-
8	Dynamic force constant at –C–C– (N/m)	377 ± 22	600 ± 15	677 ± 102	410 ± 31
9	Dynamic force constant at –C=C– (N/m)	711 ± 37	850 ± 40	987 ± 67	623 ± 89
10	Dynamic force deformation (in %)	30	43	67	21
11	Bond deformation (in %)	92	150	213	177
12	Change in crystalline length (× fold factor)	1.5	3	1	1.03
13	Change in disordered length (× fold factor)	6	4	2	~3
14	Disorder density ($\times 10^{15}$/m^2)	227	140	23	97
15	Mobility of disorder (m/s)	45 ± 0.1	20 ± 0.5	-	13 (10%)
16	Electron–phonon coupling@300 K (N/m)	33	25	41	17
17	Fermi velocity ($\times 10^6$ m/s)	11	08	03	15
18	Energy dissipation (in %)	73	65	89	46

From Table 3.3 and the comparative analysis of the mentioned parameters, we underlined a fact that the PNC may be superior in several aspects of the elastic–electronic properties; however, it is mediocre in a few cases.

3.7.3 SHOCK ANATOMY

The recorded dynamic force was transformed into the time-varying pressure and used to map the topology of pressure distribution within the PNC. The obtained data of dynamic pressure with time were subjected to the wavelet transform. Figure 3.35 shows the topology of impulse pressure exerted onto the PNC obtained via the signal transformation as presented before.

The yellow-coloured region suggests that the average pressure was of the order of ~250 N/m^2, whereas the red marker represents the echo component of the incident impulse. Our calculations suggested that ~30% of the impact of the incident impulse was left in the echo. The points from where the echo generated were found to be a point source or a continuous region. The dark region that separated the incident and reflected regions was thought to be the shear belt. In this region, an enormous frictional force might have been generated, which leads to the compression effect in the PNC by reducing its volume by 25%–30%, as confirmed by the microscopy studies.

Another noteworthy point to be discussed is the rise in the temperature (T), which may have implications for the local deformation process of the PNC matrix. By and large, the elastic deformation has no bearing on the heat evolution due to the fact that expansion is in linear proportion to the applied strain. In this process, the heat is carried away and dissipated uniformly within the matrix. The low-strain-rate (10^{-4} to 10^{-3}/s) plastic deformation is also an isothermal process. But the high-strain-rate deformation is an adiabatic process and the deformation work is transformed into heat with the consequent rise in the local temperature within the PNC. The diffusivity length (λ_L) is in proportion to $2\sqrt{\alpha \cdot t}$, where α is the coefficient of thermal expansion (in cm^2/s) and t

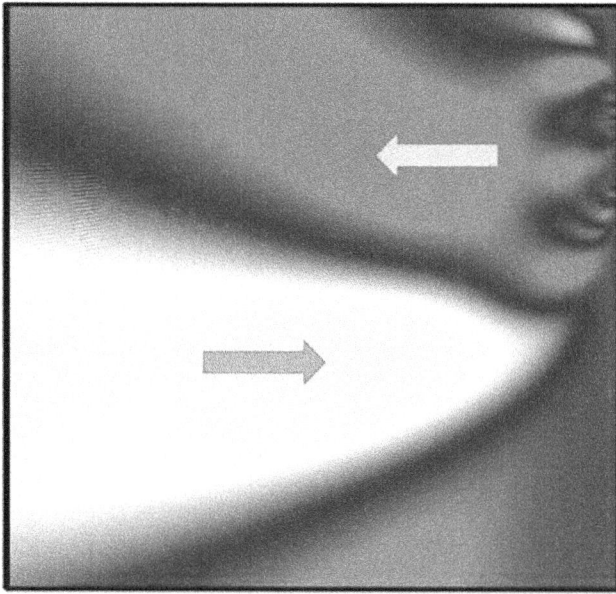

FIGURE 3.35 The zonal anatomy of the impulse pressure exerted onto the PNC. The arrows indicate the built-in (green) and echo (yellow) pressure generated due to incident impulse within the PNC matrix.

is the diffusion time. The rise in local temperature can be estimated from these variables in the above-stated equation. However, we have not carried out any explicit heat measurements using the modified constitutive Johansson–Cook equation given by: $\sigma = (\sigma_o + k\ \varepsilon^n)\cdot(1+j\log\varepsilon^*)\cdot\left(1-T^{*m}\right)$, where $j = \dfrac{1}{\rho C_P}$ and m is in the range of 0.6–0.9 for a carbonaceous medium with an effective density $(\rho) \approx 10^{-3}\,\text{kg/m}^3$ and heat capacity at constant pressure (C_P) at $300°\text{K} \approx 0.5$ J/kg/°K. The calculated temperature rise was found to be ~97°K with respect to 300°K. In comparison with other nanocarbons, the increase was negligible and has no bearing on the deformation as compared to the hydrodynamic impulse [44].

Before summarizing, we have investigated the interaction of the impulse pressure of a peak strength in the range of GPa with a time response in ~μs scale on various NC systems. Our focus was on the nature of interaction, its implication on the structural transformation, and the effective percentage attenuation of the shock energy. In general, the literature, mostly, revealed the impact of the shock wave on the fabricated NC, thereby taking the benefit of its structure to mitigate the mechanical properties. By and large, the reported work is on the advances in nanocomposites bringing the novelty by the shock treatment. However, physics-wise, pressure, density, and temperature changes in correlation with changes behind and ahead of the shock front are not reported. This work enriched our understanding about the shock and the involved hydrodynamic state of affairs of nanocarbons.

REFERENCES

[1] Sandhu, Inderpal Singh. "Design and establishment of test setups for blast evaluation and studies on blast mitigation in polyurethane and rubber foams." Ph.D. dissertation, DIAT deemed to be University, (2020).

[2] Smith, Peter, and John Hetherington. *Blast and ballistic loading of structures*. Laxtons, Oxford, 1994.

[3] Kinney, Gilbert F., and Kenneth J. Graham. *Explosive shocks in air*, Springer-Verlag, New York, 1985.

[4] Cooper, Paul W. *Explosives engineering*. John Wiley & Sons, New York, 2018.

[5] Jones, Gordon E., James E. Kennedy, and Larry D. Bertholf. "Ballistics calculations of RW Gurney." *American Journal of Physics* 48, no. 4 (1980): 264–269.

[6] Meyers, Marc A. *Dynamic behaviour of materials*. John Wiley & Sons, New York, 1994.

[7] Stolz, Alexander, Oliver Millon, and Arno Klomfass. "Analysis of the resistance of structural components to explosive loading by shock-tube tests and SDOF models." *Chemical Engineering Transactions* 48 (2016): 151–156.

[8] Sandhu, Inderpal Singh, Ankush Sharma, Meenakshi Bhatt Kala, Prashant S. Alegaonkar, Manpreet Singh, Rama Arora, and Dev Raj Saroha. "Blast mitigation study in composite materials." In *31st International Symposium on Ballistics*, 2019.

[9] Sandhu, Inderpal Singh, Ankush Sharma, Manpreet Singh, Prashant S. Alegaonkar, and Dev Raj Saroha. "Study of effect of blast wave generator parameters on blast wave strength." In *11th International High Energy Materials Conference & Exhibits (HEMCE)*, 2017.

[10] Sandhu, Inderpal Singh, Ankush Sharma, Mritunjay Kumar Singh, Rajesh Kumari, Prashant S. Alegaonkar, and Dev Raj Saroha. "Study of blast wave pressure modification through rubber foam." *Procedia Engineering* 173 (2017): 570–576.

[11] Sandhu, Inderpal Singh, Ankush Sharma, Meenakshi Bhatt. Kala, Manpreet Singh, Dev Raj Saroha, Murugan Thangadurai, and Prashant S. Alegaonkar. "Comparison of blast pressure mitigation in rubber foam in a blast wave generator and field test setups." *Combustion, Explosion, and Shock Waves* 56, no. 1 (2020): 116–123.

[12] Sandhu, Inderpal Singh, Murugan Thangadurai, Prashant S. Alegaonkar, and Dev Raj Saroha. "Mitigation of blast induced acceleration using open cell natural rubber and synthetic foam." *Defence Science Journal* 69, no. 1 (2019): 53.

[13] Sandhu, Inderpal Singh, Meenakshi Bhatt Kala, Murugan Thangadurai, Manpreet Singh, Prashant S. Alegaonkar, and Dev Raj Saroha. "Experimental study of blast wave mitigation in open cell foams." *Materials Today: Proceedings* 5, no. 14 (2018): 28170–28179.

[14] Pandya, Kedar S., and Niranjan K. Naik. "Energy absorption capability of carbon nanotubes dispersed in resins under compressive high strain rate loading." *Composites Part B: Engineering* 72 (2015): 40–44.

[15] Yu, Min-Feng, Bradley S. Files, Sivaram Arepalli, and Rodney S. Ruoff. "Tensile loading of ropes of single wall carbon nanotubes and their mechanical properties." *Physical Review Letters* 84, no. 24 (2000): 5552.

[16] Yu, Min-Feng, Oleg Lourie, Mark J. Dyer, Katerina Moloni, Thomas F. Kelly, and Rodney S. Ruoff. "Strength and breaking mechanism of multiwalled carbon nanotubes under tensile load." *Science* 287, no. 5453 (2000): 637–640.

[17] Gojny, Florian H., Malte H. G. Wichmann, Bodo Fiedler, and Karl Schulte. "Influence of different carbon nanotubes on the mechanical properties of epoxy matrix composites–a comparative study." *Composites Science and Technology* 65, no. 15–16 (2005): 2300–2313.

[18] Martinez-Rubi, Yadienka, Behnam Ashrafi, Jingwen Guan, Christopher Kingston, Andrew Johnston, Benoit Simard, Vahid Mirjalili, Pascal Hubert, Libo Deng, and Robert J. Young. "Toughening of epoxy matrices with reduced single-walled carbon nanotubes." *ACS Applied Materials & Interfaces* 3, no. 7 (2011): 2309–2317.

[19] Jindal, Prashant, Shailaja Pande, Prince Sharma, Vikas Mangla, Anisha Chaudhury, Deepak Patel, Bhanu Pratap Singh, Rakesh Behari Mathur, and Meenakshi Goyal. "High strain rate behaviour of multi-walled carbon nanotubes–polycarbonate composites." *Composites Part B: Engineering* 45, no. 1 (2013): 417–422.

[20] Pandya, Kedar S., Kiran Akella, Makrand Joshi, and Niranjan K. Naik. "Ballistic impact behavior of carbon nanotube and nanosilica dispersed resin and composites." *Journal of Applied Physics* 112, no. 11 (2012): 113522.

[21] Natsuki, Toshiaki, and Morinobu Endo. "Stress simulation of carbon nanotubes in tension and compression." *Carbon* 42, no. 11 (2004): 2147–2151.

[22] Shen, G. A., Sirish Namilae, and Namas Chandra. "Load transfer issues in the tensile and compressive behavior of multiwall carbon nanotubes." *Materials Science and Engineering: A* 429, no. 1–2 (2006): 66–73.

[23] Lee, Jae-Hwang, Phillip E. Loya, Jun Lou, and Edwin L. Thomas. "Dynamic mechanical behaviour of multilayer graphene via supersonic projectile penetration." *Science* 346, no. 6213 (2014): 1092–1096.

[24] Lahiri, Debrupa, Santanu Das, Wonbong Choi, and Arvind Agarwal. "Unfolding the damping behavior of multilayer graphene membrane in the low-frequency regime." *ACS Nano* 6, no. 5 (2012): 3992–4000.

[25] Gelfand, B. E., M. V. Silnikov, and M. V. Chernyshov. "Modification of air blast loading transmission by foams and high density materials." In *Shock waves*, pp. 103–108. Springer, Berlin, Heidelberg, 2009.

[26] Petel, Orene E., F. X. Jetté, Samuel Goroshin, David L. Frost, and Simon Ouellet. "Blast wave attenuation through a composite of varying layer distribution." *Shock Waves* 21, no. 3 (2011): 215–224.

[27] Hamilton, Brenden W., Matthew P. Kroonblawd, Chunyu Li, and Alejandro Strachan. "A hotspot's better half: non-equilibrium intra-molecular strain in shock physics." *The Journal of Physical Chemistry Letters* 12, no. 11 (2021): 2756–2762.

[28] Merrett, Robert P., Genevieve S. Langdon, and M. D. Theobald. "The blast and impact loading of aluminium foam." *Materials & Design* 44 (2013): 311–319.

[29] Kaiser, Michael Adam. "Advancements in the split Hopkinson bar test." PhD diss., Virginia Tech, 1998.

[30] Francis, D. K., Wilburn R. Whittington, W. B. Lawrimore, P. G. Allison, S. A. Turnage, and J. J. Bhattacharyya. "Split hopkinson pressure bar graphical analysis tool." *Experimental Mechanics* 57, no. 1 (2017): 179–183.

[31] Chinke, Shamal L., Inderpal Singh Sandhu, Dev Raj Saroha, and Prashant S. Alegaonkar. "Graphene-like nanoflakes for shock absorption applications." *ACS Applied Nano Materials* 1, no. 11 (2018): 6027–6037.

[32] Kai, Yun, Walter Garen, and Ulrich Teubner. "Experimental investigations on microshock waves and contact surfaces." *Physical Review Letters* 120, no. 6 (2018): 064501.

[33] Long, X. J., Bo Li, Liang Wang, Jun Yu Huang, J. Zhu, and S. N. Luo. "Shock response of Cu/graphene nanolayered composites." *Carbon* 103 (2016): 457–463.

[34] Delavignette, P., and Severin Amelinckx. "Dislocation patterns in graphite." *Journal of Nuclear Materials* 5, no. 1 (1962): 17–66.

[35] Young, Philippa J. *Radiation damage in graphite: line defects and processes.* University of Surrey, Guildford, UK, 2016.

[36] Nabarro, F. R. N. "Fifty-year study of the Peierls-Nabarro stress." *Materials Science and Engineering: A* 234 (1997): 67–76.

[37] Peierls, Rudolf. "The size of a dislocation." *Proceedings of the Physical Society (1926–1948)* 52, no. 1 (1940): 34.

[38] Nabarro, F. "Dislocations in a simple cubic lattice." *Proceedings of the Physical Society (1926–1948)* 1947, 59, 256–272.

[39] Malard, Leandro M., Marcos AssunçãoPimenta, GeneDresselhaus, and Mildred S. "Raman spectroscopy in graphene." *Physics Reports* 473, no. 5–6 (2009): 51–87.

[40] Ferrari, Andrea Carlo, and John Robertson. "Raman spectroscopy of amorphous, nanostructured, diamond–like carbon, and nanodiamond." *Philosophical Transactions of the Royal Society of London. Series A: Mathematical, Physical and Engineering Sciences* 362, no. 1824 (2004): 2477–2512.

[41] Chinke, Shamal, Rohini Gawade, and Prashant Alegaonkar. "Synthesis and characterization of graphene-like nano ribbons (GNR) using chemical vapor deposition for shock absorbent application." In *AIP Conference Proceedings*, vol. 2142, no. 1, p. 060005. AIP Publishing LLC, 2019.

[42] Chinke, Shamal L., Inderpal Singh Sandhu, Tejashree M. Bhave, and Prashant S. Alegaonkar. "Surface interactions of transonic shock waves with graphene-like nanoribbons." *Surfaces* 3, no. 3 (2020): 505–515.

[43] Chinke, Shamal L., Inperpal S. Sandhu, Tejashree M. Bhave, and Prashant S. Alegaonkar. "Shock wave hydrodynamics of nano-carbons." *Materials Chemistry and Physics* 263 (2021): 124337.

[44] Chinke, Shamal L., Inderpal S. Sandhu, and Prashant S. Alegaonkar. "Blast mitigation properties of porous nano-carbon." *Diamond and Related Materials* 120 (2021): 108691.

4 Microwave Scattering and Radar Absorption Coating Properties of Nanocarbons

Electromagnetic radiations, also popularly termed with various names such as electromagnetic field, waves, power, and energy, are known to the mankind since the post-*Gallian* era as a *scientific* discipline named waves and optics [1]. Such waves exist over a range of frequency $>10^{20}$ Hz (gamma rays) to $<10^3$ Hz (radio waves) and obey the established principles of optics, such as dispersion, interference, diffraction, and polarization. Acoustic waves are their mechanical counterpart! However, unlike acoustic waves, the transportation of energy is medium independent for electromagnetic waves. On their interaction with any medium, changes occur in their nature of rectilinear wave propagation quantified, macroscopically, in terms of reflection, transmission, absorption, or scattering, which depends upon both waves and physical properties of the material. Microscopically, such changes are associated with the modulations of electromagnetic fields. However, controlling electromagnetic field around an object by manipulating its surface properties is challenging and is of particular interest in the defence sector [2].

4.1 RADAR KNOW-HOWS: THE BACKGROUND

The word RADAR is an abbreviation of *RAdio wave Detection And Ranging.* Although popularly attached with the name *RADIO*; these are the kind of communication systems that emit microwaves and detect the objects of interest upon their reflection. In the electromagnetic spectrum, a microwave region exists between the infrared (IR) and radio (RF) waves that, characteristically, have the frequency range 0.3–300 GHz with wavelength 100–0.1 cm, power 100 mW–0.01 kW, and energy 10^{-6} to 10^{-3} eV [3]. They have become increasingly important in the defence sector and are used as a seeker probe to sense, detect, and track the anonymous flying objects that could be disruptive in nature. In other words, radars are extended eyes in military and in aviation sector. There are two types of radars: (a) the civil, which are used for the airway communication, and (b) the military, deployed for tactical purposes such as surveillance, tracking, detection, and, if required, homing the targets! Depending upon the functioning, radars are classified into two types: primary radars that are based on a reflection-based detection, and secondary radars that work on the transponder principle. All aviation communication, generally, involves primary detection, whereas the military radar, mostly, works on the secondary principle. In strategic sector, creating a clutter to a military radar is assumed to be a disruptive technology and is a subject of intense research and development, since after the World War II, called the electronic warfare and, more specifically, the electronic countermeasures [4]. Such countermeasures are a payload on missile, fighter plane, uninhabited air vehicle, and naval vessel, including upcoming soldier's camouflage suits.

4.1.1 Detection and Range Finding: The Communication Bands

The military radars are deployed on the land as well as in communication. They are commissioned with no clue about their position (i.e. latitude/longitude), power, frequency, bandwidth, dimension, polarization, elevation, range, etc., and, hence, have a major concern for the enemy in terms of identifying them or creating a clutter in their communication. The greater challenge is to gather enough

TABLE 4.1

The Threat Spectrum of a Radar and Corresponding Jammers, Shielding, and Stealth Bands [5]

	Threat	Specific Threats	Countermeasure	
Band Designation	Frequency Range (GHz)		Frequency Range (GHz)	Band Designation
UHF	0.3–1	Very long-range surveillance	0.5–1	C
L	1–2	Long-range surveillance	1–2	D
S	2–4	Surface ship radar, weather radar, communication satellites, microwave ovens, Bluetooth, GPS, WLAN, cellular phones, etc.	2–3 3–4	E and F
C	4–8	Long-range radio tracking, telecommunications	4–6 6–8	G and H
X	8–12	Short-range tracking and missile guidance (locking mode) radar, terrestrial broadband, space communications, amateur radio, satellite communications, etc.	8–10	I
Ku	12–18	High-resolution tracking/homing mode, satellite communication	10–20	J and K
K	18–27	Water vapour void	20–40	
Ka	27–70	Millimetric tracking		

intelligence about the military deployment of radars! Table 4.1 shows the threat spectrum of a radar. Depending upon the terrain, geographical location, topology, etc., a primary radar has a detection range of up to 500 km; however, a secondary radar may have 2000 km or even more [5].

Tactically, managing an object signature on tracking radar threat, from a seeker projectile, is of great tactical importance [6]. Such signature involves characteristic combination of the geometry, the physical optics, and, importantly, the reflection properties of a target pop-up onto radar dial. The combination, known as the radar cross section (RCS), could effectively be quantified in terms of the shielding effectiveness (SE) of a target. The SE is the ratio of reflected to incident power of radiation, termed as the reflection loss and measured in dB at laboratory scale and m^2 at the field. For example, the RCS of a modern jet liner is ~500–700 m^2, whereas a stealth bomber has ~0.5 m^2, practically making it impossible to detect on a surveillance radar. Physically, the amount of reflection loss depends on the correlation between the wavelength (λ) of the incident signal and the dimension of the target (d). A $d < \lambda$ yields a typical Rayleigh scattering, whereas with a $d > \lambda$, diffraction/specular scattering occurs, and at the condition $d \sim \lambda$, the Mie scattering probability is more [7,8]. It is difficult to control each component selectively and tunably because of the arbitrary shape and size of the target. However, surface coating with specific magnitudes of complex permittivity and permeability could be a viable option to enhance the absorption of incident electromagnetic power of the signals. Such a surface manages to attenuate electric or magnetic field at the coating interface with a minimum amount of reflection, causing a large amount of power loss in the reflection called the reflection loss. The extent to which field interaction exists involves complex phenomena such as polarization, motion of mobile charge carriers, and joule heating within the surface of the skin, and the incident power is mostly consumed in the absorption process. Such a technique of enhancing reflection loss of a coating is known as the electromagnetic interference (EMI) shielding or electromagnetic interference compatibility (EMC).

There exist several techniques to determine the shielding properties of a specimen. Typically, the methods used for the microwave characterization of the dielectric materials are classified into four types based on the measurement structure implemented and ease of suitability to the experimental set-up: (a) transmission/reflection line, (b) open-ended coaxial probe, (c) free space, and (d) resonant mode. Each technique has its limitations, including the frequency at which the measurements can be performed and the type of the material that can be measured.

4.1.2 S-PARAMETER MEASUREMENTS: VECTOR NETWORK ANALYSIS

In general, free space measurements are expensive to investigate SE and analyse EMI characteristics in terms of measurements of S (scattering)-parameters, making the study impractical; however, port vector analysis is a laboratory-scale testing technique to validate the quality of the shielding architecture (Table 4.2).

In the current study, the S-parameter measurements for shield architecture have been carried out using a two-port vector network analyser (VNA, Model: PNA E8364B) equipped with a software module (85071E). The instrument is able to measure S-parameters including the permittivity (ε) and permeability (μ) functions over the frequency range 2–18 GHz. Figure 4.1a shows the schematic of a VNA and 1b displays the actual set-up. The VNA is used for impedance measurements. In its basic form, both DC (resistance) and AC (impedance) level measurements could be carried out, respectively, in scalar and vector modes [10].

However, characterizations at higher-frequency regimes such as microwaves are, in general, complex. They not only are limited to the impedance measurements, but also depend on investigating the incident, reflected, and transmitted microwave powers. Thus, assimilating principle of basic impedance measurement with suitable advancements in hardware at those frequencies enable for quantification of S-parameters. Figure 4.2a shows the block diagram of a VNA, typically, representing the number of components and functioning units such as the high-frequency signal source, signal separator, signal receivers, and processor and display unit. Figure 4.2b shows the scheme of measuring S-parameters, including their definition. The S_{ij} – parameters describe the response of an N-port network to the obtained voltage signal at each port, in which the first subscript (#) refers to responding, while the second refers to the incident port, for example, S_{21} – response @2 due to signal @1. Commonly carried out N-port measurements, using the VNA, in microwave region are one-, two-, and three-port networks. Among them, two-port S-parameters are easy to model, are reliable, and possess least errors with the advancement in data collection software such as Agilent

---

TABLE 4.2
Typical Specimen Type, Measured S-Parameters, and Calculation of Magneto-Dielectric Functions Using Various Conversion Techniques [9]

Measurement Technique	Specimen Type	S-Parameters	Magneto-Dielectric Functions	Waveguide Mode[a]
Transmission/reflection line (broadband measurement method)	Coaxial line, waveguides	S_{11}, S_{21}	$\varepsilon(\omega), \mu(\omega)$	TEM
Open-ended coaxial probe	Liquids, biological specimen, semi-solids	S_{11}	$\varepsilon(\omega)$	TE
Free space	High-temperature material, large flat solid, gas, hot liquids	S_{11}, S_{21}	$\varepsilon(\omega), \mu(\omega)$	TEM
Resonant (cavity)	Rod-shaped solid materials, waveguides, liquids	Frequency response Q-factors/cavity losses	$\varepsilon(\omega), \mu(\omega)$	TE, TM

[a] Mode: TE, transverse electric; TEM, transverse electromagnetic; TM, transverse magnetic

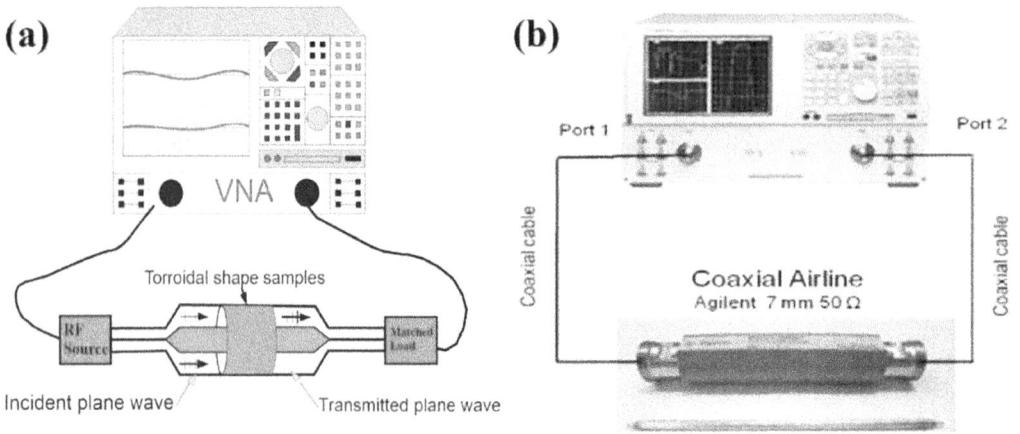

FIGURE 4.1 (a) Schematic diagrams of vector network analyser connected to the test specimen; (b) actual set-up displaying a coaxial airliner with load matching condition.

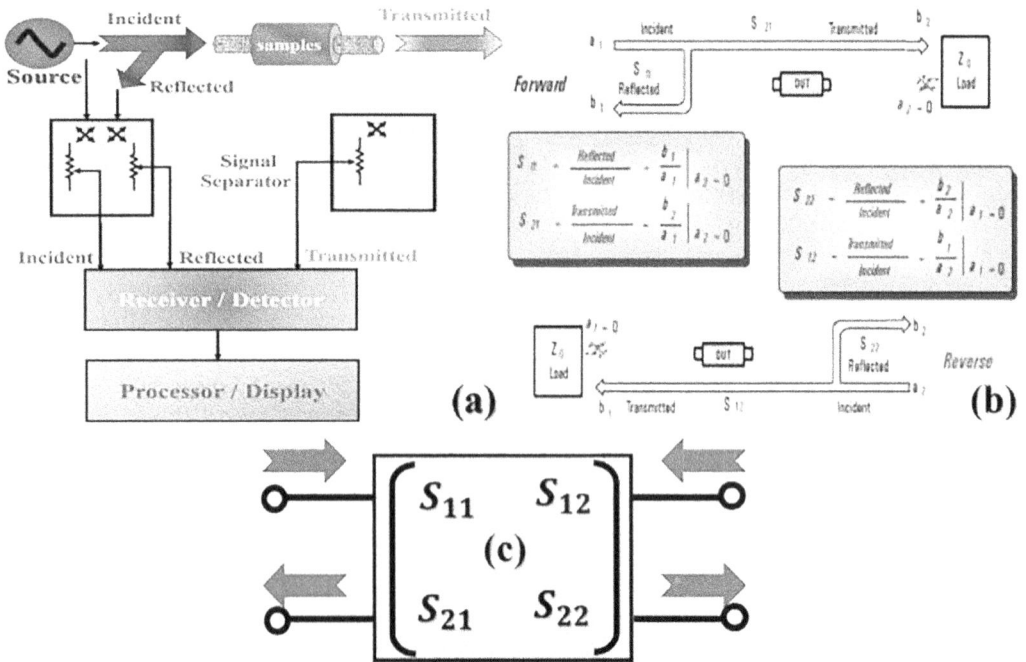

FIGURE 4.2 Schematic diagrams for (a) generalized block diagram of network analyser set-up, (b) definition of S-parameters, and (c) two-port network of S-parameter displaying various S_{ij} components, where the first subscript # refers responding, while the second refers to the incident port.

ADS [11,12]. Assuming each port terminated with Z_0 impedance, the four S-parameters: S_{11}, and S_{21}, respectively, are ratios of reflected to incidence, transmitted to incidence with same notion for S_{22}, S_{12}, as shown in Figure 4.2c, schematically. In the current study, we performed measurements in coaxial line as well as in waveguide mode (reflection loss) with TEM mode and incident power 1 μW. Prior to measurements, calibration and standardization of the instrument has been carried out. Once calibrated, the data recorded are absolute.

4.1.3 SAMPLE PREPARATION FOR S-PARAMETER AND REFLECTION LOSS STUDIES: COAXIAL AND SLAB-SHAPED SPECIMEN

4.1.3.1 Coaxial Measurements

One of the critical issues in the microwave scattering measurements is the specimen preparation for coaxial and reflection loss study. Broadly, the synthesized material has to be transformed into cylindrical- and slab-shaped articulate. The procedure is as follows.

For this purpose, the synthesized material, mostly a nanocomposite, was moulded into the desired shape using a composite blending technique. Particularly, for S-parameter measurements, performed in the coaxial mode, the requirement is a cylindrical-shaped specimen. These measurements were critical from the point of view of injecting the microwave power into the specimen.

For this, die was used, comprised of a bottom and top component with a Teflon-made coaxial prong, respectively, 7.0 mm OD and 3.00 mm ID as shown in Figure 4.3a. The prepared composite was taken as per the volume requirement of the mould. Initially, we calculated the amount of nanocomposite to be filled along with weight/volume % proportion of epoxy content. Accordingly, the weighing of nanocomposite was done and kept ready to blend with the epoxy/resin. In parallel, the preparation of epoxy was carried out, which consisted of a separate resin and hardener to be admixed in a specific weight ratio, say 10:4. After admixing, they were blended swiftly, with a continuous manual stirring. This step was carried out with utmost care for preventing epoxy matrix from pre-curing before the addition of nanocomposite. Actual addition of nanocomposite into epoxy matrix was carried out on a Teflon platform in order to prevent the wastage of material. Following this, the nanocomposite was blended thoroughly with the epoxy matrix for a period of few minutes to ensure homogenous paste formation. Prior to filling of the paste into the mould, the Teflon prong was waxed with silicone grease in order to avoid permanent sticking and, consequently, the damage of specimen shape [13,14].

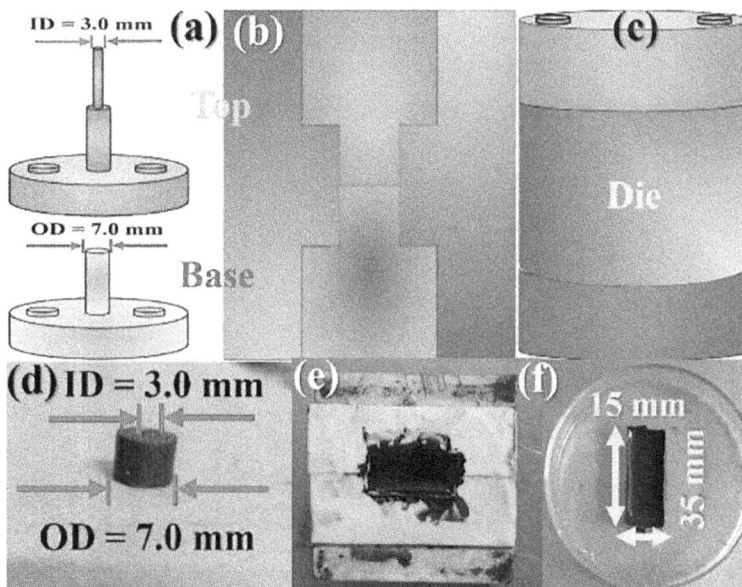

FIGURE 4.3 Top row: schematic design and drawings of (a) top and base components of mould displaying internal and external diameters of prong, respectively, 3.0 and 7.0 mm, (b) cross-sectional view of the mould, and (c) die. Bottom row: recorded digi-cam photographs of: (d) the fabricated unpolished toroidal-shaped (coaxial type) specimen displaying dimensions for S-parameter measurements, (e) nanocomposites loaded into slab-type mould, (f) unpolished slab-type specimens for reflection loss microwave measurements.

Subsequently, a metallic rod of ID 3 mm was placed at the centre of the mould. The paste was inserted into the mould with the help of a specula, and the entire die was closed and left for a period of 3–4 hours of curing at 300 K. The position of the mould and, correspondingly, the die is, respectively, shown schematically in Figure 4.3b and c. After room-temperature curing, the die was heated at 400 K for about 60 m in a hot vacuum oven for further curing and settlement of the nanocomposite. Following this, the die was cooled down and the nanocomposite specimen from the mould was separated by, first, removing the metallic rod, and with the help of a hollow cylindrical tube, the toroidal-shaped specimen was removed as shown in Figure 4.3d. The specimen was polished. A large number of specimens were fabricated and kept ready for S-parameter measurements. For reflection loss measurements, slab-shaped specimens were prepared with 3 mm height, 35 mm length, and 15 mm width. The procedure of paste preparation was identical as mentioned before. Figure 4.3e shows a loaded mould with nanocomposite, and Figure 4.3f shows the respective unpolished specimen. These specimens were also polished, subsequently, prior to measurements [15,16].

4.1.3.2 Dallenbach Scattering: Reflection/Return Loss Measurements

In another study, absorption measurements using the Dallenbach set-up has been carried out on the prepared nanocomposites, as shown, schematically, in Figure 4.4a. For this purpose, the nanocomposite absorber slab was attached to a metallic support as shown in Figure 4.4b. Mainly, the thickness of the block was optimized, approximately, to $\lambda/4$ value considering the permittivity and permeability of the composite such that the reflection component could be minimized for a preferred frequency. In general, the Dallenbach approach is a kind of remedy to erase radar signature from those metal surfaces causing high reflections of military aircraft, submarines, drones, tanks, etc. [17].

Basically, the mechanism of scattering in the Dallenbach absorber depends upon the interference (null power) condition for the incident microwave power that gets reflected from the first and second interfaces of the coating to achieve destructive interference. The reflection characteristics of microwaves at the interface of absorbers, mainly, depends upon complex permittivity and permeability functions of two media (i.e. air/composite interface) that can be adjusted to achieve a minimum reflection loss at the interface. In such case, the power nulling depends upon loss factors of μ'', ε'' as well as the thickness of the composite coating. Such coatings act as a resonant absorber and can, readily, be applied as a paint on aircraft, drones, and marine vessels, etc. These structures are sharp in their geometrical shape with highly curved surfaces or small in terms of wavelength

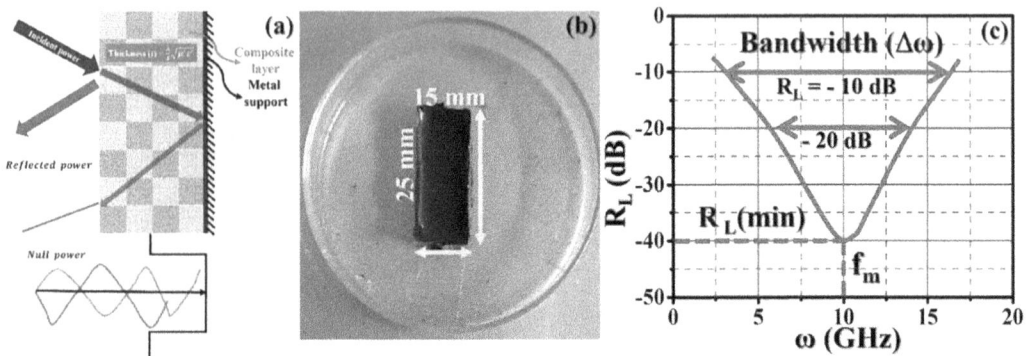

FIGURE 4.4 (a) Dallenbach set-up, i.e. single-layer absorber, for reflection loss (R_L) measurements labelling thickness condition on the composite layers, metal support, involved power losses from the first and second interfaces, and wave cancellation principle at the bottom, (b) photograph of nanocomposite slab-shaped specimen with dimensions to measure return losses, and (c) a typical performance of an absorber displayed over 2–8 GHz.

for transmitted waves. These coatings are pretty slim when compared to the other similar class of absorbers such as Salisbury or Jaumann type [17].

In the absence of bandwidth specifications, these coatings could easily be fabricated. A typical absorption/reflection loss curve as a function of frequency is shown in Figure 4.4c. An evaluation of the quality of the reflection loss curve is rather arbitrary and depends on the desired frequency range and the absorption level, which usually differ for various applications. However, three parameters are important, namely the dip in reflection loss level (R_L min), that is the level of minimum reflection or maximum absorption of the microwave power; the matching frequency (f_m) at which the dip in reflection loss occurs; and the width of the absorption band, i.e. the bandwidth ($\Delta\omega$) where reflection loss is below some level, usually −10 or −20 dB. The reflection loss of −10 dB is equivalent to 90% absorption or 10% reflection, while −20 dB represents 99% absorption or only 1% reflection of microwave power back to the incident direction [18].

Notably, the compositional investigations of the functional group involved, including estimations of particle size, were carried out using Fourier transform infrared (FTIR) spectrometer (Model: Tensor-27, Make: M/s Bruker India) over 400–4000 cm^{-1}, UV–visible spectrometer (Make: Shimadzu, Model: UV-2450) over 200–700 nm, X-ray diffraction technique (Make: Bruker, Model: 5678R), field emission scanning electron microscopy (Make: Technai, Model: 2278i), and energy-dispersive X-ray analysis (EDS).

4.2 MICROWAVE SCATTERING MECHANISM: THE MAXWELLIAN FORMULATION

In this section, we are going to review some basic interaction processes between the electromagnetic waves at the microwave region with a medium. This will help to develop an insight into the designing architecture for a shield material.

4.2.1 MAXWELL'S FORMULATION: REFLECTION AND TRANSMISSION COEFFICIENTS

The space through which the electromagnetic (EM) wave propagates is classified, fundamentally, into free space, and lossless and lossy dielectrics, including conducting medium [19–21]. There are conditions on the medium set via AC conductivity, σ_{ac}, dielectric, $\varepsilon(\omega)$, and diamagnetic, $\mu(\omega)$, functions to categorize them as a free space in which $\sigma_{ac} = 0$, lossless dielectrics for which $\sigma_{ac} \ll \omega\varepsilon$, lossy dielectrics with $\sigma \neq 0$, and good conductors as $\sigma_{ac} \approx \infty$, where ω is the angular frequency of the wave. These conditions are originated from the Maxwell's relations (typically displayed for the free space and expressed in SI units): $\nabla \cdot \mathbf{D} = \rho$; $\nabla \times \mathbf{H} = \mathbf{J} + \dfrac{\partial \mathbf{D}}{\partial t}$; $\nabla \cdot \mathbf{B} = 0$; $\nabla \times \mathbf{E} + \dfrac{\partial \mathbf{B}}{\partial t} = 0$ with a faculty equation, the Ampere's law: $\nabla \times \mathbf{J} + \dfrac{\partial \rho}{\partial t} = 0$. Mathematical treatment to these relations, further, permeates to the Helmholtz wave equation; here, ξ is $\mathbf{E}$ or $\mathbf{H}$ field and γ is a propagation constant with a condition: $\gamma^2 = \alpha^2 + \beta^2$; both α (propagation) and β (attenuation) are ω-dependent constants, and functions of μ, ε, and σ_{ac}. The bold letters denote a vector quantity. A basic feature of the Helmholtz equation for the EM field is the existence of travelling wave solutions that represent the transport of energy/power from one point to another. The equation provides the most general form of solutions: $\xi(\mathbf{r}) = |\xi_0| e^{\pm\gamma \cdot \mathbf{r}}$, where $|\xi_0|$ is amplitude and $\mathbf{r}$ is a propagation vector representing the universal forms of a medium as stated above [19–21].

Though looks trivial in its formulation, the real power of the established tool gets manifested when one treats them with the media. In general, when the incident power from air medium meets a realistic medium, say a composite material, part of the power gets reflected and partly transmission occurs at the composite/air interface. Importantly, the proportion of the incident power that is reflected or transmitted depends upon the constitutive parameters $\varepsilon'(\omega)$, $\sigma_{ac}(\omega, \varepsilon')$, and $\mu(\beta)$ of the nanocomposites, as discussed above, which get further reduced to a single electrodynamic

variable, namely the characteristic impedance, η, of the medium given by: $\eta = \dfrac{\sqrt{\dfrac{\mu'}{\varepsilon'}}}{\left[1 + \left(\dfrac{\sigma_{ac}}{\omega\varepsilon'}\right)^2\right]^{\frac{1}{4}}}$

[19–21].

Depending upon the angle of power incidence (usually taken to be normal incidence), the reflected and the transmitted power could be obtained by solving for $\xi \equiv f(\mathbf{E} \text{ or } \mathbf{H})$, which leads to the estimation of the generalized coefficient of reflection and transmission, respectively; $\Gamma = \dfrac{\eta_{nc} - \eta_{air}}{\eta_{nc} + \eta_{air}}$;

and $\varsigma = 2\dfrac{\eta_{nc}}{\eta_{nc} + \eta_{air}}$, where η_{nc} is the characteristic impedance of nanocomposites and η_{air} is that

of air, which, subsequently, reduces to a single component variable: $|\Gamma| = \dfrac{S_{ij} - 1}{S_{ij} + 1}$, where S_{ij} is the

scattering coefficient at the air (i)/composite (j) interface and given as shielding effectiveness (SE):

S_{ij} (in dB) $= 20\log_{10}\dfrac{S_{ij}\,(\text{out})}{S_{ij}\,(\text{in})}$. However, herein one has to note that this is again an oversimplification of the problem and represents a very basic formulation. In reality, the materials are far too complex in their responses when interacting with the incident EM power, as schematically shown in Figure 4.6, after the discussion presented in Section 4.2.2.

Typically, an EM absorbent medium can be termed as a perfect dielectric when σ_{ac} is null; in this case, there are no losses and the $\varepsilon(\omega)$ and $\mu(\omega)$ of the medium are the real quantities. In the case of a lossy medium, μ and ε are complex functions. A material medium that has a high σ_{ac} possesses low ability to store the energy as metals have high $\varepsilon''(\omega)$. In this case, the depth of penetration approaches zero and the material has the characteristics of a perfect reflector. In materials with low $\varepsilon''(\omega)$, the depth of penetration is larger and, as a result, little energy is absorbed by the medium, rendering the material transparent to the EM radiation. The determination of $\mu(\omega)$ and $\varepsilon(\omega)$ of a material is usually based on the measurement of complex EM parameters: the reflection coefficients (S_{11}) and transmission coefficients (S_{21}), with a measurement instrument VNA [20,21] as discussed in Section 4.1.2.

4.2.2 Microwave Interactions with Material: Losses, Absorption Factors, and Conditions

Microwave, being a small portion of the entire EM spectrum, is comprised of both electric $\mathbf{E}$ and magnetic $\mathbf{H}$ fields. To develop a suitable composite material for the microwave absorption, it is highly desirable to understand the nature of EM waves and its interaction with the material medium. Fundamentally, any time-varying $\mathbf{E}$ field is a source for $\mathbf{H}$ field and any time-varying $\mathbf{H}$ field is responsible for inducing $\mathbf{E}$ field. In free space (i.e. vacuum), for EM waves, the $\mathbf{E}$ and $\mathbf{H}$ fields are orthogonal to each other and propagate in a coupled manner in a direction perpendicular to both the fields. EM waves can propagate in free space and inside a material medium. However, in a material, the characteristics of EM waves change from that of the free space in terms of wave impedance, velocity, etc. The material responds to the field quantities by inducing electrical polarization or magnetization within surface and interfaces of the media [19–21].

4.2.2.1 Losses

There are a number of ways by which microwaves interact with a medium, and there are various means by which the microwave power can be exhausted in a material. The principal lossy channels are electric, hysteric, resonant (domain wall/interfacial polarization), eddy conduction (loop current), and electron spin resonance (ESR). However, it is difficult to separate out a single lossy channel or a combination of them exists for a particular shield architecture. The mechanism is mainly dependent on physical properties of the shield such as density (Z, atomic number) and particle size, and measurement conditions such as pressure, temperature, humidity, and frequency. The following diagram shows the losses involved in microwave absorption by material [19–21].

In a conducting medium, free electrons develop currents in response to the oscillatory electric field. The carrier flow results in the heating of the material through resistive heating (due to heating effect of the current). Microwaves are largely reflected from metal conductors, and therefore, in such conductors, heating by microwaves is minimal. In insulators, electrons are bound and an oscillatory field causes electronic reorientation or distortions of the permanent dipoles or dipole induction, resulting in heat. Such an absorber is, fundamentally, characterized by $\varepsilon(\omega)$ and $\mu(\omega)$ functions. Their real part is a measure of the extent to which the material will be polarized or magnetized by the application of an electric or magnetic field, respectively, to store the energy. And the imaginary part is a measure of energy losses incurred in rearranging the alignment of electric or magnetic dipoles according to the applied alternating fields. The loss tangent (tan δ) is indicative of the ability of the material to convert the absorbed energy into heat. For optimum coupling, a balanced combination of moderate ε' to permit adequate penetration and high loss (maximum ε'' and tan δ) is required for the exhaustion of microwave power. Two main loss channels exist for nonmagnetic materials: the conduction and the dipolar (dielectric) losses. Conduction losses are prominent in metallic (high-conductivity materials), whereas dipolar losses are dominant in dielectrics. Magnetic materials also exhibit conduction losses besides magnetic losses such as hysteresis, ESR, and domain wall resonance. Figure 4.5 shows the summary of the lossy effects operational in the shielding material on their interaction with microwaves [19–21].

4.2.2.2 Factors Affecting Microwave Absorption

There are numerous factors that influence the interaction of microwaves with an absorbent material. Particularly, radar absorbent materials (RAMs) exhaust most of the incident microwave power through heat due to molecular excitations. The most important property of a RAM comes from $\varepsilon'(\omega)$ and magnetic permeability $\mu'(\omega)$ connected with the intrinsic or characteristic impedance (Z) of the RAM given by the equation: $Z = Z_0\sqrt{\mu'/\varepsilon'}$, where $Z_0 = \sqrt{\mu_0/\varepsilon_0} = 376.7$ is the characteristic impedance of free space. Another factor for an optical shield is the Maxwell's identity $n = (\varepsilon'\mu')^{1/2}$ with wave velocity $v = \dfrac{c}{n}$, where $c = 3 \times 10^8$ m/s and n is the refractive index.

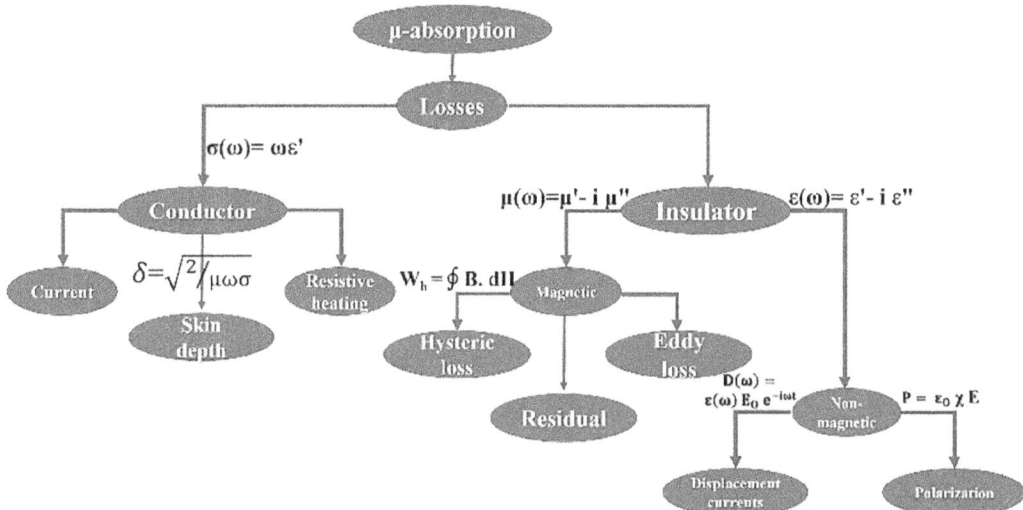

FIGURE 4.5 Origin of microwave absorption in material showing various channels available for exhaustion of power.

4.2.2.3 Conditions for Absorption

The key component of a microwave absorber is the low reflection at the interface layer with respect to the air medium and high attenuation of the incident signal within the absorber. To realize the condition, the primary requirement is that the absorbing materials should be lossy in nature. However, a high lossy material also reflects the microwave signal incident from free space due to impedance mismatch at the air–absorber interface. To obtain a low reflecting surface with attenuating characteristics, proper designing of the microwave absorber is required. The following criterion should be achieved [19–21].

4.2.2.3.1 Reflection Minimization

In case of a metal-backed single-layer absorber (Figure 4.4), the reflection phenomenon mainly occurs at two locations: first, at the air/absorber interface and, second, at the metal backing of the absorber. Technically, the reflection at the interface can be minimized by making the input impedance of absorptive layer close to the free space. At the air/absorber interface, the normalized input impedance (Z_{in}) with respect to its impedance in free space (Z_0) is determined from the equation:

$Z_{in} = \sqrt{\mu'/\varepsilon'} \, \tanh\left[\left(-i\omega/c\right)\sqrt{\mu'\varepsilon'}\, t\right]$, where t is the thickness of the coating. The expression reveals

the dependence of Z_{in} on the electrodynamic properties of the absorber, such as ε' and μ' including external parameters such as the thickness (t) of the absorber layer that are correlated with the microwave frequency (ω) of the incident radiation. In order to realize the impedance matching condition between the absorber, typically, a radar absorbing material, and the free space interface, the ratio μ'/ε' should approach unity. For a dielectric composite, $\mu(\omega) = 1 - i.0$; hence, the intrinsic tuneable parameter to achieve the impedance matching is achieved, thereby architecting ε' close to unity [20,21].

4.2.2.3.2 Enhancing Attenuation Performance

Within the composite, the microwave energy decays exponentially with distance by the factor $e^{-\alpha Z}$,

where the attenuation constant, α, is described by the equation: $\alpha\,(\varepsilon')=\omega\sqrt{\dfrac{\mu\varepsilon'}{2}\left[\sqrt{1+\left[\dfrac{\sigma}{\omega\varepsilon'}\right]^2}+1\right]}$.

The attenuation of microwave power increases with $\varepsilon(\omega)$, and since the real part (ε') is the constrained by the condition for low reflection at the interface, for increasing attenuation, the imaginary part (ε'') can only be enhanced [21].

4.2.2.3.3 Optimization of Coating Thickness

The reflected wave can, effectively, be minimized by reducing the reflected wave from the front face interface and from the back face (absorber–metal) interface, which can be achieved by phase cancellation. The principle of destructive interference is used, which is given by the equation: $t = \dfrac{\lambda}{4}\sqrt{\mu'\varepsilon'}$,

where t is the thickness of the absorber layer and λ is the free space wavelength of the incident wave. So, the thickness optimization of the RAM coating is an important condition [20].

4.2.2.3.4 Managing Reflection Loss, R_L

The transmission line technique put forwarded a condition on reflection loss, R_L, for a single-layer absorber, expressed by the equation: $R_L(dB) = -20\,log_{10}\left\{\dfrac{Z_{in} - Z_0}{Z_{in} + Z_0}\right\}$, where symbols are defined as above and carry the usual meaning.

4.2.2.3.5 Condition for Impedance Matching

With free space wave impedance, $Z_0 = \sqrt{\mu_0/\varepsilon_0} = 376.7$, input impedance, Z_{in}, $Z_{in} = \sqrt{\mu'/\varepsilon'} \, \tanh\left[\left(-i\omega/c\right)\sqrt{\mu'\varepsilon'}\, t\right]$; reflection loss condition $R_L(dB) = -20\,log_{10}\left\{\dfrac{Z_{in} - Z_0}{Z_{in} + Z_0}\right\}$.

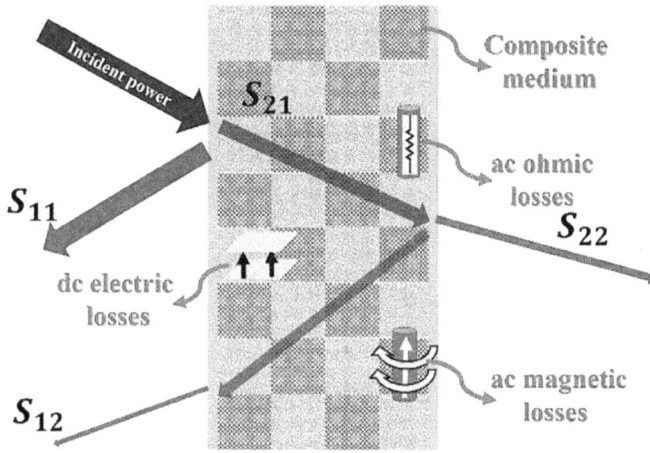

FIGURE 4.6 Schematic representation of the interaction of microwave power with a composite medium. It displays component of power reflected S_{11}, and transmitted S_{12}, S_{21}, S_{22} with respect to the incident power. Various AC and DC electric and magnetic losses are represented that are responsible for the incident power.

For minimum R_L, $\mu(\omega) = \mu' - i\ \mu''$; $\varepsilon(\omega) = \varepsilon' - i\ \varepsilon''$ are optimized for thickness, t, @ $Z = Z_0$. Ideally, $\mu' = \mu''$ and $\varepsilon' = \varepsilon''$; the impedance matching condition representing a perfectly absorbing RAM is given by: $Z_{in} = Z/Z_0 = 1$. This puts forward a demand to optimize six parameters, namely μ', μ'', ε', ε'', ω, and t. The above-presented equations, apparently, look simpler; however, the relation between the properties of the material and the minimum reflection is not that trivial to achieve. The high loss tangent in the material alone is not indicative of the minimum reflection. In the minimization problem, one has to optimize the properties for the two loosely connected phenomena, viz. the reflection from the front surface (impedance matching) and the absorption inside the absorptive layer. A typical absorption/reflection loss curve as a function of frequency is shown in Figure 4.4c. An evaluation of the quality of the reflection loss curve is rather arbitrary and depends on the desired frequency range and the absorption level, which usually differ for various applications, as discussed above. Figure 4.6 shows the actual loss encountered by a realistic medium like composites [19–21].

In general, it is difficult to comment on all these properties simultaneously and such an exercise could be cumbersome, painstaking, and sometimes repetitive.

4.2.3 S-Parameters by Transmission Line Approach: Nicolson–Ross Algorithm

The adopted transmission line approach involved the measurement of two-port complex scattering parameters using a VNA by placing the sample in the section of a waveguide or coaxial line as shown in Figure 4.7a. The calibration of the system is an essential prerequisite before performing the measurements. The technique measured port S_{11} (reflected) and S_{21} (transmitted) signals. The relevant scattering parameters relate closely to the complex permittivity and permeability of the material by equations as discussed above. There are a number of approaches for obtaining the permittivity and permeability functions from the measured S-parameters. One such approach is the Nicholson–Ross algorithm (NRA) [22,23]. The flow chart for the NRA is shown in Figure 4.7b.

The protocols for calculating S_{11} and S_{21} using the NRA come from the equations: $S_{11} = \dfrac{\Gamma\left(1-T^2\right)}{1-\Gamma^2 T^2}$ and $S_{21} = \dfrac{T\left(1-\Gamma^2\right)}{1-\Gamma^2 T^2}$; both the parameters could directly be obtained from the VNA measurements. The Γ and T are, respectively, reflection and transmission coefficients. They can be expressed in terms of S-parameters as: $\Gamma = K \pm\sqrt{K^2-1}$, where $K = \dfrac{S_{11}^2 - S_{21}^2 + 1}{2S_{11}}$ and $T = \dfrac{S_{11} + S_{21} - \Gamma}{1-(S_{11}+S_{21})\Gamma}$.

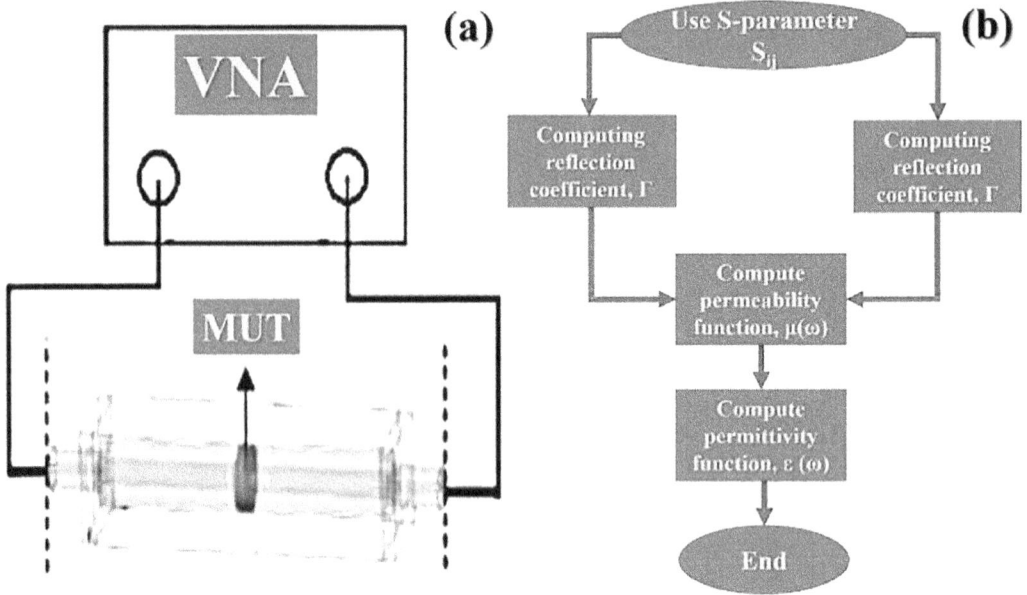

FIGURE 4.7 (a) Transmission line method for measurements of S-parameters using VNA; (b) flow chart for Nicolson–Ross algorithm (MUT, material under test).

Moreover, Γ and T are represented in terms of $\mu(\omega)$ and $\varepsilon(\omega)$ functions as: $\Gamma = \dfrac{Z_{sn}-1}{Z_{sn}+1}$, where $Z_{sn} = \sqrt{\dfrac{\mu}{\varepsilon}}$, and $T = e^{-\gamma\,t}$, where $\gamma = \gamma_0\sqrt{\mu\varepsilon}$ and $\gamma_0 = \left(\dfrac{2\pi}{\lambda_0}i\right)$ with λ_0 as free space wavelength.

Solving for ε, μ, and γ in terms of Γ and T, one can arrive at: $\sqrt{\dfrac{\mu}{\varepsilon}} = \dfrac{1+\Gamma}{1-\Gamma};\ \gamma = \dfrac{\left[Log_e\left(1/T\right)\right]}{t}$, to deduce: $\mu = \dfrac{\gamma}{\gamma_0}\dfrac{1+\Gamma}{1-\Gamma}\ ;\ \varepsilon = \dfrac{\gamma}{\gamma_0}\dfrac{1-\Gamma}{1+\Gamma}$. In a state-of-the-art measurement system, the conversion of S-parameters to permittivity/permeability functions is usually computed by MATLAB or embedded programming such as C/C++ with the application software module 85071 [22,23].

4.2.4 Coating Characteristics: The Survey

In this chapter, representative systems have been investigated, which include ferrites and their combination with nanocarbons such as multiwalled carbon nanotubes (MWCNTs), reduced graphene oxide (rGO), and carbon nano-spheres (CNS). The choice has been made from the most relevant literature, and preference is given to magnetic particle inclusions in the dielectric matrix with a high amount of electric conductivity of nanocarbon fillers.

4.3 SHIELDING PERFORMANCE OF MATERIALS ARCHITECTED

Recently, a number of attempts have been made to develop ferrite-based RAMs [8,9] due to their better EMI suppression properties such as superior permittivity/permeability functions to offer shielding of a larger bandwidth, including thinner coating geometries. Section 4.3.1 presents the preparation of variably thin Ni-Zn nano-ferrite composites for their performance evaluation for EM and microwave absorption properties.

4.3.1   MICROWAVE ABSORPTION PROPERTIES OF NI-ZN FERRITE NANOPARTICLE NANOCOMPOSITES

The composite preparation was carried out by using Ni-Zn nano-ferrite powder thoroughly mixed using non-aqueous acetone medium in two proportion polyurethane matrix consists of polyol-8 (Make: Ciba-Geigy, Switzerland) and hexamethylene diisocyanate (E-Merck, Germany) mixed in a ratio of 50:50:30 by weight Ni-Zn nano-ferrite was mixed in polyurethane (PU). The mixture was homogenized using the solid-state synthesis route in a mortar and pestle and then subjected to the mould processing as described in Section 4.1.3 to make toroidal- and slab-shaped samples with PU:Ni-Zn at a ratio of 30:100 by weight [13,14].

4.3.1.1   Morphological Properties

The SEM of virgin PU and Ni-Zn ferrite nanoparticles distribution in PU is shown in Figure 4.8a and b, respectively. The phase evolution of Ni-Zn nano-ferrite particles (in image (b)) shows amorphous distribution of Ni-Zn nano-ferrite particles within PU matrix.

Moreover, the thermogravimetric analysis (TGA) has been carried out to study the thermal stability of the prepared nano-ferrite composite. Figure 4.8c shows the recorded TGA plot of the prepared composite that exhibited multi-stage weight loss. However, it was found to have a significant thermal stability of up to ~270°C.

4.3.1.2   Microwave Characteristics

The electromagnetic parameters (ε', ε'', μ', and μ'') of the Ni-Zn nano-ferrite are displayed in Figure 4.9a and b. Figure 4.9a shows the complicated behaviour of the permittivity ε (ε', ε'') with frequency with a zero dielectric loss @ 13 GHz for composite. The inset in Figure 4.9a shows the linearly increasing magnitude of permittivity (ε), whereas Figure 4.9b shows the variation of permeability (μ) with frequency in the range of 14–18 GHz of frequency. Figure 4.9b shows that the real part of

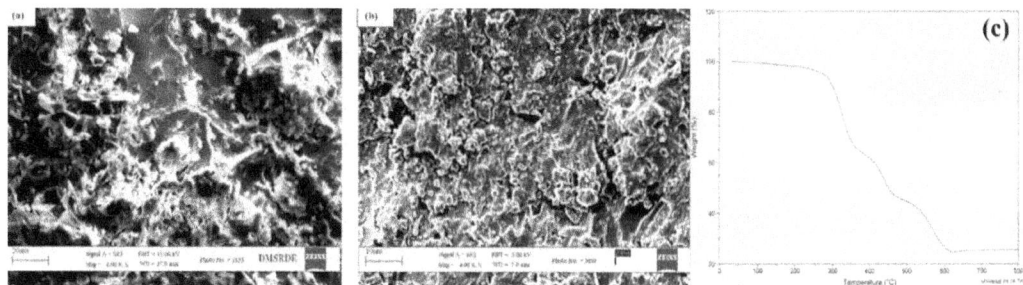

FIGURE 4.8   Recorded SEM images for (a) PU and (b) Ni-Zn/PU nanocomposite, and (c) recorded thermogram for nanocomposite.

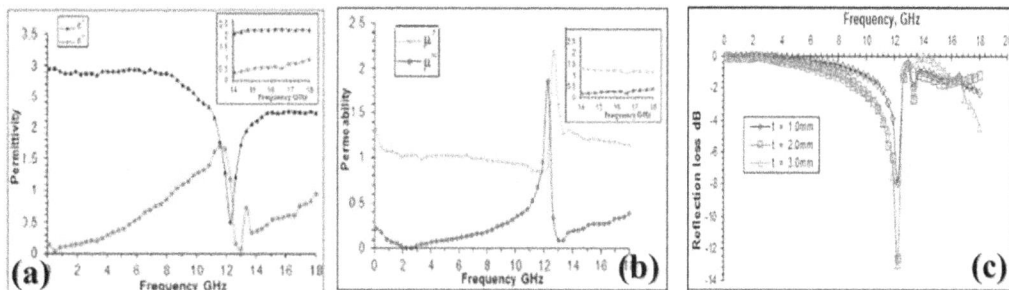

FIGURE 4.9   Recorded (a) $\varepsilon(\omega)$ and (b) $\mu(\omega)$ behaviour over 2–18 GHz; (b) reflection loss, R_L, measured in dB for various thicknesses of nanocomposite.

TABLE 4.3

Constitutive Data Corresponding to NC and 1%–4% Ni/NC Composites [16]

				$\sigma_{ac} \times 10^9$ (S/m)			
		$\sigma_{dc} \times 10^2$			Jonscher's Coefficient		
S. No.	Composition	(S/m)	ε'	Average σ_{ac}	A	a	Average β (H)
1	NC	0.76±0.20	0.0913±0.0290	0.741±0.042	0.466±0.208	0.507±0.019	16.89±0.12
2	01%	1.51±0.39	3.0293±0.0057	1.503±0.050	0.470±0.130	1.205±0.011	17.52±0.06
3	02%	1.67±0.41	4.1735±0.0095	2.956±0.074	0.430±0.013	1.174±0.031	20.56±0.11
4	03%	1.80±0.52	4.5479±0.0324	3.670±0.034	0.423±0.054	1.097±0.014	21.49±0.27
5	04%	1.93±0.55	5.5012±0.0537	4.965±0.228	0.041±0.031	0.909±0.023	23.75±0.12

permeability (μ') decreases with increasing frequency, while the magnetic loss, i.e. the imaginary part of permeability (μ''), increases with increasing frequency in the range 14–18 GHz because of the presence of Ni-Zn filler [13,14].

The reflection loss (dB) of the prepared Ni-Zn nano-ferrite-based nanocomposite sample having 30% (by wt.) Ni-Zn/PU composite matrix for various thicknesses (t = 1.0, 2.0, and 3.0 mm) were calculated using experimentally obtained values of $\varepsilon(\omega)$ and $\mu(\omega)$. The value of R_L was increased by 57.98% at a matching frequency, f_m, for the increase in the thickness from 1.0 to 2.0 mm. However, with increasing thickness to t = 3.0 mm, a marginal elevation by ~3% in measured R_L, i.e. from −12.56 to −12.93 dB, is noted as shown in Table 4.3. The prepared material was demonstrated to be utilized for EMI shielding and stealth applications.

In another study, nanocarbons were investigated for their shielding performance. Due to their superior properties, nanocarbons such as single-walled (S) and multiwalled (MW) carbon nanotubes (CNTs), and graphene in the form of polymer nanocomposites have been reported to be effective shielding materials in the X-band regime [24].

4.3.2 Impressive Transmission Mode EMI Shield Parameters of GNC/PU Nanocomposites for Short-Range Tracking Countermeasures

Herein, we have identified a peculiar interface polarization mechanism between graphene-like nanocarbon sheets (GNCs, 1–25 wt%) and PU to improve electromagnetic shielding properties of PU, in the X-band region [7]. The nature of GNCs bonding with PU has been analysed using vibration spectroscopy. The microwave scattering measurements have been carried out on toroidal-shaped samples to determine complex permittivity ($\varepsilon'-j\varepsilon''$), AC conductivity, skin thickness, transmission loss, S_{21}, and shielding effectiveness. The scattering analysis is presented in light of chemical bonding, dispersibility, and morphology of the nanocomposites. The polarization mechanism indicated that the atomic polarization associated with urethane amide rings acts as a backbone to engage incident electromagnetic field wiggles via charge transfer polarization current at doubly bonded nitrogen, oxygen, and hydrosorbed sp^3 carbon sites in GNCs. By and large, for the transmission loss of ~40%, the required PU thickness is more than a centimetre, whereas almost 99.9% loss is recorded for a millimetre thick PU, at 25 wt% loading of GNCs. Details are presented in the reference [7] and are briefly described below.

The GNCs were used as a typical nanofiller for nanocomposites preparation [25]. The precursor materials such as polyol-8 resin (Ciba-Geigy, Switzerland) and diphenylmethane 4,4′-diisocyanate ($C_{15}H_{10}N_2O$, MDI, Merck Chemical) were used to synthesize PU, along with analytical-grade reagent acetone (Alfa Aesar, India) as the solvent. The preparation protocols were same as that described in preceding sections.

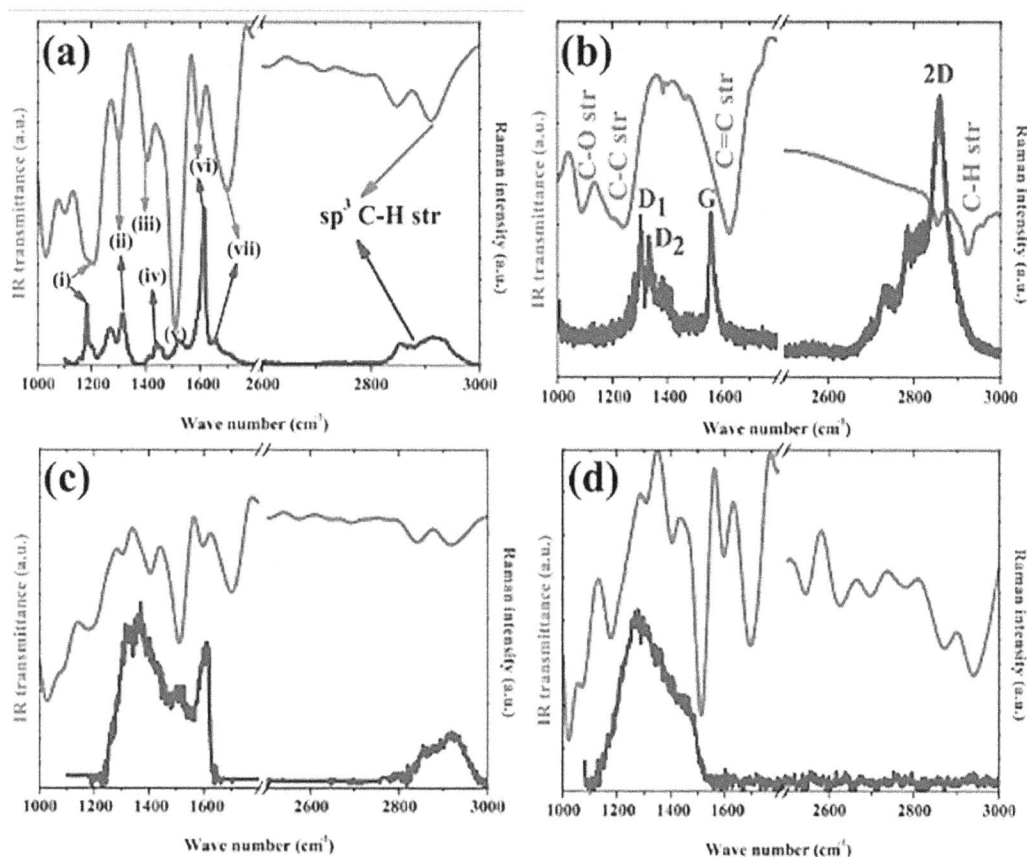

FIGURE 4.10 Typical FTIR and Raman spectra (excitation wavelength: 785 nm) recorded for the systems: (a) PU, (b) GNCs, (c) 1wt%, and (d) 25 wt% GNCs/PU nanocomposite samples. For PU, the peak indexing is (i) urethane amide (Raman active),=C−N−(IR), (ii) urethane amide III (both Raman and IR), (iii) δ(CH) urethane amide III (IR), (iv) δ(CH2) ν_{sym} N−C−O (Raman), (v) ν(Ar)-urethane amide II ν(C−N)+δ(N−H) (IR), (vi) ν (Ar) (both IR and Raman), and (vii) ν (Ar) urethane amide I ν(C=O) (both IR and Raman).

4.3.2.1 Vibration Spectroscopic Analysis: FTIR and Raman Spectroscopy

The typical trend observed for the transmittance loss of IR bands was in commensuration with the change in Raman-active modes for, mostly, all the samples. Broadly speaking, with subsequent incorporation of GNCs in PU, any change in the band-associated amide backbone, i.e. ν (Ar), is marginal. For these peaks, neither intensity nor vibration frequency is observed to be varied. This preliminarily indicates that the single-bonded backbone of the PU matrix mostly remained intact even after the incorporation of GNCs. We have summarized our observations as follows: (a) a decrease in the Raman peak intensity associated with the non-planar, hydrogen-bonded peaks of urethane amide III δ(CH) at ~1315 cm⁻¹ and 1500 cm⁻¹ (−N−H−), (b) significant intensity variations in doubly bonded oxygen (C−O at 1700 cm⁻¹), δ(CH₂) ν_{sym} N−C−O at ~1441 cm⁻¹, nitrogen (=C−N− at 1250 cm⁻¹), and (c) vanishing Raman-active sp³ C−H str (str = stretching) modes at ~3000 cm⁻¹ [12] (Figure 4.10).

This indicates that hydrogen bonding in the PU matrix is modified heavily after the incorporation of GNCs, in addition to the change in the environment at doubly bonded moieties. Among this, the nitrogen-based moieties (−N−H−, N−C−O, and =C−N−) are electron donor-loaded species that could be contributing to bonding with GNCs via hydrosorption. In this, the double-bonded site could be more reactive, due to π-conjugation, and the effect is, probably, dominant at these sites. In general, the nature of hydrogen bonding is tricky in the sense that these moieties are neither electronegative nor electropositive. Thus, the discussions lead to an intermediate conclusion that the nature of interaction could be hydrosorption in origin and prone at double-bonded sites, dominantly [26,27].

4.3.2.2 DC Conductivity

Figure 4.11 shows the recorded variations in DC conductivity (σ_{dc}) as a function of GNC weight fraction (p). One can see that, below the weight % threshold of ~5.0, the conductivity showed a dramatic decrease of ~7 orders of magnitude. This indicates that, above this threshold, the percolating network is formed in the PU matrix. The observed variations could be attributed to the increase in the number of hydrosorped conducting sites. Further, the inset in Figure 4.4 shows that the electrical conductivity obeys the exponent law given by: $\sigma \propto \left(p - p_c\right)^{\gamma}$ [28], where σ is the composite conductivity, p, the weight fraction of GNCs, pc, the percolation threshold, and γ, the critical exponent. Since the density of GNCs can only be approximately estimated, the weight fraction of GNCs is used instead of volume fraction [7,28].

For log(σ) as a function of log ($p-p_c/p_c$), the GNCs/PU nanocomposite conductivity agrees well with the percolation behaviour as predicted by equation, as stated above. The straight line with p_c ~3.0 wt% and $\gamma = 1.69$ gives an excellent fit to the obtained data with a correlation factor of 0.02. The percolation threshold is found to be at lower side, that is 3.0 wt% GNCs. This could be attributed to the efficient dispersibility of GNCs into the PU matrix, thereafter. The theoretical value of γ for a three-dimensional (3D) percolating network varies from 1.6 to 2.0, while the experimental values for carbonaceous materials such as carbon nanotube composites are reported to be varied from 0.7 to 3.1 [29].

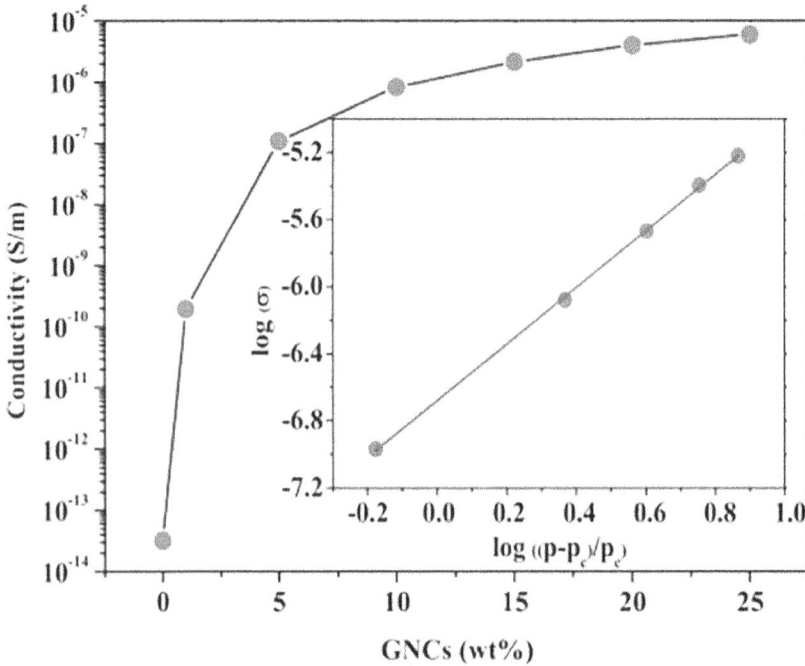

FIGURE 4.11   Variations in DC conductivity (σ_{dc}) (in logarithmic scale) as a function of weight fraction (p) of GNCs in the PU matrix. Measurements were performed using a standard two-probe technique, at room temperature. Inset: the log-log profile for σ_{dc} vs log ($p-p_c/p$). The straight line in the inset is fitted using least square methods for the obtained data using equation (4.1, above), returning the best-fit values $p_c \sim 5.0$ wt% and $\gamma = 1.69$ (correlation factor: 0.02).

Basically, three mechanisms are responsible for the attenuation of an incident electromagnetic wave: (a) reflection, (b) absorption, and (c) multiple internal reflection losses at the interface due to conductive fillers or porosity of the materials.

4.3.2.3 Analysis of Scattering Parameters: Real and Imaginary Permittivity

Figure 4.12a and b is the recorded frequency response spectra, respectively, for real (ϵ') and imaginary (ϵ'') parts of permittivity of PU and GNCs/PU nanocomposites with variable GNC wt%. A systematic increase in both real and imaginary parts of permittivity has been observed, over the frequency regime, with an increase in the GNC content.

On a relative scale, the comparison has been made across the categories of the samples by taking average at logarithmic-normal scale (indicated in Figure 4.12c). At 25 wt%, the value of real part of permittivity is 9.31 ± 0.03. In contrast, PU offers a low (real) permittivity of 1.84 ± 0.01. This increase is ~5 times. However, the imaginary part of permittivity is 1.29 ± 0.03 (for 25 wt%), which indicates an increase by a factor of ~30 with respect to the base value of PU (0.043 ± 0.006). Thus, on a relative platform, an increase in both the real and imaginary parts of permittivity is due to the increase in AC conductivity via enhancing active modes of charge transfer polarization by

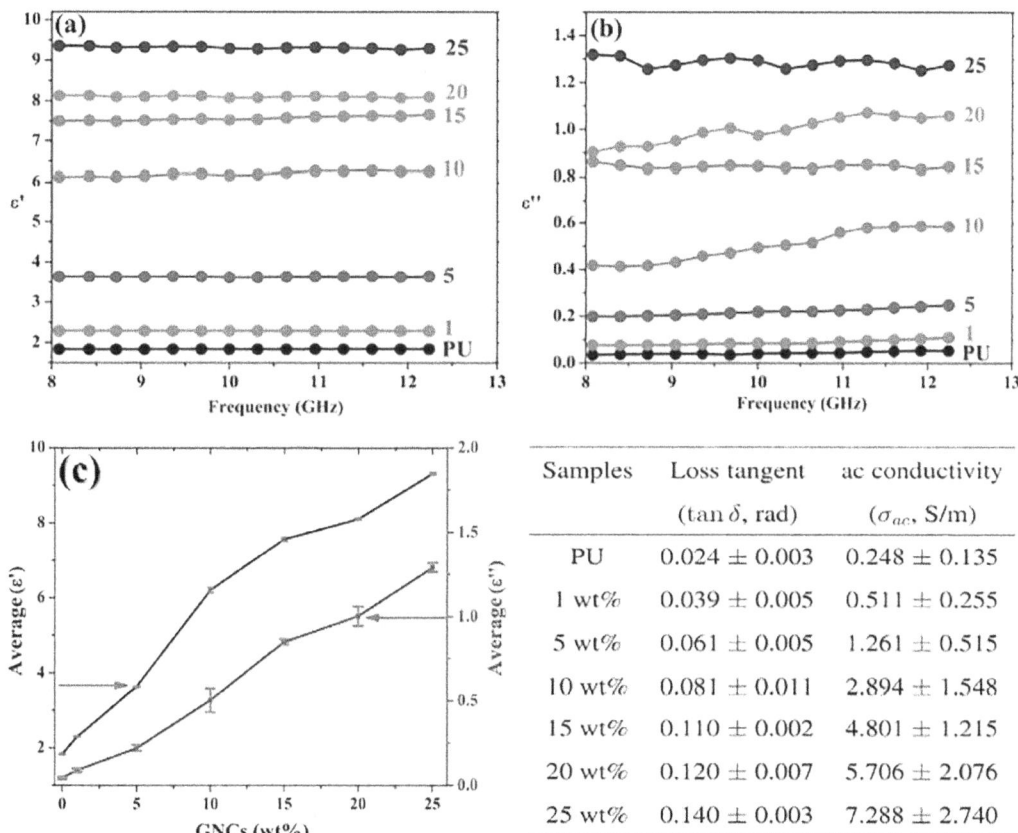

Samples	Loss tangent (tan δ, rad)	ac conductivity (σ_{ac}, S/m)
PU	0.024 ± 0.003	0.248 ± 0.135
1 wt%	0.039 ± 0.005	0.511 ± 0.255
5 wt%	0.061 ± 0.005	1.261 ± 0.515
10 wt%	0.081 ± 0.011	2.894 ± 1.548
15 wt%	0.110 ± 0.002	4.801 ± 1.215
20 wt%	0.120 ± 0.007	5.706 ± 2.076
25 wt%	0.140 ± 0.003	7.288 ± 2.740

FIGURE 4.12 Recorded complex microwave scattering data over the measured frequency regime. (a) Real (ϵ') and (b) imaginary (ϵ'') parts of the permittivity spectra for PU and GNCs/PU nanocomposites. The numbers to the right-hand side in each profile indicate wt% of GNCs. (c) Estimated magnitudes of real ϵ' and imaginary ϵ'' parts of permittivity (average at logarithmic-normal scale) as a function of GNCs wt%. Table: Magnitudes of measured loss tangent, tan δ, and AC conductivity, σ_{ac}. The σ_{ac} is increased linearly by a factor of 30 with subsequent GNCs incorporation up to 25 wt%.

GNCs in PU matrix. Further, using ϵ'' parameter, the AC conductivity (σ_{ac}) of a dielectric material could be evaluated using: $\sigma_{ac} = 2\pi f \epsilon_0 \epsilon''$, where σ_{ac} is measured in S/m, $\epsilon 0$, the free space permittivity (8.854×10^{12} F/m), and f, the applied frequency in Hz. The value of σ_{ac} for PU is 0.248 ± 0.135, which was increased linearly by a factor of 30 times with sequential incorporation of GNCs till 25 wt%. The observed increase is due to donor-loaded nitrogen sites such as –N–H–, N––C––O, and =C–N– attached to GNCs. The effect seems to be dominant at double-bonded GNCs sites, due to π-conjugation with the host matrix [30].

4.3.2.4   Efficient Microwave Absorbing Properties

The mode of measurements was a typical transmission measurement (scalar S_{21}-measurements). The dB value describes how much the level of an incident power (or power flux density) has decreased, after passing the specimen under test. The calculation of percentage values presented in Figure 4.13 refers to the power relationship. It tells us that at ~20 dB the shielding reduced the penetrating power down to 1%. To calculate the dB value, the following relation is used to compute SE: $SE = 10 \log_{10} [P_T/P_I]$ (in dB), where P_I is the incident power and P_T is the transmitted power. The scalar S_{21} has been computed as shown in Figure 4.13a. The S_{21} is the flat dispersion response, and variations are consistent with the permittivity response. The uniformity of S_{21} indicates that the operative mechanism as explained above surmises the obtained results. Further, S_{11} and S data are not presented and S_{21} alone could not be claimed for total shielding effectiveness (SE_{TO}). For this purpose, the SE due to transmission (SE_T) has been measured using equation for the effective transmittance loss and is plotted in Figure 4.13b.

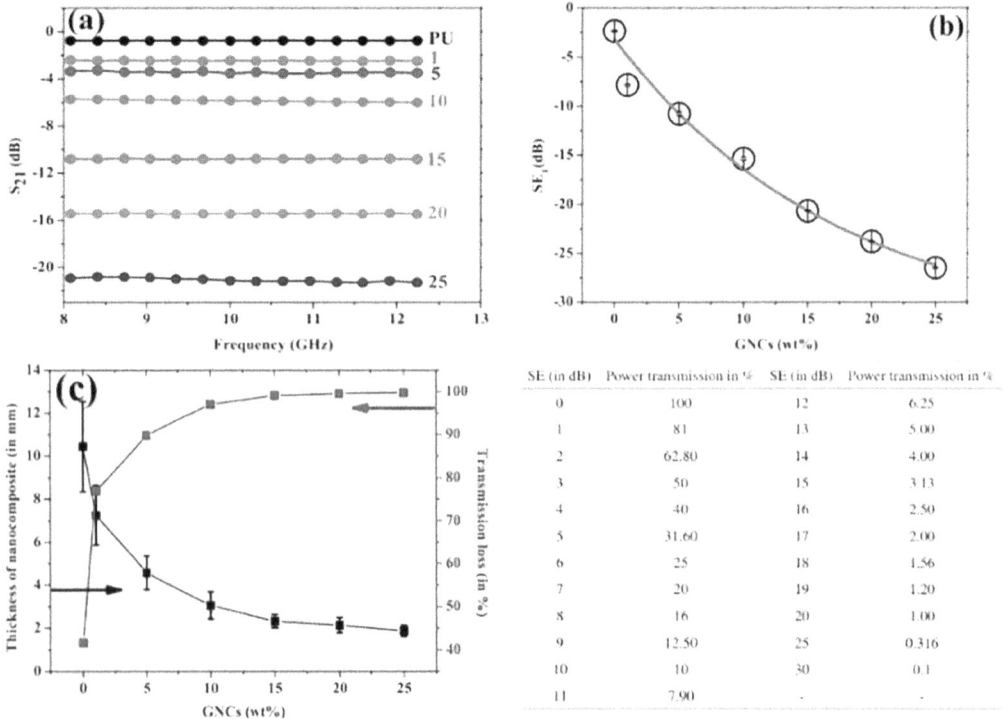

SE (in dB)	Power transmission in %	SE (in dB)	Power transmission in %
0	100	12	6.25
1	81	13	5.00
2	62.80	14	4.00
3	50	15	3.13
4	40	16	2.50
5	31.60	17	2.00
6	25	18	1.56
7	20	19	1.20
8	16	20	1.00
9	12.50	25	0.316
10	10	30	0.1
11	7.90	.	.

FIGURE 4.13 V(a)ariation in scalar S_{21} parameters measured in dB unit. Response is plotted over 8.2–12.4 GHz for PU and variable wt% of GNCs in composite, (b) computed shielding effectiveness due to transmission loss, SE_T, (c) variation in thickness measured (in mm) and transmission loss (in %) as a function of wt% of GNCs in PU. Figure also indicates measurements on base PU for comparison. Table: Relationship between shielding effectiveness (SE) and power transmission in %.

The SE_T is highest for PU and gradually decreased with filler weight fraction. The magnitude of SE_T, for PU, is −2.34 dB, whereas, on percolation threshold, the value was found to be −10.76 dB. Gradually, the value is increased up to −26.45 dB for 25 wt% GNCs/PU composites. In general, the trend of SE_T is identical to that of S_{21}. Hence, the average variation is plotted. For PU, the magnitude of SE_T is estimated to be 12 times lower with respect to 25 wt%. Hence, it is interesting to estimate percentage variation in transmission loss. Further, the extent of microwave propagation in a medium decides the amount of transmission loss. The discussions on the extent of microwave propagation with percentage transmission loss are presented in subsequent sections [7].

4.3.3 FERRITE/NANOCARBON COMPOSITES

4.3.3.1 γ- and Ni-Zn Ferrite Thermoplastic Polyurethane

In the previous section, the work on epoxy-based Ni-Zn ferrites has been presented for the stealth and EMI applications. In another study, composition wt. 40% γ-nano-ferrite and wt. 20% Ni-Zn ferrite/epoxy-based nanocomposites were investigated for their EMI performance [13]. The morphology (Figure 4.14a) and thermal behaviour (Figure 4.14b) of the nanocomposite samples were investigated using the TGA technique, including SEM imaging of the thermally treated composite. The permittivity and permeability measurements (Figure 4.14c and d) were performed in coaxial configuration over 2–15 GHz, and variations in complex electromagnetic parameters (ε, μ) and reflection loss (R_L, Figure 4.14e) with the applied frequency were studied. Further, the microwave reflection loss measurements of the nanocomposite have been analysed in the S-, C-, and X-band (2–10 GHz) frequency for various sample thicknesses for single-layer PEC backed condition (data table in Figure 4.14) [13,14].

The microwave absorption properties of γ-ferrite nanocomposites were investigated in S-, C- and lower X (8–10 GHz)-bands. The maximum reflection losses were found to be increased with increasing thickness. The EM parameters (ε, μ) were found to be varied with the applied frequency of the microwave. The analysis of magnetic and electric losses confirmed that the composition 40% (wt.) gamma ferrite + 20% (wt.) Ni-Zn ferrite in epoxy matrix was lossy in the frequency range 2–15 GHz. The SEM confirmed the microstructure of the nanocomposite. The TGA showed that the composite was thermally stable up to 190°C. The nanocomposite had the potential to be used as a RAM, electromagnetic shielding screens, coatings or jackets, and stealth in S (2–4 GHz)-, C (4–8 GHz)-, and lower X (8–10 GHz)-bands.

4.3.3.2 Stealth Properties of MWCNTs/Ferrite/PU Composites

As an alternative route, the stealth characteristics of toroidal-shaped nanocomposite samples with bi-filler nanoparticles had been investigated [14,15]. Filler 1 was chosen to be MWCNT, and filler 2 was Ni-Zn ferrite with a variable weight content that was thoroughly admixed in the thermoplastic PU (TPU) matrix. Simulation studies for metal-backed single-layered absorbers were carried out for studying the EM absorbing properties of ~2.0 mm thick stealth coating of the nanocomposites. Network analyser studies in coaxial measurements produced complex permittivity and permeability function data. The absorbing properties were examined by utilizing the measured values of complex permittivity and complex permeability functions over 2–18 GHz. Reflection loss, R_L (dB), vs. frequency variation was also determined for the 2.0 mm thick nanocomposites employing the simulation code. Experimentally, SEM (Figure 4.15a–c) and TGA measurements (data not shown) were performed to analyse the morphological and thermal behaviour of the nanocomposite. The complex permittivity and permeability of the nanocomposites were found to be frequency dependent (Figure 4.15d and e). A higher R_L was reported in the C-band (higher-frequency side) and the X-band (higher) for the 2.0 mm thick sample.

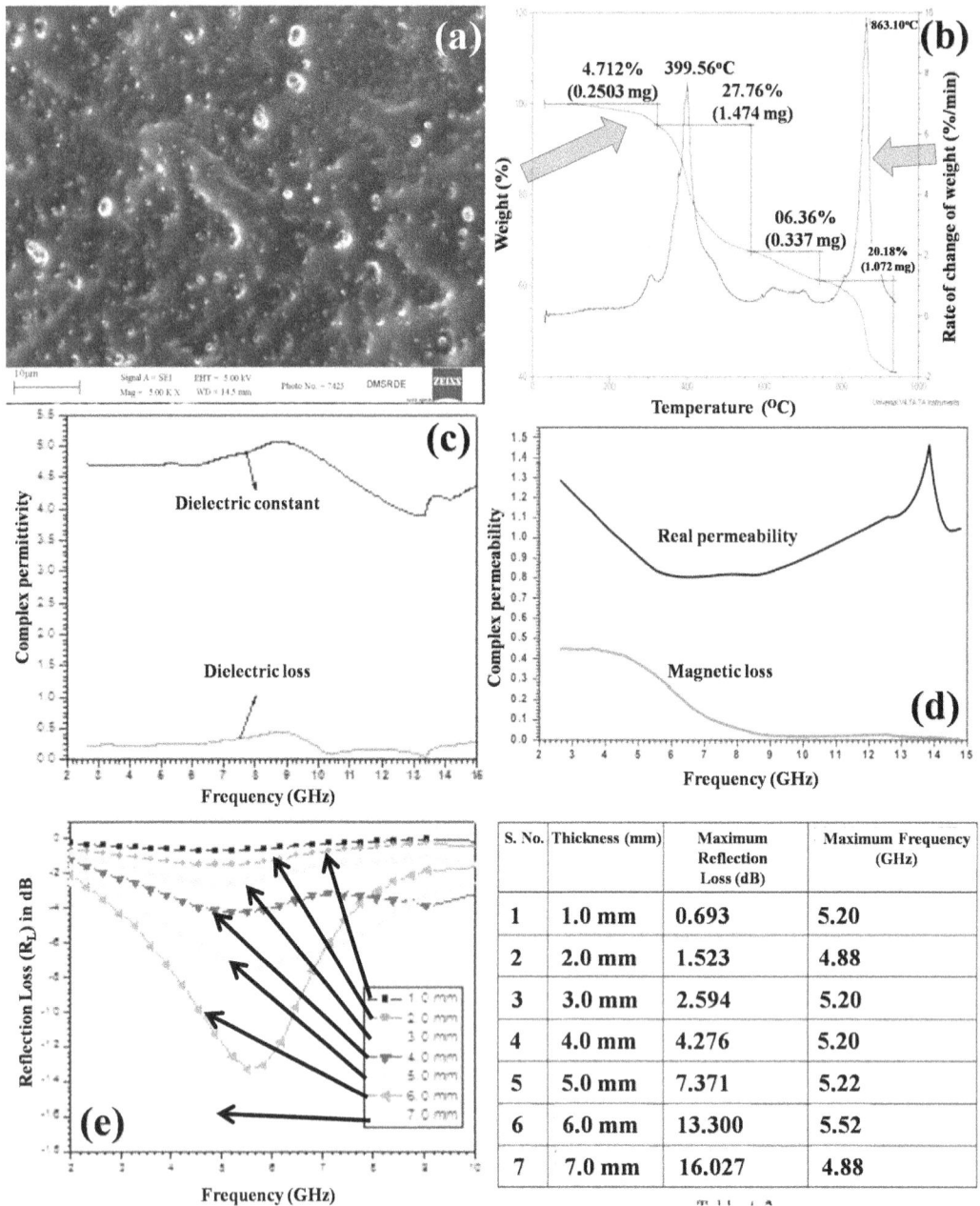

FIGURE 4.14 (a) SEM micrograph of γ-ferrite nanocomposite, (b) the recorded TGA, (c) $\varepsilon(\omega)$ and (d) $\mu(\omega)$ functions, (e) measured R_L, and (f) data table for thickness, matching frequency, and R_L as a function of thickness over 2–15 GHz.

S. No.	Thickness (mm)	Maximum Reflection Loss (dB)	Maximum Frequency (GHz)
1	1.0 mm	0.693	5.20
2	2.0 mm	1.523	4.88
3	3.0 mm	2.594	5.20
4	4.0 mm	4.276	5.20
5	5.0 mm	7.371	5.22
6	6.0 mm	13.300	5.52
7	7.0 mm	16.027	4.88

By and large, the nanocomposite sample with 35% Ni-Zn ferrite nanoparticles (filler 1) and 15% MWCNT nanocomposite showed the maximum R_L of −12 dB at a matching frequency of ~7.9 GHz (Figure 4.15f). In TGA analysis, it had been found that the RAM was thermally stable up to 250°C.

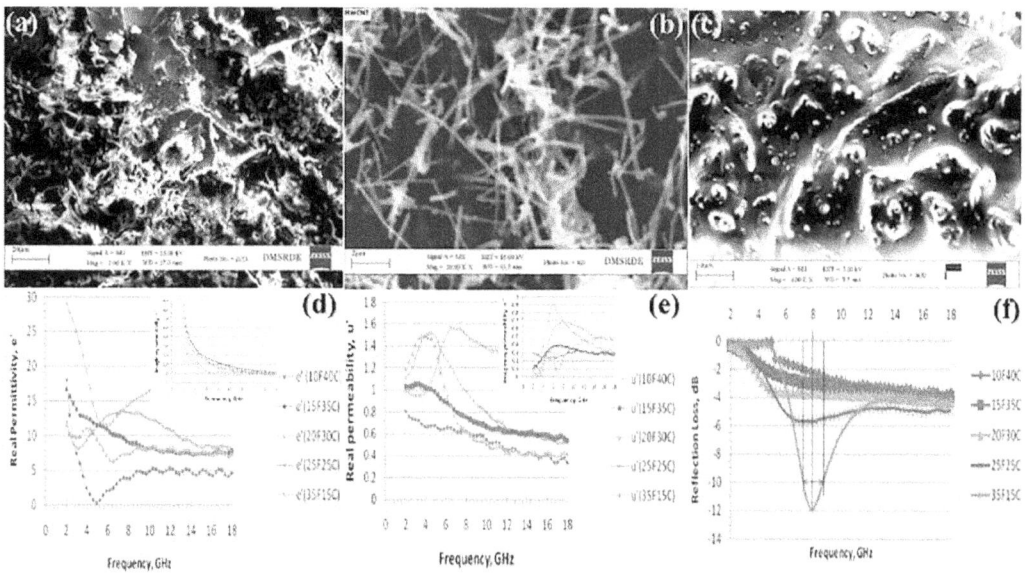

FIGURE 4.15 Recorded SEM micrographs for (a) thermoplastic PU, (b) MWCNTs, (c) Ni-Zn ferrite/TPU composite, (d) variations in ε', (inset) loss (ε'') (e) real permeability (μ') and (inset) magnetic loss (μ'') for MWCNTs/Ni-Zn/TPU nanocomposites, (f) recorded reflection loss, R_L (in dB).

4.4   GRAPHENE AND GRAPHENE DERIVATIVES FOR SHIELDING

In this work, the shielding performance of multicomponent composite materials, typically the three components: reduced graphene oxide (rGO) and strontium ferrite ($SrFe_{12}O_{19}$) (SF) bound together by polymer polyvinylidene fluoride (PVDF), was studied over the frequency range 8–12 GHz [31].

4.4.1   Performance of Multicomponent rGOSFPVDF Composite

The composite rGO/SF (rGOSF) was prepared by a facile one-pot chemical reduction technique by surface anchoring of ~500 nm of magnetic inclusions spread over the reduced graphene oxide sheets. The surface morphology, before and after composite formation, was analysed by FESEM and transmission electron microscopy (TEM). The effective crystallite size was estimated using both XRD and Raman spectroscopy that showed a specific morphological trend moving from SF to rGOSF to rGOSFPVDF. For X-band measurements, the samples were prepared mould-cast as per the requirements of the measurement, as discussed before. Magneto-dielectric functions were examined, which revealed an enhancement in the interfacial polarization and anisotropic heat losses in the PVDF created by rGOSF, which ultimately led to more scattering centres and helped in the absorption of EM radiation. The critical coating thickness of 3 mm was able to shield up to 33 dB EM power, which is more than 99.9%, accounting for absorption losses [31].

4.4.1.1   Preparation of Coating Material

For the synthesis of strontium ferrite (SF) ($SrFe_{12}O_{19}$), all the chemicals such as iron nitrate ($Fe\text{-}NO_3)_{39}H_2O$, strontium nitrate $Sr(NO_3)_2$, citric acid, and 28%–30% ammonia solution were purchased from Thomas Baker Pvt. Limited, Mumbai. Graphene oxide (GO) was synthesized from graphite flakes (natural, ~325 mesh, 99.8%, metal flakes, Alfa Aesar). Sodium nitrate ($NaNO_3$), potassium permanganate ($KMnO_4$), hydrochloric acid (HCl), sulphuric acid (H_2SO_4), hydrogen peroxide (H_2O_2), N,N-dimethylformamide (DMF), ethanol (AR), and hydrazine hydrate were purchased from SD Fine Chem Limited, Mumbai. No further purification was required. Millipore distilled water (DI) was used throughout the reaction. GO synthesis was carried out using the modified Hummers method from graphite flakes, and SF ($SrFe_{12}O_{19}$) magnetic nanoparticles were synthesized by the sol-gel technique [19].

In short, the gel precursor was prepared by taking $Fe(NO_3)_{39}H_2O$ and $Sr(NO_3)_2$ in a molar ratio of 1:12 and dissolving them in DI water in citric acid. Ammonia solution was added to maintain the pH at ~7. The as-prepared solution was then heated at 80°C and then at 200°C for the completion of reaction. The obtained xerogel powder was annealed to 1100°C for 1 hour to form SF hexaferrite phase. The required amount of GO (5, 7, and 10 mg) solution was sonicated properly before the addition of SF (100 mg) nanoparticles. Hydrazine hydrate, used as the reducing agent (GO:hydrazine=3 mg:1 mL), was added to the solution at the temperature of 90°C and stirred for more than 5–6 hours. After the completion of reaction, the final black colour rGOSF product was collected by centrifuging and drying.

To prepare rGOSFPVDF nanocomposite, both rGOSF and PVDF with a mass ratio of 4:1 were sonicated in DMF to form a homogeneous solution. The obtained solution was constantly heated and stirred for 3–4 hours to form a thick gel and poured into a glass plate to form a thick film. The obtained films were dried in a vacuum oven at 80°C for the complete removal of solvent. Samples with m(rGO):m(SF) in the ratio of 0:100, 100:0, 5:100, 7:100, and 10:100 mg were prepared and labelled as SFPVDF, rGOPVDF, 5rGOSFPVDF, 7rGOSFPVDF, and 10rGOSFPVDF, respectively. A pure PVDF film was also prepared for control experiments. The obtained films before measurement were hot-pressed at 210°C for 20 minutes to form 1, 1.5, 2, and 3 mm thickness films [31].

4.4.1.2 Morphological and Chemical Studies

Figure 4.16 shows the investigated structure and morphology of SF and its decoration on rGO. Image (a) shows the hexagonal structure of SF particles with an area of ~500×500 nm^2. Image (b) shows the surface decoration of SF nanoparticles over rGO sheet. The sheets seem to be buckled by the presence of SF clusters. At some places, SF nanostructures are intercalated in rGO. In (a), the stand-alone SF shows pretty dispersed nanoparticles, whereas in (b), these structures are coagulated in the presence of rGO [7,31].

The SF particles were distributed on the surface as well as within the layers of rGO sheets. The agglomeration in SF particles could be due to attractive magnetic interactions. Images (c–f) show TEM micrographs in which the SF inclusion in rGO layers is more evidently seen. The individual particle with its retained geometry has been accommodated within rGO. Near the SF/rGO interface, the wrinkles and exfoliation of conjugated layers are visible at few places. The HR imaging shows the nature of crystallinity, which is also likely to be retained in most of the particles ((e) dotted enclosures). The warps and wrinkles on the over-layer rGO were also visible at few places ((f) rectangles).

4.4.1.3 Shielding Character

In this work, rGOF composites with various wt% were synthesized and films were made with PVDF to analyse the effect of addition of rGO in SF on the EMI shielding properties. The morphology, crystallite size, and chemical environment of the composite were studied using FESEM, TEM, Raman analysis, XPS, and FTIR (only morphological results are presented). Magnetic studies suggested that the contribution to loss due to the anisotropic energy was maximum in 7rGOSFPVDF. One can tune the percentage of rGOSF in this hybrid composite to obtain the optimum effect for impedance matching. 7rGOSFPVDF was the best suited composition that provided more than 99.9% EMI shielding due to 3 mm thick absorber sample. From dielectric analysis, it was confirmed that with the addition of rGO in SF, both the real and imaginary parts of permittivity enhanced and it is maximum for 7rGOSFPVDF, which again supports our observed SE properties [7,31] (Figure 4.17).

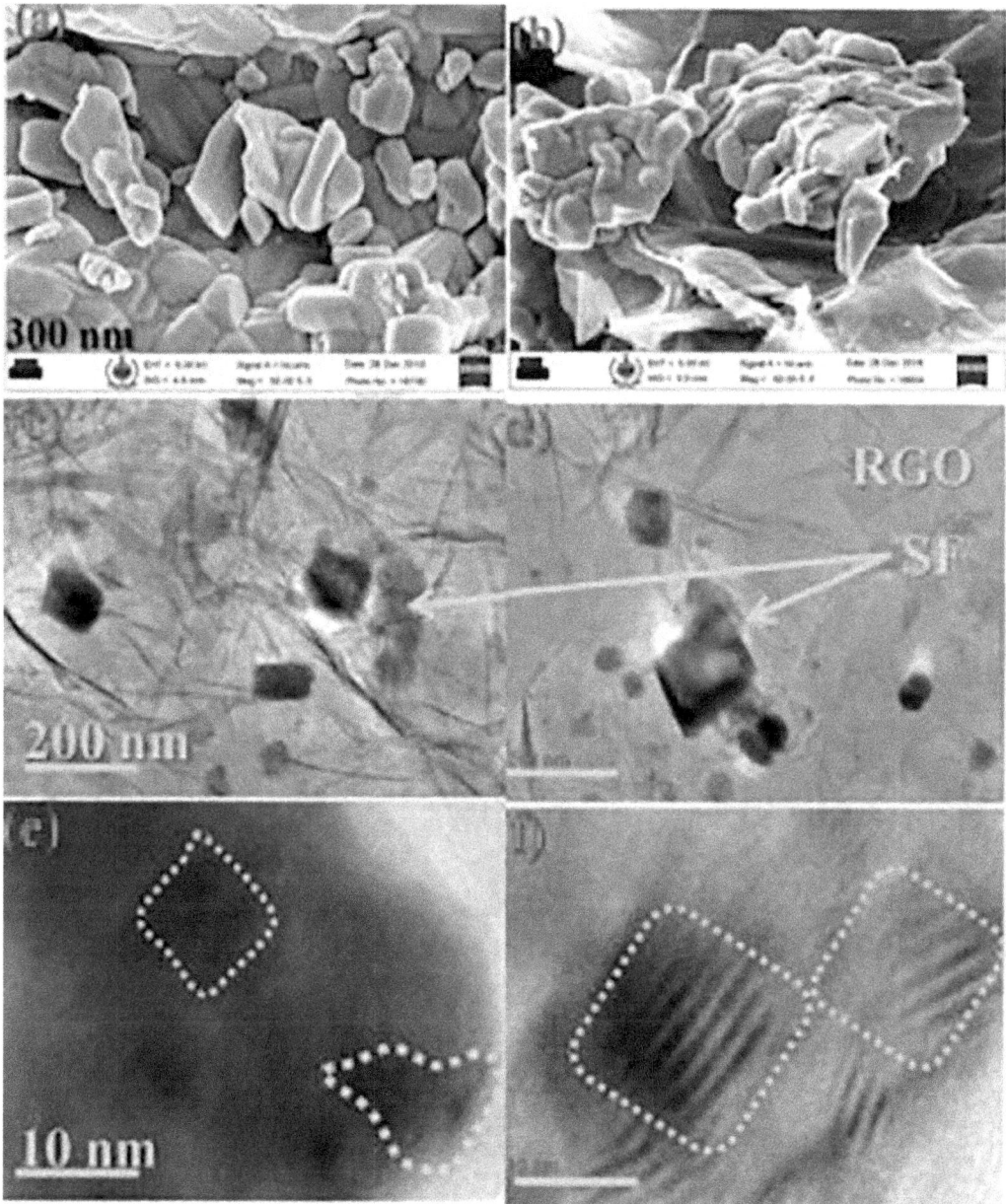

FIGURE 4.16 Recorded electron microscopic images for the samples. (a) SF nanocrystals; (b) rGOSF. HRTEM images for (c–f) rGOSF, indicating entrapment of SF within rGO and their crystallinity.

4.4.1.4 Heterostructure of $SrAl_4Fe_8O_{19}$/rGO/PVDF Composites

In another study, the fabrication of a novel RAM through one-pot chemical reduction of graphene oxide (GO) in the presence of magnetic inclusion of nano-strontium aluminium ferrite $SrAl_4Fe_8O_{19}$ (SAF) to make ternary composite films in PVDF was carried out. Two concentrations with varying ratios of rGO:SAF, 1:1 ($rGOSAF_{11}$) and 1:2 ($rGOSAF_{21}$) in PVDF, were prepared. The films were tested for RAM properties in transmission line in X-band (8–12 GHz). The morphology determined using SEM and TEM and elemental analysis using energy-dispersive X-ray analysis (EDAX) revealed that there was more non-uniformity in the dispersion as well as elemental composition in

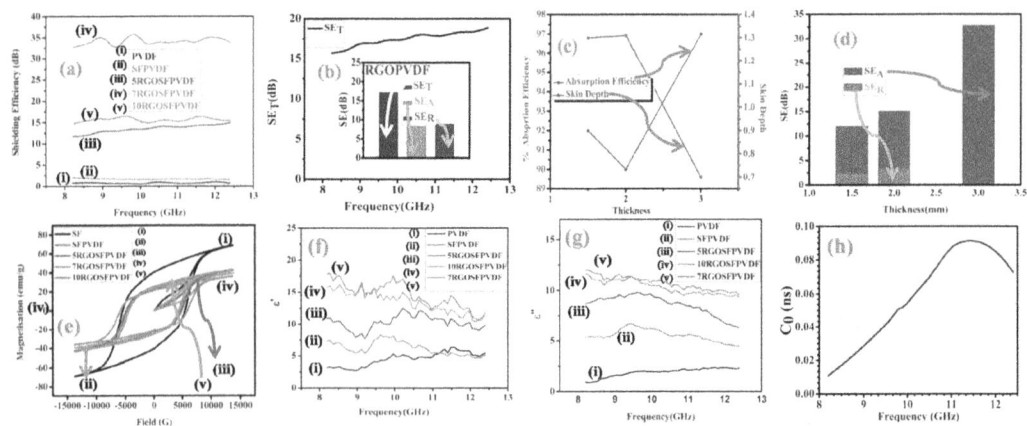

FIGURE 4.17 Calculated shielding efficiency of (a) coating films, (b) efficiency of rGOSF films and contribution to total SE from SE_A and SE_R, (c) absorption efficiency and skin depth with variations in thickness for 7rGOSFPVDF and (d) SE_A and SE_R plots for variable thickness 7rGOSFPVDF samples, (e) $B-H$ curve recorded for SF and composites @ 300 K, measured (f) real and (g) imaginary parts of permeability functions, (h) estimation of eddy current loss, C_O, over 8–12 GHz.

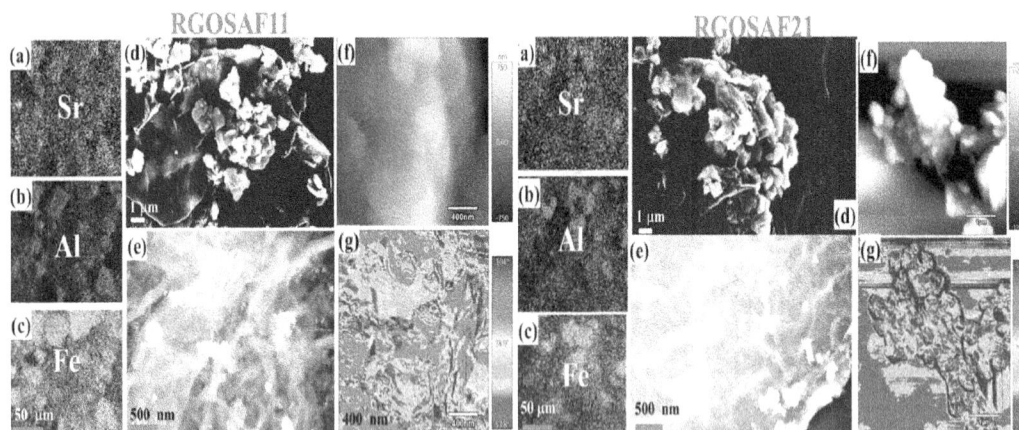

FIGURE 4.18 (a–c) Picture gallery of elemental topography, (d, e) surface morphology both by FESEM and TEM, (f, g) magnetic probe microscopy and corresponding phase contrast of rGOSAF$_{11}$ as well as rGOSAF$_{21}$.

rGOSAF$_{11}$ compared to rGOSAF$_{21}$. This was also reflected in the magnetic studies using vibrating sample magnetometer (VSM) and magnetic force microscopy (MFM). rGO played a crucial role not only in providing the conductive paths in the EM wave absorption, but also in the magnetic domain communication of magnetic entities within the rGO network. This further facilitated the effective trapping of incoming radiation and therefore exhibited shielding effectiveness of more than 40 dB. The study of dielectric properties suggested that this optimized ratio of rGO and SAF in the composite provides perfect opportunity for various effects such as interfacial polarization and scattering centres, which are responsible for the improved absorption properties. Such a material can be an impressive absorption block in emerging EMI shielding technology. The gallery of pictures for two specific composites is shown in Figure 4.18 [32,33].

The ternary composite films based on rGOSAF$_{11}$PVDF and rGOSAF$_{21}$PVDF, prepared by one-pot chemical route, showed the excellent dielectric behaviour of the composite revealing an

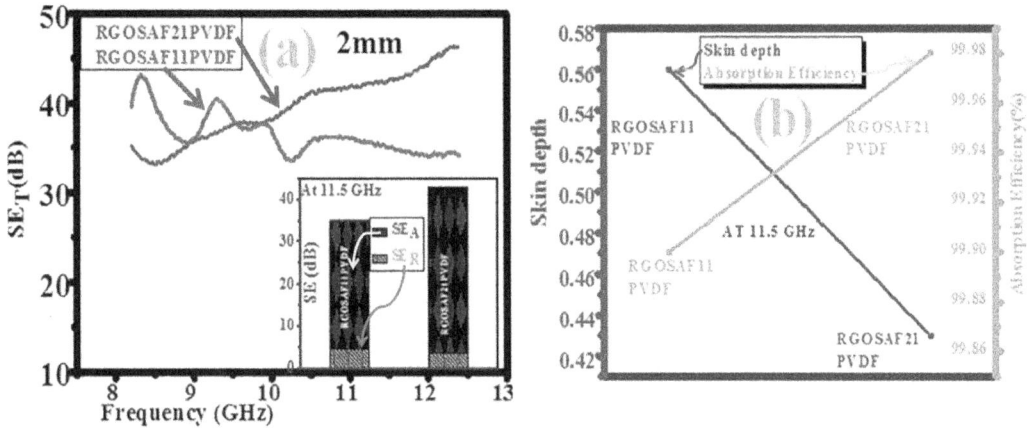

FIGURE 4.19 (a) Calculated total shielding efficiency for films at inset shows individual SE_A and SE_R parameters at 11.5 GHz and (b) calculated skin depth as well as absorption efficiency @ 11.5 GHz.

interfacial polarization within the fabricated matrix. In addition, several other lossy mechanisms were involved to arrest the incoming EM radiation. While the microwave shielding ability as well as the absorbing nature of both the as-prepared films are found to be excellent, the rGOSAF$_{21}$PVDF was found to be a more efficient shielding material as compared to rGOSAF$_{21}$PVDF. This could have been due to the uniform distribution of magnetic inclusions in rGOSAF$_{21}$PVDF as revealed from the colour code mapping in FESEM as well as MFM. Moreover, TEM showed a higher amount of SAF nanoparticles trapped within the wrinkles and warps of rGO [32,33] (Figure 4.19).

The higher value for coercivity and saturation magnetization was achieved in rGOSAF$_{21}$. MFM also showed the presence of well-connected magnetic domains with lesser degree of contrast showing uniformity of magnetic nature in the sample. Comparatively higher losses (dielectric/magnetic) and high attenuation values considering the number density along with more scattering centres further added a key reason for wave absorption [32].

4.4.2 MOLECULAR COMPOSITES FOR EMI PAINTS

In this subsection, we would be presenting the shield architectures designed and developed at molecular level by an economic route. Nanocarbon (NC) or carbon black (CB) was simply admixed with nickel (Ni), iron (Fe), and MoS$_2$ to achieve molecular composite that had been prepared by solid-state combustion route with variable wt% from 1 to 5. The fabrication is simple, of single step, involved no complex chemistry, and readily implemented as an EMI shield as a paint. The structure–property relationship has been investigated (however not discussed in here), including measurements of port S_{11}, port S_{12}, dielectric, and diamagnetic functions over 8–12 GHz (X-band) as well as return loss, R_L. The analysis of constitutive data has been carried out in light of S-parameters and correlated to the physical properties of fabricated molecular composites. A number of electrodynamic variables were calculated, such as characteristic impedance, skin depth, skin resistance, and standing wave ratio, and examined as a function of composition. It has been revealed that Ni or Fe or MoS$_2$@NC/CB induces long-range ordering in electronic polarization. The addition of such nanoparticles tunes the charge carriers to form magneto-electric dipoles. At higher particle contents such as 5%, the composite showed losses with magnitude >95% with matching frequency located in the range of ~08–09 GHz, in microwave absorption characteristics.

4.4.2.1 Nickel/Nanocarbon Composites

Microwaves interact with shielding materials at molecular level through magneto-dielectric coupling by radiative heating, generating eddy currents, and inducing polarization. It is therefore crucial to

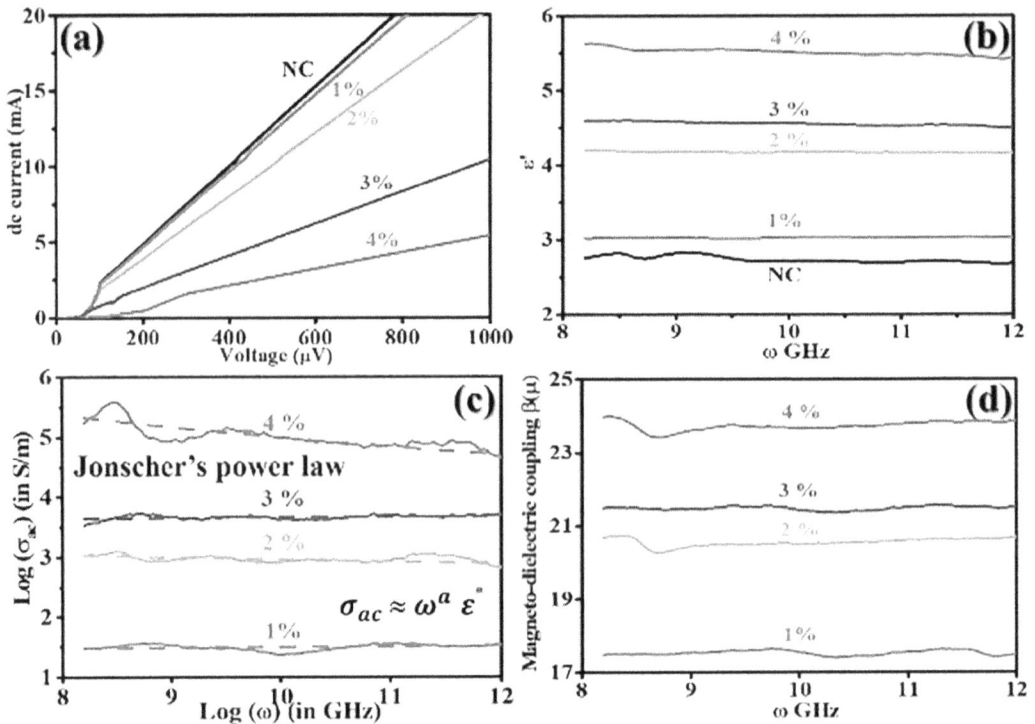

FIGURE 4.20 (a) Measured *I–V* profiles, (b) variations in ε', (c) calculated σ_{ac} (in log–log scale), and (d) estimated magneto-dielectric coupling, $\beta(\mu)$, for NC and Ni/NC nanocomposite in the X-band regime.

investigate the molecular bonding of the synthesized Ni/NC composites before quantifying X-band scattering characteristics. The presence of magnetic moiety such as Ni in nanocarbon matrix may have had an effect on the physical properties of the fabricated nanocomposites. Particularly, this would provide a clue to investigate constitutive parameters in correlating trilogy among polarization, conductivity, and signature *S*-parameters [16].

From Figure 4.20a, one can see that the *I–V* profiles are fairly linear over measured current sweep. From the measured *I–V*, values of DC conductivity, σ_{dc}, were obtained using the relation: $\sigma_{dc} = R/A \cdot l$, where *R* is the calculated resistance, and *A* and *l* are pellet area and width, respectively. Their values are provided in Table 4.3. Figure 4.20b shows the nature of ε' in the X-band for the fabricated material. It is seen that variations in ε' are nearly frequency independent over 8–12 GHz. Moreover, with a subsequent increase in Ni from 1% to 4%, a systematic increase in ε' has been observed. It seems that Ni induces long-range polarization by bridging C through oxygen resulting in almost flat dielectric function for nanocomposites [16].

Figure 4.20c shows the calculated $\log \sigma_{ac}$ in the X-band for nanocomposites. By and large, the σ_{ac} response of the composites is observed to be universal, obeying Jonscher's power law [34]. The dielectric function, $\varepsilon(\omega)$, is composed of both real and imaginary parts, and σ_{ac} is, classically, governed by a Debye equation: $\sigma_{ac} \approx \omega \varepsilon''$, which is consistent with Kramers–Kronig relations [35]. However, the departure from Debye's relation yielded almost a flat response for our systems. The response originates from the ratio of imaginary to real part of ε, which is frequency dependent and governed by σ_{ac} –Debye behaviour. It seems that, for NC, discrete polarization generates a discontinuous electrical conducting network within carbon layers yielding less magnitude of conductivity, as shown in Figure 4.21, left portion of the scheme. After the inclusion of Ni particles, the scenario changes for polar interconnects, which in turn results in a gradual increase in σ_{ac} due to long-range ordering of electronic polarization in specific directions, as shown in Figure 4.21, right-hand side.

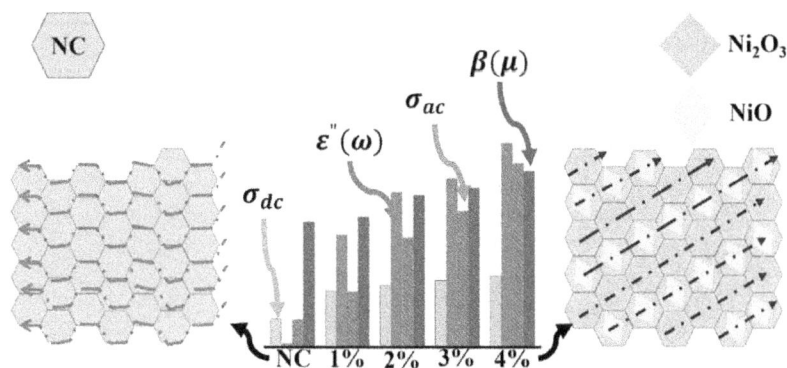

FIGURE 4.21 Schematic representation of weak, non-Debye polarization governing feeble AC conductivity in NC matrix (left side). The mid-portion shows average magnitudes of calculated constitutive parameters with increasing Ni content and order. Right side: Ni/NC molecular environment showing long-range polarization by the presence of NiO and Ni_2O_3 phases.

The behaviour of σ_{ac}, in NC, is governed by non-Debye response; consequently, it led to a fact that there existed a discrete electronic polarization of asymmetric origin. However, with nickel inclusion, the conducting path became prominently continuous due to the emergence of bigger and symmetric domains of polarization (as shown in Figure 4.21). The variations in σ_{ac} are governed by Jonscher's power law given by: $\sigma = \sigma_{dc} + \sigma_{ac} = \sigma_{dc} + \omega^a \, \varepsilon''(\omega)$, with $a \sim 0$ within experimental errors, yield frequency-independent response. Hence, with Ni content, the frequency independence of ε'' in such composite materials has been seen to follow this law: $\varepsilon''(\omega) \propto \omega^{a-1}$ over a decade of frequency (8–12 GHz), which seems to be regardless of their chemical, physical, and geometrical properties, and also regardless of the nature of electrically active species responsible for polarization, whether dipolar, ionic, or electronic. The fitting to σ_{ac} curves gave exponent values (provided in Table 4.3) confirming Jonscher's power law [36,37].

The typical characteristic of the fabricated material was noted to be lossy dielectric type. The calculated AC magnetic coupling energy $\beta(\mu)$ measured in arbitrary units is measured using: $\beta(\mu) = \omega \sqrt{\mu\varepsilon'}$. Figure 4.20d displays variations in $\beta(\mu)$ as a function of frequency for the material, and the respective average values are provided in Table 4.3. Typically, for various compositions of Ni/NC composites, there is a gradual increase in $\beta(\mu)$ with Ni content. The analysis of constitutive parameter indicated dramatic changes in ε', and marked variations in the nature of both σ_{dc} and σ_{ac}, including changes in $\beta(\mu)$ over the X-band, are significantly high for Ni/NC composites. It is thus of our interest to investigate microwave scattering parameters of them. Figure 4.22 shows port S-parameters measured at (a) S_{11} and (b) S_{12} in the X-band for 1%–5% Ni/NC composites. Correspondingly, the schematic below plots (a) and (b) shows the respective absorption mechanisms for 1% and 5% composites. In Figure 4.22a and b, the recorded spectra, respectively, for S_{11} and S_{12} are observed to be flat in shape with distinct changes with Ni inclusion. At 1%, the overall magnitude of reflected power is comparable to NC and recorded highest with maximum reflection ~99% (graphic below Figure 4.22a). It seems that Ni moieties such as FCC NiO, Ni_2O_3, and amorphous Ni clusters are responsible for introducing charge carriers that induce symmetric and anisotropic electronic polarization through symmetric and asymmetric vibration modes of Ni, C, and bridging O atoms. It amounts to large scattering of incoming microwave power mostly via absorption process, graphic shown below Figure 4.22b. Notably, an increase in Ni content downshifted the entire power at S_{11} and S_{12} in the entire X-band [16].

In another study, the microwave scattering (2–18 GHz) behaviour of the nanocomposite for architecting efficient shield composed of carbon black (CB), molybdenum disulphide (MoS_2), and cobalt (Co) has been studied and presented in the section below [38].

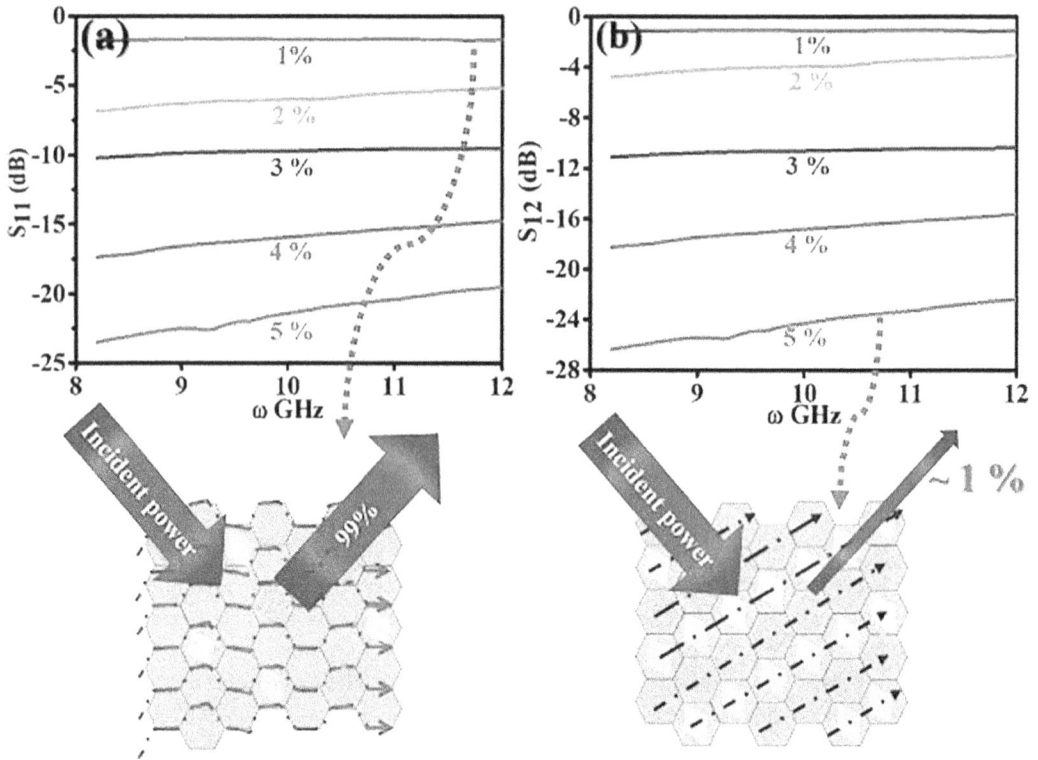

FIGURE 4.22 Recorded port S-parameters at (a) S_{11}, and (b) S_{12} over 8–12 GHz for 1%–5% Ni/NC composites (NC not given; resembles 1%). The corresponding scheme below plot (a) shows ~99% reflection@1% Ni content, whereas the graphic below plot (b) shows almost ~99% reduction in reflection@5% Ni/NC composition.

4.4.2.2 Carbon Black/Molybdenum Disulphide/Cobalt Composite

Nanocomposites, prepared by a facile, solid-state synthesis route, were characterized using infrared, X-ray diffractometry, UV–visible, and energy-dispersive X-ray spectroscopic techniques, including scanning electron microscopy. In infrared and Rayleigh analysis, the formation of –C–Mo–C, O=S=O, Mo–O, and Co–O phases generated asymmetric polarization at low % CB content that was transformed into a symmetric mode due to the formation of Co–O and Co–S–C radicals at high CB % to result into maximum power losses. The possible polarization mechanism is discussed. In constitutive analysis, a twofold increase in the dielectric function, the dual behaviour (Debye and Jonscher) of the AC conductivity, and the synergistic magneto-dielectric coupling factor influenced the scattering performance of the composite. Broadly, at 18% composition, the shielding effectiveness was recorded to be >97@14 GHz with 7.7 GHz of bandwidth and thickness of ~2.5 mm. The Dallenbach return loss is almost 75%, revealing a high-performance shield design @18% CB/MoS$_2$/Co nanocomposite [39].

Figure 4.23 displays the recorded SEM micrographs, typically, for 6%, 12%, and 18% composite. The magnification was kept at 30× for the top row and 50× for the bottom row. In Figure 4.23a and d, recorded for the 6% composition, the spread of CB is found to be somewhat non-uniform within the matrix. No prominent sign of Co or its nano-cluster has been noted. The appearance of MoS$_2$ is seen in the form of the scaffolds in which the bottom framework is observed to be broader as compared to the top layers. Perhaps, the top layers appeared to be quite small area-wise. At very few places, they are covered and surrounded by the CB particles. For 6% composite, most of the MoS$_2$/Co scaffolds are found to be free from the CB coverage, as shown, schematically, below 6% composite micrographs. In Figure 4.23b and e, the surface morphology of 12% composite is shown,

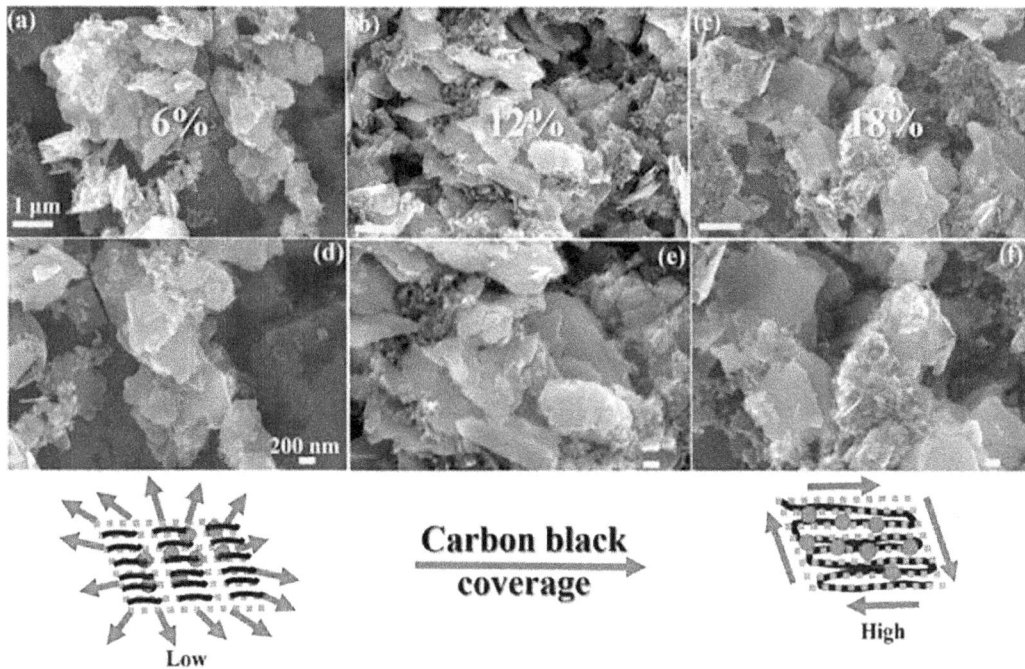

FIGURE 4.23 Typical SEM micrographs recorded for the 6%, 12%, and 18% composites. Top row: 30× and bottom row: 50× magnification. The scheme below SEM micrograph cluster shows the influence of spread and coverage of the CB within MoS_2/Co scaffolds to modify the degree of polarization.

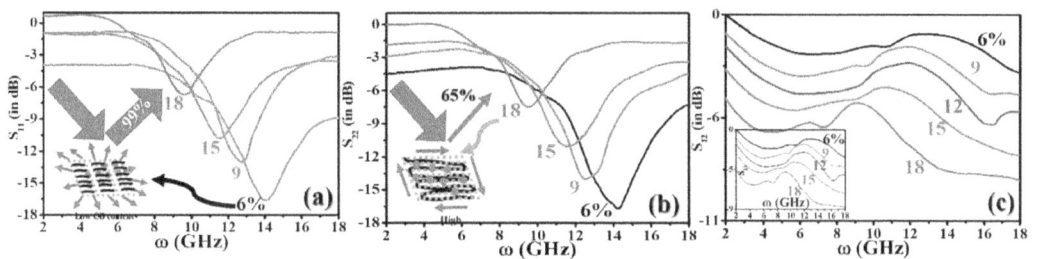

FIGURE 4.24 Recorded S-parameters at the port (a) S_{11}, (b) S_{22}, and (c) S_{12} over 2–18 GHz. Inset in (a) schematically shows ~99% reflection@6% CB content. Inset in (b) schematically shows almost ~25% reduction in the reflection@18% CB composition. Inset in (c) shows the symmetric polarization behaviour of the composite to absorb power at the two different interfaces 1 (entrance) and 2 (exit).

in which MoS_2/Co layers are seen to be enriched with the CB coverage. The CB chains are forming the interconnections between the conjugated stacks of the MoS_2/Co; however, for 12%, these interconnections are observed to be partially coupled and not fully networked. In Figure 4.23c and f, the dramatic variations in the surface morphology of 18% composite are seen. The CB is seen to almost cover the MoS_2/Co surfaces. The improvement in interconnects is quite visible as seen in the schematic below 18% composite micrographs. The enrichment in CB provides a conducting network, thereby enhancing both ε' and AC conductivities of the CB/MoS_2/Co nanocomposites. Figure 4.24 shows the measured S-parameters at port (a) S_{11}, (b) S_{22}, (c) S_{21}, and (d) S_{12} over 2–18 GHz. Correspondingly, the inset in plot (a) and (b) shows the respective absorption mechanism. The inset in (c) shows an identical behaviour of the incident microwaves at the first exit, i.e. the second interface [38,39].

TABLE 4.4

The Calculated Data of Frequency Dispersion, Reflection Loss (in %), Calculated Characteristic Impedance, η, Skin Depth, δ, Surface Resistance, R_s, and Standing Wave Ratio (SWR)

S. No.	Composition (in %)	ω (GHz)	Reflection Loss (in %)	$\eta\left(\dfrac{Z}{Z_0}\right)$	δ (nm)	R_s (Ω/m^2)	SWR (in dB%)
1	06	14.0	99	0.19 ± 0.007	395 ± 34	43.56 ± 10.89	~ 09
2	09	12.0	96	0.16 ± 0.008	440 ± 89	43.12 ± 10.66	~ 15
3	12	11.7	94	0.14 ± 0.006	470 ± 13	42.74 ± 10.51	~ 25
4	15	11.0	92	0.12 ± 0.009	490 ± 46	42.18 ± 10.47	~ 37
5	18	09.0	75	0.09 ± 0.006	530 ± 27	41.71 ± 10.38	~ 48

By and large, the recorded spectra for S_{11} and S_{22} are observed to be symmetric in their shape with the observed gradual change. In chemical analysis, the moieties such as –C–Mo–C–, O=S=O, Mo–O, and Co–O may be responsible for inducing an isotropic and asymmetric electronic polarization, which in turn resulted in a large amount of scattering for the incoming microwave power, as schematically shown in Figure 4.24a inset. However, with a subsequent increase in the CB content (9%–18%), a downshift in the frequency from ~12 to ~9 GHz is seen with a reduction in the amount of reflection from 96% to 75%. At 18%, almost 25%, or more, power is absorbed. The Co-type inclusion may favourably couple the incident time-varying magnetic field component with the matrix, effectively, resulting in an enhanced absorption of the microwave power as, schematically, shown in the inset in Figure 4.24b. The behaviour in the variations of S_{11} and S_{22} over the frequency is observed to be identical including a shift and the power losses for all the composites. In Figure 4.24c, at 6% no prominent loss feature is seen for the recorded S_{12} data and this is consistent with the inset S_{21} data recorded. There is a unique type of resemblance noted between the S_{12} and S_{21} scattering data. In Table 4.4, the calculated frequency dispersion and reflection loss values are compiled and presented including a number of other electrodynamic variables such as characteristic impedance, η, skin depth, δ, skin resistance, R_s, and standing wave ratio (SWR) for the nanocomposites.

From Table 4.4, a large frequency variation is noted ranging from higher Ku (14 GHz) to higher X (9 GHz) radar band with almost 25% gain in the power absorption over the change in carbon composition from 6% to 18% [39].

In another work, the analysis of S-parameter performance of a ferro-nanocarbon (FNC) composite tested over a frequency region of 8–12 GHz (X-band) is performed [34,40,41].

4.4.2.3 Ferro-Nanocarbon Composites

The microwave analysis showed that iron inclusion developed crystalline Fe_2O_3 (d[110]) and Fe_3O_4 (d[220]) phases that were self-dispersed well within the nanocarbon structure. Iron inclusion was responsible for creating asymmetric bond molecular environment of C–O–Fe bringing the synergistic magneto-dielectric effects in terms of long-range polarization ordering mainly through the transitions between O 2p and Fe 3d states to engage the incoming microwave field effectively. In its performance, the composite showed >98% shielding effectiveness with infinite bandwidth and >99% return loss at a matching frequency of 9.01 GHz [41].

The Dallenbach absorption technique used in the analysis provided a type of onset remedy to reduce the radar signatures of an object from the sharp angles, edges, and cones; such features exist on aircraft, submarines, drones, tanks, etc. These metallic surfaces are the origin of high reflections. Dallenbach absorption measurements were performed on NC and (1%–4%) FNC composites using a typical scattering set-up, as shown, schematically, in Figure 4.25a. In this, a metallic support is attached to the one end of the nanocomposite absorber block. Considering the extracted constitutive

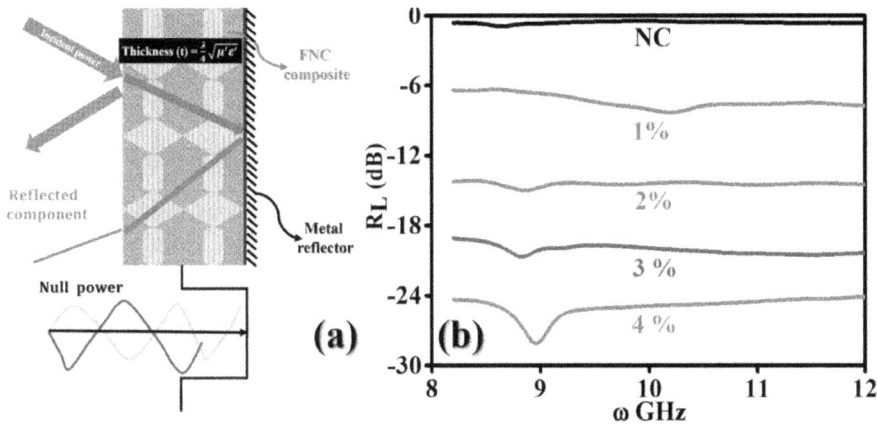

FIGURE 4.25 (a) Sketch of the Dallenbach scattering technique for the reflection loss (R_L) analysis of the NC/FNC coatings. Condition on thickness in terms of constitutive parameters is displayed on the top for the composite layers, attached metal support, schematics of power losses from first and second interfaces, and null/minimum power condition at the bottom, (b) recorded R_L spectra for NC and (1%–4%) FNC composites over 8–12 GHz.

parameters of the composite, one could minimize the reflection for a preferred frequency, thereby optimizing the thickness to $\lambda/4$. Broadly, the return loss is a typical absorption property of a coating with characteristic dip R_L (min) @ a frequency called the matching frequency, f_m, of the coat. The thickness of the absorber can be set by determining the position of f_m, since they are inversely correlated. For a value of R_L −10 dB has 90% of absorption, −20 dB shows 99% and −30 dB 99.9% value of the absorption for the transmitted electromagnetic wave reflected back to the receiver.

In Dallenbach coating, the scatterer obeys a destructive interference (null power) condition. For the incident signal that gets reflected from the second interface gets interfere with the incident one to achieve a destructive interference pattern. In such interactions, reflection, mainly, depends upon the complex permittivity and permeability functions of the air/composite interface that can be adjusted to a minimum returned power. The nulling of the power depends upon the loss factors of μ'', ε'' as well as the thickness of the coating. They are a kind of resonant absorber and can, readily, be applied as a paint on the military objects. At some places, these structures are enormously sharp in their geometrical shape with highly curved surfaces or small areas in terms of wavelength for the transmitted waves. The Dallenbach coatings are pretty slim and trim as compared to other similar classes of absorbers such as Salisbury (graded) or Jaumann (resonant) coat. Moreover, these coatings could easily be chosen, in the absence of bandwidth specifications [41].

In Figure 4.25b, R_L spectra recorded for NC and FNC (1%–4%) composites over 8–12 GHz are displayed. The return loss for NC and 1% FNC is in the range of 0–10 dB, which revealed the maximum reflection of the incident power. As the content of iron is increased, the magnitude of R_L has gradually increased. For 2%–3% FNC, R_L is observed to be in the range of 15–20 dB, which shows a significant return loss up to 50%–60%. Correspondingly, the matching frequency is seen to be quite broad covering the entire X-band. For the highest Fe composition, i.e. 4%, the recorded profile is somewhat sharp with >99% power absorption with matching frequency located at ~9.10 GHz. On incidence, electromagnetic signals interact with the fabricated nanocomposite surface. The charge carriers and polar moieties present in the matrix offer a response to the incident field and perform the Coriolis precision. This drift in turn induces magnetic field and, consequently, magnetic currents that are responsible for inducing losses to the incident field termed as the eddy current loss, which, consequently, produces a resistive heating. These loopy currents, in the surface region of the composite, lead to the emergence of a small secondary magnetic field. This field opposes the incident field, causing a reflection of the incident power. The observed increase in the loss of reflection magnitude with an increase in the iron content is, thus, attributed to larger and symmetric eddy loops associated with the polarizations of larger carbon chains within the composites. Further, the absorption of power at the air/composite interface can also be attributed to the larger values of the magneto-dielectric losses of the composites that obey an effective diffraction condition: $t = \dfrac{\lambda}{4} \sqrt{\mu' \varepsilon'}$, as shown in Figure 4.12a. Due to this, two partially reflected electromagnetic waves are generated with phase 180° apart with respect to the transmitted signal, resulting in a response as shown in Figure 4.25b. In our loss studies, for NC a very shallow peak is observed below 09 GHz, for 1% the peak is shifted down to 10.00 GHz, whereas for 2% and 3%, the peak emerges below 9.0 GHz and for 4% it is located at ~9.10 GHz [41].

Our studies revealed that the molar polarization of the composite matrix gets modified with increasing magnetic inclusions [42]. The obtained constitutive parameters played a pivotal role in analysing microwave power dissipation in which the real part represents the storage capability and the imaginary part symbolizes dissipation of the incident power [21,24,26]. Both imaginary components are observed to be increased with nanoparticle wt%, thereby maintaining a flat behaviour of R_L without any significant variations. This in turn revels that the dissipation capability of both electric and magnetic components is quite stable over the measured frequency range. The obtained values of ε'' and ε' are observed to be in proportion to the AC conductivity and obey Jonscher's exponent law. The individual nanoparticles in NC are connected randomly, resulting in a low magnitude of electric conductivity. However, iron inclusion enhanced the values of ε'' and ε' quite dramatically. We speculate that magnetic inclusion is responsible for enhancing NC interconnections that in turn optimized the impedance matching condition, thereby prompting the electromagnetic power to enter into composite absorbers to dissipate to a greater extent. Briefly, for permeability behaviour, the gradual increase in μ'' and μ' reveals strengthening of the magnetic coupling with dielectric component that favours impedance match at the high iron content. As a result, magneto-dielectric polarization seems to be oriented along the iron phases in a charge-symmetric mode, which influenced the behaviour of the scattering for the composites at higher iron content. Broadly, these modes seem to have enhanced the return losses, making higher wt% composites more promising candidates for shield paints [43–45].

REFERENCES

[1] Hecht, Eugene. *Optics*, 4th edition. Addison Wesley Longman Inc, England 1998.
[2] Geetha, Shielding, K. K. Satheesh Kumar, Chepuri R. K. Rao, M. Vijayan, and D. C. Trivedi. "EMI shielding: methods and materials—a review." *Journal of Applied Polymer Science* 112, no. 4 (2009): 2073–2086.
[3] Pozar, David M. *Microwave engineering*. John Wiley & Sons, Hoboken, NJ 2011.
[4] Pillet, V. Mártínez, Antonio Aparicio, and Francisco Sánchez. "Payload and mission definition in space sciences." In *Payload and mission definition in space sciences*, Cambridge University Press, Cambridge, UK 2011.
[5] Knott, Eugene F., John F. Shaeffer, and Michael T. Tuley. *Radar cross section*, 2nd edition, SciTech, Raleigh, NC, 2004.
[6] Ra'di, Younes, Constantin R. Simovski, and Sergei A. Tretyakov. "Thin perfect absorbers for electromagnetic waves: theory, design, and realizations." *Physical Review Applied* 3, no. 3 (2015): 037001.
[7] Kumar, Arvind, and Prashant S. Alegaonkar. "Impressive transmission mode electromagnetic interference shielding parameters of graphene-like nano-carbon/polyurethane nanocomposites for short range tracking countermeasures." *ACS Applied Materials & Interfaces* 7, no. 27 (2015): 14833–14842.
[8] Tripathi Krishna Chand, Sayaad M. Abbas, Prashant S. Alegaonkar, and Rishi B. Sharma. "Microwave absorption properties of Ni-Zn ferrite nano-particle based nano composite." *International Journal of Advanced Research in Science, Engineering and Technology* 2, no. 2 (2015): 463–468.
[9] Tripathi Krishna Chand. "Preparation and performance evaluation of nano-composites for electromagnetic and microwave absorption." PhD diss., DIAT deemd to be University, 2017.

[10] Ghodgaonkar, D. K., V. V. Varadan, and Vijay K. Varadan. "Free-space measurement of complex permittivity and complex permeability of magnetic materials at microwave frequencies." *IEEE Transactions on Instrumentation and Measurement* 39, no. 2 (1990): 387–394.

[11] Lee, Thomas H., and Thomas H. Lee. *Planar microwave engineering: a practical guide to theory, measurement, and circuits*, vol. 1. Cambridge University Press, Cambridge, UK 2004.

[12] Agilent, PAN Microwave Network Analyzer, Catalogue and Product Note E8364B, 2009.

[13] Tripathi Krishna Chand, Sayaad M. Abbas, Rishi B. Sharma, Prashant S. Alegaonkar, and Manish Verma. "Microwave absorption studies of γ-ferrite & Ni–Zn ferrite/epoxy based nano composites." *International Journal for Science and Advance Research in Technology* 1 (2015): 16.

[14] Tripathi Krishna Chand, Sayaad M. Abbas, Prashant S. Alegaonkar, and Rishi B. Sharma. "Microwave absorption properties of Ni-Zn ferrite nano-fillers/TPU based nano composites." *International Journal of Advanced Research in Science, Engineering and Technology* 3, no. 3 (2016): 1598–1604.

[15] Alegaonkar, Ashwini P., and Prashant S. Alegaonkar. "Nano-carbon/polymer composites for electromagnetic shielding, structural mechanical and field emission applications." *Thermoset Composites: Preparation, Properties and Applications* 38 (2018): 128.

[16] Alegaonkar, Ashwini P., Krishna C. Tripathi, Himangshu B. Baskey, Satish K. Pardeshi, and Prashant S. Alegaonkar. "X-band scattering characteristics of nickel/nano-carbon composites for anti-tracking application." *ChemNanoMat* 8, no. 2 (2022): e202100301.

[17] Yusoff, A. N., M. H. Abdullah, S. H. Ahmad, S. F. Jusoh, A. A. Mansor, and S. A. A. Hamid. "Electromagnetic and absorption properties of some microwave absorbers." *Journal of Applied Physics* 92, no. 2 (2002): 876–882.

[18] Wang, Tao, Rui Han, Guoguo Tan, Jianqiang Wei, Liang Qiao, and Fashen Li. "Reflection loss mechanism of single layer absorber for flake-shaped carbonyl-iron particle composite." *Journal of Applied Physics* 112, no. 10 (2012): 104903.

[19] Jackson, John David. "Classical electrodynamics." *American Journal of Physics* 67 (1999): 841–842.

[20] Chen, Lin-Feng, C. K. Ong, C. P. Neo, V. V. Varadan, and Vijay K. Varadan. *Microwave electronics: measurement and materials characterization*. John Wiley & Sons, Chichester, UK 2004.

[21] Hayt Jr, W. H., and J. A. Buck. *Engineering electromagnetic*, 1988.

[22] Bennett, C. Leonard, and Gerald F. Ross. "Time-domain electromagnetics and its applications." *Proceedings of the IEEE* 66, no. 3 (1978): 299–318.

[23] Weir, William B. "Automatic measurement of complex dielectric constant and permeability at microwave frequencies." *Proceedings of the IEEE* 62, no. 1 (1974): 33–36.

[24] Alegaonkar, Ashwini P., and Prashant S. Alegaonkar. "Nano-carbons: preparation, assessments, and applications in structural engineering, spintronics, gas sensing, EMI shielding, and cloaking in X-band." In *Nano-carbon and its composites*, pp. 171–285. Woodhead Publishing, Sawston, UK 2019.

[25] Kumar, Arvind, Sumati Patil, Anupama Joshi, Vasant Bhoraskar, Suwarna Datar, and Prashant Alegaonkar. "Mixed phase, sp^2–sp^3 bonded, and disordered few layer graphene-like nano-carbon: synthesis and characterizations." *Applied Surface Science* 271 (2013): 86–92.

[26] Stuart, Barbara. "Infrared spectroscopy." In *Analytical techniques in forensic science*, pp. 145–160, John Wiley & Sons, Inc., Hoboken, NJ 2021.

[27] Larkin, Peter. *Infrared and Raman spectroscopy: principles and spectral interpretation*. Elsevier, Amsterdam 2017.

[28] Stauffer, Dietrich, and Ammon Aharony. *Introduction to percolation theory*. Taylor & Francis, London, UK 2018.

[29] Regev, Oren, Paul N. B. ElKati, Joachim Loos, and Cor E. Koning. "Preparation of conductive nanotube–polymer composites using latex technology." *Advanced Materials* 16, no. 3 (2004): 248–251.

[30] Ounaies, Z., C. Park, K. E. Wise, E. J. Siochi, and J. S. Harrison. "Electrical properties of single wall carbon nanotube reinforced polyimide composites." *Composites Science and Technology* 63, no. 11 (2003): 1637–1646.

[31] Acharya, Sanghamitra, Chinnakonda S. Gopinath, Prashant Alegaonkar, and Suwarna Datar. "Enhanced microwave absorption property of reduced graphene oxide (RGO)–strontium hexaferrite (SF)/poly (vinylidene) fluoride (PVDF)." *Diamond and Related Materials* 89 (2018): 28–34.

[32] Acharya, Sanghamitra, Prashant Alegaonkar, and Suwarna Datar. "Effect of formation of heterostructure of SrAl4Fe8O19/RGO/PVDF on the microwave absorption properties of the composite." *Chemical Engineering Journal* 374 (2019): 144–154.

[33] Acharya, Sanghamitra, Prashant Alegaokar, and Suwarna Datar. "SrAl4Fe8O19 hexaferrite and reduced graphene oxide: for microwave absorption application." In *AIP Conference Proceedings*, vol. 2115, no. 1, p. 030145. AIP Publishing LLC, 2019.

[34] Tripathi, Krishna Chandra, Sayaad M. Abbas, Rishi B. Sharma, and Prashant S. Alegaonkar. "Microwave absorbing properties of MWCNT/Carbon black-PU Nano-composites." In *2017 IEEE International Conference on Power, Control, Signals and Instrumentation Engineering (ICPCSI)*, pp. 489–496. IEEE, 2017.

[35] Joshi, Anupama, Anil Bajaj, Rajvinder Singh, Anoop Anand, P. S. Alegaonkar, and Suwarna Datar. "Processing of graphene nanoribbon based hybrid composite for electromagnetic shielding." *Composites Part B: Engineering* 69 (2015): 472–477.

[36] Jonscher, A. K. "Dielectric relaxation in solids (Chelsea Dielectric, London, 1983). KL Ngai, AK Jonscher and CT White." *Nature* 277 (1979): 185.

[37] Jonscher, Andrew K. "The 'universal'dielectric response." *Nature* 267, no. 5613 (1977): 673–679.

[38] Yadav, Akshita, Krishna C. Tripathi, Himangshu B. Baskey, and Prashant S. Alegaonkar. "Microwave scattering behaviour of carbon black/molybdenum di sulphide/cobalt composite for electromagnetic interference shielding application." *Materials Chemistry and Physics* 279 (2022): 125766.

[39] Yadav, Akshita, and Prashant S. Alegaonkar. "MoS$_2$-carbon black nanocomposite for EMI shielding purpose." In *Abstracts of International Conferences & Meetings*, vol. 1, no. 5, pp. 17–17. 2021.

[40] Acharya, Sanghamitra, Jay Ray, Tutiki Umasankar Patro, Prashant Alegaonkar, and Suwarna Datar. "Microwave absorption properties of reduced graphene oxide strontium hexaferrite/poly (methyl methacrylate) composites." *Nanotechnology* 29, no. 11 (2018): 115605.

[41] Alegaonkar, Ashwini P., Himangshu B. Baskey, and Prashant S. Alegaonkar. "Microwave scattering parameters of ferro–nano-carbon composites for tracking range countermeasures." *Materials Advances* (2022).

[42] Joshi, Anupama, Anil Bajaj, Rajvinder Singh, Prashant S. Alegaonkar, Kanda Subramanian Balasubramanian, and Suwarna Datar. "Graphene nanoribbon–PVA composite as EMI shielding material in the X band." *Nanotechnology* 24, no. 45 (2013): 455705.

[43] Zhang, Yang, Zhangjing Yang, and Bianying Wen. "An ingenious strategy to construct helical structure with excellent electromagnetic shielding performance." *Advanced Materials Interfaces* 6, no. 11 (2019): 1900375.

[44] Hammani, H., W. Boumya, F. Laghrib, A. Farahi, S. Lahrich, A. Aboulkas, and M. A. El Mhammedi. "Electrocatalytic effect of NiO supported onto activated carbon in oxidizing phenol at graphite electrode: application in tap water and olive oil samples." *Journal of the Association of Arab Universities for Basic and Applied Sciences* 24, no. 1 (2017): 26–33.

[45] Uran, S., A. Alhani, and C. Silva. "Study of ultraviolet-visible light absorbance of exfoliated graphite forms." *AIP Advances* 7, no. 3 (2017): 035323.

5 Heat Transfer and Thermodynamics in Micrographitic Nozzles

In this chapter, we have described the quality assessment of micrographitic rocket nozzles in terms of its heat transfer and thermodynamic capabilities. The chapter is divided into to two parts: First, investigations on thermo-physical parameters of the variable density graphite are presented for designing an optimal jet propulsion nozzle; second, its thermal testing and performance is presented.

Initially, graphite specimens were fabricated by hot isostatic pressing foundry processing to achieve densities of 1611, 1744, 1811.7, and 1910.4 kg/m^3. For thermal property, American Society for Testing and Materials (ASTM)-grade samples were prepared to examine specific heat capacity at constant pressure, C_P, volume, C_V, thermal conductivity, λ, diffusivity, α, and expansivity, α_L, over 300–1300°K. C_P and C_V examinations revealed energy contributed in mechanical (lattice) vibrations, and at surfaces/interfaces of graphite micro-matrix. Notably, for 1811.7 kg/m^3, this played a dominant role in showing a larger entropy change with a higher enthalpy value. Debye temperature is found to be 575°K; however, the corresponding thermal activation mechanism is not so clear. Parametric λ variations obeyed $\sim T^{0-2}$ law and provided more insights into thermo-structural aspects. Enthalpy-derived heat flux characteristics showed efficient heat transfer characteristics of 1811.7 kg/m^3 graphite compounded with >90% resemblance to reported α_L at 300°K. Data reduction technique is presented for diffusivity, conductivity, and expansivity data. X-ray diffraction analysed for d_{002} reflex showed peculiar values of stacking, in-plane crystallinity, layer density per stack, degree of graphitization, and α_L for 1811.7 kg/m^3 graphite advantageous over others. Raman analysis showed the same electron–phonon coupling strength at high density with no impact on Fermi velocity of conduction electrons in any graphitic system. Microscopy study of graphite texture is presented. Favourable thermo-physical properties offered by 1811.7 kg/m^3 graphite prompted in designing of a prototype nozzle.

5.1 THERMO-PHYSICAL ASSESSMENTS OF VARIABLE DENSITY GRAPHITE

5.1.1 GRAPHITIC CARBON IN MISSILE ENGINEERING

Carbon is a wondrous element in the periodic table [1]. Due to its ability to form covalently bonded long chains, it has a number of natural and synthetic allotropes. Natural ones such as diamond and graphite are well known to the mankind, whereas artificially synthesized fullerene (C60), carbon nanotubes, and graphene are the focus of many recent research investigations [2–6]. Among them, graphite has widespread applications ranging from simple pencil lead to DC motor brushes, some traces of electronic circuitry to water filter purifier, from battery carbon electrodes to heat exchangers, from aerospace to nuclear reactor engineering and several more [7].

Particularly, in missile technology, graphite is used as the propulsion gas exhaust nozzles to maintain constant pressure inside the rocket motor during the mission period. Jet propulsion system is a complex engineering machinery comprised of motor casing, solid rocket propellants, thermal insulation, and exhaust nozzle at the rear end. From nozzle, high-pressure and high-temperature gases are ejected, which are generated due to the combustion of propellants and transform chemical/heat energy into kinetic energy. Although the performance of the rocket motor depends, integrally,

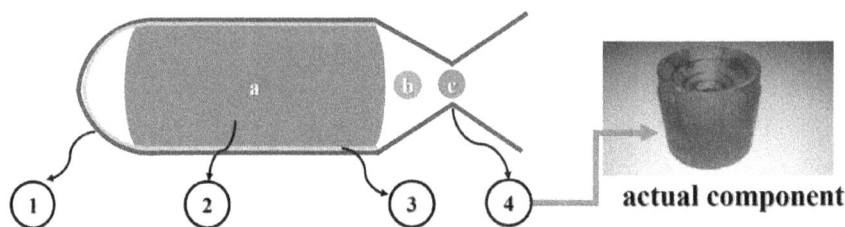

actual component

FIGURE 5.1 Schematics of rocket motor system: (1) motor casing, (2) solid rocket propellants, (3) thermal insulation, and (4) nozzle, including the actual photograph of the component. Heat flux experienced by the projectile surface at (a) 02–04, (b) 04–09, and (c) >20 MW/m².

on all the components, the nozzle is a crucial component. Figure 5.1 schematically shows a rocket motor system including the actual photograph of a prototype graphite nozzle. The choice of graphite over other materials such as molybdenum and tungsten comes from its rich abundance, cheap cost, excellent chemical and corrosion resistance properties, superior mechanical stability, less char (porous, sooty carbon deposits) yield, and remarkable thermal stability. Moreover, choosing low-density nozzles that offer a lower payload on weight is advantageous for aerospace applications.

5.1.1.1 Nozzles: System Engineering in Short-Range Missiles

Geometrically, a nozzle consists of a convergent cone, a throat, and an exit cone. Any flaw in the designed component may result in thermal stresses and fatigue to the nozzle and, consequently, result in the failure of the mission. Hence, precision designing of a nozzle is an important issue, particularly from the thermal performance point of view. Fundamentally, thermal characteristics of any solid are gauged by four major parameters: its capability to absorb heat, i.e. specific heat capacity (C_P, measured in J/kg/K), ability to transfer it efficiently by means of thermal conductivity (λ, W/m/K) as well as diffusivity (α, m²/s), and its degree of deformation on heating, i.e. coefficient of thermal expansion (α_L, K^{-1}). Crystalline graphite, being a layered material, exhibits thermal anisotropy.

There are several investigations carried out to explore one or the other thermal properties of several categories of graphite (laboratory synthesized, commercial, nuclear reactor grade, foams, etc.) over 40–3500 K; however, they lack in mutual agreement in thermal properties [8–12]. Reasons are several [13–17]. First and foremost, the physical chemistry of graphite is very complex. Second, crystalline microstructure, grain orientation, stacking, packing, bulk density, intrinsic void, distribution of porosity, surfaces/interfaces, degree of graphitization, etc., play a vital role in influencing thermal properties of graphite [17–29]. Notably, thermo-physical assessments on aerospace-grade graphite of variable densities have rarely been reported in the literature.

5.1.1.2 Density Effects on Thermodynamics of Nozzles

The purpose of this communication is to examine the thermal parameters (C_P, λ, α, and α_L) of variable density graphite (VDG), over 300–1300 K, and to choose a right candidate for jet propulsion nozzle design. VDG specimens of densities in the range of 1600–1900 kg/m^{-3} were fabricated in foundry by HIP and subjected to thermo-physical measurements. The effects of density variations on thermal properties and intricately derived thermodynamic quantities such as Debye temperature, θ_D, change in entropy, ΔS, enthalpy, ΔH, and heat flux were analysed. Correspondingly, X-ray diffractometry, Raman spectroscopy, and scanning electron microscopy measurements extracted numerus involved structural parameters: A_x, stacking thickness, L_a, in-plane crystallinity, L_a, stacking density, ρ_s, electron–phonon coupling, γ_{e-p-c}, etc., to shed light on thermo-physical correlations. Notably, VDG with density 1811.7 kg/m³ is found superior in its thermal, physical, and textural performance over others.

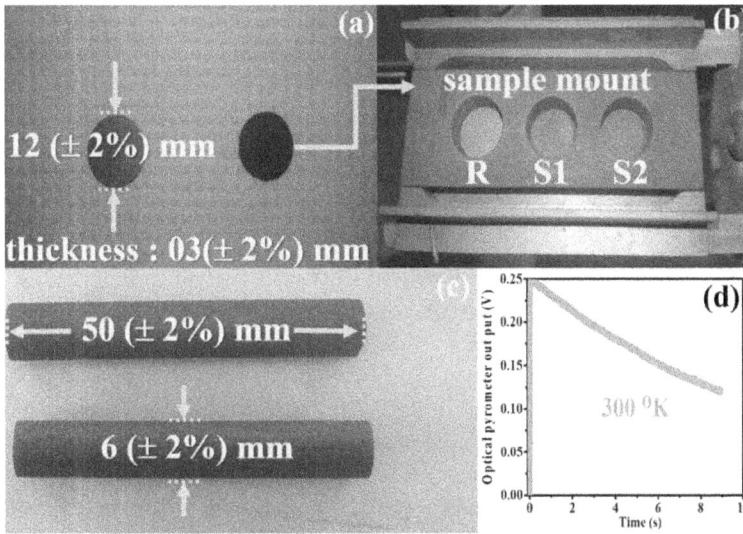

FIGURE 5.2 (a) Disc-shaped (ASTM E1461) specimens for measurements of C_P, λ, and α, (b) assembly for sample (S1, S2) and reference (thermographite) mount, (c) cylindrical (ASTM E228) samples for measurement of α_L, and (d) typical calibration thermogram recorded for standard thermographite at 300 K.

5.1.2 THERMAL MEASUREMENTS

The obtained VDG was subjected to thermal measurements. For these measurements, two types of specimens were prepared by meeting specifications laid down by American Society for Testing and Materials (ASTM). Figure 5.2 shows (a) a disc (ASTM E1461)-shaped specimen for the measurement of C_P, λ, α, (b) a mount assembly indicating reference (R) and two sample (S1, S2) positions, and (c) a cylinder (ASTM E228) for the estimation of α_L [30]. Dimensions for ASTM E1461 specimens were the following: diameter: 12% ±2% mm and thickness: 3% ±2% mm, whereas dimensions of ASTM E228 samples were the following: length: 50% ±2% mm and diameter: 6% ±2% mm.

5.1.2.1 Transient Flash Technique: Measurement of C_P, λ, and α

The transient flash technique, also popularly known as the laser flash method, is one of the most accurate methods for determining thermal properties of any material under investigation. It has number of advantages in terms of accessing high temperature range, non-contact sensing, and easy and swift measurement provisions. One can perform absolute thermal diffusivity measurements using this technique. The amount of sample required is quite small as compared to other similar classes of measurement set-ups such as guarded hot plate and transient hot wire methods. In its modular architecture, our thermal properties analyser (FlashLine 3000, M/s Anter Corporation, the USA) is having an infrared (IR) heating furnace equipped with complete temperature control (−200 to +1100°C, ramp rate ~ 100°C/m, accuracy: ±1°C). The instrument consists of data acquisition subsystems, a high-speed xenon discharge (HSXD) pulsed light source (peak power ~500 W; pulse width 200–300 μs), it's power supply compounded with control electronics, safety interlocks, optical pulse delivery components, along with IR optics and liquid nitrogen-cooled IR InSb (indium antimonide) detector to record rear face thermogram of the specimens, at ambient temperatures.

The physical principle involved in the technique is the following: The emitted radiation from the source falls onto the specimen surface and gets absorbed in the form of heat. Incidentally, the absorbed heat propagates through the specimen and changes the thermal signature at the rear surface that is picked up by an IR detector. The detector signal is further amplified, conditioned, denoised, and processed through data acquisition system to obtain thermogram data that enable to estimate of thermal parameters. In order to absorb maximum heat energy, sometimes, graphite spray is coated depending upon emission characteristics of the sample.

As such, no calibration is required for flash technique due to its primitive sensing mechanism. However, random or systematic errors that originated during measurements were standardized by performing thermogram measurements onto a standard specimen (thermographite): A typical spectrum is shown in Figure 5.3d.

5.1.2.2   Push-Rod Dilatometry: Measurement of α_L

In expansivity measurements, a push-rod dilatometer was used. The involved instrumentation and measurement principle is identical to that of transient flash technique, except the light source component. The dilatometer consisted of Nd:glass laser (class 1 laser) capable of producing a pulse energy of ~15 J (max.) and a pulse length of 0.33 ms. The system (MicroFlash, Make: NETZSCH LFA 457) is able to perform thermal measurements between −125°C and 1100°C. Measurement tests at a particular temperature value (say room temperature) take, notably, ~ few minutes, whereas the entire temperature range can be spanned in a period of few hours, typically four to five hours. The tool is non-destructive and could be used for materials with thermal diffusivities ranging from 0.01 to 1000 mm²/s. Like transient technique, the dilatometer method is an absolute test technique and needs no calibration. However, standardization operations could be performed on the system in accordance with the international/national standards such as ASTM E1461, DIN EN821, and DIN 30905. The accuracy to measure thermal diffusivity is determined to be ±3% and thermal conductivity within ±5%. In our study, the system was calibrated with a standard sapphire specimen for the full temperature range 300–1300 K. The corrections generated with calibration curve were applied in the actual sample scans.

5.1.2.3   Other Characterization Techniques

For the structure–property relationship, X-ray diffractometry (XRD), Raman spectroscopy, and scanning electron microscopy measurements were performed. All measurements were carried out at room temperature.

- **X-ray diffractometry:** XRD measurements were performed using an instrument from M/s Bruker D8 Advanced, the USA, with a gun voltage kept at 40 kV and a beam current of 30 mA to obtain Cu K_α line of wavelength $(\lambda) = 1.54$ Å. The scan rate was fixed to 0.02°/s, and the range for 2θ scan was 10°–80°.
- **Raman spectroscopy:** Raman spectra were obtained using LABRAM HR 800 from M/s Jobin Vyon, France. The excitation wavelength $\lambda = 532$ nm was used to record spectra over the wavenumber regime 200–3000 cm⁻¹.
- **Electron microscopy:** Field emission scanning electron microscopy (FESEM) was carried out using an instrument from M/s Zeiss ΣIGMA, the USA, at an electron beam potential of 30 kV with a fixed working distance for various magnifications, under high vacuum conditions (10⁻⁶ Torr). A few typical micrographs have been presented.

5.1.3   HEAT CAPACITY, C_P, AT CONSTANT PRESSURE AND VOLUME

Figure 5.3a shows the measured C_P profiles fitted with Butland's equation, (b) the corresponding $C_P - C_V$ curves, (c) the obtained d-factors, and (d) change in entropy, Δs, over 300–1300 K for VDG. Butland et al. had, extensively, measured and evaluated C_P for nuclear reactor-grade graphite over 200–3500 K and fitted with a mathematical relation [11]: $C_P = d + e\,T + f/T + g/T^2 + h/T^3 + i/T^4$, as shown in Figure 5.3a. The bulk C_P has contributions from pure crystalline phase, relative lattice movement (i.e. quantum of work done by heat in enhancing the lattice vibrations), and change in

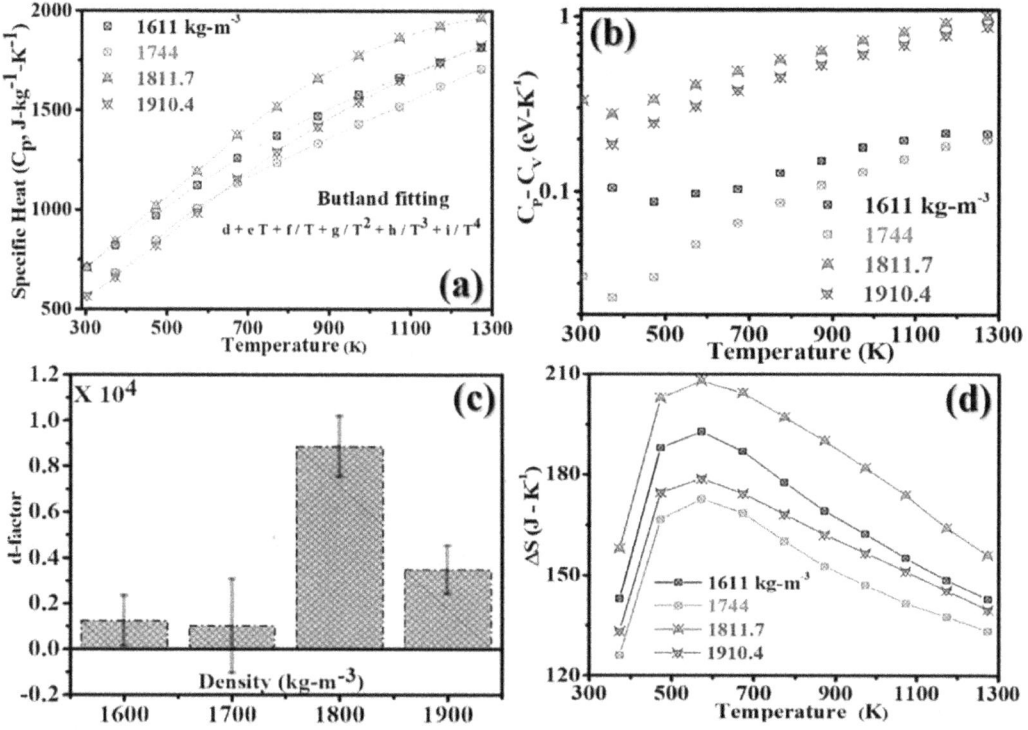

FIGURE 5.3 (a) Estimated C_P, (b) fundamental C_P-C_V variations, (c) lattice contribution d-coefficient from Butland's equation, and (d) change in entropy, Δs, for VDG over 300–1300 K.

the internal energy of the crystal microstructure in terms of modifications in its surface/interface energy. The first term in the equation is attributed to the lattice contribution. From Figure 5.3a and c, one can clearly see that for VDG with 1811.7 kg/m³, lattice contribution played a dominating role over other classes of VDG. The Butland fitment Table 5.1 shows contributions of various other terms.

Perhaps for VDG@1811.7, the d-factor is observed to be multi-fold high over other samples. Moreover, from the experimentally determined C_P, one can derive C_V (@constant volume), which is a more fundamental physical entity connected by a relation: $C_P - C_V = 9\,\alpha_L^2\,B\,V\,T$, where B is the elastic modulus of VDG, which is $\sim 18\,\text{GPa}$, and V is the measured percentage volumetric deformation. The $C_P - C_V$ behaviour is displayed in Figure 5.4b for VDG, indicating superior performance of VDG@1811.7 in terms of C_P and C_V over other graphite specimens.

The C_V has connection to coupling of phonons to lattice by a relation: $C_V / 3R = \int_0^\infty \rho(\omega') \dfrac{\beta'^2 \exp(\beta')}{\left[\exp(\beta') - 1\right]^2}\,d\omega'$ that is evaluated by Raman spectroscopy to extract electron–phonon coupling, γ_{e-p-c}, as presented further [13]. The exchange variable β' is the ratio of $\dfrac{\hbar\omega'}{k_B T}$, where ω' is the phonon frequency (in rad/s) associated with C–C and C=C molecules. The C_V is canonically connected to ΔS by Loschmidt's expansion at $T = \theta_D$:

$$\Delta S = 1 + C_V / \left(\frac{T}{\Delta T}\right)^1 + C_V / \left(\frac{T}{\Delta T}\right)^2 + C_V / \left(\frac{T}{\Delta T}\right)^3 + \ldots,$$ where θ_D is the Debye temperature. By

neglecting higher-order terms and setting $1 \ll C_V / \left(\dfrac{T}{\Delta T}\right)^1$, we obtained variations in ΔS for VDG

that are displayed in Figure 5.3d. Interestingly, ΔS vs T plot suggested only one value of $\theta_D \sim 575\,\text{K}$ for VDG. The value does not agree with the reported $\theta_D = 420\,\text{K}$ for graphite [23]. Moreover, usually,

TABLE 5.1

Fitment Data for Butland's Equation Showing Contributions of Various Thermodynamic Energies Such as Crystalline Phase, Lattice Mechanical Vibration, And Internal Energy/ Surface Energy Change for Microstructure of VDG

Density (kg/m³)	d	$\pm\Delta d$	e	$\pm\Delta e$	$f\times10^6$	$\pm\Delta f\times10^6$	$g\times10^8$	$\pm\Delta g\times10^8$	$h\times10^{11}$	$\pm\Delta h\times10^{11}$	$i\times10^{13}$	$\pm\Delta i\times10^{13}$
1611.00	1246.27	1093.37	0.553	0.329	0.5	1.36	−3.8	8.02	1.54	2.22	−1.1	2.33
1744.00	1019.17	2054.80	0.858	0.559	−1.1	2.82	1.1	1.95	−5.2	6.31	8.24	7.81
1811.70	8871.06	1334.29	−1.95	0.402	−8.0	1.66	3.65	9.8	−8.1	2.71	7.16	2.84
1910.40	3481.11	1056.06	−0.07	0.318	−2.7	1.32	1.05	7.75	−2.0	2.14	1.63	2.25

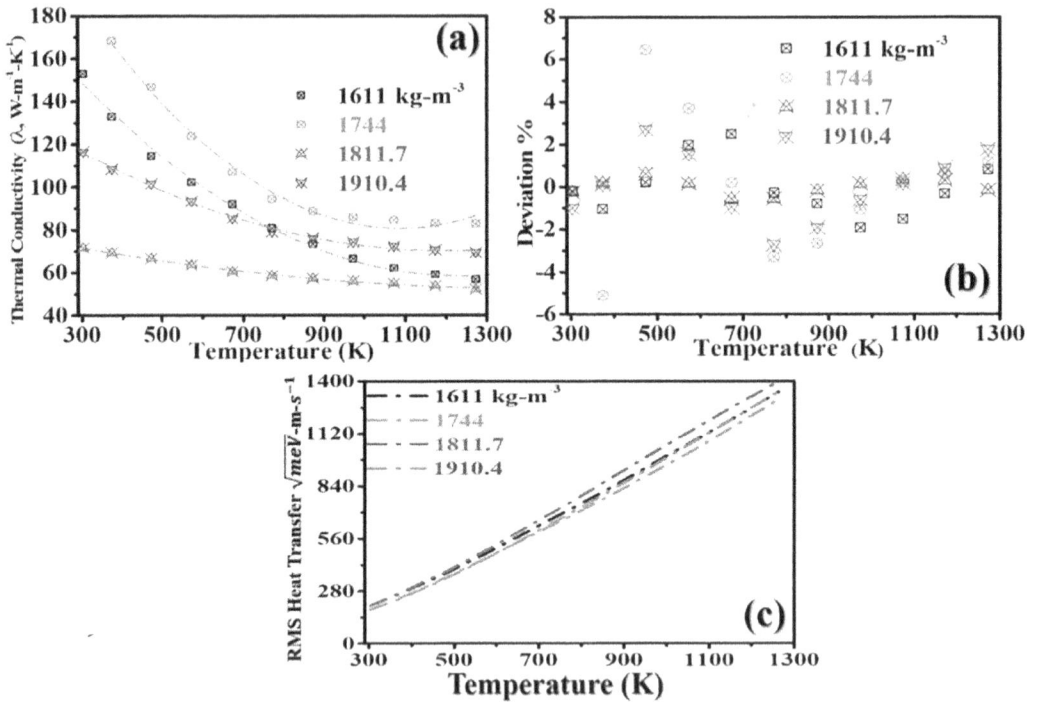

FIGURE 5.4 (a) Measured λ, (b) % deviation between measurements and correlations, and (c) root mean square heat transfer for VDG over the measured temperature regime.

graphite is represented via two Debye temperatures – one for the basal plane vibration and the other for the prismatic plane vibrations [24]. Thermodynamically, $C_V \approx T \sum \left(\dfrac{\partial S}{\partial T}\right)_V \approx \left(\dfrac{\partial S}{\partial H}\right)_V$, where H is the enthalpy of the system. Table 5.2 shows the estimated values of H obtained at four typical temperatures 375, 575, 1075, and 1275 K, including θ_D.

Hence, H is associated with the thermal reactivity of any material, which seems to be high for VDG@1811.7 over others.

5.1.4 Thermal Conductivity, Diffusivity, and Heat Flux

Further, the specific heat capacity, Cp, together with thermal conductivity, λ, and thermal diffusivity, α, is given according to a relationship: $\lambda = \alpha \cdot C_P \cdot \rho$. Both the parameters λ and α for graphite

TABLE 5.2

Estimation of H (in kJ) for VDG at Four Typical Temperatures Including θ_D

Density (kg/m³)	Temperature in K			
	375	575 (θ_D)	1075	1275
1611	55	110	160	180
1744	45	100	140	170
1811.7	65	125	180	200
1910.4	50	100	150	180

TABLE 5.3

Fitting Parameters for Quadratic Equation

Density (kg/m³)	McEligot's Thermo-Structural Coefficients		
	J_0	J_1	J_2
1611	216.37±5.15	0.255±0.014	$1.03\times10^{-4}\pm9.2\times10^{-4}$
1744	281.4±8.48	0.369±0.022	$1.64\times10^{-4}\pm1.33\times10^{-5}$
1811.7	82.99±0.92	0.041±0.002	$1.43\times10^{-5}\pm1.64\times10^{-6}$
1910.4	152.49±2.20	0.136±0.006	$5.68\times10^{-3}\pm3.92\times10^{-5}$

were reported to be varied over the measured temperature range 40–2500 K. Their values were found to be different along parallel and perpendicular directions of the basal plane [9]. Particularly, for λ, the nature of exponent followed a number of functional forms such as T^{-n} with $n \leq -1/2$ over 300–1700 K.

In a study, λ perpendicular to the basal plane followed an inverse temperature relationship over 300–900 K and was found to be changed over 1000 K obeying T^3 relationship [15]. For temperature above 1500 K, the deviation from inverse temperature law, of the form $(A+BT)^{-1}$, was observed to be critically high >100%. Broadly speaking, the magnitude of $\lambda_\parallel$ and $\lambda_\perp$ may vary by a factor of 1.5–3.5, has functional behaviour between $T^{-1/2}$ and T^{+3}, and varies by 10%–30% depending upon the preparation conditions of graphite. However, such a behaviour was reported to be independent of degree of graphitization of the sample. Figure 5.4a shows the exponent fitting for the superimposed $\lambda_\parallel$ and $\lambda_\perp$ components. Fitting matches well with McEligot's quadratic equation of the form: $= J_0 - J_1T + J_2T^2$, where J_i's are McEligot's thermo-structural coefficients [26]. Their value and strength are given in Table 5.3.

The values of λ obtained are purely the outcome of the measurements on VDG, and any structural variations are not taken into account. Figure 5.4b shows the % deviation between measurements and correlations over 300–1300 K, which was recorded for two samples at a time for 10+ data points. The smallest % deviation was found to be 1%–3%.Figure 5.4c shows the derived root mean square (RMS) heat flux H estimated via interaction potential formulism. It is expressed as:

$$H\left(D_o + x\right) = H_{D_o} + (\partial H/\partial D)_{D_o} + \frac{1}{2!} \cdot (\partial^2 H/\partial D^2)_{D_o} \cdot x^2 + \frac{1}{3!} \cdot (\partial^3 H/\partial D^3)_{D_o} \cdot x^3 + \dots, \quad \text{where}$$

D_o is the equilibrium position of carbon atoms and x is thermally assisted perturbation [29]. It leads to an expression: $H \approx k_B T^2 C_V$, where k_B is the Boltzmann constant providing values of RMS heat flux for VDG. This expression gives root mean square heat flux values. From Figure 5.4c, one can see that the RMS heat flux capability of VDG@1811.7 is far efficient over other VDGs.

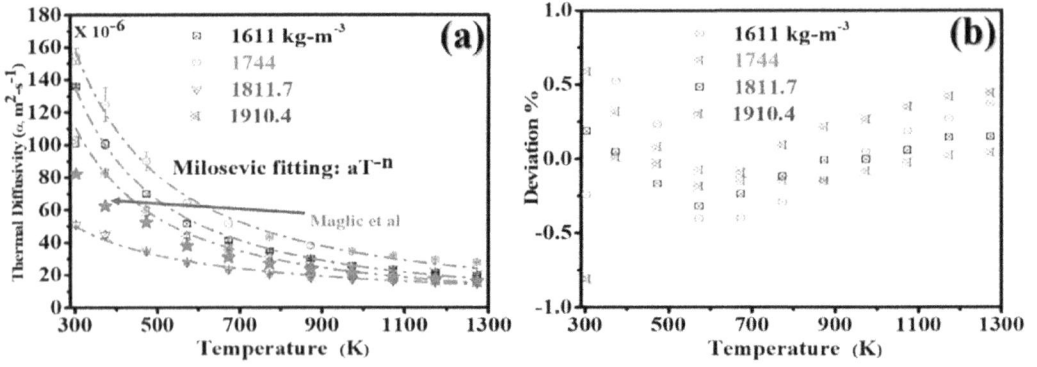

FIGURE 5.5 (a) Estimation of α, indicating Milosevic fitting; the arrow shows data recorded by Maglic et al. on thermographite [27]. (b) Recorded % deviation in measurements and correlations.

Next, we studied thermal diffusivity, α, characteristics of VDG. Maglic et al., extensively, carried out thermal diffusivity and conductivity investigations on AXM-5Q and thermographite specimens. They reported α in the range of 60–15($\times 10^{-6}$) m^2/s over 300–1230 K. The density of AXM-5Q and thermographite was, respectively, 1722 and 1755 kg/m^3. We referred Maglic's work from the point of view of density effects that are close to VDGs [27].

The behaviour of α resembles the recorded λ features. The diffusivity was observed to be varied as T^{-n} and found to be consistent with the reported work of Maglic et al. [27]. The data correlations of % deviation were in agreement with 1% variation, as seen in Figure 5.5b. The data table for Milosevic's fitting parameters is provided in Appendix A (Table A.5.1 and Figure A.5.2).

5.1.5 THERMAL EXPANSIVITY EFFECTS

The coefficient of thermal expansion plays an important role in estimating internal stresses in a material. Particularly, from rocket nozzle viewpoint, the onset combustion of propellants and gas ejection, during the mission period, could result in enormous temperature and pressure flux gradient at nozzle, along radial and axial directions. This may lead to compositional modifications at nozzle surface due to oxidation, ablation, charring, etc., causing high amount of internal thermal stress generation. The coefficient of thermal expansion (CTE), α_L, is given by an expression: $\dfrac{\partial L_o}{L_o} \cdot \dfrac{1}{\partial T} = \alpha_L$, where $\dfrac{\partial L_o}{L_o}$ is the % length deformation at temperature T with reference to some reference temperature, T_o.

For a graphitic medium, the estimation of α_L, particularly, for crystalline graphite is somewhat complex. This is, mainly, due to the anisotropy offered by the matrix within basal planes of sp^2 hybridized network, referred to as a-axis, and conjugated planes connected by weak van der Waals interactions, referred to as c-axis. Therefore, at room temperature, for a crystalline graphite, α_L is composed of parallel (a-axis, α_{L_a}, $\sim -1.5 \times 10^{-6}$/K) and perpendicular (c-axis, α_{L_c}, $\sim 27.0 \times 10^{-6}$/K) values. The negative a-axis value is originated for parallel elongation, which is attributed to shrinkage in the magnitude of Poisson's ratio associated with large expansion perpendicular to basal planes. But this is a story of single crystal graphite with density >2000 kg/m^3. For practical purposes, such as HIP-processed VDG, they are polygranular graphite. Therefore, the α_L for VDG could be modelled via an ensemble of several single crystals nearly isotropic in their graphite orientations in which α_L attains an arbitrary direction, α_{L_x}. Figure 5.6a displays α_L variations recorded for VDG over 300–1300 K, and Figure 5.6b, the corresponding % deviation. At 300 K, VDG@1811.7 shows >90% resemblance with a theoretical value of α_{L_x} [22]. Further, α_{L_x} is empirically connected to structure factor, A_x, and could be examined by physical characterization such as XRD. In general,

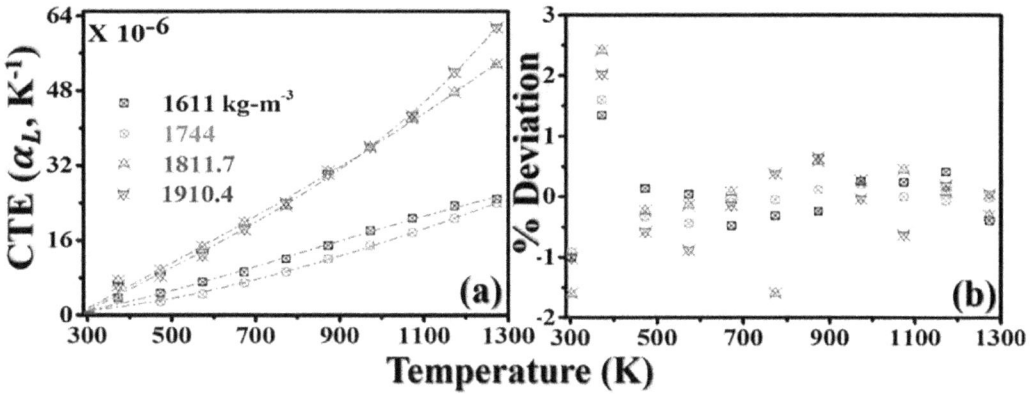

FIGURE 5.6 (a) Measured α_L for VDG; (b) % deviation in correlations recorded. VDG@1811.7 resembles theoretical value, α_{L_x}, with >90%.

TABLE 5.4

Estimated Pre-Factors for α_L Relation

Density (kg/m³)	K_1	K_2	K_3	K_4
1611	3.151 ± 3.871	0.0072 ± 0.0176	$-2.42 \times 10^{-5} \pm 2.4 \times 10^{-5}$	$-9.72 \times 10^{-9} \pm 1.01 \times 10^{-8}$
1744	0.543 ± 3.92	-0.0014 ± 0.017	$-2.19 \times 10^{-5} \pm 2.44 \times 10^{-5}$	$-4.47 \times 10^{-9} \pm 1.03 \times 10^{-8}$
1811.7	13.31 ± 6.92	0.05 ± 0.031	$6.16 \times 10^{-6} \pm 4.31 \times 10^{-5}$	$6.27 \times 10^{-9} \pm 1.18 \times 10^{-8}$
1910.4	15.78 ± 5.44	0.065 ± 0.024	$4.05 \times 10^{-5} \pm 3.39 \times 10^{-5}$	$2.91 \times 10^{-8} \pm 1.43 \times 10^{-8}$

A_x is a parameter in the interpretations of scattering patterns obtained by XRD and depends on the degree of orientation of graphitic planes and the way crystal expansion is accommodated within the microstructure [29,31].

Broadly, α_{L_x}, at 300 K, is correlated to α_{L_a} and α_{LC} by a relation: $\alpha_{L_x} = A_x \, \alpha_{LC} + (1 - A_x)\alpha_{L_a} \approx (\alpha_{LC} + 2\,\alpha_{L_a})/3 \sim 8.1 \times 10^{-6}/\text{K}$. Herein, for avoiding complexity α_{L_x} is designated as α_L. The pre-factors shown in Table 5.4 are obtained using the equation proposed by Tsang, which is given by: $\alpha_L = -K_1 + K_2 T - K_3 T^2 + K_4 T^3$, which fits well to all data with % deviation 1%–3% and χ^2 fitting yielding ~99% of the convergence value [29,32].

5.1.6 XRD: THE STRUCTURE CHANGE

The XRD analysis was carried out to obtain A_x for VDG. VDG was basically characterized by an interplane distance of d_{002} or d_{004}, dimensions of structural components (L_a, L_c), and degree of graphitization. Figure 5.7 shows (002) reflex used to extract information details of VDG. The inset shows the corresponding full 2θ scan indicating various phases of VDGs.

On the basis of the recorded XRD pattern, the following parameters were extracted [33,34]. The interlayer distance d_{002} for VDG sample was computed using Bragg's law: $d_{002} = \lambda/2 \sin\theta_{002}$, where λ is the wavelength of X-ray, i.e. 1.55 Å, and θ_{002} is the angle of diffraction. The dimensions of structural components such as the thickness of graphitic layer, L_c, are given by the equation: $L_C = 0.9 \, \lambda/\zeta_{002} \cos\theta_{002}$, where ζ_{002} is the full width at half maximum (FWHM) of (002) reflex measured in radians. The lateral/parallel crystalline extent, L_a, is computed using: $L_a = 0.9 \, \lambda/\zeta_{100} \cos\theta_{100}$, where ζ_{100} is the FWHM of reflex (100). The stacking density of VDG layers is given by: $\rho_s = 0.762/d_{002}$, and the number of layers in stacks was estimated by: $N = (L_C/d_{002}) + 1$. Crystallographically, the

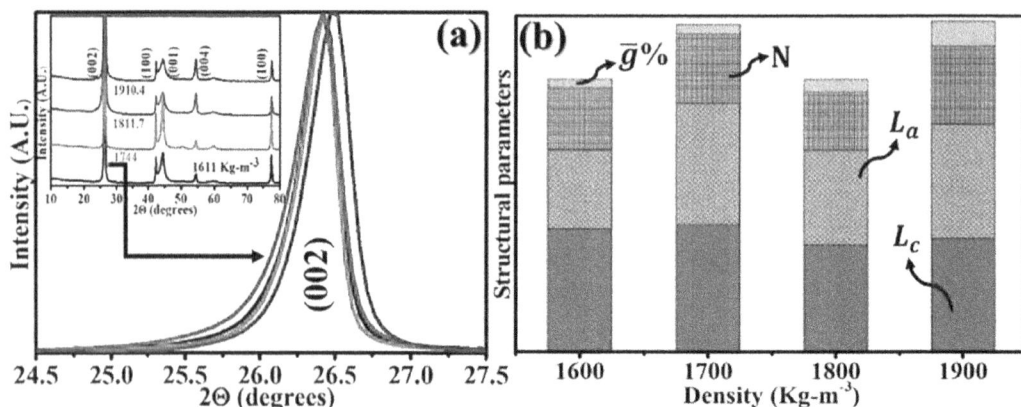

FIGURE 5.7   (a) Recorded XRD spectra. Reflex (002) is used to analyse various structural aspects, whereas inset shows the full scan recorded over 2θ, $10°$–$80°$. (b) Structural parameters: percentage graphitization, $\bar{g}\%$, the number of layers in stack, N, in-plane crystalline length, L_a, and the thickness of graphitic plane, L_c.

TABLE 5.5
Estimated Structure Factors, A_x, for VDG by XRD (Some Are Displayed in Figure 5.8b)

Density (kg/m³)	d_{002} (A°)	L_c (A°)	L_a (A°)	ρ_s (kg/m³)	N	$\bar{g}$ %	$\alpha_L \times 10^{-6}$ (K^{-1})@300K
1611	4.941	265	170	1602	135	17	~2.10
1744	4.831	275	260	1749	150	21	~1.50
1811.7	4.620	230	205	1810	125	27	~7.10
1910.4	4.893	245	245	1920	168	55	~3.00

bonding distance between C–C, a_o, in the honeycomb sheet is $a/\sqrt{3}$. For an ideal single crystal at 300 K, $c = 6.708\,\text{Å}$, $a = 2.4614\,\text{Å}$, $a_o = 1.4211\,\text{Å}$, and $c/a = 2.725$. The degree of graphitization, $\bar{g}$, was estimated by analysing d_{002}, where $\bar{g}$ becomes 1 for $d_{002} = 3.354\,\text{Å}$ and 0 for 3.44 Å. The governing equation is $\bar{g} = \dfrac{3.440 - d_{002}}{3.440 - 3.354}$. Further, we have estimated α_L, at room temperature, using the d_{002} line, by using the equation: $\alpha_L = \left(d_{002}\,(300°K) - d_{002}\,(0K)/T\,(300\ K) - T\,(0\ K\,)\right)$, where the value of $d_{002}\,(0\ K)$ is estimated to be 3.44 Å. Experimental errors in determining d_{002} and α_L magnitudes were, respectively, 0.001 nm and 0.01×10^{-5}/K. In general, for computations of structure parameters such as d_{001}, L_c, L_a, ρ_s, N, and $\bar{g}$, one can use reflex (002) or even higher-order reflexes such as (004) and (006). In our study, we have chosen (002) to explore them. Table 5.5 shows the magnitudes of structure factors, A_x, estimated.

5.1.7   RAMAN: THE MOLECULAR-LEVEL THERMAL TRACE

The objective of performing Raman studies was to look into the bond molecular environment of VDG. Raman analysis was carried out at λ (excitation)$= 532$ nm, by recording VDG spectra at 300 K over 200–3500 cm^{-1}. A number of features were revealed associated with the appearance of D, G, and 2D bands that appeared, respectively, at 1320–1350, 1540–1590, and 2640–2670 cm^{-1}. Notably, a prominent presence of amorphous carbon was observed in the low-wavenumber regime below 850 cm^{-1}. Raman is a resonant spectroscopy operative at molecular level and provides rich information on L_c, L_a, sp^3, sp^2 content, etc. Since in-plane and out-of-plane crystalline parameters were already examined and presented using XRD analysis, we have not performed this exercise again.

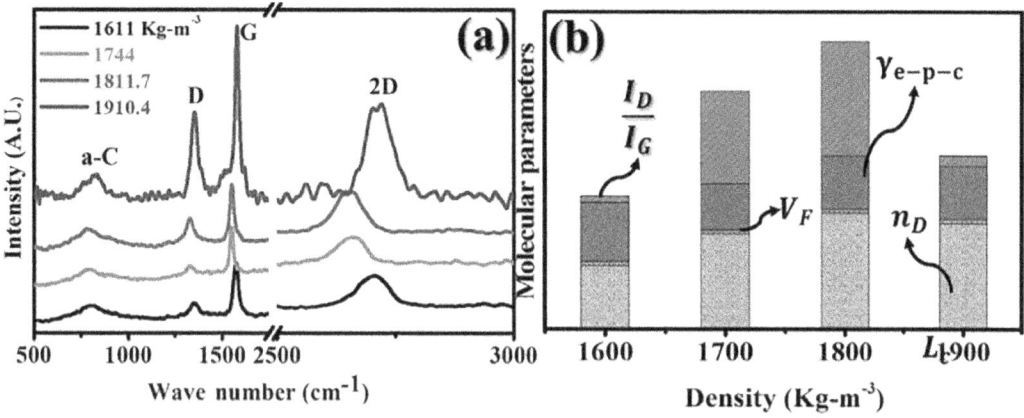

FIGURE 5.8 (a) Recorded Raman spectra at λ (excitation) = 532 nm for VDG, over 200–3500 cm^{-1} at 300 K, (b) extraction of molecular parameters such as integrated I_D/I_G (%), electron–phonon coupling, γ_{e-p-c}, Fermi velocity, V_F, and n_D (others are shown in Table 5.6).

TABLE 5.6

Estimated Molecular Parameters, Electron–Phonon Coupling, γ_{e-p-c}, Disordered Length, L_D, and Defect Density, n_D, Fermi Velocity, V_F, and Dynamic Force Constant, k, for VDG Obtained from Raman Spectroscopy

Density (kg/m³)	I_D/I_G (%) (Integrated)	γ_{e-p-c} (cm^{-1})	L_D^2 (nm²)	$n_D \times 10^{18}$ (m^{-2})	$V_F \times 10^6$ (m/s)	k_G (–C=C–) (N/m)	k_D (–C–C–) (N/m)
1611	30	28	340	0.00290	~2.0	723±30	1020±70
1744	45	22	230	0.00434	~1.8	671±17	970±44
1811.7	55	25	185	0.00540	~1.8	678±53	1101±23
1910.4	50	25	200	0.00500	~2.0	855±22	1007±89

However, there are interesting extractions associated with electron–phonon coupling, γ_{e-p-c}, which in turn depends on a number of intricate molecular dynamic parameters such as natural broadening of G-peak; Fermi velocity V_F (m/s) given by: $V_F^2 = \dfrac{\sqrt{3}}{2} \cdot \dfrac{a_o^2}{\mu} \dfrac{\Gamma_{e-p-c}^2}{\gamma_{e-p-c}}$, where μ is the reduced mass of C–C and Γ_{e-p-c}^2 is the theoretical Γ-phonon branching factor (50 (eV-Å^{-1})^2); and defect density n_D given by: $n_D \left(m^{-2} \right) = 10^{18}/L_D^2$, where L_D is the disordered length estimated from integrated I_D/I_G ratio given by: $L_D^2 = C(\lambda_L)/\left(\dfrac{I_D}{I_G} \right)$, where $C(\lambda_L)$ is the laser beam parameter (102 nm²). These formulations are given in detail in [31,35]. Figure 5.8a shows the recorded Raman spectra, and Figure 5.8b, the extracted molecular parameters. From the recorded spectra, one can see that the appeared peaks are dispersed with respect to their original position.

The Raman study, broadly, revealed that I_D/I_G (%) was maximum for VDG@1811.7, showing higher sp^2 content over others. The value of γ_{e-p-c} was found to be almost identical, i.e. ~0.5 meV, for higher VDG. The obtained V_F had no impact on the molecular compaction of VDG and, consequently, on the density of graphite matrix. Other molecular parameters displayed in Table 5.6 may have canonical impact on density.

5.1.8 MICROSCOPIC ANALYSIS

SEM investigations were carried out on VDG specimens at a suitable electron beam working distance of ~6mm to record the texture of graphite, clearly and legitimately, at various magnifications from 2000 to 30,000. The recorded images brought out distinct features of matrices that include but are not limited to planes, stakes, flakes, warps, exfoliated and eroded surfaces, micro- and macro-textures, and amorphous carbon deposit chunks. Morphology observations at the edge of the flakes to examine their thickness, mostly, resemble the estimation of L_c obtained from XRD analysis (presented in Table 5.5).

As such, it is not possible to display all recorded images, herein. Representative images are shown in Figure 5.9, in which the top row, (a)–(d), shows the morphology captured (with hierarchy in density from 1611 to 1910.4 kg/m) at 2000 magnification. The surface textures seen in (a)–(d) were entirely distinct. Particularly in Figure 5.10a, a larger, smoother, planar, and less warpy surface texture was observed; however, as one moves from Figure 5.9a to Figure 5.9d, this trend gets trade-off. Perhaps for image in Figure 5.9c and, more prominently, in Figure 5.9d, flake squashing was observed. The reason is somewhat not clear, but this could be attributed to HIP processing of graphite, where enormous compression may result into redistribution of surface as well as bulk of the matrix. In fact, for VDG@1744 and VDG@1811.7, we observed a large number of amorphous carbon particles on flakes and within graphitic layers (Figure A.5.1 is provided in Annexure C). However, there was a trade-off between these samples and VDG@1910.4. It seems that HIP not only changed density, but also modified microstructural features of graphite, significantly.

The bottom row in Figure 5.9e–h, correspondingly, displays images taken with the same protocol at 30,000 magnification, portraying an identical trend. The analysis presented, broadly, establishes a correlation between the recorded microstructural features and their investigated thermo-molecular properties.

Our study further prompted to design a batch of prototype nozzle components for static field testing. Figure 5.10 shows HIP-manufactured VDG@1811.7 that underwent thermo-physical

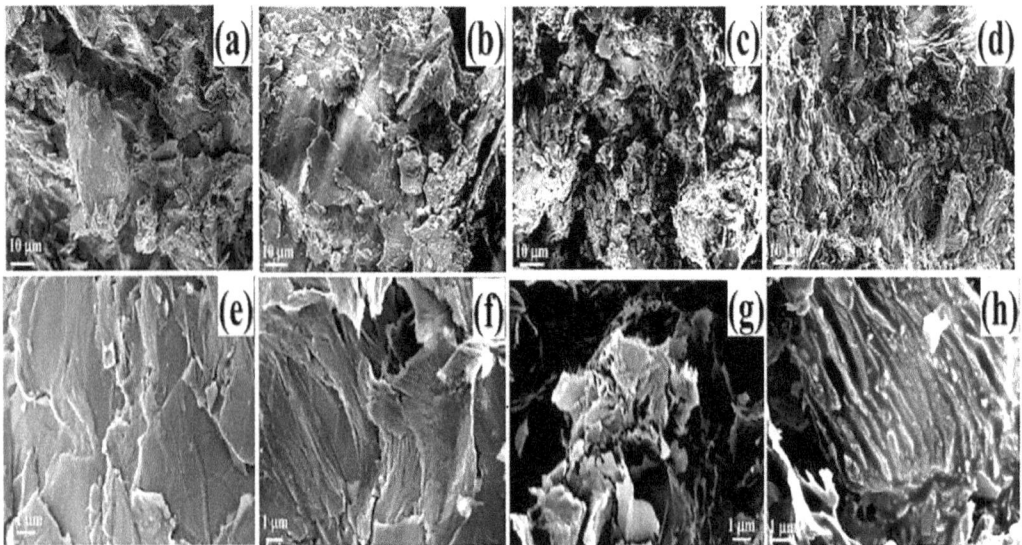

FIGURE 5.9 Recorded SEM micrographs for VDG. Top row: (a)–(d) images recorded at 2000 magnification (10 μm scale bar). Bottom row: (e)–(h) at 30,000 (1 μm). Increase in density value from left column (1611 kg/m^3) to right (1910.4 kg/m^3).

FIGURE 5.10 Photograph of HIP-fabricated VDG@1811.7 that qualified thermo-physical testing. Inset shows a machined nozzle getting ready for static firing test.

assessments and qualified. The inset displays a photograph of a machined nozzle that is getting ready for a static firing test of 40 kg propulsion fuel load. It is planned to investigate post-fired nozzle as a future endeavour to extract thermal changes experienced.

The work presented, herein, is an attempt to make thermo-physical assessments on variable density graphite and identify graphite material of appropriate density that will suitably be used for high-thermal-performance rocket nozzle engineering. Nozzles are an important system component in rocket propulsion technology. Graphite nozzles are advantageous over molybdenum or tungsten due to their superior thermo-physical properties.

Further, a graphitic rocket nozzle is an important material component in aerospace engineering. Particularly, nozzle quality in terms of high-temperature stability plays a vital role in the mission accomplishment of a missile. Herein, we assessed high-temperature molecular thermodynamics in graphitic nozzles for providing solution to designing an efficient jet propulsion exhaust system. Graphite specimens of desired density ($1811.7 \, \text{kg/m}^3$) are fabricated by HIP foundry processing. Nozzles are examined for static firing carried out for a realistic payload using composite and double base propellants. Firing created a transient temperature rise to 3000+ K, leading to thermal stresses in the nozzle. Fired nozzles are examined for their thermal properties such as specific heat capacity (at constant pressure (C_P) and volume (C_V)), thermal conductivity, λ, diffusivity, α, and expansivity, α_L, over 300–1300 K, and compared with the pristine. C_P and C_V examinations revealed an entropy increase by $3\%2\pm3\%$ compounded with an enhancement in lattice specific heat by $20\%\pm3\%$ and dissociation enthalpy by $18\%\pm3\%$. However, the overall electronic specific heat is reduced by $25\%\pm3\%$ for fired nozzles. Stacking density, in-plane crystallinity, specific layer density, degree of graphitization, and α_L are examined by X-ray diffraction. However, surface graphitic content, disordered length, density of areal disorder, Fermi velocity, electron–phonon coupling, and molecular force constant are studied by Raman studies. The microscopy study of graphite texture is also presented. Efficient molecular thermodynamics, in terms of heat flux characteristics, is responsible for an effective high thermal performance of $1811.7 \, \text{kg/m}^3$ density nozzles.

5.2   HIGH-TEMPERATURE THERMODYNAMICS IN ROCKET MOTOR NOZZLES

A physical insight into any material comes up with investigating that material at the molecular level. Particularly, when the material is treated under high temperature and pressure conditions, it's the root of thermodynamics that could act as the eigen-probe to look through.

A missile is a multicomponent complex aerodynamic system comprised of guidance, flight, warhead, and engine. The engine, being a sensitive element, has a propulsion (fuel) motor equipped with an exhaust nozzle. The conversion of a solid/liquid propellant into gases takes place, in a nozzle, by transforming chemical/heat energy into kinetic energy accompanied by high amount of temperature and pressure generation. The nozzle is, thus, an important part in trajectory acquisition to accomplish the target by the projectile. Any failure in this component may right away result in the failure of the mission in toto! A nozzle design, geometrically, consists of an exit cone, a throat, and a convergent cone. Graphite is a well-suited material over others such as molybdenum and tungsten. The choice has several attributes such as cheap cost, excellent chemistry, superior mechanical capability, and extraordinary thermal properties [1,3,5,36]. Precision designing of a nozzle is an important issue, particularly, from high thermal performance point of view.

5.2.1   GENERAL THERMODYNAMIC CONSIDERATIONS

The thermal characteristics of any solid could be gauged, fundamentally, by analysing four major parameters such as specific heat capacity (C_P, measured in $J/kg/K^{-1}$), thermal conductivity (λ, W/m/K), diffusivity (α, m²/s), and expansivity (α_L, K^{-1}). Graphite, being an orthotropic material, exhibits thermal anisotropy [4,7]. Further, there are several investigations carried out to explore one or the other thermal properties of laboratory-synthesized, industrially fabricated, naturally available, specifically made for nuclear reactor engineering, and several types of other forms of graphite over 40–3500 K, with a wide density range of 1500–2000 kg/m³. Tsang et al. carried out theoretical and empirical studies on $\alpha_L \sim 10^{-6}/°C$ of nuclear-grade graphite (IG-110) over 20°C–700°C and compared their model with nuclear industries in the USA and Japan [32]. Morgan et al. established α_L and C_P relationship on the same model over 300–3000 K, which was estimated to be from 0.75 to 6.5 J/m/K for crystalline graphite of density 2000 kg/m³ [8]. Null et al. carried out the estimation of diffusivity (α) of pristine and annealed pyrolytic graphite over 300–2700 K reporting $\alpha \sim 1.94$ m²/s (@300 K) and ~ 0.32 m²/s (@2700 K) for the as-deposited graphite and ~ 9.54 m²/s (@300 K) and ~ 0.401 m²/s (@2700 K) for annealed pyrolytic graphite [9]. In a study, Taylor et al. carried out C_P investigations on carbon/carbon composites over 340–750 K and compared these values with industrially fabricated (POCO) AXM-5Q graphite, which were in agreement by 1% variation over 350 K; moreover, the investigation established a correlation between λ and α for such a class of graphite [10]. Butland et al. performed measurement and evaluation of C_P for graphite over 200–3500 K by fitting a mathematical model using higher-order polynomial equations. In this study, a phonon frequency spectrum at 1800 K was used to obtain C_V [11]. In another investigation, Neumann et al. determined the conductivity (λ) of graphite and other types of high-temperature alloys by laser flash technique, in which λ was found to be varied from 100%±3% to 40%±5% over 600–1500 K [12]. Kellet et al. investigated the thermal expansion of graphite within plane of layers over −200°C to +1200°C, in which α_L was found to be varied, respectively, from $+1.2 \times 10^{-6}/°C$ to $-1.6 \times 10^{-6}/°C$ [14]. Hacker et al. studied two kinds of nuclear graphite: one isotropic and another non-isotropic graphite to investigate the influence of oxidation on α_L, in which it was found that the isotropic graphite showed no change in α_L values till 50%–60% oxidation over 20°C–600°C [22]. McEligot et al. studied the thermal properties of G-348 nuclear-grade graphite from 300 to 1300 K, in which α was found to be varied between 100 and 20 mm²/s, α_L, $1-6 \times 10^{-6}/°C$, and C_P, 700–2000 J/kg/K [26]. Earlier, Kelly et al. carried out extensive theoretical as well as experimental work on crystalline graphite with density ~ 1900–2000 kg/m³. The studies involved thermal expansion of parallel and perpendicular basal planes, which revealed that at a high temperature above 600 K, magnitudes of α_L are less with the formation of warp to graphitic planes. Theoretically, Kelly et al. obtained values of $\alpha_L \sim +0.9 \times 10^{-6} - K^{-1}$ and $+1.64 \times 10^{-6} - K^{-1}$, respectively, at 1000 and 2000 K, which were marginally in agreement with experimental values recorded, respectively, to be $\sim +0.7 \times 10^{-6}/K$ and $+1.3 \times 10^{-6}/K$. Kelly et al. also studied the thermal expansion of nuclear-grade graphite below 400 K using Komatsu–Nagamiya lattice dynamic theory [15–17]. Interestingly, Martin et al. studied the thermal expansion of graphite over 100–700 K to underline the correlation between the elastic model and bulk thermal expansion of polycrystalline graphite to its degree of graphitization. Their study was in good mutual agreement with pyrolytic graphite [18]. The crystallographic thermal expansion of graphite was studied experimentally as well as theoretically by Nelson et al. and Riley et al. [20,37]. The study revealed that the a-axis C–C bonding got contracted up to ~400 K and, above it, the expansion takes place, whereas the C–C bonding behaved quite complexly and was very large over this temperature regime. In recent years, Seehra et al. have reported the thermal expansion parameters of a number of graphites obtained from solvent-extracted coal feedstocks. Their study revealed that the c-axis α_L varied from $+28.7 \times 10^{-6}/K$ to $+22 \times 10^{-6}/K$ @ 300K, whereas the a-axis α_L from $-0.95 \times 10^{-6}/K$ to $0 \times 10^{-6}/K$ @ 300 K [37]. Maglic et al. carried out thermal diffusivity studies on AXM-5Q graphite used in nuclear reactor engineering. The values of coefficient obtained by them are found to be matching with our current work, as discussed further [21]. Nagano et al. studied the thermo-physical parameters of spacecraft-quality graphite of ultralow density ~840 kg/m³, using the laser calorimetric technique from 100 to 350 K, in which the average values of $\lambda_\parallel$ and $\lambda_\perp$ were, respectively, found to be ~460+ and ~7.0+ W/m/K [28]. The survey shows that graphite lacks in mutual agreement in thermal parameters. Reasons are several. First and foremost, the physical chemistry of carbon compounds is quite complex and, second, crystalline microstructure, grain orientation, stacking, packing, bulk density, intrinsic void, distribution of porosity, surfaces/interfaces, degree of graphitization, etc., play a vital role in influencing the thermal performance of different classes of graphite. In our earlier studies, the thermo-physical properties of variable density graphite (1611, 1744, 1811.7, and 1910.4 kg/m³) were examined, over 300–1300 K, to choose a well-suited candidate for nozzle design. Thermal, physical, and textural investigations revealed the superior behaviour of 1811.7 kg/m³ density graphite over others [38]. Further, prototype nozzles were fabricated and subjected to the static firing test at 5000 kg/N thrust for a period of 3–5 seconds to achieve transient temperature and pressure rise, respectively, of ~3000+ K and 1–3 GPa.

The purpose of this communication is to revisit and investigate the thermo-physical properties of the nozzle that underwent firing and compare them with the before firing conditions. We addressed the fundamental questions involved in the study, such as: What happened to their thermal properties? Is there any change in the estimated entropy and enthalpy once the nozzle goes to such a high thermal as well as pressure stress? Does the nozzle effectively deform? If it deforms, what is the effective % volume deformation? What is the thermal transport mechanism? What dictates heat flow? And how the underlying phonon model works? The details are presented.

5.2.1.1 Static Firing Test of Nozzles

For nozzle fabrication and thermo-physical property analysis, graphite blocks (density = 1811.7 kg/m³) were prepared by HIP processing. The starting material was obtained from M/s Graphite India Pvt. Ltd. HIP is an industrial process in which the material of appropriate density could be fabricated by isothermal compression. In this, an appropriate amount of working substance (graphite powder, CAS registry number: 7782-42-5) was loaded into the basket followed by lowering it into the hot furnace. The zone consisted of a high-pressure, hot Ar gas injection from the top. The loaded substance was placed onto a firm platform, surrounded by heaters. In a few hours of waiting time, the high-pressure cylinder exerts an enormous pressure of 250–300 MPa including a temperature rise of ~2000 K on graphite to achieve a block of desired density [38]. A large number of blocks were prepared and cast into the nozzle to subject them to static fire test.

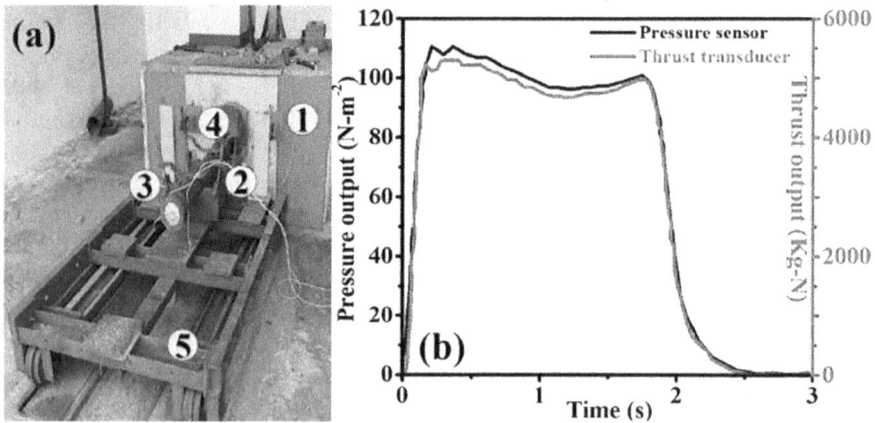

FIGURE 5.11   (a) Photograph of motor testing bay for nozzles, displaying (1) thruster block, (2) pressure sensor, (3) rocket motor, (4) load cell, and (5) rail trolley installations; (b) sensor output recorded for the generated pressure and thrust with time, around the nozzle.

A missile motor stores solid propellant, as an energy source, which works as a confined chamber for combustion. The propellant, after ignition and onset of combustion, generates hot gases that are released via the nozzle at an extremely high speed in backward direction resulting in the acceleration of the projectile in forward course by imparting thrust. The static test was simulated as follows: An assembled rocket motor was fastened, as shown in Figure 5.11a, on a frictionless trolley. Pressure sensor, load cell, and strain gauge were mounted on the motor as per the requirement. The pressure sensor used was of strain gauge type (Make: Honeywell Inc., the USA), with pressure range: 50,000–100,000 psi, ±0.50% FS BFSL, output sensitivity: 1 mV/V, 04–20 mA, 0–5 Vdc output. For thrust sensing, a load cell sensor was used (Make: Sensotech Pvt. Ltd., India), with a maximum load capacity 40 tf, sensitivity of 3 mV/V, and input/output impedance of $350\pm5/350\pm3$. Heat flux measurements were carried out using water-cooled Gardon make sensors with measurement range up to 1 GW/m^2, cooling rate ~0.014 L/s, nominal response time ~100 ms (63%@ 1 GW/m^2), output DC with range > 5 mV, and sensitivity 1 mV/V. No direct flame measurements were performed, since the combustion temperature was >3000K. The estimation of temperatures was carried out using a computer program NASA 1D based on thermo-chemistry of propellant combustion. These sensors, respectively, enabled to measure the pressure inside the rocket chamber, thrust generated by the propellant on combustion, surface temperature of motor casing, and strain generated on the motor casing during the test.

Sensor output signals were communicated through coaxial cables to the control room, approximately 100–200 m away from the test location equipped with a high-speed data acquisition unit and controllers to carry out firing. Firing of propellant was triggered by a predetermined current (~4 A). After ignition, while the onset of propellant combustion, physical parameters such as pressure, thrust, temperature, and strain were recorded and subsequently analysed with the help of computer software. The recorded pressure–time and thrust–time profiles for rocket propellant combustion test are combinedly shown in Figure 5.11b. Figure shows representative images for smaller propellant payloads. A materials summary is given in Table 5.7.

5.2.1.2   Measurement Details

Both before and after the firing test (here onwards abbreviated as BF and AF), graphite specimens were subjected to thermal measurements over 300–1300 K. For this purpose, both types of samples were transformed into (a) disc (ASTM E1461) for diffusivity to measure C_P, λ, and α and (b) cylinder (ASTM E228) for expansivity to estimate α_L using ASTM configuration [30]. The dimensions of ASTM E1461 samples were diameter: 12%±2% mm and thickness: 3%±2% mm, whereas ASTM E228 samples' dimensions were length: 50%±2% mm and diameter: 6%±2% mm. The

TABLE 5.7
Source and Purity of Materials

Chemical Name	CAS RN®	Final Mass Fraction Purity	Method	Source
Graphite	7782-42-5	≥0.997[a]	Before firing (BF)	Sigma-Aldrich; M/s Graphite India Pvt. Ltd., India
Graphite	7782-42-5	≥0.99[b]	After firing (AF)	Sigma-Aldrich; M/s Graphite India Pvt. Ltd., India

[a] In mass fraction, as stated by the supplier.
[b] In mass fraction (via analysis by X-ray diffraction and Raman spectroscopy. Details can be found in Section 5.2.3)

basic principle involved in such a measurement system is the emission of radiation energy from the light source that falls onto the sample surface and gets absorbed in the form of heat. The absorbed heat energy from the entrance surface propagates through the sample and changes thermal signature at the rear (exit) surface, which could be picked up by the infrared (IR) detector. The obtained signal from the detector was processed further through a data acquisition system by its amplification, conditioning, and denoising to obtain a thermogram. The thermogram data enable the estimation of fundamental thermal parameters such as C_P, λ, α, and α_L. In order to absorb maximum heat energy, sometimes, a graphite spray can be coated depending upon the emission characteristics of the sample.

For the measurement of C_P, α, and λ, the laser flash method, also known as the transient flash technique, was used. It is one of the most accurate methods to determine the thermal performance of a material. The method is advantageous in easily accessing high temperature range above 500 K in a non-contact sensing mode. The measurement provisions are simply integrated with swift data acquisition system. The system is able to perform absolute thermal diffusivity measurements with the requirement of smaller sample quantity when compared to other similar techniques such as guarded transient hot wire or hot plate methods. Our thermal property analyser has a modular architecture (Make: FlashLine 3000; M/s Anter Corporation, the USA). It has provisions of IR heating furnace equipped with complete temperature control ranging from −200°C to +1100°C with a ramp rate of ~ 100°C/m and accuracy in measuring temperature ±1°C. The system consists of data acquisition subsystems and a high-speed xenon discharge (HSXD) pulsed light source with peak power ~500W and pulse width 200–300 μs. The system is powered with corresponding control electronics, safety interlocks, optical pulse delivery components, along with IR optics, and a liquid nitrogen-cooled IR indium antimonide (InSb) detector to record the rear face thermogram at ambient temperatures.

For the measurement of α_L, the instrumentation and involved measurement principle in push-rod dilatometer is identical to that of the laser flash method. However, in dilatometer, the light source consists of a class 1 laser (Nd:glass laser) that produces a maximum pulse energy of ~20J with a pulse width >0.30 ms. Our dilatometer system (Model: MicroFlash, Make: NETZSCH LFA 457) is able to perform thermal expansivity measurements in the range of −125°C to 1100°C. At a particular temperature value (say 300K), measurements take, notably, ~01−02 minutes and the entire range can be covered, typically, in four to five hours of time. The technique is non-destructive and could be used for materials ranging in thermal diffusivities from 0.01 to 1000mm²/s. The dilatometer method is an absolute test technique for measuring thermal properties such as transient technique and requires no calibration for determining thermal parameters. However, one can standardize the operations by performing measurements on the system in accordance with the international/national standards such as ASTM E1461, DIN EN821, and DIN 30905. The accuracy to measure thermal diffusivity is determined to be ±3% and thermal conductivity within ±5%. Such experimental investigations have previously been reported by co-workers [11,21,26,27].

Further, to investigate the structure and property relationship in BF and AF specimens, X-ray diffractometry (XRD), Raman, and scanning electron microscopy measurements were performed.

In XRD, the beam potential was kept @40 kV with 30 mA beam current and Cu K_α wavelength, $\lambda = 1.5406$Å, with a scan rate of $0.02°$–s^{-1}@ $2\theta = 10°$–$80°$. In Raman, the excitation wavelength, λ, was 532 nm to record spectra over 200–3000 cm^{-1}. Electron microscopy was carried out at a beam potential of 30 kV for various magnifications. Typical micrographs are presented.

5.2.2 COMPARISON OF THERMAL PROPERTIES BEFORE AND AFTER FIRING

Figure 5.12a displays the estimated C_P, which was correspondingly fitted by Butland's equation: $C_P = a + bT + c/T + d/T^2 + e/T^3 + f/T^4$, where a–f are pre-exponent factors. Butland measured and fitted the C_P curve for graphite used in the reactor over 200–3500 K and has given a mathematical relation [11]. From the obtained C_P values, one can determine C_V (heat capacity at constant volume) given by: $C_P - C_V = 9 \alpha_L^2 B V T$, where B is the elastic modulus, which is ~ 18 GPa for graphite [18], and V is the estimated volume deformation in %. The behaviour of C_V is displayed in Figure 5.12b both BF and AF specimens over 300–1300 K, indicating a change in thermodynamic performance.

5.2.2.1 Estimation of Change in Entropy

Figure 5.12c shows a change in entropy, ΔS, over 300–1300 K, obtained from: $\Delta S = C_V / \dfrac{T}{\Delta T}$. Several orders of magnitude change are noted in ΔS for the measured temperature range. The accuracy to measure C_P was determined to be $\pm 3\%$. Since ΔS was derived from C_P and from C_P, the values of

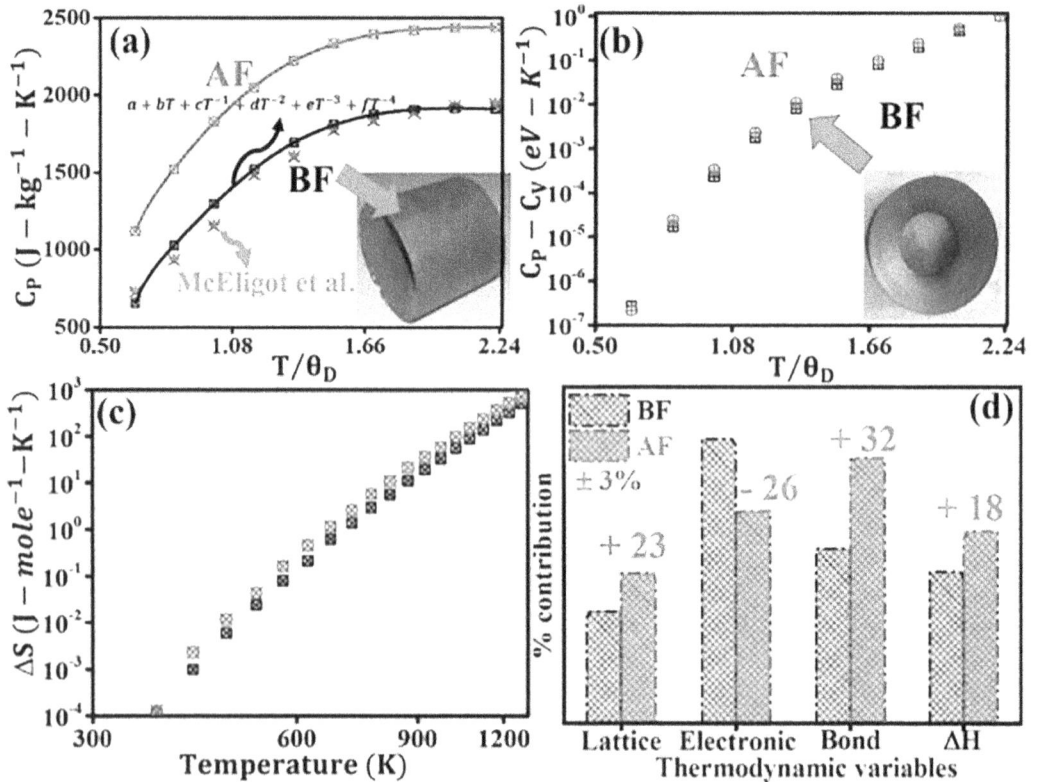

FIGURE 5.12 (a) Recorded C_P fitted for Butland's polynomial equation, including reference data from McEligot et al. [26], (b) derived $C_P - C_V$ (insets: nozzle condition), (c) change in entropy, ΔS, and (d) contributions to thermodynamic variables estimated from fittings of C_P and $C_P - C_V$ variations for BF and AF specimens over 300–1300 K. The unit conversion between C_P and $C_P - C_V$ is J/kg/K = 10.36 meV/K. The calculated Debye temperature $\theta_D = 575 \pm 0.1$ K.

C_V were obtained, so accuracy remains of the same order as that of C_P. ΔS was calculated using the following assumptions: (a) The amount of heat Q_{xy} absorbed by the system resulted in a change in the macrostate of the system from x to y, (b) the obtained macrostates were the result of a large number of combinations of unidentifiable microstates, (c) only the microstate number was somewhat quantifiable, but that had too many ambiguities, (d) for such a system, the heat equation was given by: $Q_{xy} = (\bar{E}_y - \bar{E}_x) + W_{xy}$, where W_{xy} is the macroscopic work done (total internal energy) and $E_i's$ are internal energies of x and y levels, and (e) in general, the quantification of W_{xy} provided a clue about ΔS. Experimentally, the configuration of carbon–carbon molecular vibrations, as it absorbs the heat, was going to modulate in a quantized fashion. Moreover, C–C comes in the form of sp^2, sp^3, and their inherent bond order (this is again a free and an uncontrollable variable). This could be subtly realizable once we have estimation of C_V. Thus, W_{xy}, ΔS, and C_V were on the same footy; i.e., they were in order of each other. But, W_{xy} is a canonical variable, ΔS is estimable, and C_V is derivable. Hence, we worked on obtaining C_V. Further, the temperature rise T was a definitive, measurable quantity and connected to the derivable Debye temperature, θ_D, and C_V. For solids, one can

obtain it by series expansion of the ratio of C_V to specific temperature, $\dfrac{T}{\Delta T}$ @ $T = \theta_D$. Analytically,

$$C_V \ k \int_0^{\omega_D} \frac{e^{\beta'\hbar\omega'}(\beta'\hbar\omega')^2}{(e^{\beta'\hbar\omega'} - 1)^2} \rho \ \omega^2 d\omega,$$ can however, experimentally be obtained. C_V is connected to the

coupling of phonons to lattice by a relation: $C_V / 3R = \int_0^\infty \rho(\omega') \dfrac{\beta'^2 \exp(\beta')}{\left[\exp(\beta') - 1\right]^2} \ d\omega'$ that can be

evaluated by Raman spectroscopy to extract electron–phonon coupling, γ_{e-p-c}, as presented in further analysis [35]. The exchange variable β' is the ratio of $\dfrac{\hbar\omega'}{k_B T}$, where ω' is the phonon frequency (in rad/s) associated with C–C and C=C molecules. Thus, the C_V–ΔS correlation by Loschmidt's

expansion at $T = \theta_D$ is given by: $\Delta S = 1 + C_V / \left(\dfrac{T}{\Delta T}\right)^1 + C_V / \left(\dfrac{T}{\Delta T}\right)^2 + C_V / \left(\dfrac{T}{\Delta T}\right)^3 + ...$; neglecting

the higher-order terms and setting $1 \ll C_V / \left(\dfrac{T}{\Delta T}\right)^1$, we obtained the variations in ΔS for BF and AF specimens.

5.2.2.2 Enthalpy Calculations

From the C_P–C_V plot, the suggested θ_D was 575K for graphite, which is not in agreement with the reported $\theta_D = 420$K [23]. The emergence of C_P in graphite is attributed to the degree of crystalline phase, the lattice vibrations, the internal energy stored in the microstructure, and the surface/interface energies. In Butland's equation, the higher-order terms are attributed to the contributions from lattices, whereas the intermediate- and lower-order terms are, respectively, from electronic and bond disorder. From Figure 5.12a and c, one can observe that there is a significant amount of change in the graphitic matrix after firing. In Figure 5.12d, lattice vibration and bond dissociation energy is enhanced, respectively, by 23%±3% and 32%±3%, whereas the electronic component decayed by 26%±3% with reference to the original material matrix. Thermodynamically, $C_V \approx \dfrac{(\partial S)}{(\partial H)}$, where H is the displayed enthalpy estimated over the temperature regime.

The C_P, λ, and α are correlated by: $\lambda = \alpha \cdot C_P \cdot \rho$. Both λ and α were found to be varied for graphite. Due to orthotropic nature, their values were reported to be different along the parallel and perpendicular directions of the basal (graphene layer) plane. Particularly, λ has an exponent dependence fitted from −1/2 to +3 over the range of 40–2500 K. In our case, the exponent fitting was carried out with an assumption that λ has superimposed both $\lambda_\parallel$ and $\lambda_\perp$ components. In

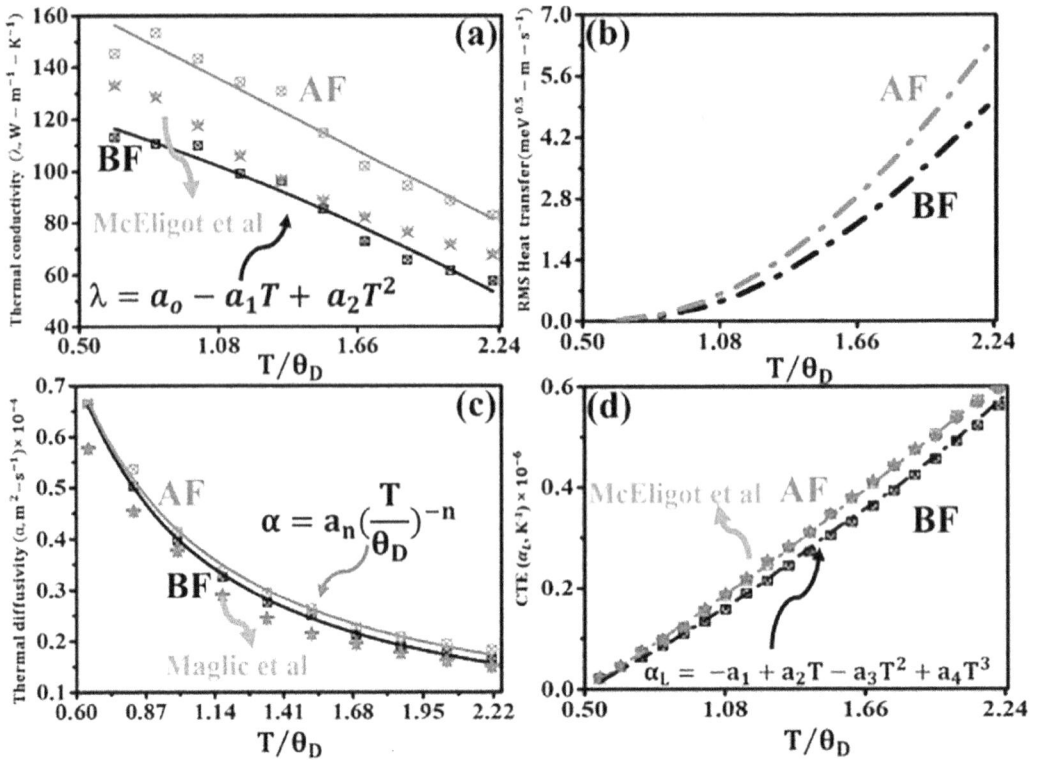

FIGURE 5.13 (a) Behaviour of conductivity, λ [15], (b) root mean square heat flux capability, (c) diffusivity, α [27], and (d) expansivity, α_L, for BF and AF specimens over the measured temperature range [26].

Figure 5.13a, the fitting for λ with T matched well with McEligot's quadratic equation of the form: $\lambda = a_o - a_1 T + a_2 T^2$, where a_i's are thermo-structural coefficients [26]. Figure 5.13b shows the derived root mean square (RMS) heat flux for both BF and AF matrices. The estimated heat flux is a function of molecular interaction potential, which led to an expression: $\langle H \rangle \approx k_B T^2 \langle C_V \rangle$, where k_B is the Boltzmann constant [25]. In Figure 5.13b, one can see that the RMS heat flux capability of AF was found to be efficient over BF specimen. In general, the solid rocket propellant combustion that takes place inside a rocket motor chamber results in gas ejection compounded with the development of high pressure and temperature gradients. The components of a propulsion system such as casing, thermal insulation, nozzles, and unburned propellant surface experience such a harsh environment during the operation of the system. Intricately, such an extreme environment depends on a number of physical factors such as the nature of propellants, thermal/heat energy imparted to different components, and ablation character of thermal insulators and graphitic nozzles. The heat energy experienced is generally expressed in terms of heat flux and quantified by root mean square (RMS) heat transfer function. Notably, heat transfer is somewhat different parameter than that of temperature. It's the rate of areal heat transfer to the direction normal to heat flux (given in $\sqrt{meV} - m/s$) that is termed as RMS heat transfer. There are a number of studies reported on heat transfer measurements: A relevant instrumentation review was done by Diller in 1993 [39]. Combustion applications and RMS heat transfer measurements were carried out on various rocket motor components by Arai et al. [40]. The effect of adiabatic cooling was investigated by Keltner [41]. Childs et al. and others carried out RMS heat transfer measurements on the motor system using different types of sensors and instrumentation [42,43]. Experimentally, the RMS heat transfer could be estimated by enthalpy, H, which is a function of molecular interaction potential expressed as:

$H\left(D_o + x\right) = H_{D_o} + (\partial H / \partial D)_{D_o} + \dfrac{1}{2!} \cdot (\partial^2 H / \partial D^2)_{D_o} \cdot x^2 + \dfrac{1}{3!} \cdot (\partial^3 H / \partial D^3)_{D_o} \cdot x^3 + \ldots$, where D_o is the equilibrium position of the carbon atoms and x is the thermally assisted perturbation [29]. It leads to an expression: $H \approx k_B T^2 C_V$, where k_B is the Boltzmann constant. The higher heat transfer capability of AF is indicative of efficient heat flow.

In general, λ of a material is the rate of change of steady heat flow passing through unit length of an infinite, isotropic, homogeneous, continuous medium and is a directional quantity. It flows in an orthogonal direction to the medium surface, induced by the uniform temperature difference. The property is referred with the specific mean temperature, due to its temperature dependence. Meanwhile, thermal diffusivity, α, is the property that combinedly arose due to conductivity, heat capacity, and density. Therefore, it is the manifestation of thermal conductivity, λ. Next, we observed the α characteristics of BF and AF specimens. By and large, the behaviour of α, resembles the recorded λ features. The diffusivity varies as x^{-n} over the measured temperature regime and was found to be consistent with the report of Maglic et al. on the graphite material [27]. Table 5.8 shows the actual data.

The coefficient of thermal expansion (CTE), α_L, plays an important role in estimating internal stresses in the graphitic matrix. Particularly, from the viewpoint of rocket nozzle, combustion and gas exhaustion results in a huge rise in temperature and gradient of heat flux within the nozzle, radially as well as axially. The nozzle may get gasified due to oxidation during its mission period. The thermal expansion of a medium is expressed in terms of α, the instant thermal expansion of a material can be expressed using α, and the instantaneous linear CTE is given by an expression: $(\partial L_0)/L_0 \, 1/\partial T = \alpha_L$, where $(\partial L_0)/L_0$ is the % length deformation at temperature T with reference to some reference temperature, T_0. Figure 5.13d displays α_L variations recorded for BF and AF specimens over 300–1300 K. Importantly, α_L for AF and BF specimens was imaged as the collection of several single-crystal graphites oriented isotropically, resulting in polygranular graphite in which α_L leads to the access of arbitrary direction and is related to the structure factor. Such a factor is easily quantifiable by performing physical characterization on BF/AF specimen using techniques such as X-ray diffractometry, Raman spectroscopy, and electron microscopy. Figure 5.14a shows the recorded XRD spectra for BF and AF specimens (inset: full 2θ scan range). Figure 5.14b is a comparison of the extracted structural parameters from d-200 reflux for BF and AF specimens.

5.2.3 COMPARISON OF PHYSICAL PARAMETERS

Using XRD d_{002} reflex, the following parameters were determined for BF and AF specimens [35,36]: The interlayer distance d_{002} for the specimens was calculated by Bragg's law: $d_{002} = \lambda_x / 2 \sin\theta_{002}$, where λ_x is the X-ray wavelength (1.5406Å) and θ_{002}, the diffraction angle. The dimensions of structural components such as the thickness of graphitic layer, L_c, is given by: $L_C = 0.9 \, \lambda_x / \beta_{002} \cos\theta_{002}$, where β_{002} is the full width at half maximum (FWHM) for (002) peak. The in-plane crystalline extent, L_a, was computed using: $L_a = 0.9 \, \lambda_x / \beta_{100} \cos\theta_{100}$, where β_{100} is FWHM of (100) peak. The stacking density was calculated: $\rho_s = 0.762 / d_{002}$, and the number of layers in stacks was estimated by: $N = (L_C / d_{002}) + 1$. Crystallographically, the bonding distance between C–C, a_o, in the honeycomb sheet is $a/\sqrt{3}$. For an ideal single crystal at 300 K, $c = 6.708$ Å, $a = 2.4614$ Å, $a_o = 1.4211$ Å, and $c/a = 2.725$. The degree of graphitization, $\bar{g}$, was determined by analysing d_{002}, where $\bar{g}$ is 1 for $d_{002} = 3.354$ Å (crystalline) and 0 for 3.44 Å (exfoliation/amorphization). The governing equation used was the following: $\bar{g} = \dfrac{3.440 - d_{002}}{3.440 - 3.354}$. Moreover, α_L near room temperature was estimated, using d_{002} line, by an equation: $\alpha_L = \left(d_{002}(300K) - d_{002}(0K)/T(300K) - T(0K)\right)$, where the value of $d_{002}(0K)$ is estimated to be 3.44 Å. Experimental errors in determining d_{002} and α_L magnitudes were, respectively, 0.001 nm and 0.01×10^{-5} K. In general, for computations of structure parameters such as d_{002}, L_c, L_a, ρ_s, N, and $\bar{g}$, one can use (002) peak or even higher-order peaks such as (004) and (006); however, we have chosen (002) to carry out the analysis.

TABLE 5.8

Experimental Specific Heat, C_P, Thermal Conductivity, λ, Thermal Diffusivity, α, and Coefficient of Thermal Expansion, α_L, Respectively, Shown in Figures 5.12a, 5.13a, c, and d Over 300 ± 0.05 to 1300 ± 0.05 K with Calculated Debye Temperature, θ_D, 575 ± 0.05 K, and Pressure, p, 101.325 ± 0.00625 kPa (Standard Absolute Transducer)

$\dfrac{T}{\theta_D}$ [a]	[b]Specific Heat, C_P, (J/kg/K)		[b]Thermal Conductivity, λ, (W/m/K)		[b]Thermal Diffusivity, α, (m²/s) (×10⁻⁴)		[c]Coefficient of Thermal Expansion α_L, K⁻¹ (×10⁻⁶)	
	BF	AF	BF	AF	BF	AF	BF	AF
0.56	–	–	–	–	–	–	0.02118	0.01937
0.64	0652.45	1115.73	113.05	145.28	0.6648	0.6645	0.04322	0.04397
0.73	–	–	–	–	–	–	0.06278	0.07439
0.82	1026.22	1521.96	110.62	153.36	0.5026	0.5361	0.08594	0.09865
0.90	–	–	–	–	–	–	0.10953	0.12312
0.99	1295.08	1825.90	109.93	143.37	0.3966	0.4138	0.13388	0.15429
1.08	–	–	–	–	–	–	0.15776	0.18367
1.17	1519.12	2047.21	099.19	134.42	0.3254	0.3367	0.18905	0.21429
1.25	–	–	–	–	–	–	0.21526	0.24495
1.34	1690.71	2222.29	096.13	130.81	0.275	0.2943	0.24461	0.27949
1.43	–	–	–	–	–	–	0.27153	0.30956
1.51	1808.74	2334.42	085.34	114.82	0.2494	0.2629	0.30403	0.34669
1.60	–	–	–	–	–	–	0.33168	0.37736
1.69	1873.22	2395.40	072.94	102.02	0.2122	0.2261	0.36303	0.40634
1.77	–	–	–	–	–	–	0.39369	0.44118
1.86	1906.01	2420.98	065.70	094.30	0.1901	0.2077	0.42457	0.47362
1.95	–	–	–	–	–	–	0.45563	0.50613
2.04	1912.56	2437.70	061.81	088.62	0.1776	0.1943	0.49115	0.54224
2.12	–	–	–	–	–	–	0.52247	0.57419
2.21	1910.38	2439.67	057.74	083.05	0.1651	0.1833	0.56141	0.59995

[a] Standard uncertainties u are u(T)=0.05 K and u(p)=0.00625 kPa (level of confidence=0.68).

[b] For thermal diffusivity and, consequently, heat capacity as well as thermal conductivity, the expanded uncertainty, U, is of 95% confidence level, i.e. ±4% over 300–1300 K.

[c] For coefficient of thermal expansion, the expanded uncertainty, U, is of 95% confidence level, i.e. ±2.25%.

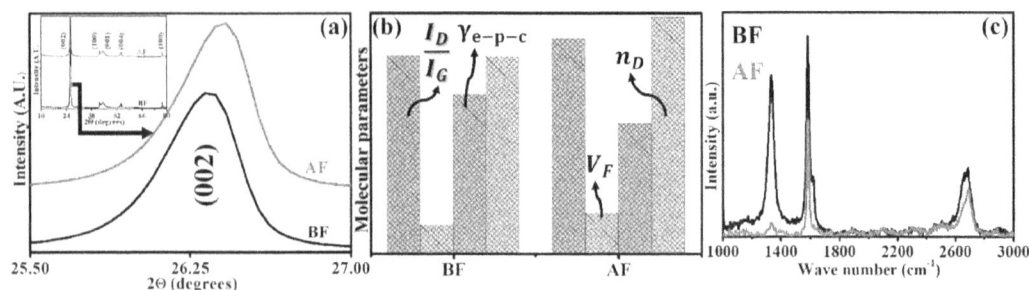

FIGURE 5.14 (a) Recorded XRD pattern and corresponding (002) peak, (b) variations in derived molecular parameters such as I_D/I_G, electron–phonon coupling, γ_{e-p-c}, Fermi velocity, V_F, disorder density, n_D from Raman compared for BF and AF specimens, and (c) recorded Raman spectra at 532 nm excitation displayed over 1000–3000 cm⁻¹.

The change in bond molecular environment before and after firing was investigated using Raman spectroscopy. The profiles are shown in Figure 5.14c, which displays a number of features recorded, such as D, G, and 2D peaks, recorded at 1320–1350, 1540–1590, and 2640–2670 cm^{-1}. The prominent presence of amorphous carbon was also noted below 850 cm^{-1} regimes. In general, the Raman phenomenon is based on resonant spectroscopy operative at molecular level and provides rich information on crystalline length, L_c, amorphous length, L_a, sp^3, sp^2 content, etc. Since in-plane and out-of-plane crystalline parameters have already been examined and presented using XRD analysis, we did not perform this exercise. There are interesting extractions associated with electron–phonon coupling, γ_{e-p-c}, which in turn depends on a number of intricate parameters such as natural broadening of G-peak; Fermi velocity, V_F (m/s), which is given by: $V_F^2 = \dfrac{\sqrt{3}}{2} \cdot \dfrac{a_o^2}{\mu} \dfrac{\Gamma_{e-p-c}^2}{\gamma_{e-p-c}}$, where μ is the reduced mass of C–C and Γ_{e-p-c}^2 is the theoretical Γ-phonon coupling factor (50 (eV-A$^{0-1}$)^2); and defect density, n_D, given by: $n_D \left(m^{-2} \right) = 10^{18}/L_D^2$, where L_D is the disordered length estimated from integrated I_D/I_G ratio measured and given as: $L_D^2 = C(\lambda_L)/\left(\dfrac{I_D}{I_G} \right)$, where $C(\lambda_L)$ is the laser beam parameter (102 nm^2). The related details are provided in [31,35,38]. Table 5.9 displays the structure factors estimated for BF and AF specimens. It is noted that there is a significant difference observed between the values of the α_L recorded @ 300 K by the two techniques, i.e. thermal and XRD, as seen in Table 5.3 (~0.02 × 10^{-6}/K) and Table 4 (~0.07 × 10^{-6}/K; ~0.11 × 10^{-6}/K).

Morphological investigations were carried out on BF and AF specimens using SEM at a suitable working distance of 5–6 mm. The texture of graphite, before and after firing, was clearly and legitimately recorded at magnifications ranging from 3000 to 30,000. Micrographs displayed distinct features of graphite that include but are not limited to amorphous carbon deposit chunks, micro- and macro-textures, exfoliated and eroded surfaces, warps, planes, flakes, and stakes. The morphology at the flake edges was examined to get a rough idea about the change in thickness by the onset of erosion during combustion. Results, mostly, matched well with the estimation of L_c obtained through XRD analysis, as displayed in Table 5.9.

Further, static testing was carried out for two different types of propellants, namely composite and double base propellants. These propellants have different combustion properties in terms of specific ignition, generated pressure, backward thrust, and temperature. Their combustion effect on the surface morphology was also examined. Typical micrographs are displayed in Figure 5.15, in which the first row (i.e. (a) and (b)) belongs to the nozzle surface before firing, whereas the second ((c) and (d)) and third ((e) and (f)) rows are nozzles that underwent firing using composite and double base propellants, respectively.

The texture of the graphite surface before firing was observed to be entirely different as seen in images (a) and (b) in Figure 5.15. After cryo-breaking of the surface, due to the extremely pure form of graphite, one can see coagulation of the graphene layers. On the sample surface, a large amount of amorphous carbon chunks was seen; they might have originated during cryo-breaking or due to the hipping and densification process while fabrication. For the recorded images (c) and (d), we observed quite different types of surface texture. These are the nozzle surfaces that experienced firing by composite propellants. The pristine fluffiness of the flakes seems to be disappeared, dramatically, due to firing and the thermal/pressure shock nozzle surface experienced during propulsion. The layers were larger, not broken, and straightened with less wrinkles and warps. Perhaps, this is seen more prominently in micrographs (e) and (f) recorded for double base propellants. The layers were observed to be more expanded and became large; the degree of warpiness (though not quantified) was seen to be still reduced further. Not all images are provided herein due to constraints on the content; only representative images have been presented.

Broadly, our analysis revealed that after firing, the thermal as well as physical properties measured were found to be improved for graphite in all respect over the temperature regime. It seems that, firing, equivalently, acted like an annealing treatment to improve the overall thermal

TABLE 5.9

Calculated Structural Parameters for BF and AF Specimens by XRD and Raman Spectroscopy[a]

XRD

Type	d_{002} (Å)	L_c (Å)	L_a (Å)	ρ_s (kg/m³)	N	$\bar{g}$ %	$\alpha_L \times 10^{-6}$ (K^{-1}) @300K
Parameter	Lattice distance	Crystalline length	In-plane amorphization	Stacking density	Number of stacking layers	Degree of graphitization	CTE
BF	4.620 (2%)	230±10	205±10	1810 (5%)	125±5	27 (3%)	0.071±0.03
AF	5.017 (2%)	301±10	144±10	2014 (5%)	259±5	55 (3%)	0.1137±0.03

Raman

Type	I_D/I_G (%) (Integrated)	γ_{e-p-c} (cm⁻¹)	L_D (nm²)	$n_D \times 10^{18}$ (m⁻²)	$V_F \times 10^6$ (m/s)	k_G (–C=C–) k_D (–C–C–) (N/m)
Parameter	Graphitic content ratio	Electron–phonon coupling	Disordered length	Disordered density	Fermi velocity	C–C molecular force constant
BF	55 (3%–5%)	25±4	185±10	0.0540	~1.8	678±53 1101±23
AF	78 (3%–5%)	14±4	257±10	1.2010	~2.3	814±45 1231±56

[a] Standard uncertainties u are u(T) = 0.05 K and u(p) 0.00625 kPa (level of confidence = 0.68), and the combined expanded uncertainty U_c (ρ) = 0.1 kg/m³ with 0.95 level of confidence (k≈2).

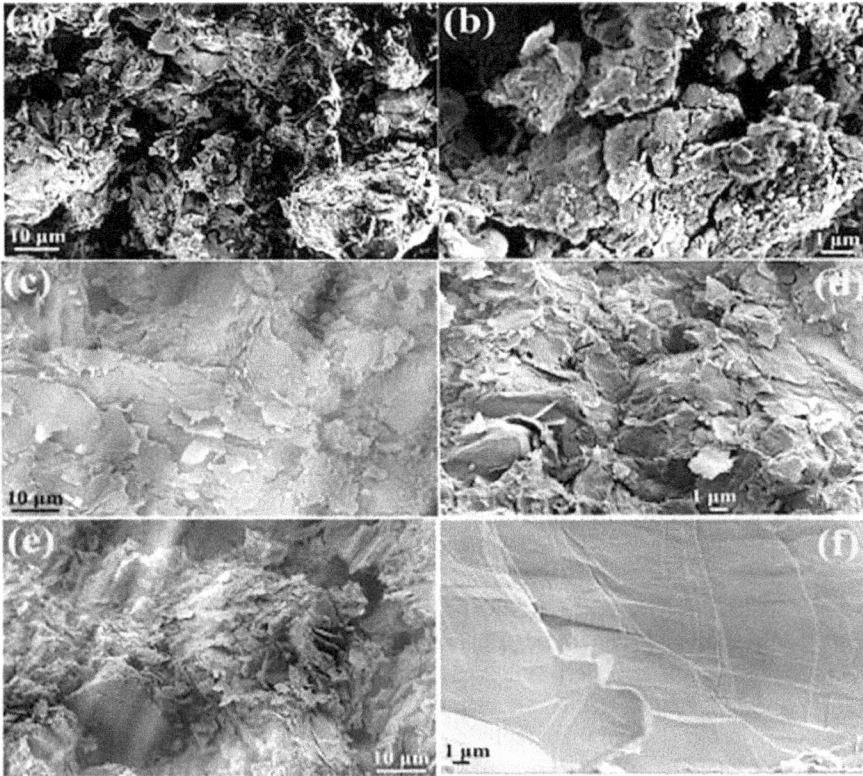

FIGURE 5.15   Typical scanning electron micrographs recorded for (a), (b) pristine nozzles, before firing; (c), (d) firing using composite propellants; and (e), (f) firing using double base propellants. Scale bar: left images: 10 μm; right: 1 μm.

performance of graphite. Efficient thermal properties mean favourable temperature flow and minimal thermal gradient. But the question is: Why such a behaviour emerges in AF specimens? As discussed earlier, as light energy gets absorbed by both types of graphite, heat energy propagates from entrance and exits from the rear surface and is recorded as a thermogram, which is analysed extensively. But, fundamentally, the process of thermal energy transfer is a random process. The energy does not simply enter one end of the specimen and proceed directly in a straight path to the other end. The heat energy diffuses through the specimen and suffers frequent collisions. Such an energy propagation through a specimen with deflection of heat sets a thermal flux, $\vec{J}$, which necessarily depends upon the temperature gradient, $\vec{\nabla} T$. Thus, the random nature of thermal conductivity process brings the temperature gradient, $\vec{\nabla} T$, and mean free path, l, of heat propagation that result in the temperature difference ΔT across the length of the specimen. So basically, it's the competition between $\vec{J}$ and $\vec{\nabla} T$ or $\dfrac{dT}{dx}$ $(for\ 1D)$. But graphite is made up of carbon molecules. The amount of heat absorbed by the specimen will, locally, be absorbed by the C–C molecule, which will transfer it to its nearest neighbour. However, this process depends on the heat capacity of C–C (at constant volume), i.e. $C_V \Delta T$. If l is the length of C–C molecule, which is of the order of phonon mean free path, l, then $C_V \Delta T \approx \dfrac{dT}{dx} \cdot l\ \dfrac{dT}{dx} \cdot v \cdot \tau$, where τ is the average time between the collisions and v represents the velocity of sound $\sim 5 \times 10^3$ m/s. This led to an expression for phonon mean free path (l): $\dfrac{1}{\rho} \cdot \dfrac{\lambda}{C_V} \cdot \dfrac{1}{v} = l$ (in nm). Thus, phonon mean free path, l, dictates the free flow of heat within the graphite. The path, l, becomes limiting when phonon encounters anharmonic lattice vibrations and

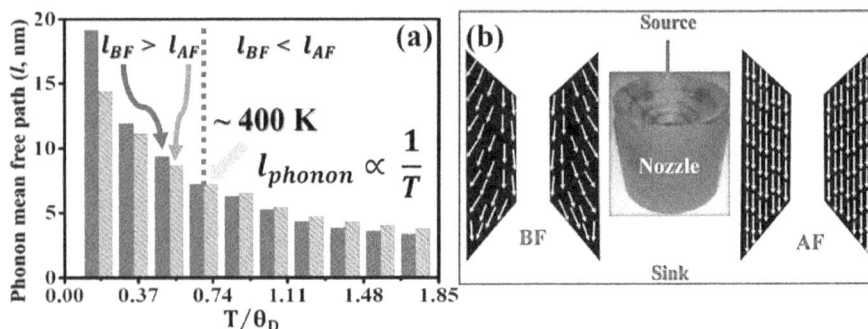

FIGURE 5.16 (a) Calculated variations in phonon mean free path, l, with $\dfrac{T}{\theta_D}$ and crossover temperature ~400K; (b) phonon flow model indicating $l \propto T^{-1}$ behaviour in BF and AF nozzles.

gets scattered, whereas geometrical scattering of phonon takes place due to interaction with microstructure boundaries and lattice imperfections.

From Figure 5.16, one can see that, below the crossover temperature, l_{BF} is higher than l_{AF}, which shows that up to 400 K, lattice imperfections and scattering of phonon from microstructure is smaller in BF as compared to AF. However, at 400 K, both are comparable and thereafter phonon flow becomes, markedly, efficient in case of AF. During firing, the combustion of propellants generates hot (2000–4000 K) and pressurized (01–03 GPa) gases resulting in enormous stresses and pyrolysation of nozzles. This has implication on the microstructural properties of graphite, estimated using XRD and Raman techniques. It showed that graphitic content was improved after firing, with a reduction in in-plane amorphization by 30%. The overall degree of graphitization was observed to be increased with a simultaneous increase in crystalline length by 5% as well as an increase in stacking density by 3%–5%. The observed reduction in electron–phonon coupling by 15% or so and increase in Fermi velocity indicated a reduction in the scattering of phonon for about 10% by the electronic ensemble in graphite. This in turn shows that the resistivity to heat flow offered by the lattice and electrons is somewhat reduced after firing test, particularly above the crossover temperature ~400 K. Nelson and Riply et al. suggested that, below 400 K, a-axis (in-plane) contraction dominates over c-axis (out-of-plane) expansion. For pristine graphite, this was fairly prominent; however, over 400 K, c-axis expansion seems to be dominant for post-fired graphite [20,37–43].

By and large, our study gain insights into the high-temperature thermodynamics in a rocket propulsion nozzle. The study is important to underline the high-temperature performance of the component for its quality assessment and assurance in order to qualify the test for accomplishment of the mission [44–48].

REFERENCES

[1] Lide, David R. *Handbook of chemistry and physics*, 86th edition, pp. 8–21. CRC Press, Boca Raton, FL, 2005.

[2] Pavlasek, P., C. J. Elliott, J. V. Pearce, S. Duris, R. Palencar, M. Koval, and G. Machin. "Hysteresis effects and strain-induced homogeneity effects in base metal thermocouples." *International Journal of Thermophysics* 36, no. 2 (2015): 467–481.

[3] Bochvar, D. A., and E. G. Galpern. "Hypothetical systems-carbododecahedron, s-icosahedrone and carbo-s-icosahedron." *Doklady Akademii Nauk Sssr* 209, no. 3 (1973): 610–612.

[4] Stankevich, I. V., Nikerov, M. V. E., and Bochvar, D. A. E. "The structural chemistry of crystalline carbon: geometry, stability, and electronic spectrum." *Russian Chemical Reviews* 53, no. 7 (1984): 640.

[5] Zhang, Qing-Ling, Sean C. O'Brien, James R. Heath, Yichang Liu, Robert F. Curl, Harold W. Kroto, and Richard E. Smalley. "Reactivity of large carbon clusters: spheroidal carbon shells and their possible relevance to the formation and morphology of soot." *The Journal of Physical Chemistry* 90, no. 4 (1986): 525–528.

[6] Iijima, Sumio. "Helical microtubules of graphitic carbon." *Nature* 354, no. 6348 (1991): 56–58.

[7] Chung, D. D. L. "Review graphite." *Journal of Materials Science* 37, no. 8 (2002): 1475–1489.

[8] Morgan, W. C. "Thermal expansion coefficients of graphite crystals." *Carbon* 10, no. 1 (1972): 73–79.

[9] Null, M. R., W. W. Lozier, and A. W. Moore. "Thermal diffusivity and thermal conductivity of pyrolytic graphite from 300 to 2700 K." *Carbon* 11, no. 2 (1973): 81–87.

[10] Taylor, R. E. *Specific heat of carbon/carbon composites.* Purdue University Research, Thermophysical Properties Research Laboratory, Inc., West Lafayette, IN 1980.

[11] Butland, A. T. D., and R. J. Maddison. "The specific heat of graphite: an evaluation of measurements." *Journal of Nuclear Materials* 49, no. 1 (1973): 45–56.

[12] Neumann, W., and K. Wallisch. "Determination of the thermal conductivity of graphite and high-temperature alloys by the laser-flash method." *Measurement* 1, no. 4 (1983): 204–208.

[13] Pavlov, T. R., M. Lestak, M. R. Wenman, L. Vlahovic, D. Robba, A. Cambriani, D. Staicu et al. "Examining the thermal properties of unirradiated nuclear grade graphite between 750 and 2500 K." *Journal of Nuclear Materials* 538 (2020): 152176.

[14] Kellett, E. A., and B. P. Richards. "The thermal expansion of graphite within the layer planes." *Journal of Nuclear Materials* 12, no. 2 (1964): 184–192.

[15] Kelly, B. T. "The thermal expansion coefficient of graphite parallel to the basal planes." *Carbon* 10, no. 4 (1972): 429–433.

[16] Kelly, B. T. "The thermal expansion coefficients of graphite crystals at low temperatures." *Carbon* 28, no. 1 (1990): 255–257.

[17] Kelly, B. T. "The thermal expansion coefficients of graphite crystals—the theoretical model and comparison with 1990 data." *Carbon* 29, no. 6 (1991): 721–724.

[18] Martin, W. H., and Miss F. Entwisle. "Thermal expansion of graphite over different temperature ranges." *Journal of Nuclear Materials* 10, no. 1 (1963): 1–7.

[19] Megaw, Helen D. "Crystal structures and thermal expansion." *Materials Research Bulletin* 6, no. 10 (1971): 1007–1018.

[20] Nelson, J. B., and D. P. Riley. "The thermal expansion of graphite from 15 c. to 800 c.: part I. Experimental." *Proceedings of the Physical Society (1926–1948)* 57, no. 6 (1945): 477.

[21] Seehra, Mohindar S., and Arthur S. Pavlovic. "X-Ray diffraction, thermal expansion, electrical conductivity, and optical microscopy studies of coal-based graphites." *Carbon* 31, no. 4 (1993): 557–564.

[22] Hacker, Paul J., Gareth B. Neighbour, and Brian McEnaney. "The coefficient of thermal expansion of nuclear graphite with increasing thermal oxidation." *Journal of Physics D: Applied Physics* 33, no. 8 (2000): 991.

[23] Flubacher, P., A. J. Leadbetter, and J. A. Morrison. "The limiting low-temperature behavior of the heat capacity of graphite." *Journal of Physics and Chemistry of Solids* 13 (1960): 43–53.

[24] Pavlov, T., L. Vlahovic, D. Staicu, R. J. M. Konings, M. R. Wenman, P. Van Uffelen, and R. W. Grimes. "A new numerical method and modified apparatus for the simultaneous evaluation of thermo-physical properties above 1500 K: a case study on isostatically pressed graphite." *Thermochimica Acta* 652 (2017): 39–52.

[25] Konings, Rudy, and Roger Stoller. *Comprehensive nuclear materials.* Elsevier, 2020.

[26] McEligot, Donald, W. David Swank, David L. Cottle, and Francisco I. Valentin. *Thermal properties of G-348 graphite.* No. INL/EXT-16–38241. Idaho National Lab., Idaho Falls, ID, 2016.

[27] Maglić, K. D., and Nenad D. Milošević. "Thermal diffusivity measurements of thermographite." *International Journal of Thermophysics* 25, no. 1 (2004): 237–247.

[28] Nagano, H., A. Ohnishi, and Yuji Nagasaka. "Thermophysical properties of high-thermal-conductivity graphite sheets for spacecraft thermal design." *Journal of Thermophysics and Heat Transfer* 15, no. 3 (2001): 347–353.

[29] Zhang, W., A. Li, B. Reznik, and O. Deutschmann. "Thermal expansion of pyrolytic carbon with various textures." *ZAMM-Journal of Applied Mathematics and Mechanics/Zeitschrift für Angewandte Mathematik und Mechanik* 93, no. 5 (2013): 338–345.

[30] ASTM E1461-13. (2013). Standard test method for thermal diffusivity by the flash method.

[31] Alegaonkar, Ashwini, Prashant Alegaonkar, and Satish Pardeshi. "Exploring molecular and spin interactions of Tellurium adatom in reduced graphene oxide." *Materials Chemistry and Physics* 195 (2017): 82–87.

[32] Tsang, D. K. L., B. J. Marsden, S. L. Fok, and G. Hall. "Graphite thermal expansion relationship for different temperature ranges." *Carbon* 43, no. 14 (2005): 2902–2906.

[33] Popova, A. N. "Crystallographic analysis of graphite by X-ray diffraction." *Coke and Chemistry* 60, no. 9 (2017): 361–365.

[34] Khokhlova, G. P., Ch N. Barnakov, A. N. Popova, and L. M. Khitsova. "Influence of carbon additives on the thermal transformation of coal pitch." *Coke and Chemistry* 58, no. 7 (2015): 268–274.

[35] Ferrari, Andrea C. "Raman spectroscopy of graphene and graphite: disorder, electron–phonon coupling, doping and nonadiabatic effects." *Solid State Communications* 143, no. 1–2 (2007): 47–57.

[36] Fu, Li, Samy Merabia, and Laurent Joly. "Understanding fast and robust thermo-osmotic flows through carbon nanotube membranes: thermodynamics meets hydrodynamics." *The Journal of Physical Chemistry Letters* 9, no. 8 (2018): 2086–2092.

[37] Riley, D. P. "The thermal expansion of graphite: part II. Theoretical." *Proceedings of the Physical Society (1926–1948)* 57, no. 6 (1945): 486.

[38] Kalal, Rakesh Kumar, Himanshu Shekhar, Prashant Alegaonkar, Rekha Sangtyani, and Arvind Kumar. "Thermo-Physical Properties and Combustion Wave of Nitramine Based Composite Propellant Compositions." In *Proceedings of the International Conference on Atomic, Molecular, Optical & Nano Physics with Applications*, pp. 371–380. Springer, Singapore, 2022.

[39] Diller, Tom E. "Advances in heat flux measurements." In *Advances in heat transfer*, vol. 23, pp. 279–368. Elsevier, UK 1993.

[40] Arai, Norio, Aritalca Matsunami, and Stuart W. Churchill. "A review of measurements of heat flux density applicable to the field of combustion." *Experimental Thermal and Fluid Science* 12, no. 4 (1996): 452–460.

[41] Gromov, Gennadi. "Thermoelectric modules as efficient heat flux sensors." In *Thermoelectric energy conversion: basic concepts and device applications*, pp. 233–282, Wiley-VCH, Germany 2017.

[42] Childs, P. R. N., J. R. Greenwood, and C. A. Long. "Heat flux measurement techniques." *Proceedings of the Institution of Mechanical Engineers, Part C: Journal of Mechanical Engineering Science* 213, no. 7 (1999): 655–677.

[43] Han, Je-Chin, Sandip Dutta, and Srinath Ekkad. *Gas turbine heat transfer and cooling technology.* CRC Press, Boca Raton, FL, 2012.

[44] Kalal, Rakesh Kumar, Shamal Chinke, Balesh Ropia, Himanshu Shekhar, and Prashant Alegaonkar. "Effect of rocket propulsion exhaust on thermophysical properties of graphite nozzle." In *AIP Conference Proceedings*, vol. 2115, no. 1, p. 030441. AIP Publishing LLC, 2019.

[45] Kalal, Rakesh Kumar, Balesh Ropia, Himanshu Shekhar, and Prashant Sudhir Alegaonkar. "Studies on heat flux imparted on thermal insulation inside rocket motor containing double base propellant." *Journal of Aerospace Technology and Management* 11 (2019): 1434–1441.

[46] Kalal, Rakesh Kumar, Balesh Ropia, Prashant S. Alegaonkar. "High temperature thermodynamics in rocket motor nozzles" *International Journal of Nanomaterial Molecular Nanotechnology* 4, no. 1 (2022): 123–130.

[47] Kalal, Rakesh Kumar, Balesh Ropia, Himanshu Shekher and Prashant S. Alegaonkar. "Thermo-physical property assessment of variable density graphite: improving jet propulsion nozzle designs for aerospace applications" *International Journal of Nanomaterial Molecular Nanotechnology* 4, no. 1 (2022): 136–147.

[48] Kalal, Rakesh Kumar, Himanshu Shekhar, Prashant Sudhir Alegaonkar, and Shrikant Pande. "Propellant combustion wave studies by embedded thermocouple and imaging method at ambient pressure." *Journal of Aerospace Technology and Management* 12 (2020): 1720–1727.

6 Electrochemistry and Energy Storage Devices Made Up of Carbon Nanoparticles

6.1 HIGH-PERFORMANCE TELLURIUM–REDUCED GRAPHENE OXIDE PSEUDO-CAPACITOR

In this section, we present the work on tellurium–reduced graphene oxide (Te–rGO) supercapacitors. A simple protocol was implemented, in which tellurium (0.5–11 w/w%) was incorporated, in situ, while reduction of GO and evaluated for its electrochemical performance. The electrode material was developed in a single step without any pre- or post-treatment and tested in an aqueous electrolyte. The structure–property relationship of rGO and Te–rGO was evaluated using Raman, infrared, X-ray diffractometry, and electron spectroscopy for chemical analysis, while surface compositional studies were done by energy-dispersive X-ray imaging, and scanning and transmission electron microscopies. In electrochemical studies, cyclic voltammetry and galvanostatic charge–discharge measurements were carried out at three- and two-electrode configurations, as a function of scan rate and current density, respectively, in addition to the electrochemical impedance spectroscopy. Analysis revealed that 1%Te–rGO showed *pseudo-capacitive* performance due to *chemistry* of Te-O and rGO layers to have superior cell characteristics over high % Te content and rGO base. Proof of concept is demonstrated [1].

6.1.1 Demand and Thrust for Energy: The Global Scenario

The annual world energy consumption demand is around 150+ TWh ($\sim 10^{22}$ J) and increasing every year by 10%–15%. To date, there have been several sources of energy generation, such as fossil fuels (67%), hydro (16%), nuclear (11%), and renewable (6%), to satisfy the need by nearly 100 TWh/y, but we still have deficiency in the supply by ~35%. Moreover, they have implication on their production in terms of effectiveness, efficiency, sustainability, and ecology of the globe [2]. To cope with the rapidly increasing demand, the development of alternative high-power sources, energy storage devices (ESD), and delivery systems is of significant particular interest. Traditional ESD provide available energy for constant active power request, which strongly depends on the type of storage such as batteries, flywheels, superconducting magnetic energy storage devices, pressure storage devices, and capacitors. They are particularly characterized by the energy, E_D (in Wh/kg), and the power density, P_D (W/kg), being available for a variable load. There are two types of storage: capacitive and inductive. In the former, E_D decreases, and in the latter, it increases as a function of P_D. Typically, batteries have high E_D (277.8), but only low P_D (~10), while capacitors have rather high P_D (10^6), but low E_D (2.7×10^{-3}). That makes battery to provide long-term energy supply, while capacitors are of instantaneous power discharge. The charge–discharge character of ESD, fundamentally, depends on the energy-to-power ratio, which ideally has to vary linearly to effectively bridge the gap between high specific energy batteries and high specific power capacitors.

Several materials such as transition metal oxides (RuO_2, Fe_3O_4, NiO, and MnO_2) [3–6], conducting polymers (polyaniline, polypyrrole, and polythiophene) [5–8], and different forms of carbon have been reported for active electrodes. These electrodes have been used with various aqueous and non-aqueous molar electrolytes such as HCl, H_2SO_4, KOH, tetraethylammonium tetrafluoroborate ($TEABF_4$), lithium chlorate ($LiClO_4$), 1-butyl-3-methylimidazolium tetrafluoroborate ([BMIm]

DOI: 10.1201/9781003317258-6

BF$_4$), and 1-ethyl-3-methylimidazolium tetrafluoroborate ([EMIm]BF$_4$). We would be discussing, mainly, the literature of electrochemical parameters of carbon-based combinations.

The activated carbon showed a specific capacitance, C_{sp}, of 200–320 F/g in 2 M H$_2$SO$_4$ with stability up to 2500 cycles @ current density 0.16 A/g. The surfactant-treated mesoporous carbon with the same electrolyte, after calcination at 600°C, showed $C_{sp} \sim 200$, E_D, ~ 8, and P_D, ~ 0.3. The para-toluene sulphuric acid-treated mesoporous carbon/SiO$_2$ templates in KOH achieved 200–220, 3–6, and 1–3, respectively [9–11]. In another study, porous carbon/PANI in TEABF$_4$ was used to achieve C_{sp}, E_D, and P_D, respectively, 160, 30, and 10. The post-treated hierarchical porous carbon in KOH and TEABF$_4$ showed E_D and P_D of 10 and 24, and 70 and 17, respectively [12,13].

Single-walled carbon nanotubes, obtained by the floating chemical vapour deposition technique, showed high electrochemical cell parameters for TEABF$_4$ (80, 69, and 43.3) than for LiClO$_4$ (10, 43, and 197) [14].

Reduced graphene oxide (rGO) in KOH achieved $C_{sp} \sim 200$, $E_D \sim 10$, and $P_D \sim 28$, while rGO treated by microwave exfoliation followed by calcination in 1M ([BMIm]BF$_4$) attained respectively, 166, 20, and 75, having stability up to 1000 cycles and in 1M [EMIm]BF$_4$ attained, respectively, 273, 151, and 776.8 with stability as high as 10,000 cycles @ 100 A/g [15,16]. The rGO/PANI/H$_2$SO$_4$ combination yielded parameters $C_{sp} \sim 210$, $E_D \sim 20$, and $P_D \sim 10$ with stability up to 800 cycles, and rGO/RuO$_2$, 570, 20, and 10, respectively, with 1000 cycles, both @3 A/g [17,18]. The performance of the above-mentioned supercells was between EDLC and SC.

By and large, the literature analysis prominently reveal two facts: first, extensive pre- or post-treatment on active carbon electrode in terms of heating, chemical functionalization, and pressure employment, and second, the use of non-aqueous/organic electrolyte. Although the main advantage of organic electrolytes is their wide electrochemical stability window, compared with aqueous alternatives, they are expensive, flammable, and in some cases, toxic. Moreover, aqueous electrolytes have a narrower electrochemical stability window, but are non-flammable, are inexpensive, have higher ion migration, and often give rise to higher C_{sp} due to smaller ions.

6.1.2 BASIC FORMULATION FOR THE ESTIMATION OF ELECTROCHEMICAL PARAMETERS

The electrochemical parameters such as values of C_{sp}, E_D, and P_D were estimated using the following set of equations [19]:

$$C_{sp} = \frac{\int iv.dV}{2\mu \ m \times \Delta V} \quad \left(@\,\text{three electrodes}\right) \tag{6.1}$$

$$C_{sp} = \frac{\int iv.dV}{\mu \times \Delta V} \times 2 \quad \left(@\,\text{two electrodes}\right) \tag{6.2}$$

$$C_{sp} = 4 \times \frac{I\Delta t}{m\Delta V} \quad \left(@\,\text{two electrodes}\right) \tag{6.3}$$

$$E_D = \frac{1}{2} \times C_{sp} \times \Delta V^2 \times \frac{1}{4} \times \frac{1}{1.36} \tag{6.4}$$

$$P_D = \frac{E}{\Delta t} \tag{6.5}$$

$$P_D = \frac{\Delta V^2}{4 \times ESR \times V} \tag{6.6}$$

where C_{sp} (F/g) is the specific capacitance, μ is the scan rate (mV/s), $\int iv.dV$ is the area under the I–V curve, I is the discharge current (A), Δt is the discharge time (s), m is the mass of active electrode (g), $\Delta V(V)$ is the potential window (V), Energy density (E_D) (Wh/kg), Power density (P_D) (W/kg), and ESR is the equivalent series resistance of a cell (Ω). Using EIS data, the C_{sp} of the electrodes was determined using equation (6.7):

$$C_{sp} = -\frac{1}{\pi \, f_l \, Z_l'' \, m} \tag{6.7}$$

where f_l is the lowest frequency and Z_l'' is the imaginary impedance at f_l. The EIS data as a function of the frequency were analysed using equations (6.8)–(6.10):

$$C(\omega) = C'(\omega) + C''(\omega) \tag{6.8}$$

$$C''(\omega) = \frac{Z'(\omega)}{\omega \, |Z(\omega)|^2} \tag{6.9}$$

$$C'(\omega) = -\frac{Z''(\omega)}{\omega \, |Z(\omega)|^2} \tag{6.10}$$

where $Z(\omega) = 1/j \, C(\omega)$; $C'(\omega)$ and $C''(\omega)$ are real and imaginary capacitance; and $Z'(\omega)$ and $Z''(\omega)$ are real and imaginary impedance.

6.1.3 Preparation of Te-Based Electrodes and Electrochemical Measurements

The electrochemical performance of rGO and Te–rGO cells was studied using cyclic voltammetry (CV), galvanostatic charge–discharge (CD) measurements, and electrochemical impedance spectroscopy (EIS) at $0.1–10^6$ Hz. The measurements were performed by Auto Lab (PGSTAT 30, Eco Chemie). All of the measurements were carried out at room temperature (300 K).

In two electrodes, they were prepared by coating (1 ×1 cm²) a slurry of 80 w/w% active material (rGO and Te–rGO), 10 w/w% carbon black, and 10 w/w% polyvinylidene difluoride (PVDF) in N-methylpyrrolidone (NMP) on a carbon paper electrode of 1×4 cm² as shown in Figure 6.1a and b. The specimen was dried at 80°C, for 24 hours. Two such electrodes were prepared and configured into a cell by sandwiching a quartz Whatman paper (as a separator) immersed in 1 M H_2SO_4 as an electrolyte and, subsequently, dried as shown in Figure 6.1c and d.

Of the three electrodes, carbon paper strip was used as a working electrode, Pt, as a counter electrode, and Ag/AgCl, as a reference electrode in the same electrolyte.

For both configurations, CV measurements were performed in a potential window of 0–1 V at scan rates of 5, 10, 20, 50, and 100 mV/s and current densities of 0.25, 0.50, 1.00, and 2.00 A/g.

In order to understand the various impedance factors experienced by the electrode/electrolyte interface and charge transfer process, EIS measurements were carried out on the specimen. The cyclic stability of the cell was investigated up to 2500 cycles at 1 A/g.

6.1.3.1 Structure and Property Relationship

Figure 6.2a shows the recorded Raman spectra for rGO and 1% Te–rGO at 532 nm photo-excitation. For rGO, D- and G-peaks were present at 1355 and 1595 cm^{-1}. After Te incorporation, a slight shift in vibration modes is observed. The D-peak is shifted up to 1359 cm^{-1}, whereas G, downshifted to 1590 cm^{-1}. The value of I_D/I_G for rGO was 1.11 and increased to 1.41 in Te–rGO. The change is

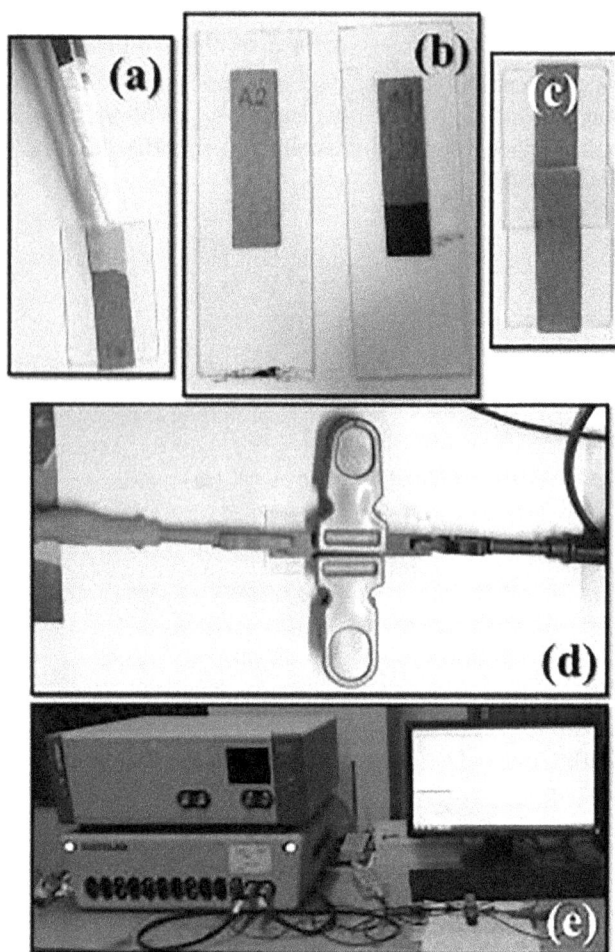

FIGURE 6.1 Sequence for supercell development for two-electrode measurements. (a) and (b) Deposition of active material, (c) sandwiching electrodes with electrolyte (1M H_2SO_4) separator, (d) and (e) mounting and system details, respectively.

attributed to lowering down the defect concentration and the increase in the number of sp^2 rings in Te–rGO. The inset shows the corresponding FTIR spectra.

In IR, peaks at 664 and 1525 cm^{-1} are associated with C–H (in-plane bending modes) and C=C at 2312 cm^{-1}, which remained unchanged for both the systems. The small shoulder at 615 cm^{-1} overlapping with the broad C–H mode indicates the presence of Te-O in Te–rGO. For rGO, the peaks related to C–O–C (epoxides), C=O (carbonyl), and C=O (carboxylate), respectively, at 1210, 1370, and 1730 cm^{-1} disappeared completely in Te–rGO. The ratio of C/O seems to be modified, which is more evident by the ESCA.

Figure 6.2b and f (lower pan) shows the edges of rGO and Te–rGO recorded using HRTEM (details including ESCA and HRTEM analysis in Appendix B, respectively, in Figures B.6.2.1 and B.6.2.2). The edge width is somewhat enhanced from 5 (rGO) to 10 nm (1% Te). It seems that Te-O might have modified the nature of conjugation and folding of rGO sheets by precipitating at the edges as indicated by arrows in (b). This is even reflected in Raman spectrum of Te–rGO (inset in (b)) with the emergence of shake-up D' at 1620 cm^{-1} to G showing stretching of edge bonds.

Figure 6.2c shows the X-ray powder diffraction patterns for both samples. The broad peak at $2\theta \sim 25°$ is assigned to hexagonal (002) C (JCPDS card no. 75-1621) plane of stacked rGO sheets

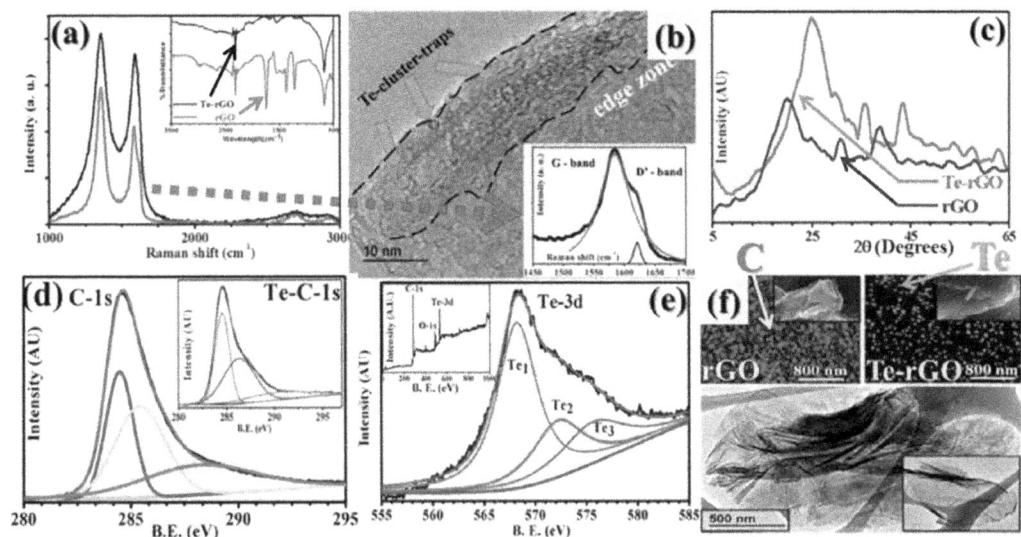

FIGURE 6.2 (a) Raman spectra for rGO and Te–rGO (@532 nm); inset shows variation in IR bands. (b) HRTEM image showing modification in edges of rGO by Te; inset shows the emergence of D' band (Raman spectrum). (c) Recorded XRD patterns, (d) ESCA spectra of C-1s and inset Te-C-1s, (e) inset general scan, Te-3d chemical states in rGO, (f) EDAX mapping for rGO and Te–rGO (1%) (upper pan); (lower) HRTEM image for rGO and Te–rGO (inset).

[20]. For Te–rGO, the peak of Te-O is not clearly visible due to mutual overlap with rGO position and small amount (1%); however, a marginal shift in (200) ~ 20° indicates exfoliation of rGO sheets by Te-O. The interlayer distance, d, is increased from 3.54 (rGO) to 4.35 Å (Te–rGO) indicating intercalation of Te-O within conjugated rGO layers and is consistent with the above discussions. For higher Te%, no significant change has been observed. The XRD patterns for 3%–11% Te–rGO and d values are provided in Appendix B supporting information.

Figure 6.2d shows the ESCA spectrum of C-1s of rGO (inset 1%Te–rGO). For rGO, the peaks at 284.48, 285.46, and 288.55 eV represent non-oxygenated C–C ring, C=C, carboxylate carbon, and C(O)O, respectively. The inset for Te-C-1s spectrum shows a reduction in the oxygen functional groups present within rGO. Mainly, the peak for C(O)O in rGO is expanded significantly and shifted to 291.23 eV, after adding Te, indicating substantial rearrangement of C(O)O group. The sharpening and increase in the intensity of C–C peak and the broad asymmetric tail of the peak indicate the emergence of high sp^2 carbon content as compared to rGO. We can say that ratio of C/O is decreased in Te–rGO, due to the detachment of oxygen moiety from graphitic planes with partial oxidation of Te. Figure 6.2e inset shows a general elemental scan along with deconvoluted peaks for Te-3d. The peak at 568.23 eV is attributed to the C–Te bonding, which has 54 at % contribution. The other two peaks at 575.63 and 572.22 eV show the formation of TeO. However, no peaks of bare Te atom have been observed. The 3d component corresponds to spin–orbit splitting character of the bond. For pure Te atoms, a doublet of 3d$_{5/2}$–3d$_{3/2}$ is reported with binding energy separation ~10.4 eV; however, in our case, no such splitting is observed. It seems that the bonding between Te and C quenched the charge coupling character, significantly [1,21].

In Figure 6.2f, the upper panel indicates EDAX mapping imaging of rGO (left) and 1% Te–rGO (right). The corresponding inset shows surface morphology of the layers captured by FESEM in active electrode materials. The scale bar is uniform ~800 nm for both images including insets. For rGO, the pink colour represents the existence of carbon, whereas the small black portion distributed within the carbon domain is allocated for oxygen and native nitrogen that exist on rGO surface. We have not quantified the distribution of these moieties; however, they seem to be negligibly small

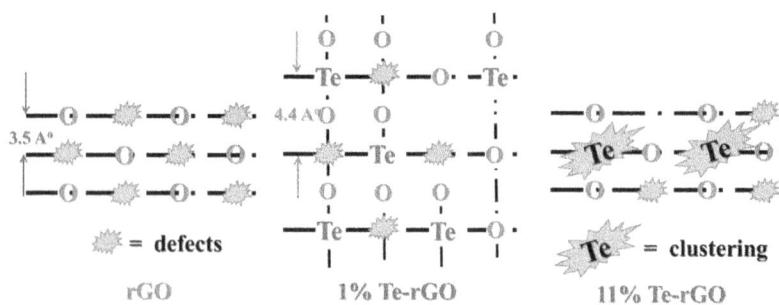

FIGURE 6.3 Schematic representation of Te incorporation and probable effects in rGO layers (at low and high %).

compared with carbon. For Te–rGO, the orange dots show the presence of Te clusters onto rGO surface. The spread of clusters seems to be uniform across the mapped area. The dark zone around the cluster consists of carbon and other elements. The lower panel shows the morphology of rGO, and the inset shows 1% Te–rGO. Both images are recorded on the same scale of 500 nm by HRETM. For rGO, the flakes seem to be thicker and consist of several number of layers, including wrinkles, warps, and folding planes. The dark zone is indicative of non-transparent region to incident electron beam for rGO. The flake thickness is seen to be reduced due to the intercalation of Te, resulting in the decreasing number of layers and increasing transparency of the incident electron beam through the flakes of Te–rGO. Figure 6.3 shows the scheme of Te incorporation and its effect on rGO, at 0.5 (low) and 11(high) %.

The analysis of active material, broadly, revealed that facile addition of Te atoms exfoliates conjugated layers [21], selectively attacks C(O)O and disorder sites, modifies the nature of local hybridization, and alters symmetry and spatial C/O arrangement. This results in the bond frustration, at Te position, and stress surrounding sp^2 superlattice, consequently modifying the electronic density distribution at Te–C and interlayer Te–O–C sites. The presence of Te seems to be acting as a buffer component that enhances the crystalline and conjugation length by a factor of two and decreases inter-defect distance, by enriching the overall sp^2 content. However, at higher w/w%, mostly clustering of Te is observed without much modification in rGO chemical environment [1,21].

6.1.3.2 Electrochemical Analysis

The representative results of three electrode measurements are shown in Figure 6.4. Figure 6.4a–c shows CV profiles of rGO, and 1% and 11% Te–rGO, respectively. From the plots, the magnitude of C_{sp} for rGO and Te–rGO is computed using equation (6.1), and data are provided in Table 6.1. After Te doping, change in nature and area of the curves are observed. The shape of CV curve is almost rectangular with slight pseudo-capacitive contribution. This may be due to heteroatoms Te-O-C molecular environment [1,21]. Figure 6.4d shows comparative profiles, at 100 mV/s, for rGO, and 1% and 11% Te–rGO. At 100 mV/s, the increase in C_{sp} for 1% Te–rGO was ~96%, whereas for 11% electrode, it was ~20% with respect to the base rGO value. In a similar fashion, a deviation in C_{sp} is observed, practically, for all scans; however, it is highest for 50 mV/s, 130% (1%), and 35% (11% Te). Further, the values of C_{sp} decreased gradually with an increase in % of Te in rGO. The highest values are obtained for 1% incorporation of Te and seem to be an ideal compositional content in rGO. For 11% Te–rGO, oxidation and reduction potential peaks are observed, respectively, at ~0.6 and 0.4 V. But, it seems that they have no favourable impact on C_{sp}.

Figure 6.4e shows CD profiles of 1% Te–rGO, and, correspondingly, the inset is for rGO at current densities of 0.5, 1, and 2 A/g. Plot (f) shows the comparison of CD profiles for rGO and 1% Te–rGO at 1.00 A/g, and the inset shows a decrease in the values of C_{sp} with scan rate. It is clear that the CD time is retarded, at this composition, by a factor of six with respect to rGO. This indicates that the surface area available is more than rGO, enhancing the accessibility for the solvated ion

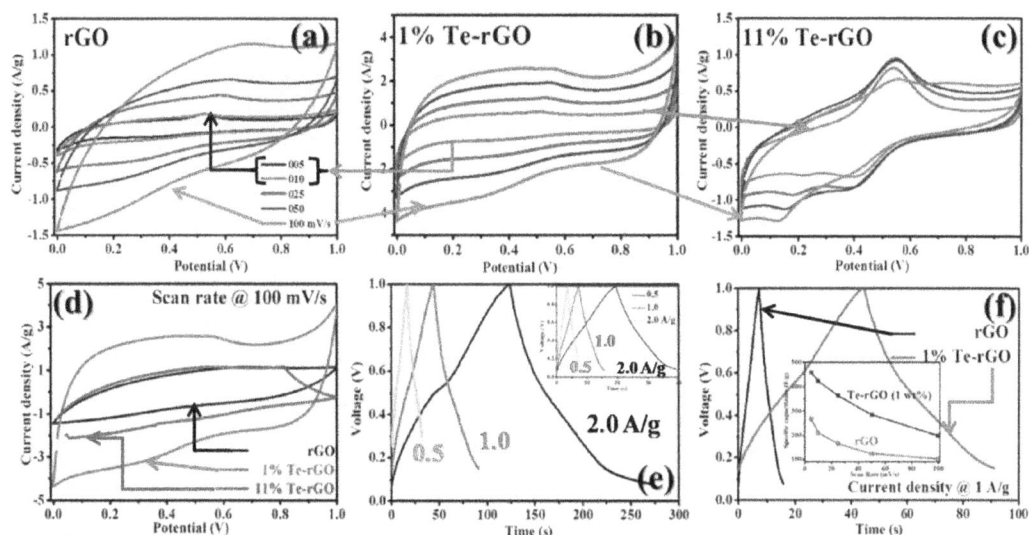

FIGURE 6.4   Three-electrode measurements. (a) rGO, (b) 1% Te–rGO, (c) 11% Te–rGO, (d) CV @100 mV/s for three systems, (e) CD profiles for 1% Te–rGO (inset: rGO @ 0.5, 1.0, and 2.0 A/g), (f) comparison of CD @1 A/g (inset: C_{sp} as a function of scan rate for rGO and 1% Te–rGO with measurement error ±3%–5%).

TABLE 6.1

Values of C_{sp} Computed for rGO and Te–rGO Using Three-Electrode System

Scan Rate (mV/s)	C_{sp} (F/g) by CV		
	rGO	1% Te–rGO	11% Te–rGO
5	268.14	458.33	272.55
10	210.78	423.33	268.14
25	165.90	362.84	210.78
50	123.40	284.95	165.93
100	101.29	198.50	123.40

migration. The presence of Te-O moieties between adjacent rGO sheets might lead to the formation of channels for ion transport, improving the rate capability. Moreover, at 1% Te, the layered sandwich structure seemed to be acting as an ideal strain buffer to accommodate changes in the volume of the exfoliated rGO layers and, thus, had better structural stability as well as superior electrochemical parameters. At higher %, Te clustering might have maximized restacking of rGO sheets, largely hindering the superior performance needed for practical applications. Due to this peculiar behaviour, other compositions of high %Te–rGO seem to be redundant; hence, in subsequent sections, only the results of 1% Te–rGO are discussed [1].

From device perspective, the practical implementation of symmetric electrode system is essentially important. The results of the two-electrode measurements are presented in Figure 6.5. Figure 6.5a shows a comparison of CV for rGO and Te–rGO @ 50 mV/s. For both profiles, a nearly rectangular shape is obtained, indicating a good rate performance of the cell, important for practical use. For Te–rGO, CV (@ 5–100 mV/s) and CD (@0.25–2.00 A/g) are shown in Figure 6.5b and c, respectively. The CV profiles have maintained their shape at a higher scan rate. The higher area under the curves shows the enhanced electrochemical performance of Te–rGO than rGO. The

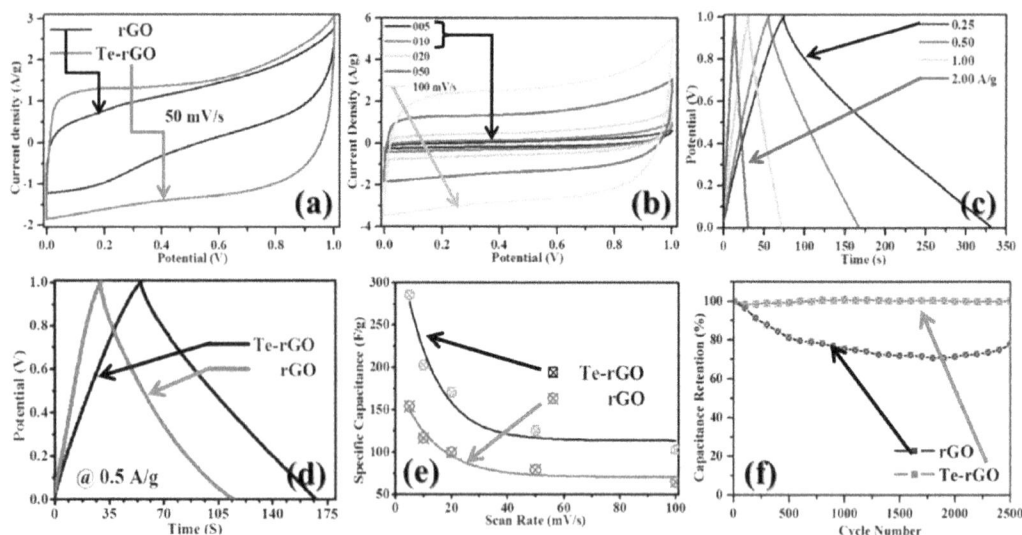

FIGURE 6.5 Two-electrode measurements. (a) CV @50 mV/s, (b) CV @5–100 mV/s, (c) CD @0.25–2.0 A/g for Te–rGO, (d) comparison of CD @0.5 A/g, (e) C_{sp} vs. scan rate, and (f) % capacitance retention.

TABLE 6.2

C_{sp} for rGO and Te–rGO Using Two-Electrode System

	C_{sp} (F/g) by CV			C_{sp} (F/g) by CD		
Scan Rate (mV/s)	rGO	Te–rGO	Current Density (A/g)	rGO	Te–rGO
5	154.02	285.71	0.25	182.02	259.53
10	117.00	202.85	0.50	170.23	226.89
20	136.40	170.50	1.00	108.66	168.44
50	79.85	125.80	2.00	88.98	132.13
100	64.92	103.80	—	—	—

nonlinear CD profiles are indicative of good capacitive action. In case of Te–rGO, the increase in CD time confirms superior capacitance properties. A comparison of CD @ 0.5A/g for both systems is shown in Figure 6.5d. The values of C_{sp} are calculated from equations (6.2) and (6.3). Figure 6.5e shows a comparison of C_{sp} for rGO and Te–rGO obtained using equation (6.1). Overall, the value of C_{sp} is higher for Te–rGO than for rGO and, over the scan rate, C_{sp} is decreased for both. This is due to the fact that a higher scan rate limits the accessibility to ions inside the pores of the electrode and only the outermost portion of the electrode is utilized for ion diffusion [1].

From CD curves, the highest C_{sp} is obtained at 0.25 A/g ~ 260 F/g with greatest CD time ~250 seconds; see Table 6.2. With current density, the C_{sp} reduces and the behaviour is similar to that observed in increasing scan rate voltage. The decrease in C_{sp} is due to limited accessible areas to electrolyte ion for diffusion with an increase in current density. At lower current density, ions can easily penetrate into the innermost regime of electrode material through almost every available pore and channel resulting in a higher capacitive performance. The estimated values of C_{sp} obtained from CD and CV are in well agreement. Overall, the values of C_{sp} are less compared to the three-electrode system. The deviation in C_{sp} for two-electrode capacitance to three-electrode one is due to the asymmetry in the adsorption of positive and negative ions. With superior electrolyte and counter

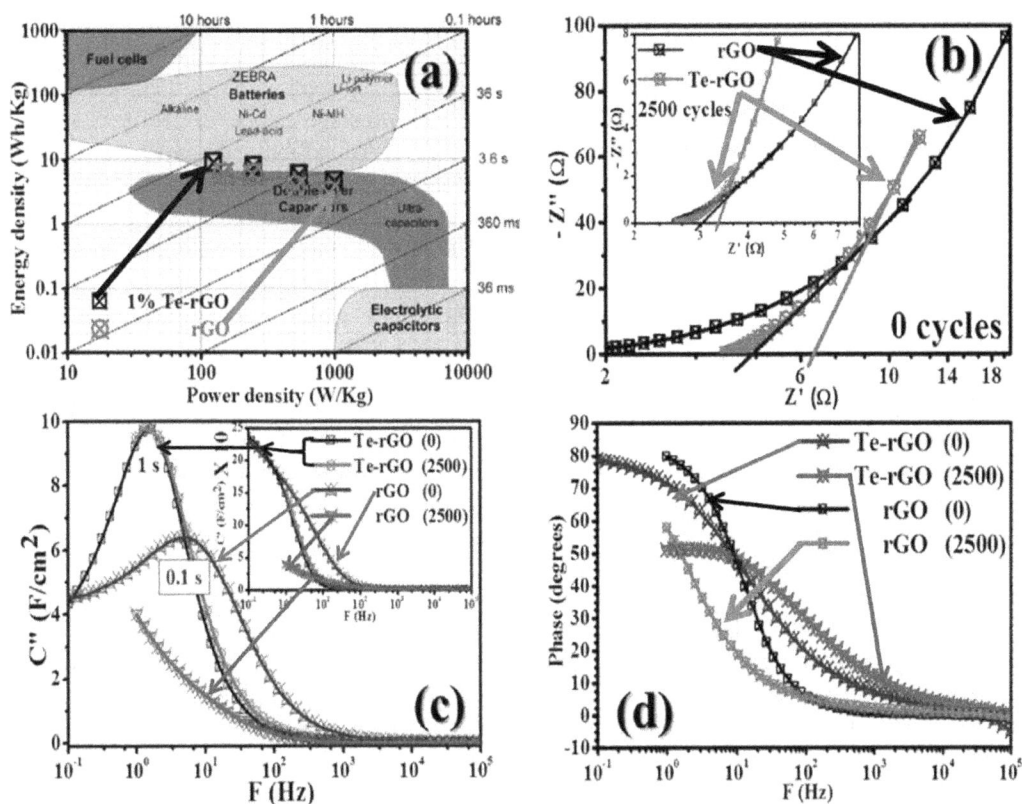

FIGURE 6.6 (a) Ragone plot of the asymmetric supercapacitors consisting of rGO (red circle) and Te–rGO (black square) as cathode and Pt as anode in comparison with various electrical energy storage devices. Times shown are the time constants of the devices as obtained by dividing E_D by P_D, (b) Nyquist plots over $0.1–10^6$ Hz after 0 (inset: 2500) cycles, (c) complex AC capacitance Bode plots, and (d) electrochemical phase impediment P_θ vs. time obtained from EIS data.

electrodes, the maximum C_{sp} can be obtained from the active material. For the electrochemical stability of the electrodes, CD cycles were carried out up to 2500 cycles at 0.5A/g. Figure 6.5f shows capacitance retention. It is observed that almost 100% capacitance is retained at the end of 2500 cycles for Te–rGO; however, there is a decrease up to 80% for rGO. Table 6.2 shows the comparison of C_{sp} obtained from CV and CD measurements [1].

Further, rGO and Te–rGO electrodes are characterized by the energy and the power being available for a load. This has been achieved by estimating E_D and P_D using equations (6.5) and (6.6), respectively. In general, ESD are located in characteristic regions in the power–energy plane. The efficiency of these devices is usually dependant on the working point; a single device belongs to a whole curve in the energy–power plane, the so-called Ragone plots, presented in log–log scale as seen in Figure 6.6a. They provide the limit in the available power, and the optimum region of working where both energy and power are high for the device. The maximum E_D obtained for Te–rGO is 9 Wh/kg with P_D 125 W/kg @ 0.25A/g. The values of E_D and P_D at 0.5, 1, and 2A/g are 8, 6, and 5 Wh/kg and 250, 501, and 1002 W/kg, respectively. Similar estimations are made for rGO, and the values of E_D and P_D at 0.5, 1, and 2A/g are 6, 4, and 3 Wh/kg and 248, 520, and 918 W/kg, respectively. In Figure 6.6a, the location of Te–rGO and rGO electrodes is, broadly, at the interface of battery and EDLC domain. Specifically, Te–rGO at low P_D and high E_D lies in the battery region, whereas at high P_D and low E_D, its behaviour is similar to that of the EDLC. For rGO, at both high and low E_D–P_D regime lie in the EDLC. The appearance of rGO and Te–rGO at the boundaries of the battery and capacitor regions is indicative of internal losses and leakage within the interface of *supercell*. To investigate cell interface, a further analysis has been carried out by EIS.

The useful presentation of EIS data is to plot Z' against Z'' to obtain a Nyquist curve recorded by using two-electrode AC cells of rGO and Te–rGO, at two-stage cycles (0, 2500), as seen in Figure 6.6b. For both electrodes at these stages, the EIS curve exhibits no semicircle over the high-frequency range, but a curved locus in the lower-frequency region. The absence of semicircle pattern, observed from the Nyquist plot, is indicative of low charge transfer resistance that leads to excellent electrode conductivity of the material. The low-frequency curved region shows that the device is largely departed from the ideal capacitor behaviour, at initial stage. Moreover, its charge storage has strong frequency dependence, having intrinsic and non-linear pseudo-capacitive contribution. Both electrodes seem to have no limiting capacitance values and possess low limiting (Warburg) impedance, Z_p. The estimated values of Z_p for Te–rGO are 6.5 (@0.4Hz) and 3.8 (@3.2Hz) Ω and the corresponding values for rGO are 4.5 (1.5Hz) and 3.0 (1.4kHz) Ω, respectively, at 0 and 2500 cycles. This indicates that the reactant ions move very far in Te–rGO, whereas in rGO, the movement of ions is restrictive. The contribution of electrolytic ionic impedance, substrate resistance, and resistance of active material/current collector interface seems to be negligibly small. The resistance offered by Te–rGO surface is high and reduced with the number of cycles, and a similar trend is observed for rGO. However, Te–rGO offers high impedance, relatively. The absence of high-frequency semicircle confirms the pseudo-capacitive action for both the electrode systems initially and after 2500 cycles. Since the behaviour is pseudo-capacitive, faradic reactions might also contribute to the impedance. However, it has been observed that, after 2500 cycles, due to more vertical line at low frequencies, Te–rGO behaves like an ideal capacitor (inset in Figure 6.6b) [1,21].

The Bode plot of complex AC capacitance (Figure 6.6c) is presented for rGO and Te–rGO for both the stages. The values of C' and C'' are calculated using equations (6.9) and (6.10), respectively. In general, when the frequency decreases, $C'(\omega)$ sharply increases and tends to be less frequency dependent. The low-frequency value of $C'(\omega)$ corresponds to the capacitance of the supercapacitor cell measured during a constant current discharge. For rGO and Te–rGO, no frequency-independent regime has been found for $C'(\omega)$. In parallel, the evolution of $C''(\omega)$ vs frequency, i.e. imaginary part of capacitance, goes to maximum at frequency f_0, which defines a time constant of $t_0 = 1/f_0$. This time constant is described by a characteristic relaxation time of the whole system (the minimum time to discharge all of the energy from device with an efficiency of more than 50%). Thus, a smaller value indicates higher rate capability. However, the minimum time to discharge 50% of energy is ~0.2 seconds (0 cycles) for rGO and ~0.7 seconds (0, 2500 cycles) for Te–rGO. After incorporating Te, the discharge time is increased by a factor more than three. As a result, rGO shows the tendency of EDLC, whereas Te–rGO behaves more like a pseudo-capacitor [1]. Thus, the impediment of electrolytic species to store charge is more in rGO compared to Te–rGO, as revealed by P_θ vs frequency curve seen in plot (d). The values of ESR are calculated using equation (6.6) [1,21] and found to be ~1.7 mΩ for rGO and ~2 mΩ for Te–rGO @ 0.25 A/g.

6.1.3.3 Fully Sealed Device Characteristics

By and large, using the horizon scan for similar studies, for instance, Jia et al. [22] showed C_{sp} ~ 79 F/g for rGO/carbon black, whereas Jang et al. [8] showed maximum C_{sp} between 100 and 250 F/g @ 1A/g for a similar system. Yoo et al. [20] showed C_{sp} ~ 247.3 F/g @ 0.17A/g for multilayer graphene, whereas Kenar et al. [23] achieved C_{sp} ~ 4 mF/cm^2 @ 1.5A/g for laser-scribed rGO, showing that the Te–rGO electrode is facile to fabricate, is easy to integrate, possesses superior electrochemical parameters, is cost-effective, and is favourable towards practicality on an industrial scale. To the best of the authors' knowledge, the report on supercapacitive behaviour of Te-incorporated rGO is not seen in the literature.

Further, both the active electrode materials are implemented for laboratory-scale demonstration of the *supercell* action. For this purpose, the electrode is developed as described in the experimental Section 2.3; Figure 6.7a shows the corresponding apparatus. The assembly is packed with a Teflon

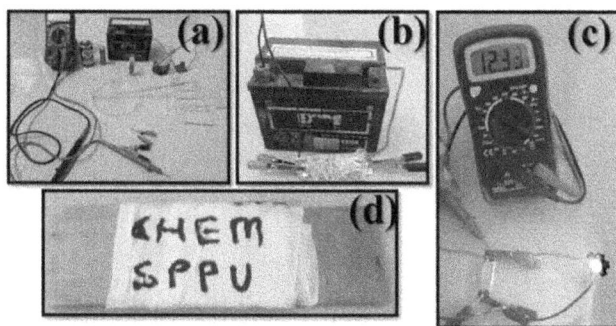

FIGURE 6.7 Laboratory-scale demonstration of 1% Te–rGO electrode showing illumination of LED. (a) Components such as substrate, digi-multimeter, 12.4 V battery, active electrode material, carbon powder, and PVDF, (b) charging of the *supercell* with a commercially available automobile battery, (c) LED illumination after charging indicating voltage attained by the fabricated assembly, and (d) fully sealed Te–rGO electrode assembly.

tape as seen in image (d). Prior to charging, the voltage between the two electrodes was measured and found to be negligibly small ~0.05 mV or so. The assembly is connected to two *crock-clips* having conducting wires. The cell is charged for about 1 minute with a normal 12.50 V automobile battery. After charging, the voltage measured, instantaneously, is found to be ~12.33 V, as seen in image (c) in Figure 6.7. The voltage is observed to be falling rapidly indicating the recombination of developed charges. So the Helmholtz layer dissipation time is found to be very small for such a primitive assembly. These charging measurements are performed repetitively, and in one case, we connected the assembly to commercially available LED equipped with current-carrying wires. The LED is found to be illuminated, as seen in image (c) of Figure 6.7 for a period of 4.5 minutes or so. The result shows the laboratory-scale performance of the fabricated electrode using Te–rGO. It is interesting to note that no such effect is observed for rGO specimen prepared and tested in an identical fashion [1].

The capacitance retention property of Te–rGO was investigated by repeating CV measurement at 2.5 mV/s up to 10,000 cycles as shown in Figure 6.8. The inset shows a comparison of FTIR and Raman spectra of pre- (0 cycle) and post-treated (10,000) Te–rGO. The capacitance retention was observed to be retained up to ~80%, indicating a high electrochemical stability. There is no change in active electrochemical sites, after cycling. From the FTIR comparison of the samples (before and after cycles), the molecular environment of Te–rGO remains almost unaltered, which is indicative of the fact that the chemical bonding between Te-C and Te-O remained invariant, after cycling. The peaks at 664, 1525, and 2312 cm^{-1} are assigned to C–H (in-plane bending modes, the first two) and C=C; later, the band remained unchanged for both the systems. Further, the small shoulder that appeared at 615 cm^{-1} is somewhat prominently emerged showing the presence of Te-O in Te–rGO even after 10,000 cycles. The chemical bonding between Te-C and Te-O remained unaltered. In FTIR, additional important peaks are recognized for C–OH and C=C functionalities (~3200–3300 cm^{-1}) corresponding to PVDF and NMP materials of the electrode. In Raman analysis, the shift and enhancement in band intensity at 2700 cm^{-1} is indicative of more exfoliation of Te–rGO sheets, compared to rGO. The corresponding FTIR and Raman details are provided in Appendix B (Figure B.6.8.1a and b). The comparison of pre- and post-cycled Te–rGO is also made using electron microscopy, typical images shown in Figure 6.8b–e. By and large, no dramatic changes are seen in the morphology of Te–rGO; however, the number of layers seems to be reduced and shows consistency with Raman analysis presented above.

By and large, at 1% concentration Te enhanced the integrated density of states by three times at adatom site, which has implications on the reduction in chemical potential from 2.4 to 1.8 eV.

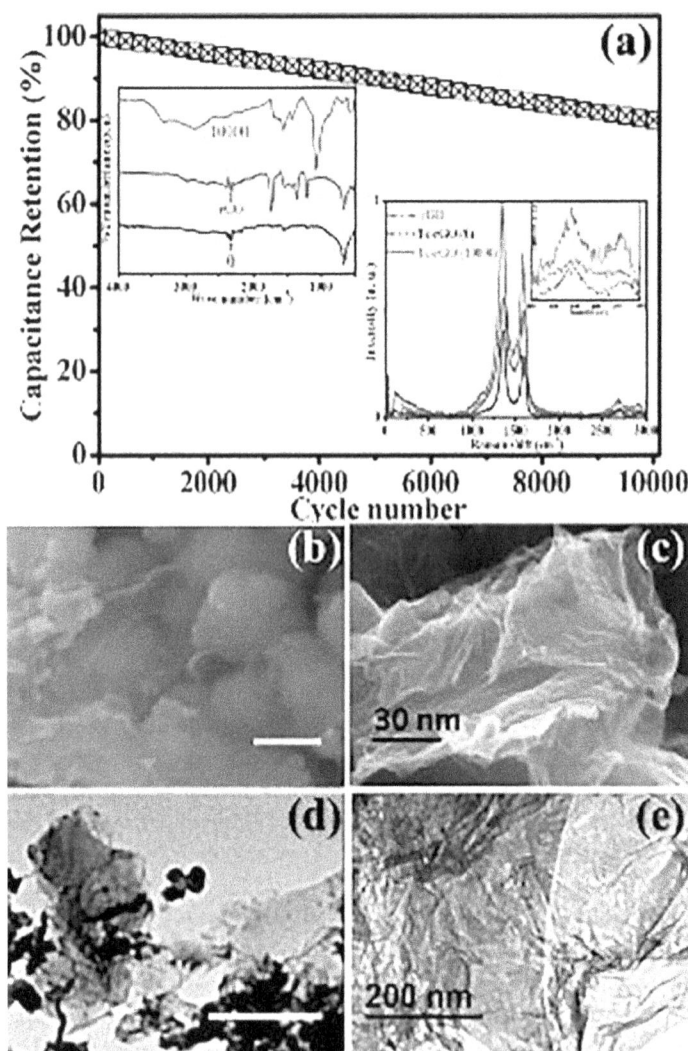

FIGURE 6.8 (a) Capacitance fading data up to 10,000 cycles (inset: pre- and post-electrode material comparison using FTIR and Raman), FESEM image @10,000 (b), pre-Te–rGO (c), HRTEM images after cycling (d), and at initial condition (e).

Moreover, it has reduced the oxygen proportion, recovered the sp^2 content, and enriched the π molecular environment, which swayed the electrical branching of transport, electron–phonon coupling, carrier Fermi velocity, conductivity, and mid-gap states in electronic band. Particularly, for supercapacitors, the energy is stored due to reversible surface or near-surface faradic reactions.

In overall analysis, we found that 1% presence of Te, particularly, offers controllable electronic properties and favourable morphology at inter- and intra-layers of rGO to meet the desired requirements for supercapacitor applications. But chemistry doesn't favour at low and higher Te wt% in rGO. In our previous communication [19], we have demonstrated that the chemisorbed Te modified the hybridization nature, altered the spatial arrangement of carbon, and resulted in electron density redistribution at C-Te sites. In this context, using Raman and electron microscopy analysis together, we have reported that it has generated bond frustration, at Te position, and stress surrounding the sp^2 superlattice, consequently altering the electronic properties. The optimum presence of Te enhanced the crystalline length by a factor of two and decreased the inter-defect distance from 7 to 10 nm, by enriching the overall sp^2

content. The defect density was reduced, by one order of magnitude, and electronic bandwidth, by 50%, due to the increase in π-π^* character. The reduction in resistivity, by Te, was responsible for offering the strength to couple the mobile charge carriers to applied electric field [1].

6.2 FABRICATION OF FLEXIBLE AND DURABLE SUPERCELL MADE UP OF CARBON NANO-SPHERES

In this section, the work on carbon nano-spheres (CNS) supercell is presented [7]. The fabricated cell was able to deliver 1.5 Wh of total energy at 38.4 V and 20.0 mA. CNS was obtained, eco-friendly, from the camphor precursor without any additional treatment; the details of synthesis are provided in Section 2.1.2. Analyses were carried out using Raman, X-ray photospectroscopy (XPS), UV–visible spectroscopy, scanning electron microscopy (FESEM), transmission electron microscopy (HRTEM), and BET (Brunauer–Emmett–Teller); CNS showed monodispersed, 3D heterostructure, low crystalline network of sp^2/sp^3 with dilute presence of oxygen, advantageous for achieving superior electrode performance. The electrochemical CV and galvanostatic charge–discharge (CD) measurements were performed, respectively, with variable scan rates (10–1000 mV/s) and current densities (A/g) (1.0–4.0), appropriately, in two- and three-electrode systems. Notably, in the two-electrode system, the value of CSP is ~560.0 with E_D, ~20.0, P_D, 250.0 @ 1 A/g with C_A, ~0.70 having high cyclic stability of ~86% @ 20,000. Post-electrode analysis showed excellent material stability. Importantly, the fabricated supercell is flexible, durable, and obtained ecologically, having nearly double-layer charge storage mechanism with no presence of any redox reaction [24].

6.2.1 BACKGROUND INFORMATION

Several routes have been adopted for the fabrication of nanocarbons by different methods and bio-precursors to obtain C_{SP}, E_D, and P_D, ranging ~74.0–400.0, ~10.0–55.0, and ~560.0–78,000.0, respectively [25]. In a study, Kalpana et al. used camphor black precursor to obtain carbon nano-beads activated @ 900°C using KOH, having S_A of ~79.6 that had provided C_{SP} ~77.0, E_D ~1.4, P_D ~568.0 with cyclic stability of 100,000 cycles in 0.1 M H_2SO_4 @ 1 mA/cm^2 [25]. Using the solvothermal in situ polymerization technique, the polyaniline/reduced graphene oxide composite was synthesized by using camphor sulfonic acid having S_A ~376.0 and C_{SP} ~438.0 with 76% cyclic stability after 1000 cycles in 1 M H_2SO_4 @ 2 Ag^{-1} [26]. In another work, Moreira et al. used the mixture of camphor and ferrocene processed in chemical vapour deposition (CVD) to prepare MWCNTs and, subsequently, functionalized with PVA having S_A ~40.8 with C_{SP} ~1.0, E_D ~0.28, P_D ~7.0, and 83% cyclic performance in 1 M Na_2SO_4 @ 100 mV/s [27]. Further, the decomposition of camphor over nickel nanoparticles generated nanocarbon with S_A ~520.0 having electrode parameters C_{SP} ~117.0, E_D ~31.9 in 1 M H_2SO_4 @ 100 mV/s [28].

In another study, carbon xerogels/carbon nanotubes obtained using the combustion of camphor precursor in CVD were achieved by phosphoric acid. The product was having S_A ~215.0, which provided C_{SP} ~151.0 in 1 M H_2SO_4 @ 2.5 A/g [29]. Moreover, Joseph et al. obtained nano-beads of carbon by pyrolysing camphor in air that has been blended in aqueous solution with 1 w% polyvinyl alcohol binder and variable concentration of sodium dodecyl sulphate surfactant. In electrochemical impedance spectroscopy analysis, the highest concentrated blend showed RC time constant ~319 µs with phase angle Φ −78° @ 120.0 Hz and impedance phase angle −45° at 4.2 KHz having areal capacitance (CA, µFcm^{-2}) ~269.0 and by cyclic voltammetry (CV) measurements ~487.0 in 1 M tetraethylammonium tetrafluoroborate electrolyte [30]. In addition, several studies were carried out on camphor blocks admixed with MnO_2^-, Ni/NiO nanowires, lithium titanate spinel, MnO_2 powder dispersed in methanol @ variable weight compositions and pyrolysed having S_A ~40.0–520.0. Their asymmetric electrochemical performance parameter, i.e. C_{SP}, E_D, and P_D were, respectively, in the range of 117.0–1950.0, 0.3–96.0, and 2.8–75,000.0 with cyclic stability 85%–90% after 1000–2000;

see [7,24] and references cited therein. The details in terms of E_D and P_D for the reported material, in the charted form, are provided in Appendix B (Table B.6.2). The reported electrodes were mostly prepared using campho-carbon as a precursor with exhaustive pre- and post-treatment to evaluate hybrid capacitor characteristics.

6.2.1.1 Pre-Analysis of CNS

Raman scattering is a remarkable tool to probe sp^2 and sp^3 fractions in the carbon medium. Figure 6.9a shows a typical Raman spectrum recorded for CNS. The spectrum consisted of two peaks, G at $1598.0\,cm^{-1}$ and D at $1375.0\,cm^{-1}$, respectively, that were assigned to sp^2 and sp^3 phases of CNS. In addition to their position, FWHM and integrated area peak ratio (I_D/I_G) disclose crystalline length, L_a, and the amount of amorphous carbon and provide information about disorder. As a broad feature, the intensity of D peak was comparable to G. This is an indication of a large amount of disorders in the obtained CNS [7,24]. The FWHM of D-peak is quite high, i.e. $\sim 323.0\,cm^{-1}$. The contribution comes from the non-sp^2 fraction in terms of topological disorder within carbon shell, Stone-Waller fraction, and sp^3 carbon phase, all contributing to disorder, dominantly. The presence G shows sp^2 fraction rich in π-electron. The I_D/I_G ratio is estimated to be 1.3, resulting in $L_a \sim 3.0\,nm$. This shows that the order of crystallinity is quite low in CNS. It seems that the sp^2 fraction is distributed within the disorder zone. In such a heterostructured environment, itinerant electrons can favourably participate in charge accumulation process. XPS analysis was carried out to gain more insights into the surface composition and the chemical state of CNS. The survey scan in Figure 6.9b (inset) shows the presence of C 1s and O 1s with the binding energies ~ 284.4 (~ 88.0 at %) and $\sim 532.3\,eV$ (~ 12.0 at %), respectively. The atomic concentration (at %) of elemental CNS is estimated using equation [16]: $x_i = \left\{ \dfrac{I_i^m}{S_i} \Big/ \sum_i^j \dfrac{I_{i,j}^m}{S_{iJ}} \right\} \times 100.$

Here, x_i is the atomic concentration of the element, $\dfrac{I_i^m}{S_i}$, the corresponding intensity, S_i, the sensitivity factor, $I_{i,j}^m$, the total composition of CNS, and S_{iJ}, the respective linear combinations of corresponding sensitivity factors. For intensity-weighted components (estimated in %), S_i is eliminated. The presence of dilute oxygen moiety may catalyse charge transfer process favourably for achieving better electrochemical parameters [24]. The details of O-1s analysis (Figure B.6.9.1) and tabulated elemental composition of CNS (Table B.6.3) are provided in Appendix B. The C-1s core-level spectrum, which appeared at 284.4 eV, was calibrated with respect to the standard carbon spectrum [24]. The spectrum showed a broad peak, which was deconvoluted into five subcomponents in Figure 6.9b. The peaks at 282.9, 284.6, 287.1, 289.6, and 292.5 eV were attributed to different phases of carbon [7,24]. The main peak at 282.9 was assigned to sp^2 hybridized state of graphitic origin, while the peak at 284.6 was due to sp^3 carbon [31]. The presence of sp^2 ($\sim 27.3\%$) and sp^3 ($\sim 45.9\%$) phases were comparably in near proportion, and such a carbon structure is advantageous for effective charge transport within CNS and their conjugated network, isentropically. They may offer reasonably efficient conduction as compared to orthotropic sp^2 superlattice. Moreover, the deconvolutions reveal some amount of oxidized carbon ascribed to C=O (287.1) and C(O)O (289.6). The peak that appeared at 292.5 is associated with π-π^* transition, responsible for charge donation process. This was further evident by UV–visible spectrum obtained between 200.0 and 800.0 nm as seen in Figure 6.9c. The sharp peak that emerged at $\sim 210.0\,nm$ was indicative of the nearly monodispersed nature of CNS. The peak indicates donor-loaded sites in CNS.

The FESEM images recorded for two different magnifications are shown in Figure 6.10a and b. The morphology of the CNS shows a coagulated, interconnected 3D network of nano-spheres. The formation of amorphous zone and discontinuity in the shells may provide a rapid access for the electrolyte to the bulk phase of the electrode. As a result, the C_{SP} of the electrode material was expected to be aggrandized. Moreover, HRTEM images recorded for CNS showed the detailed structure of individual spheres; a couple of images, representatively, are shown in Figure 6.10c and d. The

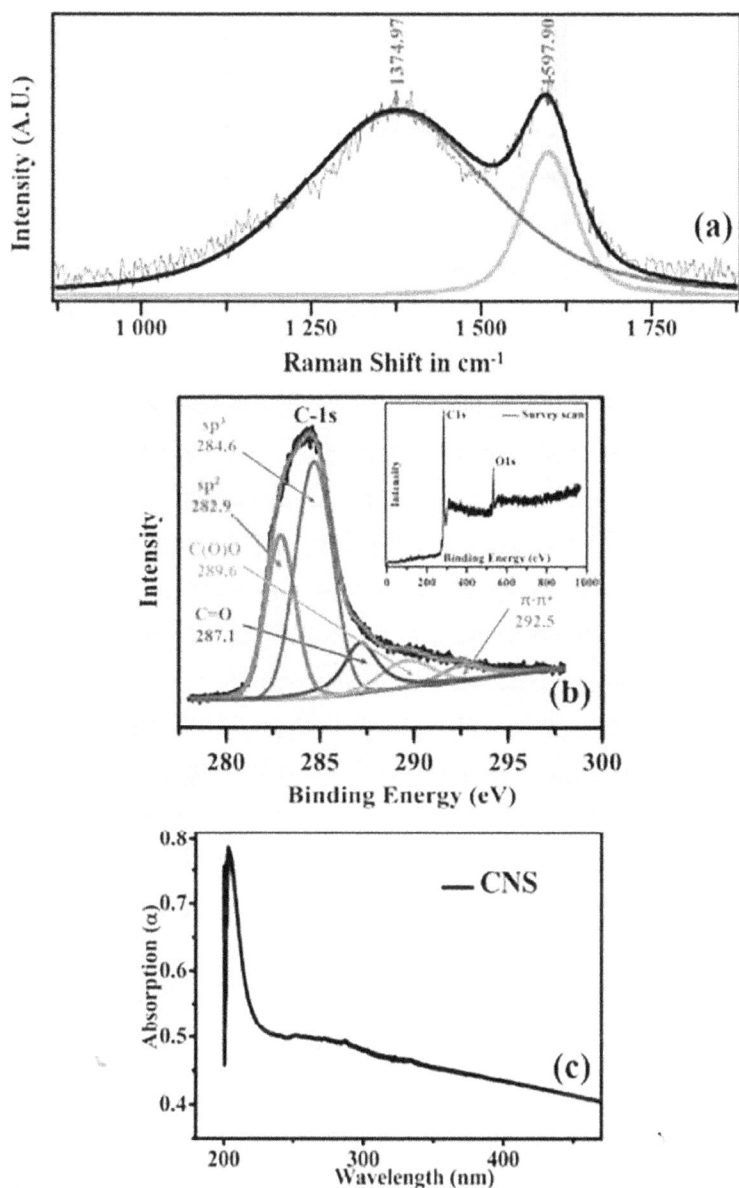

FIGURE 6.9 Recorded (a) Raman spectrum @ 633 nm, (b) XPS spectrum for C-1s, and (c) UV–visible spectrum for CNS.

concentric carbon shells are rarely seen. Due to the observed brightness, the CNS surface seems to be electrostatically charged, indicating electron-rich carbonaceous phases. A large number of sites have been examined to estimate the size of nano-spheres, which was found to be between 40.0 and 50.0 nm. The selected area electron diffraction (SAED) pattern shows bluer rings, indicative of amorphous zones present within CNS (inset (d)). For electrochemical applications, the S_A of carbon material was one of the important parameters characterized by the N_2 sorption technique at 77 K. Basically, sorption isotherms show two adsorbent regions: one at low-pressure concave isotherms and the other at relatively higher pressures, convex toward the P/P_0 axis. These curves provide information about the amount of adsorbent condensed by capillarity action on the surface under

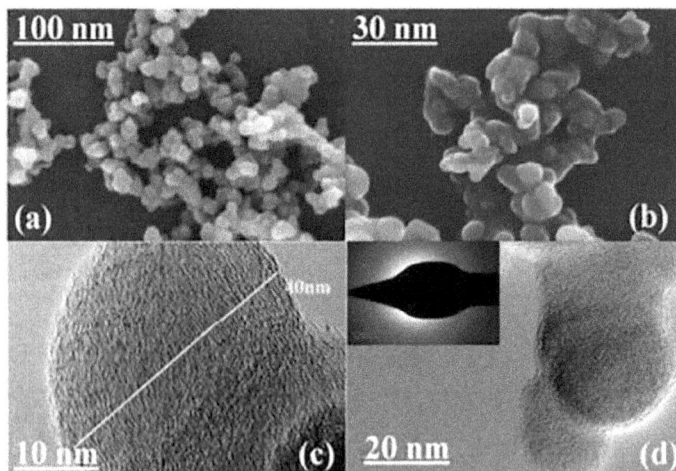

FIGURE 6.10   (a) and (b) FESEM images indicating interconnected 3D spherical carbon network; (c) and (d) details of morphology recorded by HRTEM (inset showing SAED pattern) for CNS.

examination. The action depends on the relative pressure, molecular dimensions of adsorbent to surface area available, the nature of adsorbed layers (mono-/multimolecular) condensed, etc. [31,32]. Investigating the electrode material thus becomes important to extract surface electrode parameters such as S_A, average pore size, and pore size distribution (PSD), using such techniques [4]. CNS at standard temperature and pressure (STP) was exposed to N_2 for about 1 hour, and the average S_A was estimated to be 790.0, which was quite higher than the previously reported values [5].

In Figure 6.11a, the hysteresis curve has been observed in sorption isotherms at a relatively low pressure $(P/P_0) \sim 0.30$, with an estimated average pore size ~ 3.42 nm. As per the IUPAC categorization, the porosity of the medium is classified into three channels: micropores (<2 nm), mesopores (2–50 nm), and macropores (>50 nm) [24]. Among them, micropores do not contribute effectively to EDLC due to the inaccessibility of electrolyte ions to the available pore area. However, ions can penetrate easily into the surface with size >2 nm (meso-/macropores). The obtained value indicated that CNS consists of widely distributed, heterogeneous, multi-phase inter-particle networks of carbon with high amount of porosity. Such networks compounded with high volume of surface electrons (as discussed early), appreciably high S_A and multi-channel size distribution (Figure 6.11b) is interesting to provide excellent electrode parameters, thereby utilizing macroscopic area over active surface area. The schematic representation of multi-phase surface porosity of CNS is shown in Figure 6.11c. It is noteworthy that such nanocarbons are obtained without any additional activation or treatment.

6.2.1.2   Electrochemical Performance Parameters for CNS Electrodes

CV is an effective tool to reveal the electrochemical properties of a given electrode material. Using two-electrode (Appendix B, Figure B.6.12.1) and three-electrode (Figure 6.12a) configurations, the supercapacitive parameters for CNS specimens were evaluated in 1 M electrolytes of $C_3H_4O_2$ (acrylic acid), HCl, H_2SO_4, and KOH. For a given potential window, the plot exhibited a nearly rectangular shape profile for all electrolytes, except for KOH. For nanocarbons, a perfect rectangular-shaped profile originates due to several reasons. The absence of redox (oxidation–reduction) process leading to an ideal ELDC behaviour in which the sorption of solvated ions occurs reversibly is perfect. In this, again, the porosity of the electrode surface plays a crucial role. As presented before, bi-phasor state of hybridization, the presence of native oxygen, carbon–oxygen bonding, and multi-channel surface porosity may play a pivotal role, cooperatively, in the hybrid behaviour. Thus, the

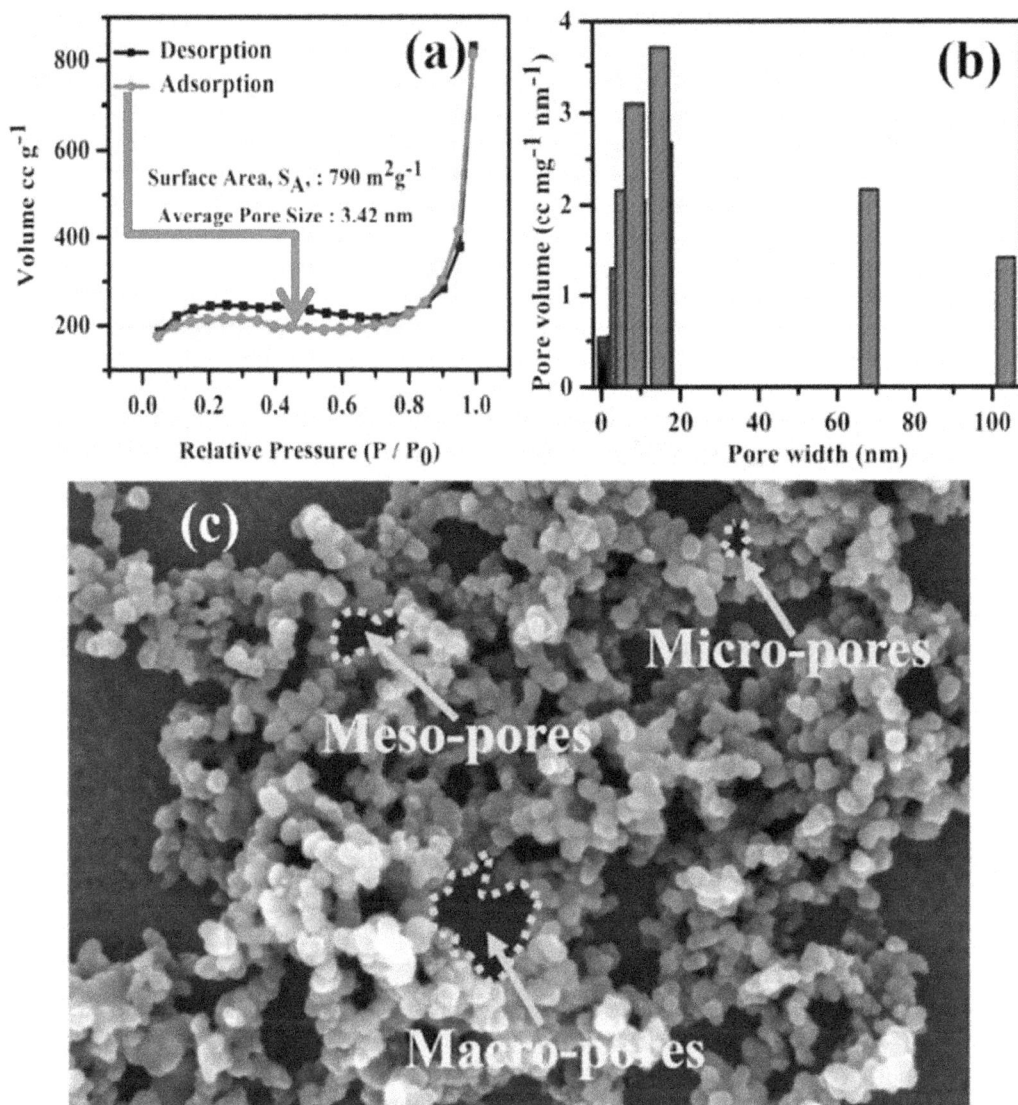

FIGURE 6.11 (a) N_2 sorption isotherms indicating estimated $S_A \sim 790 \, m^2/g$ and average pore dimension $\sim 3.42 \, nm$, (b) pore size distribution, and (c) typical SEM micrograph, schematically, indicating multi-channel surface porosity in CNS.

near-rectangular-shaped profile, obtained for CNS electrodes, is indicative of ELDC in addition to some other types of storage performance. In HCl, higher values of capacitance are obtained yielding $C_{SP} \sim 1080.0$ (two) and 570.0 (three) @ 10 mV/s (Appendix B, Table B.6.4, including all electrolytes). Although the main advantage of an organic electrolyte is its wide electrochemical stability window, compared with aqueous alternatives, they are expensive, flammable, and in some cases, toxic.

In contrast, an aqueous electrolyte has a narrower electrochemical stability window, but is not flammable and inexpensive. They have a higher ion migration, often giving rise to higher C_{SP} due to smaller ions as observed for HCl and hence chosen for further investigations. Figure 6.12b shows CV results at different scan rates for three electrodes, showing nearly rectangular-shaped profiles, which is indicative of a reasonably good capacitive behaviour. The response seems to be strongly dependent on the scan rate, and there is no evidence of redox current on both positive and negative

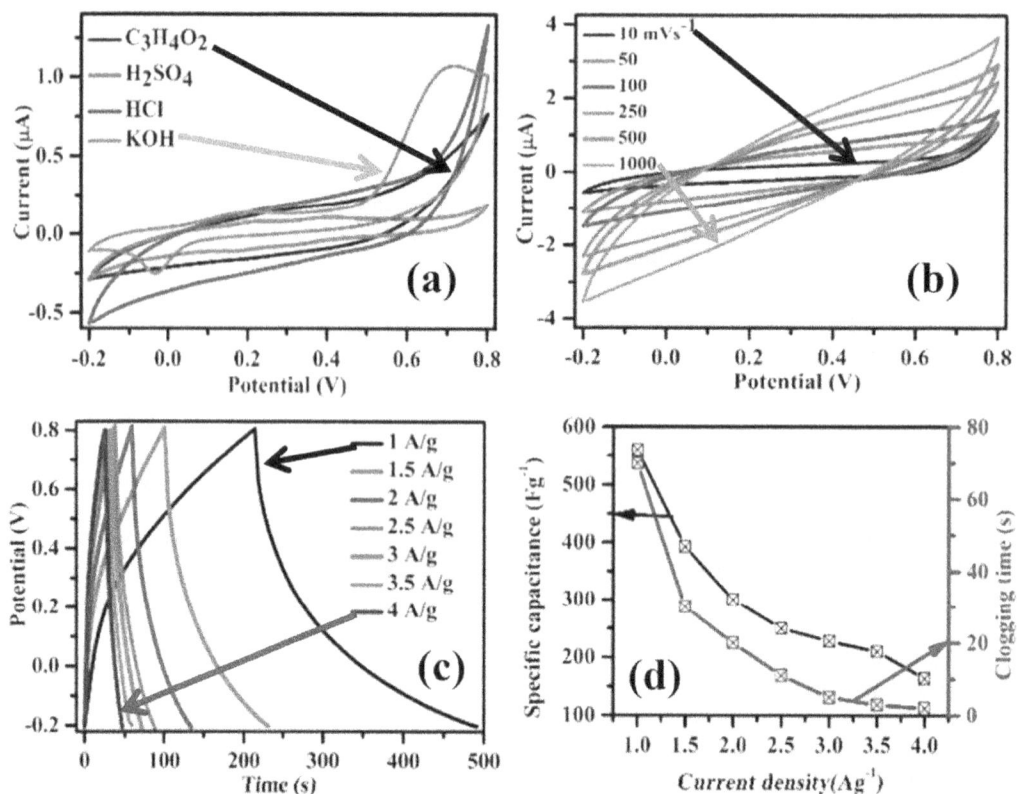

FIGURE 6.12 (a) Three-electrode CV curves, at 10 mV/s at different electrolyte [C₃H₄O₂: acrylic acid], (b) three-electrode CV curves at different scan rates, (c) CD profiles (two electrodes), and (d) for 1M HCl.

voltammetric sweeps. A similar trend is observed for the two electrodes displayed in Figure B.6.12.2 in Appendix B. Further, C_{SP} as a function of scan rate is found to have reached a minimal value of 65.0 (two) and 33.0 (three) @ 1000 mV/s (Appendix B, Figure B.6.12.3). For a lower scan rate, solvated ions penetrate deeper into the pores of CNS accessing larger S_A of pores, thereby achieving higher C_{SP}; however, at higher rates, ions can access only some portion of CNS attaining lower values of CSP [33,34]. For the two electrodes, the CD measurements were performed between 1.0 and 4.0 A/g to investigate the time for charging and discharging at a constant current sweep, as shown in Figure 6.12c. All curves exhibited a nearly triangular isosceles shape with discharge time greater than that of the charge time. The CD time was ~500 seconds at 1.0 A/g and found to be reduced to 50 s (4.0 A/g). The CNS in HCl electrolyte is behaving like a battery in low P_D, while hybrid in high P_D regime (discussed further in detail). The basic physical process involved in CD is electrolysis, i.e. the chemical charges that are produced at the electrodes at the expense of electric energy. This process depends on the polarization of the electrolyte under applied potential, electrolytic decomposition, and reverse decomposition potential for the electrolyte. In case of HCl, the decomposition and reverse decomposition potentials are, respectively, 1.31 and 1.37 V in the standard electrode configuration [34].

The product of electrolysis obtained on charging diffuses slowly back into electrolyte during the discharging cycle to recombine. The diffusion is continuous; however, the recombination depends on the concentration gradient of polarization and intrinsic overvoltage phenomenon. The overvoltage is a polarization potential whose source lies in the process at electrode, which takes place perfectly reversibly and depends on the reverse decomposition potential of electrolyte and the chemical

TABLE 6.3

Estimated Parameters for Binary CNS Electrode

Current Density	1.0	1.5	2.0	2.5	3.0	3.5	4.0
C_{SP}	560.0	390.0	300.0	250.0	230.0	210.0	165.0
C_A	0.70	0.50	0.40	0.32	0.30	0.25	0.20
P_D	250.0	375.0	500.0	625.0	745.0	875.0	1000.0
E_D	20.0	15.0	10.0	9.0	8.0	7.0	6.0

nature of electrode. The excess charge arrived at electrode (anode) due to intrinsic overvoltage creates a hindrance to the ions that have already participated in the charging process and are, now, ready to discharge, during the discharge cycle. The arrived excess charges, at that instant, are neither readily diffusible into electrolyte, nor able to participate in the recombination process. This leads to anodic clogging that dilates discharging time [34]. In our case, a similar effect is observed and the corresponding profile is shown in Figure 6.12d, indicating variation in current density (A/g) with clogging time(s). In our opinion, this could be an advantageous characteristic under the condition that an appropriate cell voltage is designed for a variable known load, driving sufficient amount of E_D and P_D for longer time. From the CD data, the estimated values of CSP were varied from 560.0 (1.0) to 165.0 (4.0) and are displayed in Figure 6.12d. For a specific current density, the CNS surface gets charged electrostatically due to electrolytic ions that move towards the nano-spheres. They get accumulated on the surface of CNS to form Helmholtz bilayer and participate in capacitive action. At a lower current density, by slow charging, the CSP attains a maximum value due to the accumulation of a large amount of solvated ions nearly accessing all pores of CNS. However, at a higher current density, all micropores are not reachable, due to relatively swift charging, resulting in inadequate accrual of ions, consequently reducing C_{SP} [24]. To evaluate the electrochemical stability, initially, cyclic efficiency has been measured (for both electrodes) at a lower number of cycles (max. 1000), at 10 mV/s. Figure B.6.12.4 shows the variation in C_{SP} as a function of cycle number, indicating a higher stability of ~98.5% and ~97.0% at 500 and 1000 cycles, respectively. Table 6.3 shows several electrochemical parameters estimated for CNS. Among them, C_A is an important device parameter in supercapacitor application. At miniaturized level, a high gravimetric capacitance is, essentially, required by loading low active mass fraction [23]. In the reported literature, the values of C_A were found to be between 0.50 and 0.32 as obtained from various carbonaceous materials [4,24]. Interestingly, our result showed a highest value of $C_A \sim 0.70$. Further, we have superimposed the estimated E_D and P_D onto the Ragone plot.

In general, the Ragone plot is used to identify an ESD domain against the E_D–P_D performance of an active electrode material. The increase in P_D compromises the value of E_D and vice versa [25,26]. Figure 6.13a shows the battery-like behaviour at highest E_D, 20.0 with low P_D, 250.0 (@ 1.0 A/g), and E_D was reduced to ~6.0 with increased in P_D, ~1000.0 (@ 4.0 A/g), revealing the battery/EDLC-like hybrid characteristics, for CNS electrodes. At a current drain time of 36 seconds, the values of E_D and P_D were found to be ~8.0 and 750.0, respectively. Figure 6.13b shows a comparison of Nyquist plots after 1 and 1000 cycles. In general, the Nyquist plot is separated in two regions: low- and high-frequency regions (inset). The transition point of two sections was known as the knee frequency given by R_B (base impedance)+R_E (electrolytic impedance) [27]. For CNS, the knee frequency was, initially, 60.0 Hz and found to be reduced to ~40.0 Hz. It shows that the change in the frequency of capacitive behaviour resembles that reported in the carbon literature [28,29]. At a low frequency, it shows nearly linear features, indicating the performance of CNS as an ideal capacitor. The parallel lines revealed that the behaviour is cyclically invariant. The extrapolation of the line on x-axis gave a value of impedance $R_B+R_E+R_L$ (limiting impedance) ~9.0–10.0 Ω @ 1 and 1000 cycles. The combination of ionic resistance of electrolyte, intrinsic substrate impedance,

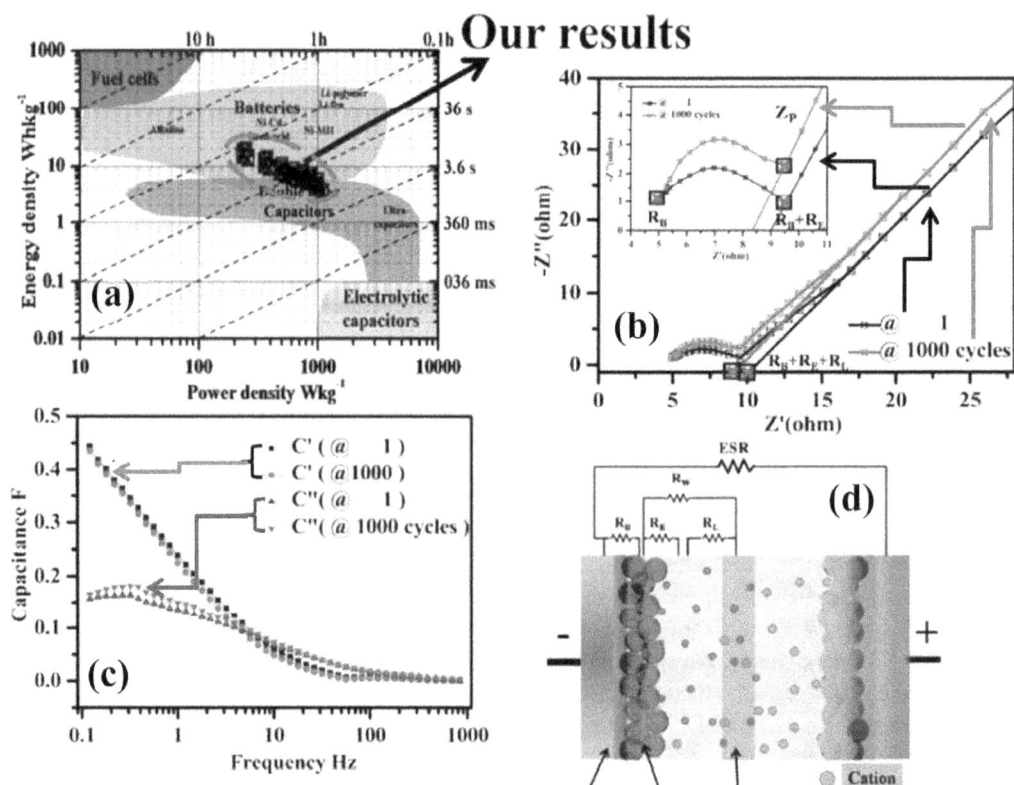

FIGURE 6.13 (a) Plot of E_D vs P_D (Ragone plane), ratio of E_D to P_D provides time scale as shown, (b) Nyquist plot for CNS electrode showing R_B (base), knee frequency ($R_B + R_L$ (electrolytic)), $R_B + R_E + R_L$ (limiting), and Z_P (Warburg impedance). Inset shows magnified higher-frequency region, (c) frequency-dependent complex capacitance ((b) and (c) recorded after 1 and 1000 cycles over 1 kHz–100 mHz), and (d) CNS electrode showing electrochemical impedance scheme.

interface contact resistances, charge transfer EDLC impedance, and limiting impedance is almost unaltered [30]. Mid-frequency band (inset), connected to charge transfer rate, quantified in terms of Z_P (Warburg impedance), which revealed the typical features of electrode, such as porosity and thickness, influences charge conduction. From Table 6.4, we see that, although there is no appreciable change in Z_P, after cycling, the corresponding frequency response is reduced by ~50%. This indicates retardation in charge transfer process attributed to modification in effective porous structure, consequently affecting the migration of ions in CNS electrodes [31,32]. The high-frequency semicircle pattern (inset) indicated that the capacitor action is in effect towards the lower-frequency region, however, the impediment to electrolyte ionic conductivity at electrode–electrolyte interface is high [34].

After 1000 cycles, the shift of semicircle towards the higher impedance side with an increase in radius is suggestive of further enhancement in ionic impedance [30,31]. Figure 6.13c shows the frequency-dependent complex capacitance, in which C' is real and C'' is the imaginary component. In a typical curve, when frequency decreases, C' sharply increases and then tends to be frequency independent. But, in our case, no such region is identified, and hence, C' curve is extrapolated onto the y-axis to obtain supercapacitive value of the electrode, at constant current discharge. The value was found to be between 0.50 and 0.49 $\mu F/cm^2$ @ 1 and 1000 cycles. The imaginary part of capacitance goes through a maximum at frequency f_0, which defines time constant $t_0 = 1/f_0$. This time constant is described as a characteristic relaxation time of the electrode to discharge all of the E_D

TABLE 6.4

Impedance Parameter Obtained by EIS for CNS

Parameters	At 1 Cycle		At 1000 Cycles	
	Z (ohm)	Freq (Hz)	Z (ohm)	Freq (Hz)
R_B	5.0	830.0	5.0	820.0
$R_B + R_E$	9.5	60.0	9.5	40.0
$R_B + R_E + R_L$	9.0	70.0	10.0	30.0
Z_P (Delivery of P_D)	9.0	80.0	9.0	40.0
C'_{SP}	0.5		0.5	
C''_{SP}	0.2		0.3	
t_0 (Delivery of E_D)		~5.0–5.5 seconds		

R_B, base impedance; R_E, electrolytic impedance; R_L, limiting impedance; Z_P, Warburg impedance.

from the device with an efficiency more than 50.0%. The value of t_0 is found to be ~5.0–5.5 seconds, in the course of the cycles. The Bode phase plot shown in SI, Figure S6, specified that in the lower-frequency region, the ionic phase shift was ~800 and reduced, insignificantly, to ~770, respectively, recorded @ 1 and 1000 cycles. The nature of the plot reveals that a double-layer charge storage phenomenon has taken place in the electrode with no redox (faradic) charge storage type due to the absence of a shallow curve [28].

Broadly, our electrochemical impedance spectroscopy (EIS) analysis revealed that the high S_A of electrode generated electrostatic charges on the surface of CNS in the presence of migrating electrolytic ions. The ions diffused through the interconnected pores of CNS and seem to have formed an electrostatic Helmholtz layer on respective electrodes, as seen in the schematic in Figure 5.13d. The characteristic impedance of the substrate to the CNS is high, generating an inductive coupling at the substrate/CNS interface. With subsequent electrical ageing/cycling, the interface coupling became stronger with further strengthening of the Helmholtz layer. This has implication on the transfer impedance (Z) of ED and PD, which is found to be almost invariant even after 1000 cycles as seen in Table 6.4 [24].

6.2.2 Post-Material Analysis, Cell Fabrication, and Performance Evaluation

6.2.2.1 CNS: Post-Analysis

From the application point of view, the cyclic efficiency is an important and essential parameter for EDLC over rechargeable batteries to evaluate the electrochemical stability [4]. Figure 6.14 shows the percentage stability of CNS electrode measured up to 20,000 cycles and compared with lower cycles (5000), at different scan rates, which indicated the robust nature of CNS having 86.0% (20,000) of performance stability.

Moreover, material stability has been revealed by post-analysis of CNS electrodes using CV and Raman measurements. One can see that the shape of the CV curve remained almost invariant, suggestive of nearly no change in electrochemical parameters. In Raman, the I_D/I_G ratio was found to be 1.05 having L_a ~2.2 nm, which is almost unchanged with respect to the pre-analysed CNS material. The material remained unaltered even up to molecular level though aged for several thousand cycles. However, the slight departure from 100% performance seems to be due to irreversible charge consumption, at initial stage, because of faradic reactions associated with possible oxidation and reduction of loosely bound surface particles. After few initial cycles, the nature of interactions between electrode and electrolyte seems to be stabilized and remain unchanged, thereafter. This shows that the ability to transfer charges remained almost constant as the cycle progresses further [7,24].

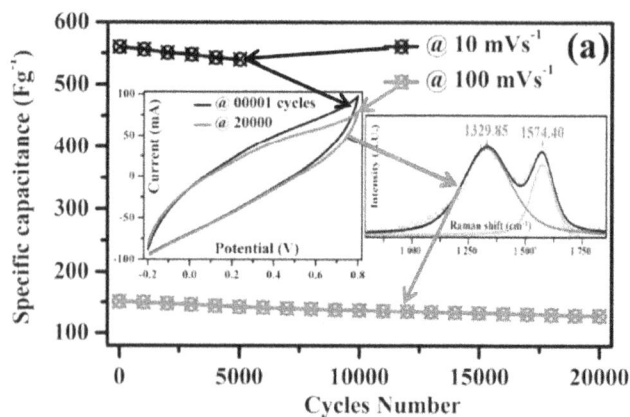

FIGURE 6.14 Stability of CNS @5000 and @20,000 cycles at different scan rates. [Inset: CV and Raman data of CNS after cycling.]

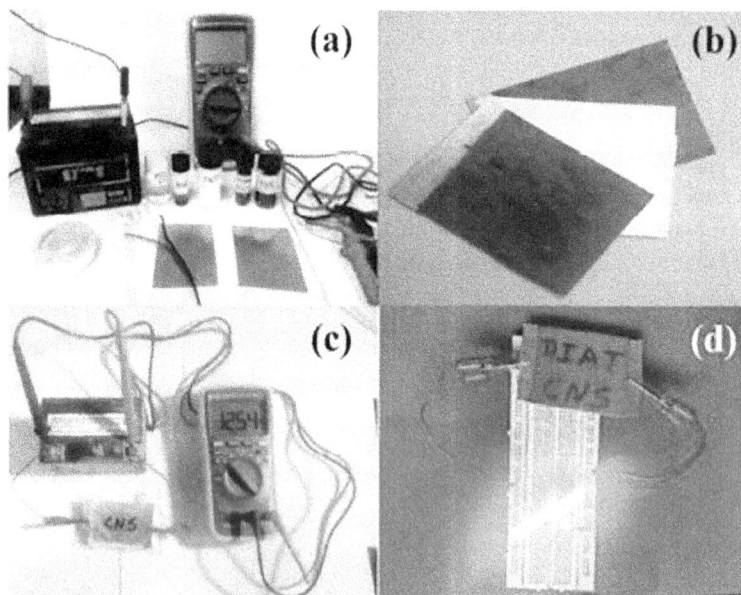

FIGURE 6.15 (a) Apparatus for fabrication of flat cell, (b) CNS slurry-coated electrodes and HCl-soaked separator, (c) charging of cell with automobile battery, (d) LED illumination by flat cell.

6.2.2.2 Flat-Cell Characteristics

We have implemented CNS for the fabrication of a flat cell, as seen in Figure 6.15. Figure 6.15a shows the photograph of apparatus required, such as (in bottles) CNS powder, PVDF, N-methyl-2-pyrrolidone (NMP), and 1 M HCl-soaked Whatman paper, automobile battery (12.5 V, DC), 1.0 mm thick aluminium (Al) sheets, connecting wires, crock-clips, digital multimeter, and 4–5 LEDs (1.5 V, each). Initially, CNS:PVDF in the ratio 9:1 were mixed in NMP solution of volume ~10 mL to form a slurry. Two Al electrodes were taken, and the slurry was coated onto the area ~4×5 cm^2 on each electrode. In parallel, the Whatman paper was soaked into 1 M HCl solution and kept for drying. The painted electrodes and the electrolyte-soaked separator can be seen in Figure 6.15b.

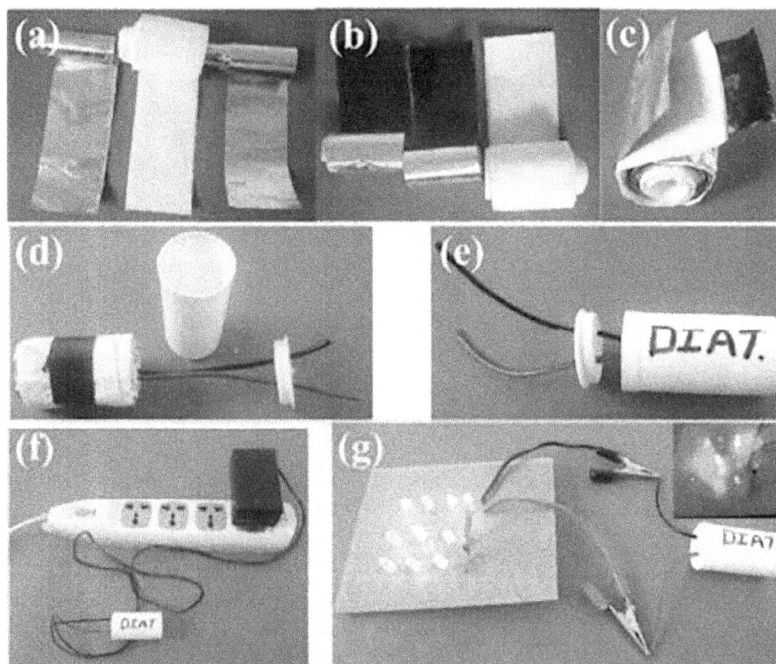

FIGURE 6.16 Production sequence for stand-alone CNS supercell: (a) Al electrodes ($4 \times 90 \times 0.05\,cm^3$), separator ($4 \times 100\,cm^2$), (b) CNS:PVDF: 90:10 (w/w%) in NMP-coated electrodes and HCl-soaked separator, (c) sandwiched Swiss roll assembly, (d) and (e) assembly installation in plastic encasing, (f) supercell charging ~5 m, and (g) power delivery ~0.8 W 45 m.

The electrode/separator/electrode junction was packaged with Teflon tape, and the two ends of the cell were connected to the battery and digital multimeter. The cell was subjected to charging, as seen in Figure 6.15c. The time required to charge the flat cell was ~30 seconds. Notably, the voltage of the cell decreased gradually. The charging of the cell was carried out for several times. In one charging cycle, 4 LEDs in series were connected to the cell and found to be illuminated for about 5 minutes as seen in Figure 6.15d. The total energy delivered was 3.0 Wh with total power 0.26 W [7,24].

6.2.2.3 Flexible CNS Electrodes: Supercell Device

In continuation with the flat-cell device, further development was made by fabricating a stand-alone CNS supercell. For this, a paper separator ($4.0 \times 100.0\,cm^2$) (Figure 6.16a) was sandwiched between the two Al electrodes ($4.0 \times 90.0\,cm^2$, thickness ~50 µm) (Figure 6.16a), in a Swiss roll configuration, and assembled into a cylindrical plastic container (inner diameter 2.5 cm and height 5.0 cm). For the development of the electrodes, a similar recipe was adopted, in which CNS and PVDF in the ratio of 9: 1 (w/w %) were taken and mixed into 10 ml of NMP to make the slurry. The slurry was coated onto the electrodes, and the coating was allowed to dry partially for a period of 15–20 minutes (Figure 6.16b).

Following this, a small amount of amorphous charcoal powder was sprinkled onto the semi-dried electrodes and, subsequently, allowed to dry fully. The separator was soaked into HCl and sandwiched between the coated electrodes, and this configuration was physically transformed into a Swiss roll structure and encased into the container to form a cell (Figure 6.16c–e). The cell was charged with a DC adapter for about 5 m and connected to several LED series (Figure 6.16f). Typically, 12 LEDs were illuminated with the supercell to deliver a total power of ~0.8 W to sustain for about 45 m or so (Figure 6.16g). Thus, from a laboratory scale (three electrodes) to a device level

TABLE 6.5
Performance Sheet: CNS Electrodes to Supercell Device

| | Electrodes | | | |
Parameter/Configuration	3	2	Flat Cell	Supercell
C_{SP} (@1 A/g)	–	560.0	–	
E_D (@1 A/g)	–	020.0	–	
P_D (@1 A/g)	–	250.0	–	
Cyclic stability @500	98.5%	–	–	
01000	97.0%			
05000	93.0%			
20,000	86.0%			
Total power delivered			0.26 W	0.80 W
Total energy delivered			3.0 Wh	1.5 Wh
Charging time			30 seconds	05 minutes
Discharging time			05 minutes	45 minutes

(supercell), CNS showed reliable performance characteristics as, comparatively, shown in Table 6.5. The results presented herein are of technological importance in energy storage devices, obtained via an environmentally friendly route, in which CNS is obtained by a facile pyrolysis of the camphor. The process is one step, involved complete combustion of precursor, does not yield any toxic by-products, and is time- and cost-effective [34]. The fabricated electrode material required no pre-/post-treatment, was flexible, and was readily integrable into the cell; cells are durable, having reliable performance characteristics [24].

6.3 SELF-ASSEMBLED TWO-DIMENSIONAL HETEROSTRUCTURE OF rGO/MoS$_2$/h-BN (GMH)

The focus of this current work was to investigate the electrochemical parameters for rGO/h-BN/MoS$_2$ heterostructure. The self-assembled composite was prepared by a simple addition of their equal volume ratios. The blend, after characterization, showed successful formation of composite made up of 2D materials. The electrochemical properties were studied to realize the effect of blend that attended the effective $C_{sp} > 650$ F/g (@1 A/g) with highest E_D 90 + Wh/kg and $P_D \sim 1700$ W/kg. The combination studied, herein, has not been reported thus far for its electrochemical performance [35].

6.3.1 2D HETEROJUNCTION: THE SURVEY

Several two-dimensional (2D) materials such as graphene, molybdenum disulphide, hexagonal boron nitride, transition metal dichalcogenides (TMDs), g-C$_3$N$_4$, phosphorene layered double hydroxides, and MXenes [36–39] are of great potential for energy storage applications due to their outstanding structure–property relationship. In addition to graphene, transition metal dichalcogenides (TMDs), post-graphene contenders, transition metal oxides (TMO)/hydroxides (TMH) are considered as promising candidates for energy storage. The structures of TMO and TMH in the form of films, flakes, platelets, petals, belts, etc., have significantly altered the inherent properties of supercapacitors attending excellent E_D. This is a huge class, among which we have specifically chosen reduced graphene oxide (rGO), MoS$_2$, and h-BN in the form of a blend to investigate their electrochemical performance. Basically, the use of carbon compounds in energy storage devices has its own advantages in the form of superior performance characteristics, high surface area, low cost, affordability to large-scale applications, and easy preparation protocols. There are number of reports on the use of nanocarbons such as rGO, single-walled carbon nanotubes (CNTs), CNTs/polymer, rGO-TMO, activated carbon, and multiwalled CNTs to achieve superior specific capacitance ($C_{sp,}$ in F/g). Particularly, rGO showed $C_{sp} \sim 200@1200$ cycles in electrolyte KOH [39–41], >300@1000 cycles in PVA [35], ~600@1000 with electrolyte H_2SO_4 [32], respectively.

Similarly, MoS_2 layers alone show the ability to store electrostatic charge by electric double-layer capacitance (ELDC) mechanism. However, they suffer from inferior electrochemical performance in terms of poor cyclic life, inherently low charge transport, large volume change during cycling, and restacking [7,24,32] yielding $C_{sp} \sim 100$ F/g @ 1 mV/s (scan rate) [15]. In order to overcome these issues, it has been mixed, wrapped, or deposited with highly conductive/electroactive materials such as carbonaceous materials or conducting polymers using various top-down/bottom-up synthetic as well as combinatorial approaches [24].

Pristine h-BN, being an electrical insulator, does not suitably fit in this branch of application. But due to structural similarity with graphene and comparatively weak van der Waals forces, it offers a strong ionic bond and is expected to provide superior C_{sp} if combined with graphene and MoS_2 [23].

6.3.1.1 rGO/MoS₂/h-BN (GMH Composite) Heterojunction: Fabrication and Analysis

Initially, graphene oxide (GO) was prepared from the natural graphite precursor by modified Hummers' route. GO has been washed with DI water for several times in order to make the obtained GO free from acidic moieties. Hexagonal boron nitride (h-BN) was prepared according to the protocols given in [32]. For this, around 3 g of boric acid (HBO_3) and approximately 9 g of urea were mixed in acetone and stirred for 15 m or so. Following this, acetone was allowed to evaporate under natural conditions and the obtained powder was heated for ~ 700°C for 5 hours, under nitrogen atmosphere. In a round-bottom flask, subsequently, GO (500 mg), h-BN (500 mg), ammonium molybdate ((NH_4)$_6Mo_7O_{24}$-$4H_2O$, 1.25 g), and thiourea (CH_4-NH_2S, 2.31 g) were added into the DI water [25]. In the reaction mixture, hydrazine hydrate (10 mL) was added and the entire system was kept for heating at ~80°C in an oil bath for about 4 hours. After this, the system was allowed to cool down to room temperature. During the synthesis of h-BN, the small amount of B_2O_3 formed may be reduced in the presence of hydrazine hydrate. Moreover, we prepared MoS_2, in situ, in the reduction process of GO. The water was separated using vacuum filtration, and the *black-coloured* residue was washed using hot DI water and ethanol, sequentially. The residue was sonicated, further, for about 4 hours to exfoliate the interlayers. Finally, the product was assumed to be multilayered rGO/MoS_2/h-BN and *termed as* GMH composite material. It was further characterized for the structure–property relationship and for its electrochemical application studies.

Electron spectroscopy for chemical analysis (ESCA) measurements were performed using an Omicron ESCA probe (Omicron Nanotechnology). X-ray powder diffraction studies were carried out using Rigaku instrument with Cu Kα radiation (1.5406 Å) over 2θ range 10°–90° @ scanning rate 2°m⁻¹. FTIR measurements were performed at 400–4000 cm⁻¹, using Bruker Tensor 37 and Raman over 200–3500 cm⁻¹ using LABRAM HR 800 (@ $\lambda \sim 533$ nm). Nanostructure imaging was carried out by field emission scanning (FSEM, S-4700), high-resolution transmission electron microscopy (HRTEM, JEOL-2100F), and energy-dispersive X-ray (EDAX) elemental mapping. ESR measurements were performed at 9.4 GHz (X-band) with microwave input 950 μW. Due to space limitations, we have discussed a few results.

6.3.1.2 Surface Chemical and Morphological Investigations

In order to understand the surface chemical composition and the valence state of the GMH blend, XPS analysis has been carried out. Figure 6.17 shows the corresponding spectra, and the inset in (a) shows a survey spectrum, indicating the coexistence of the elements such as B, S, Mo, C, N, and O in GMH.

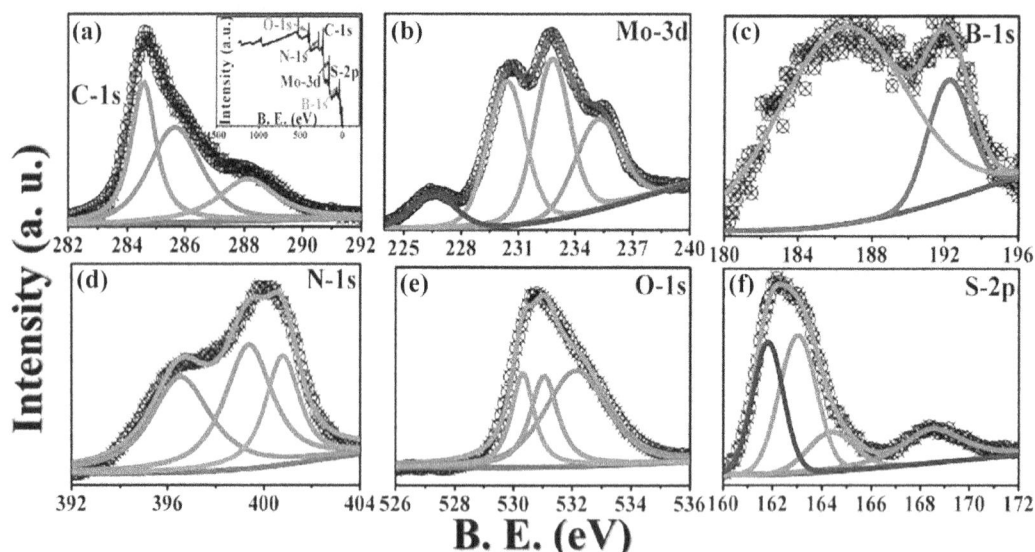

FIGURE 6.17 Recorded XPS spectra for GMH composites in which (a) C-1s (inset: elemental survey scan), (b) Mo-3d, (c) B-1s, (d) N-1s, (e) O-1s, and (f) S-2p.

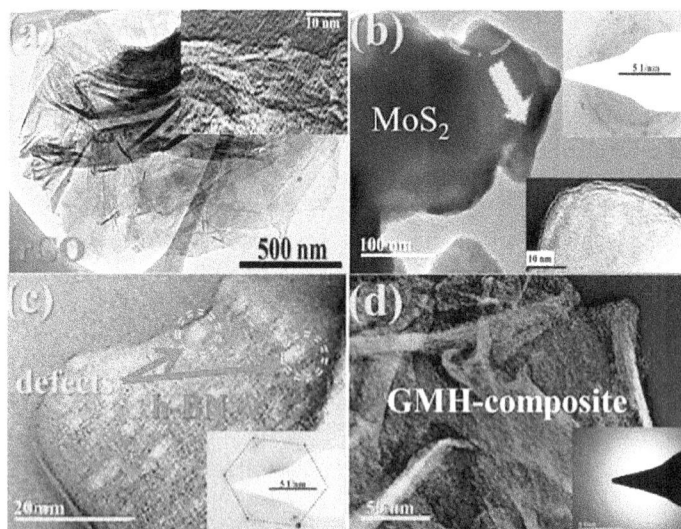

FIGURE 6.18 Typical HRTEM images for (a) rGO (inset: higher resolution of layers), (b) MoS$_2$ (inset: higher resolution and SAED pattern), (c) h-BN arrows and double circles show defects (inset: the corresponding SAED), and (d) GMH composites with SAED.

In the surface analysis of GMH composites, broadly, the excess of oxygen-containing groups were significantly reduced and only a small amount of oxygen was observed to reside in the 2D multilayers. From rGO, h-BN, and MoS$_2$ sheets, the unwanted traces of, respectively, hydroxide/epoxide, B$_2$O$_3$, and MoO$_3$ were removed due to the action of hydrazine hydrate as a reducing agent. The blend was found to be rich in C–B–C, Mo–S, C–N, B–N, Mo–C, etc., molecular bonding forming composite formation of composite layers.

Figure 6.18a image shows a typical TEM image of rGO, indicating well-defined graphitic planes stacked, and the respective inset displays higher-resolution imaging with lattice fringes. It indicates

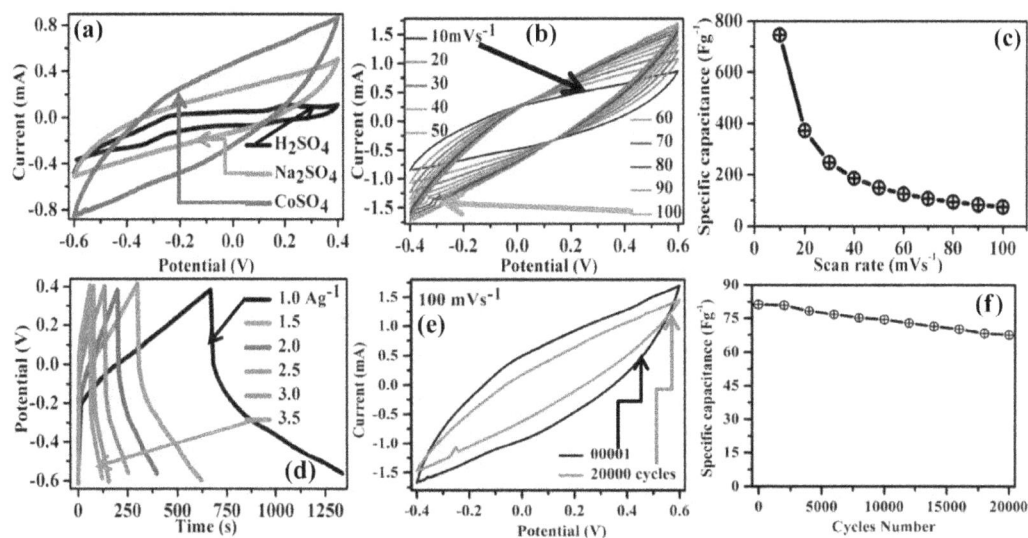

FIGURE 6.19 CV measurements on GMH composite for (a) different aqueous electrolytes (scan rate @10 mV/s) with superior performance in $CoSO_4$, (b) variable scan rates between 10 and 100 mV/s, (c) change in C_{SP} (specific capacitance) with scan rate, (d) stability curve up to 20,000 cycles, (e) CV curves recorded after the first and the last cycles, (f) charge–discharge (CD) curves at variable current density (A/g). Plots (b)–(f) are CV measurements in $CoSO_4$ electrolyte.

notable crystallization of the sheets. Figure 6.18b shows two layers of MoS_2, and it seems that due to stacking fault, the crystallinity was somewhat lost. In Figure 6.18c, crystalline h-BN is seen with some structural defects shown by yellow double circles. The defects could be attributed to the presence of B_2O_3. They were passivated as more amount of hydrazine hydrate was added, as discussed previously. The corresponding inset shows a SAED pattern, which confirmed the crystallinity of h-BN. Figure 6.19d shows stacked compact layers of rGO, MoS_2, and h-BN and a SAED pattern that indicated the amorphous nature of the blend. The supercapacitor performance of the GMH composite was investigated by cyclic voltammetry (CV) studies [32].

6.3.1.3 Electrochemical Studies

Figure 6.19a shows the three-electrode CV curves obtained for GMH composites in different aqueous electrolytes at a scan rate of 10 mV/s. The values of specific capacitance, C_{SP}, measured in F/g, were recorded to be 302.67, 564.18, and 745.18 for electrolytes H_2SO_4, Na_2SO_4, and $CoSO_4$, respectively. Figure 6.19b represents profiles (three electrodes) at various scan rates from 10 to 100 mV/s in $CoSO_4$ electrolyte. The CV curves exhibited a nearly rectangular shape and was indicative of an ideal electrode double-layer capacitor. From the device perspective, it is important to investigate two-electrode CV measurements. In Figure 6.19c, variations in C_{SP} with a change in scan rate have been observed to study the effective interaction between the ions and the electrode. Further, Figure 6.19d shows the galvanostatic charge discharge curves at various current densities from 1 to 3.5 A/g. The increased scan rate as well as current density shows a decrease in specific capacitance. Table 6.6 shows the specific capacitance, C_{SP}, for the two-electrode technique with corresponding power, P_D, and energy density, E_D. The values of C_{SP} in two-electrode systems are usually less as compared to three-electrode configurations and were calculated using the formulae stated in Section 6.1.2.

Figure 6.19e and f, respectively, shows I–V curves, after 1 and 20,000 cycles. After 20,000 cycles, a slight decrease in specific capacitance is observed by ~25% or so with a reduction in area under the curve. The good cyclic stability can be attributed to their excellent chemical stability, mechanical strength, and flexibility of GMH composites [33].

TABLE 6.6

Calculated C_{SP}, E_D, and P_D for GMH Composites for Two-Electrode System

Current Density (A/g)	1.0	1.5	2.0	2.5	3.0	3.5
C_{SP} (F/g)	683	482	401	296	224	214
E_D (Wh/kg)	92.3	65.1	54.2	40.0	30.0	28.9
P_D (W/kg)	486	729	972	1213	1458	1701

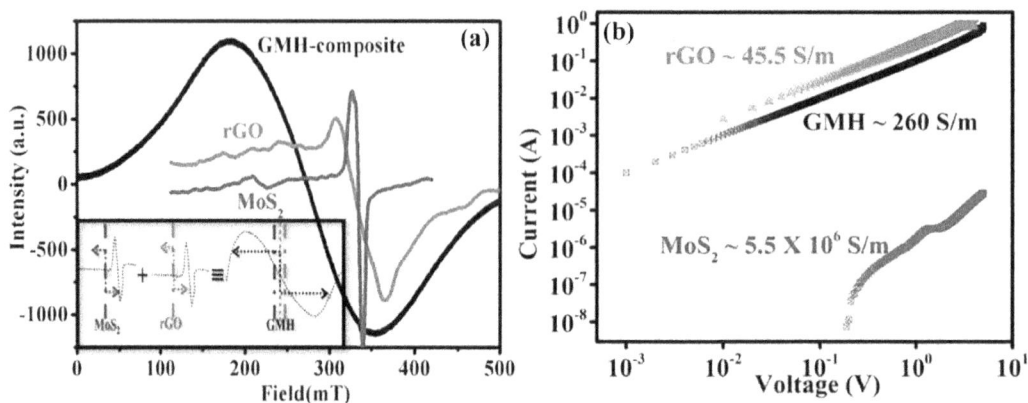

FIGURE 6.20 (a) Recorded electron spin resonance profiles for rGO, MoS$_2$, GMH composites @ 300 K. Inset shows the scheme for electron diffusion for the individual 2D materials and their composites. Arrows indicate, schematically, electron diffusion length, D_L, during charging in MoS$_2$, rGO, and GMH composites, respectively, 200, 600, and 1500 μm. (b) Recorded I–V profiles for rGO, MoS$_2$, and GMH composites @ 300 K indicating specific conductance. In both measurements, profiles for h-BN are not obtained.

Further, electron spin resonance (ESR) spectroscopy reveals dynamics of radical electron in a solid system resonating para-magnetically by the absorption of microwave energy. The purpose of ESR analysis was to estimate diffusion dynamics of such radical electrons in GMH blends vis-a-vis their individual counterparts.

As a first step towards computing electron transport parameters, the magnitude of the line width was determined by peak-to-peak distance, ΔH_{pp}, and the shape of the ESR spectra is investigated. Figure 6.20a shows the variation in intensity with applied magnetic field for rGO, MoS$_2$, and GMH composites at room temperature. The ESR spectrum of h-BN was not obtained. In case of rGO and MoS$_2$, the first absorption derivative dY/dH as a function of applied field shows unsymmetrical nature, while for GMH composites, symmetric and homogenous peaks were observed. Further, the principal parameter governing spintronics usability is the spin–lattice relaxation time T_{sl}, which characterizes non-thermal spin states around the lattice. From the measured values of ΔH_{pp}, and using $1/T_{sl} = (28.0\,\text{GHz})/\text{T (in K)} \times \Delta H_{pp}$, the measures of T_{sl} and electron diffusion length, D_L, have been carried out with the assumption: The Fermi velocity is ~10^6 m/s in all 2D materials [39,40]. For MoS$_2$, the D_L is estimated to be ~200 μm, whereas for rGO, it is ~600 μm. The effective D_L was found to be ~1500 μm for GMH composites. This indicates that on charging, electron diffusion is quite high for GMH compared with their individual components. Further, this also supports the conductivity measurements.

Figure 6.20b shows the I–V plots for rGO, MoS$_2$, and GMH using two-probe conductivity measurements at room temperature. In case of h-BN, no such curve is obtained. The calculated

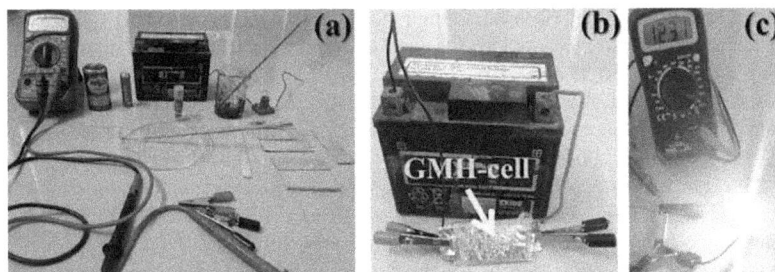

FIGURE 6.21 Laboratory-scale demonstration of GMH electrode demonstrating LED illumination. (a) Components such as substrate, digital multimeter, 12.4 V commercial automobile battery, active electrode material, carbon powder, and PVDF, (b) charging of the *blend-cell* with battery, and(c) LED illumination after charging, indicating voltage attained by the fully sealed fabricated blend electrode assembly.

values for the specific conductivity of the prepared pellets of rGO, MoS_2, and GMH are 45.5 ± 3.5, $5.5 \pm 15 \times 10^6$, and 260 ± 30 S/m, respectively. The specific conductivity of rGO is very high as compared to GMH and MoS_2. The observed decrease is due to the presence of h-BN in the composite, due to which the magnitude of T_{sl} and implications of D_L prove the composite material good for energy storage. Therefore, further we have prepared the cell for GMH blend and the performance is tested [32].

6.3.1.4 GMH Cell: Performance

GMH blend has been implemented for active electrode material for laboratory-scale demonstration of the *blend-cell* action. The electrodes are developed as described in [31]; Figure 6.21a shows the corresponding apparatus, starting materials, flat electrodes, etc. The assembly is packed with an aluminium foil as seen in Figure 6.21b. Prior to charging, the voltage between the two electrodes was measured and found to be negligibly small ~0.01 mV or so. The assembly is connected to *crock-clips* having conducting wires. The cell is charged for about 1.5 minutes with nominal 12.50 V automobile battery. The voltage measured, instantaneously, is found to be ~12.30 V, after charging, as seen in Figure 6.21c.

The voltage fallen down gradually with time, indicating recombination of the developed charges across 2D materials. The Helmholtz layer degradation time is found to be very small for such primitive assembly. The charging measurements are performed, repetitively. In one case, we connected the assembly to a commercially available LED equipped with current-carrying wires. The LED is found to be illuminated, seen in Figure 6.21c for a period of 5 minutes or so. The result shows the laboratory-scale performance of the fabricated electrode using GMH molecular blend. No such effect is observed for rGO specimen prepared and tested in an identical fashion [42,43].

REFERENCES

[1] Alegaonkar, Ashwini P., Manjiri A. Mahadadalkar, Prashant S. Alegaonkar, Bharat B. Kale, and Satish K. Pardeshi. "High performance tellurium-reduced graphene oxide pseudocapacitor electrodes." *Electrochimica Acta* 291 (2018): 225–233.

[2] Nguyen, Thu Thuy. "CO_2 emissions, financial development, and renewable energy consumption (REC): a metadata analysis." In *Industrial transformation*, edited by Om Prakash Jena, Sudhansu Shekhar Patra, Mrutyunjaya Panda, Zdzislaw Polkowski, S. Balamurugan, pp. 251–270. CRC Press, Boca Raton, FL, 2022.

[3] Brousse, Thierry, Mathieu Toupin, Romain Dugas, Laurence Athouël, Olivier Crosnier, and Daniel Belanger. "Crystalline MnO_2 as possible alternatives to amorphous compounds in electrochemical supercapacitors." *Journal of the Electrochemical Society* 153, no. 12 (2006): A2171.

[4] Wang, Da-Wei, Feng Li, and Hui-Ming Cheng. "Hierarchical porous nickel oxide and carbon as electrode materials for asymmetric supercapacitor." *Journal of Power Sources* 185, no. 2 (2008): 1563–1568.

[5] Cottineau, Thomas, Mathieu Toupin, Timur Delahaye, Thierry Brousse, and Daniel Bélanger. "Nanostructured transition metal oxides for aqueous hybrid electrochemical supercapacitors." *Applied Physics A* 82, no. 4 (2006): 599–606.

[6] Mastragostino, Marina, Catia Arbizzani, Luca Meneghello, and Ruggero Paraventi. "Electronically conducting polymers and activated carbon: electrode materials in supercapacitor technology." *Advanced Materials* 8, no. 4 (1996): 331–334.

[7] Haladkar, Sushant, and Prashant Alegaonkar. "Preparation and performance evaluation of carbon-nano-sphere for electrode double layer capacitor." *Applied Surface Science* 449 (2018): 500–506.

[8] Liu, Chenguang, Zhenning Yu, David Neff, Aruna Zhamu, and Bor Z. Jang. "Graphene-based supercapacitor with an ultrahigh energy density." *Nano Letters* 10, no. 12 (2010): 4863–4868.

[9] Raymundo-Piñero, Encarnacion, Fabrice Leroux, and François Béguin. "A high-performance carbon for supercapacitors obtained by carbonization of a seaweed biopolymer." *Advanced Materials* 18, no. 14 (2006): 1877–1882.

[10] Fuertes, Antonio B., Fernando Pico, and Jose M. Rojo. "Influence of pore structure on electric double-layer capacitance of template mesoporous carbons." *Journal of Power Sources* 133, no. 2 (2004): 329–336.

[11] Sevilla, M., S. Alvarez, Teresa A. Centeno, A. B. Fuertes, and Fritz Stoeckli. "Performance of templated mesoporous carbons in supercapacitors." *Electrochimica Acta* 52, no. 9 (2007): 3207–3215.

[12] Wang, Da-Wei, Feng Li, Jinping Zhao, Wencai Ren, Zhi-Gang Chen, Jun Tan, Zhong-Shuai Wu, Ian Gentle, Gao Qing Lu, and Hui-Ming Cheng. "Fabrication of graphene/polyaniline composite paper via in situ anodic electropolymerization for high-performance flexible electrode." *ACS Nano* 3, no. 7 (2009): 1745–1752.

[13] Weng, Zhe, Feng Li, Da-Wei Wang, Lei Wen, and Hui-Ming Cheng. "Controlled electrochemical charge injection to maximize the energy density of supercapacitors." *Angewandte Chemie International Edition* 52, no. 13 (2013): 3722–3725.

[14] Niu, Zhiqiang, Weiya Zhou, Jun Chen, Guoxing Feng, Hong Li, Wenjun Ma, Jinzhu Li et al. "Compact-designed supercapacitors using free-standing single-walled carbon nanotube films." *Energy & Environmental Science* 4, no. 4 (2011): 1440–1446.

[15] Stoller, Meryl D., Sungjin Park, Yanwu Zhu, Jinho An, and Rodney S. Ruoff. "Graphene-based ultracapacitors." *Nano Letters* 8, no. 10 (2008): 3498–3502.

[16] Zhu, Yanwu, Shanthi Murali, Meryl D. Stoller, K. J. Ganesh, Weiwei Cai, Paulo J. Ferreira, Adam Pirkle et al. "Carbon-based supercapacitors produced by activation of graphene." *Science* 332, no. 6037 (2011): 1537–1541.

[17] Wu, Qiong, Yuxi Xu, Zhiyi Yao, Anran Liu, and Gaoquan Shi. "Supercapacitors based on flexible graphene/polyaniline nanofiber composite films." *ACS Nano* 4, no. 4 (2010): 1963–1970.

[18] Kang, Yu Jin, Sang-Jin Chun, Sung-Suk Lee, Bo-Yeong Kim, Jung Hyeun Kim, Haegeun Chung, Sun-Young Lee, and Woong Kim. "All-solid-state flexible supercapacitors fabricated with bacterial nanocellulose papers, carbon nanotubes, and triblock-copolymer ion gels." *ACS Nano* 6, no. 7 (2012): 6400–6406.

[19] Childres, Isaac, Luis A. Jauregui, WonjunPark, HelinCao, and Yong P. Chen. *New developments in photon and materials research.* Nova Science Publishers, Switzerland, 2013.

[20] Yoo, Jung Joon, Kaushik Balakrishnan, Jingsong Huang, Vincent Meunier, Bobby G. Sumpter, Anchal Srivastava, Michelle Conway et al. "Ultrathin planar graphene supercapacitors." *Nano Letters* 11, no. 4 (2011): 1423–1427.

[21] Rani, Pinki, Ashwini P. Alegaonkar, Santosh K. Mahapatra, and Prashant S. Alegaonkar. "Tellurium nanostructures for optoelectronic applications." *Applied Physics A* 128, no. 4 (2022): 1–13.

[22] Wang, Yaming, Junchen Chen, Jianyun Cao, Yan Liu, Yu Zhou, Jia-Hu Ouyang, and Dechang Jia. "Graphene/carbon black hybrid film for flexible and high rate performance supercapacitor." *Journal of Power Sources* 271 (2014): 269–277.

[23] Kerner, Ross A., and Barry P. Rand. "Linking chemistry at the $TiO_2/CH_3NH_3PbI_3$ interface to current–voltage hysteresis." *The Journal of Physical Chemistry Letters* 8, no. 10 (2017): 2298–2303.

[24] Haladkar, Sushant A., Mangesh A. Desai, Shrikrishna D. Sartale, and Prashant S. Alegaonkar. "Assessment of ecologically prepared carbon-nano-spheres for fabrication of flexible and durable supercell devices." *Journal of Materials Chemistry A* 6, no. 16 (2018): 7246–7256.

[25] Biswal, Mandakini, Abhik Banerjee, Meenal Deo, and Satishchandra Ogale. "From dead leaves to high energy density supercapacitors." *Energy & Environmental Science* 6, no. 4 (2013): 1249–1259.

[26] Kalpana, Dharmalingam, Kaliyappan Karthikeyan, Nedumaram Gopalan Renganathan, and Yun-sung Lee. "Camphoric carbon nanobeads–a new electrode material for supercapacitors." *Electrochemistry communications* 10, no. 7 (2008): 977–979.

[27] Mi, Hongyu, Jiapan Zhou, Qingxia Cui, Zongbin Zhao, Chang Yu, Xuzhen Wang, and Jieshan Qiu. "Chemically patterned polyaniline arrays located on pyrolytic graphene for supercapacitors." *Carbon* 80 (2014): 799–807.

[28] Moreira, João Vitor Silva, Evaldo José Corat, Paul William May, Lays Dias Ribeiro Cardoso, Pedro Almeida Lelis, and Hudson Zanin. "Freestanding aligned multi-walled carbon nanotubes for supercapacitor devices." *Journal of Electronic Materials* 45, no. 11 (2016): 5781–5788.

[29] Martindale, Benjamin C. M., Georgina A. M. Hutton, Christine A. Caputo, Sebastian Prantl, Robert Godin, James R. Durrant, and Erwin Reisner. "Enhancing light absorption and charge transfer efficiency in carbon dots through graphitization and core nitrogen doping." *Angewandte Chemie* 129, no. 23 (2017): 6559–6563.

[30] Joseph, Jickson, Anjali Paravannoor, Shantikumar V. Nair, Zhao Jun Han, Kostya Ken Ostrikov, and Avinash Balakrishnan. "Supercapacitors based on camphor-derived meso/macroporous carbon sponge electrodes with ultrafast frequency response for ac line-filtering." *Journal of Materials Chemistry A* 3, no. 27 (2015): 14105–14108.

[31] Joshi, Anupama, Arvind Kumar, Prashant S. Alegaonkar, and Suwarna Datar. "Graphene-like-nanocarbon—polyaniline composite as supercapacitor." *Energy and Environment Focus* 2, no. 3 (2013): 176–180.

[32] Adoor, Rashmi S., Sushant A. Haladkar, Prashant S. Alegaonkar, and Narasimha H. Ayachit. "Study of electrochemical parameters of carbon-nano-spheres/polyaniline nano-composite." In *AIP Conference Proceedings*, vol. 2244, no. 1, p. 110010. AIP Publishing LLC, 2020.

[33] Gangwar, Rajesh K., Vinayak A. Dhumale, Arvind Kumar, Prashant Alegaonkar, Rishi B. Sharma, and Suwarna S. Datar. "Gold-graphene nanocomposite based ultrasensitive electrochemical glucose sensor." In *2012 1st International Symposium on Physics and Technology of Sensors (ISPTS-1)*, pp. 282–285. IEEE, 2012.

[34] Iqbal, Muzahir, Nilesh G. Saykar, Anil Arya, Indrani Banerjee, Prashant S. Alegaonkar, and Santosh K. Mahapatra. "High-performance supercapacitor based on MoS_2@ TiO_2 composite for wide range temperature application." *Journal of Alloys and Compounds* 883 (2021): 160705.

[35] Iqbal, Muzahir, Nilesh Gokul Saykar, Prashant Alegaonkar, and Santosh K. Mahapatra. "Synergistically modified WS_2@ PANI binary nanocomposite based all-solid-state symmetric supercapacitor with high energy density." *New Journal of Chemistry* 46 (2022) 7043–7054.

[36] Alegaonkar, Ashwini P., Prashant S. Alegaonkar, and Satish K. Pardeshi. "Electrochemical performance of a self-assembled two-dimensional heterostructure of $rGO/MoS_2/h$-BN." *Nanoscale Advances* 2, no. 4 (2020): 1531–1541.

[37] Chhowalla, Manish, Zhongfan Liu, and Hua Zhang. "Two-dimensional transition metal dichalcogenide (TMD) nanosheets." *Chemical Society Reviews* 44, no. 9 (2015): 2584–2586.

[38] Tian, Yue, Huide Wang, Haonan Li, Zhinan Guo, Bining Tian, Yanxia Cui, Zhanfeng Li, Guohui Li, Han Zhang, and Yucheng Wu. "Recent advances in black phosphorus/carbon hybrid composites: from improved stability to applications." *Journal of Materials Chemistry A* 8, no. 9 (2020): 4647–4676.

[39] Zhang, Yu, Yun Zheng, Kun Rui, Huey Hoon Hng, Kedar Hippalgaonkar, Jianwei Xu, Wenping Sun, Jixin Zhu, Qingyu Yan, and Wei Huang. "2D black phosphorus for energy storage and thermoelectric applications." *Small* 13, no. 28 (2017): 1700661.

[40] Zhang, Shilin, Feng Yao, Lan Yang, Fazhi Zhang, and Sailong Xu. "Sulfur-doped mesoporous carbon from surfactant-intercalated layered double hydroxide precursor as high-performance anode nanomaterials for both Li-ion and Na-ion batteries." *Carbon* 93 (2015): 143–150.

[41] Lipatov, Alexey, Mohamed Alhabeb, Maria R. Lukatskaya, Alex Boson, Yury Gogotsi, and Alexander Sinitskii. "Effect of synthesis on quality, electronic properties and environmental stability of individual monolayer Ti_3C_2 MXene flakes." *Advanced Electronic Materials* 2, no. 12 (2016): 1600255.

[42] Wang, Yan, Zhiqiang Shi, Yi Huang, Yanfeng Ma, Chengyang Wang, Mingming Chen, and Yongsheng Chen. "Supercapacitor devices based on graphene materials." *The Journal of Physical Chemistry C* 113, no. 30 (2009): 13103–13107.

[43] Victor L. Pushparaj, Manikoth M. Shaijumon, Ashavani Kumar, Saravanababu Murugesan, Lijie Ci, Robert Vajtai, Robert J. Linhardt, Omkaram Nalamasu and Pulickel M. Ajayan, "Flexible energy storage devices based on nanocomposite paper." *Proceedings of the National Academy of Sciences of the United States of America* 104 (2007): 13574–13577.

7 Magnetism in Otherwise Non-Magnetic Nanocarbons and Their Derivatives

7.1 SPIN TRANSPORT AND MAGNETIC CORRELATION IN GNCs DOPED WITH NITROGEN

Herein, we have investigated magneto-spin correlations in GNCs that were doped with nitrogen, using an organic compound tetrakis(dimethylamino)ethylene (TDAE). The spin transport measurements, performed by electron spin resonance technique, showed that both spin–spin and spin–lattice relaxation times are increased by doping nitrogen. The magnetic correlations, measured using vibrating sample magnetometer, showed that ordering parameters are reduced for nitrogen-loaded GNCs. Chemical analysis, carried out using electron spectroscopy, revealed that nitrogen atoms, in TDAE, donate electrons to carbon network and exchange holes. I–V measurements, performed on the system, showed that higher-order resistance is appreciably decreased for nitrogen-doped GNCs. The observed decrease could be due to the increase in nonbonding states, having small local density. After doping, states in this region may be localized π spin populated around the doped region. By and large, around 20% magnetization, which exists in GNCs, is found to reduce to 5% by introducing nitrogen [1].

7.1.1 NON-MAGNETIC CARBON: THE SURVEY

Magnetic materials are omnipresent component in today's technology. Currently used metal magnets involve partially filled d- or f-band atoms. Pristine carbon is strongly diamagnetic and consists of s- and p-electrons. However, in recent years, the existence of ferro-, antiferro-, and paramagnetic ordering in carbon has become the focus of several investigations. The issue of *carbon magnetism* is controversial, is intriguing, and originates due to size reduction of the system. A large number of experimental attempts have been made to demonstrate allotropes of carbon, such as fullerenes, highly ordered pyrolytic graphite, carbon nano-foams, and nano-diamonds, as magnetic materials [2–5]. Graphene is not an exception [6]. The magnetic moments, in graphene, emerge due to zig-zag edge states, topological disorders, unsaturated dangling bonds, mixed sp^2/sp^3 interconnected phases, etc. [2,5,7,8]. Moreover, multi-shaped graphene fragments such as triangular and hexagonal nano-islands, ribbons, nano-flakes, and fractal carbon have shown high spin ground state and behaved as artificial ferromagnetic atoms [9–12]. To provide an atomic-level understanding of the observed magnetism, in carbon, numerous theoretical studies have been performed [13–15]. The study showed that isolated vacancies and chemisorption of foreign atoms near vacancies could induce strong local magnetic moments [16]. The atomic origin of magnetic moments has three principal sources: (a) the spin with which electrons are endowed, (b) their orbital angular momentum about the nucleus, and (c) the change in the orbital momentum induced by external perturbations. The first two give spin–spin and spin–orbit interactions, and the third measures the strength of spin–orbit coupling. In graphene, spin–orbit interactions couples π and σ bands. The principal parameter governing usability of graphene in magnetic applications is spin–lattice relaxation time. The relaxation of spin coupled to its lattice depends on broken inversion symmetry and the presence of heterostructure in two-dimensional graphene superlattice. The Elliott–Yafet mechanism explains the former case, and the latter is based on the Dyakonov–Perel (DyP) theory [17–21].

DOI: 10.1201/9781003317258-7

7.1.1.1 Methods for Nitrogen Doping in GNCs

GNCs were synthesized according to the protocol presented in Section 2.1.1. The material obtained was a class of graphene nano-sheets and not graphene. It contains mixed sp^2+sp^3 phase rather than pure sp^2 bonded graphene network comprised of, typically, two–five layers with heavy local disorder therein [1,6,22,23]. Nitrogen doping was carried out using tetrakis(dimethylamino)ethylene (TDAE) compound. Initially, the suspension of GNCs was prepared in 25 mL of tetrahydrofuran (THF) and 0.1 mg of TDAE was added in GNCs suspension. After adding TDAE, the suspension was sonicated for 30 minutes followed by room-temperature stirring for 8–10 hours. The suspension was allowed to settle for about 5 hours. The vacuum filtration was carried out using PTFE filter (pore size ~1.2 μm). The GNCs treated with TDAE were termed as N-GNCs. The powder obtained was used for further preliminary characterization using Fourier transform infrared and UV–visible spectroscopies.

7.1.2 Magneto-Spin Investigations

7.1.2.1 ESR

To explore spin transport, electron spin resonance (ESR) measurements were carried out using a standard ESR set-up equipped with electromagnet, microwave bridge, resonant cavity, waveguide circuitry, and spectrometer consol. Initially, a known amount of sample, under investigation, was placed in the rectangular cavity consisting of cylindrical sample transfer ports. The mode of the cavity was TE_{102}. The sample was positioned at cavity centre where the magnetic component of the microwave standing wave attains the local maxima. This configuration provides maximum sensitivity to the measurement. The microwave energy was injected and coupled via an iris screw. The critical coupling achieved using an iris screw controlled the amount of incident and reflected microwave radiation in the cavity. The measurements were performed at a microwave frequency of ~9.1 GHz (X-band). The maximum microwave output was varied from sample to sample in the range 985–1000 μW. Further, for receiver mode, phase shifter enables one to match the phase of microwave signal reflected from cavity with the phase of microwave injected in the reference arm of the circulator. The value of variable phase shifter was kept fixed at zero. The quality factor of the cavity was 12,000. In general, the ESR signal is weak and submerged in the background level noise. By enhancing the sensitivity of spectrometer, one can amplify the obtained ESR signal. In the present measurements, the modulation frequency was kept constant at 100 kHz by adjusting the band-pass filter parameter of the lock-in amplifier. The static magnetic field was swept slowly at a spectrum–point time constant 0.1 seconds over the range 300–370 mT with the amplitude of modulation frequency kept at 6 kHz. The field centre was 336 mT and ESR line width, $\Delta H = 0.05$ mT. The signal-to-noise ratio was computed to be 20 at 300 K with 1 second per spectrum–point constant. The sensitivity of the system was 7.0×10^9 spins/0.1 mT and resolution 2.35 μT. All measurements were carried out at room temperature. And the first derivative of the paramagnetic absorption signal was recorded for the GNCs and N-GNCs. The highest concentration samples were taken from the batch of N-GNCs [24].

7.1.2.2 Magnetometry

A vibrating sample magnetometer (VSM), based on the principle of mutual inductance, is used to measure the magnetic behaviour of magnetic materials. The sensitivity of VSM system is crucially important and depends on the calibration as well as background noise level. Hence, prior to VSM measurements, calibration and estimation of noise measurements were carried out. To perform calibration, saturation magnetization of nickel was recorded as a function of field and temperature (300–77 K) for different masses of nickel. The calibration constant was found to be 0.580 ± 0.001 A–m²/V. The sensitivity of the system was 5×10^{-6} A–m² above 2T and 1×10^{-6} A–m² below 2T at 77 K. At room temperature, sensitivity of the system was found to be $1–2.7 \times 10^{-7}$ A–m² below 2T.

In order to estimate diamagnetic background signal, the *null scan* of tufnol/PTFE sample holder was performed before every sample scan. The sample was poured into sample holder and attached firmly to head driver in order to vibrate, harmonically, in synchronization with driver at a frequency 60 Hz. The isotherm plots of magnetization (M) as a function of applied magnetic field (H) were recorded for GNCs and N-GNCs samples. For both samples, M–H loops were obtained after linear subtraction of the diamagnetic background signal. The measurements were performed at room temperature, and field was swept from -1.5×10^4 to $+1.5 \times 10^4$ Oe. For all measurements, the transimpedance amplifier course gain was fixed at 0.1 and a lock-in amplifier at window gain 0.35 [25].

In another study presented in Section 7.4, magnetic measurements were carried out on a Quantum Design MPMS-XL-1 SQUID magnetometer with a 1 T magnet over 2–300 K with sensitivity better than 5×10^{-8} emu. For thermo-magnetic measurements, both field-cooled (FC) and zero-field-cooled (ZFC) curves were obtained at 100 Oe over 2–300 K. The susceptibility, χ, measurements were performed at 1 T magnetic field at 2 and 300 K. Prior to analysis, the diamagnetic corrections were performed on the collected data [24,25].

7.1.3 ELECTRONIC TRANSPORT PROPERTIES

Both GNCs and N-GNCs were weighed from 0.1 to 0.5 mg and sonicated well in THF for a period of 15 minutes to obtained better dispersion. These suspensions were used for *I–V* measurements. To perform *I–V* measurements, a standard cyclic voltmeter set-up was used [1,6,22–25].

Figure 7.1a shows the photograph of *I–V* measurement set-up, and the schematic of assembly is shown in Figure 7.1b. The facility consists of a computer-controlled potentiostat and galvanostat having a four-electrode assembly. For *I–V* measurements, one of the electrodes was disabled and another was used as a reference electrode connected to potentiostat via current meter and voltmeter. The key features of the system are the following: The sweep voltage can vary from −10 to +10 V, having compliance voltage ± 30 V with maximum current ± 2A. The current can vary from 1 A to 10 nA with voltage accuracy ± 0.2%. The voltage resolution was 0.3 mV with current accuracy ± 0.2%. The resolution for current was 0.0003% over the measured current sweep range. The input impedance was greater than 1 TΩ with potentiostatic bandwidth of 1 MHz. The *I–V* measurements were performed on all N-GNCs samples.

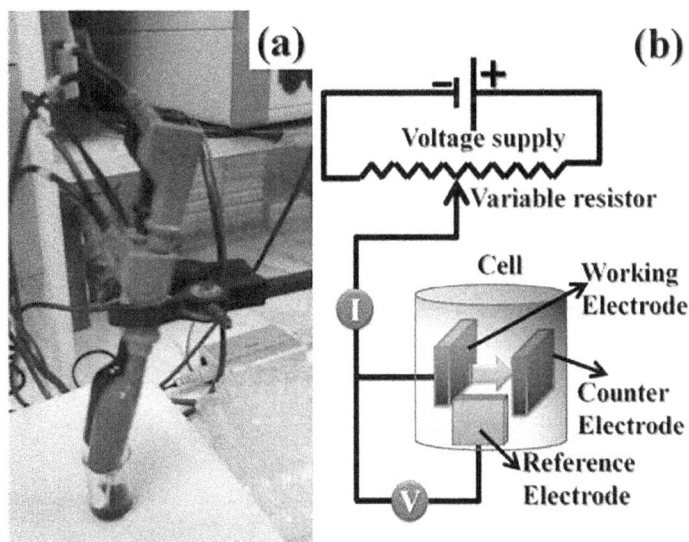

FIGURE 7.1 (a) Photograph of *I–V* measurement set-up. (b) Schematic assembly details. It shows cell arrangement that consists of electrodes, circuitry arrangements for current, voltage meter, potentiostat, etc.

7.1.4 MAGNETIC CORRELATIONS: GNCs vs N-GNCs

7.1.4.1 Spin Transport

Figure 7.2a shows the room-temperature ESR spectrum for GNCs. The first absorption derivative, dY/dH, as a function of applied field shows a symmetric absorption peak. The broadening seems to be symmetric and homogeneous for GNCs. The ESR line shape, in general, gives information about magnetic interaction in the measured sample. The line width is determined by measuring peak to peak distance, ΔH_{pp}. The peak width ΔH_{pp} has an azimuth angular dependence, θ, i.e. orientation of applied field with respect to the orientation of GNCs planes. Since measurements were conducted on the bulk powder specimen, $\Delta H_{pp}(\theta) = \sin^2(\theta) H_{\parallel} + \cos^2(\theta) H_{\perp}$. The computed width is found to be 0.6756 mT for GNCs. The measured ΔH_{pp} consists of contribution from $H_{\parallel}$ as well as $H_{\perp}$. The magnitude of ΔH_{pp} bears important information about the spin dynamics of the system, specifically, T_{ss}, which corresponds to electron spin–spin relaxation time. And ΔH_{pp} and T_{ss} are correlated by the equation: $\Delta H_{pp} = \dfrac{1}{\gamma_e T_{ss}}$, where γ_e is the gyromagnetic ratio for electrons having magnitude 1.760859×10^{11}/sec–T. The computed value of T_{ss} is found to be 0.8406 psec [1,6,22–26].

The value of g-factor, at which the resonance has occurred under applied microwave frequency, characterizes the magnetic moment and gyromagnetic ratio associated with unpaired electrons in the material. The magnitude of g-factor is estimated to be 1.99685 for GNCs. The value of g-factor is observed to be less than the g-factor for a typical non-degenerated Pauli gas, 2.0023. This indicates that, in GNCs, the local magnetic environment is distinctly different than that exists in a conventional magnetic material. The small values of line width and the small deviation in g-factor suggest that spin does not originate from transition metal impurities, but from only carbon-inherited spin species in GNCs. Furthermore, the principal parameter governing spintronic usability is the spin–lattice relaxation time, T_{sl}, which characterizes the variation in non-thermal spin state around the lattice. The magnitude of T_{sl} is computed using relation: $\dfrac{1}{T_{sl}} = \dfrac{28.0 \text{ GHz}}{T \ (in \text{ K})} \times \Delta H_{pp}$ and found to be 1.586 nsec. For spintronic applications, the theoretical estimate for T_{sl} is 1–100 ns, whereas the experimental spin transport measurements on graphene showed that T_{sl} is as short as 60–150 psec

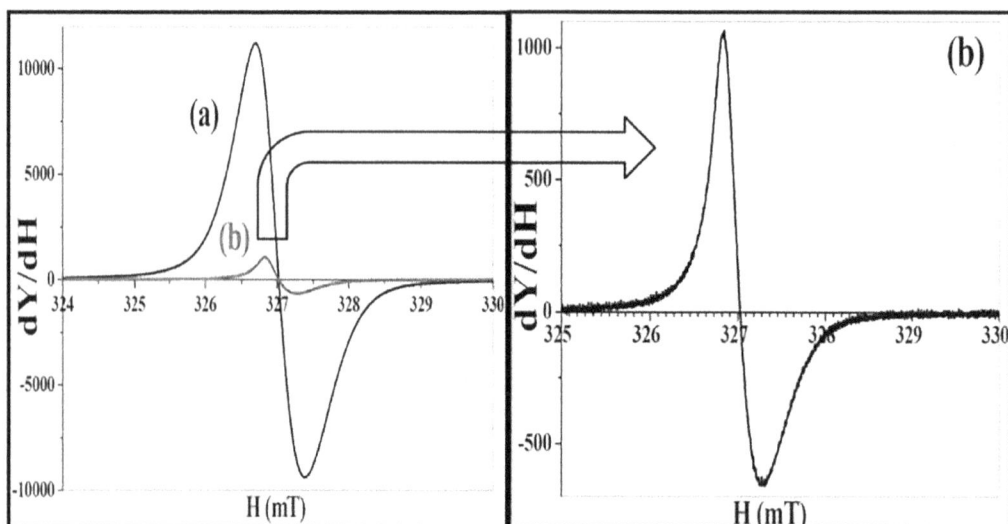

FIGURE 7.2 A typical electron spin resonance spectra recorded, at room temperature. The left panel shows the comparison of (a) GNCs and (b) N-GNCs line widths, and the right panel shows enlarged feature for N-GNCs, indicated by an arrow.

TABLE 7.1

Comparison of Spin Transport Parameters for GNCs and N-GNCs (Maximum Concentration ~ 0.5 mg/mL) Computed Using Theory of Spin Relaxation from the Experimentally Obtained ESR Lines

Parameters	GNCs	N-GNCs
Resonance magnetic field (H_r)	327.163 mT	327.020 mT
g-Factor	1.99685	1.99856
Peak-to-peak line width (ΔH_{pp})	0.6756 mT	0.4618 mT
Spin–spin relaxation time (T_{ss})	0.8406 psec	1.2297 psec
Spin–lattice relaxation time (T_{sl})	1.586 nsec	2.3201 nsec
Δg	5.45×10^{-3}	4.24×10^{-3}
Band width (Δ)	4.06 eV	3.79 ev
Spin–orbit coupling constant (L_i)	22.12 meV	16.07 meV
Spin relaxation rate (Γ_{spin})	7.81×10^{-8} eV	5.35×10^{-8} eV
Momentum relaxation rate (Γ)	2.63 meV	2.97 meV
Pseudo-chemical potential (.)	1.94 eV	1.79 eV
Density of states (ρ)	0.07469 stats/eV-atom	0.06892 stats/eV-atom
Pauli-spin susceptibility (χ at 300 K)	4.80×10^{-7}	4.43×10^{-7}
ESR line limit of detection	$514.81 \bullet L_D$	$580.55 \bullet L_D$

[22]. Using the theory of spin relaxation, spin transport parameters were obtained, analytically. A comprehensive exercise, performed on GNCs, is provided in Appendix C (Table C.7.1.4 and Figure C.7.2.1), and in a similar fashion, parameters were obtained for N-GNCs over 123–473K, including graphene and N-graphene. Figure 7.2b shows the ESR spectrum for N-GNCs. The comparison of obtained parameters is shown in Table 7.1 [1,6,22–25].

The comparison shows that the nature of spin transport in N-GNCs is modified and reflected in variations in measured relaxation parameters. Both T_{ss} and T_{sl} are increased, respectively, to 1.2297 psec and 2.3201 nsec after doping GNCs with nitrogen. Thus, electron spin takes more time to regain its original state. The hindrance could be offered by donor-loaded nitrogen group. The magnitude of spin–orbit coupling constant, L_i, is decreased for GNCs from 22.12 to 16.07 meV, after nitrogen doping. In graphene, there are three principal sources of spin–orbit coupling constant (SOCC): intrinsic, Bychkov–Rashba (BR, related to structural symmetry break), and ripples (related to inevitable wrinkles/folding edges). The theoretical estimate for intrinsic SOCC ranges between 1 µeV and 0.2 meV and BR 10–36 µeV/V/nm. For the intrinsic component, when $\tilde{\mu} \gg \Gamma$, Γ (intrinsic) $\approx \dfrac{L_i^2}{\Delta^2} \Gamma$, which is an Elliot–Yafet-like result [19]. Here, $\tilde{\mu}$ is the pseudo-chemical potential, Γ is the momentum relaxation rate, and Γ (intrinsic) is the intrinsic momentum relaxation rate. Further, the ripple relaxation contribution becomes dominant only when $\Gamma \gg \tilde{\mu}$. The computed value of Γ and $\tilde{\mu}$ for our systems indicates that intrinsic and BR could be the operative channels. But one can neglect intrinsic term since disorder is present in GNCs. Once vacancy is generated, the carbon atom which is the nearest neighbour to a vacancy is displaced from its equilibrium position. As a result, the system with vacancy undergoes a Jahn–Teller distortion and breaks the honeycomb symmetry. The amount of such sites with broken spatial inversion symmetry is high in GNCs compared to pure graphene. Further, in pure graphene, the SO interaction couples π and σ band under the condition $\Delta \gg L_i$, where Δ is band gap [1,6,22–26].

Thus, for GNCs, the dominant component in L_i is BR with π–σ band distortion. After nitrogen doping, the observed reduction in magnitude of L_i could be attributed to the introduction of mid-gap between π and σ bands due to modification at the broken symmetry site. The perturbation could

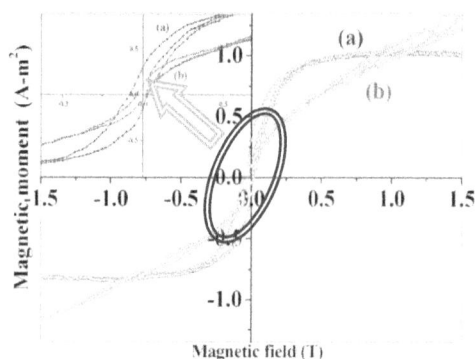

FIGURE 7.3 Vibrating sample magnetometer spectra typically recorded for (a) GNCs and (b) N-GNCs at 300 K. Inset shows zoomed in portion of $M–H$ loop as indicated by an arrow. Open ellipse shows the region of magnification.

be offered by nitrogen-loaded donor moieties. They have strong tendency to donate an itinerant electron via charge transfer interaction with sp^2 network. The spin relaxation time, T_{ss}, and momentum scattering rate, τ, are connected [23] as $T_{ss}=N_{col}\cdot\tau$, where $\Gamma\sim 1/\tau$. The obtained N_{col} is the spin-flip scattering probability (meV–psec) of spin density wave propagating in crystal. For GNCs, the magnitude of N_{col} is estimated to be 2.22 meV–psec, and after nitrogen doping, the collisional probability is enhanced to 3.65 meV–psec. The variations observed for N_{col} are inconsistent with the changes obtained for density of states (DOS) ρ ($\tilde{\mu}$, Γ_{spin}) in both systems. It is interesting to note that large variations have been observed between theoretically and experimentally obtained values of spin transport parameters for GNC systems [1,6,22–26].

The values of Pauli spin susceptibility, χ_{Pauli}, obtained using the ESR technique (given in Table 7.1) were reconfirmed by the data obtained using VSM measurements.

Figure 7.3 shows the VSM spectrum for (a) GNCs and (b) N-GNCs. Saturation magnetic moment (M_s), effective magnetic moment (μ_{eff}), coercive field (H_c), remanence field (M_r), and squareness ratio (M_r/M_s) were obtained, compared, and studied using VSM. A shift in hysteresis loop is expected when field is applied in reverse direction due to the presence of exchange anisotropy. The hysteresis curve recorded, for GNCs, exhibit saturation behaviour. In contrast, no clear evidence is obtained for saturation trend in N-GNCs. The magnitude of M_s is computed to be ~1.02 A–m^2 for GNCs. The recorded value is found to be three orders of magnitude more than that reported in the literature [24]. For N-GNCs, in order to estimate M_s, geometrical projection of $M–H$ is taken on the Y-axis magnetic moment. This methodology is followed due to the absence of saturation magnetic moment feature. The value of M_s is estimated to be ~0.27 A–m^2. The value of M_s is reduced by 3.8 times, compared to GNCs, after doping with nitrogen [27].

7.1.4.2 Magnetometry Studies

The effective magnetic moment, μ_{eff}, is computed using values of M_s obtained for both samples. The Bohr magneton, μ_B, is used as a basic physical constant for magnetic moment of electrons per single atom. The μ_B expresses an intrinsic electron magnetic dipole moment, which is ~1 Bohr magneton for electrons [25,26]. The study of μ_B of a single carbon atom is critical because it verifies how many μ_B a single carbon atom contains. For the computation of μ_B, two basic assumptions were made: (a) Deformation in unit cell for both systems is neglected, and (b) post-doped flake area remains constant. The concentration of carbon atoms, C_c, is computed by relation: $C_c=\dfrac{4}{\sqrt[3]{3^2}}\times a^{-2}$, where a is the C–C bond length in graphene and the value of a is 1.421 Å. The value of C_c is estimated to be 3.81×10^{15} atoms/cm^2. The effective magnetic moment, μ_{eff}, is computed by the use of relation:

$$\mu_{eff} = \frac{M_s}{W\ (in\ gm)}\times\frac{V}{N_c},$$ where W is the weight of samples (in gm) subjected to VSM, N_c is the number

TABLE 7.2

Magnetic Correlation Parameters Obtained after Analysing $M-H$ Loop of VSM Recorded for GNCs and N-GNCs

Parameters	GNCs	N-GNCs
Saturation magnetic moment (M_s)	1.02 A-m^2	0.27 A-m^2
Effective magnetic moment (μ_{eff})	0.198	0.053
χ_{pauli} (at 300 K)	1.63×10^{-5}	1.17×10^{-6}
Coercive field (H_c)	50.477 mT	22.849 mT
Remanence field (M_r)	0.302 A-m^2	0.098 A-m^2
Magnetic hardness (M_H)	423.0 mT/A-m^2	481.5 mT/A-m^2
Squareness ratio (M_r/M_s)	0.296	0.363

The measurements were performed at 300 K.

of carbon atoms in active layer, and V is the effective volume of the GNCs layer. The thickness of carbon layer is taken as 3.5 Å. The value of N_c is given by $N_c = C_c \times A$, where A is the area of the flakes. Our scanning electron microscopy study revealed that the area of the flake is roughly 20 μm^2 [21]. And hence, N_c is estimated to be 7.62×10^{10} carbon atoms. The weight of the samples, W, taken was in the range 47–51 mg. Thus, the magnitude of volume, V, is computed $\sim 7 \times 10^{-13}$ cm^3. The value of μ_{eff} is found to be 0.198 μ_B and 0.053 μ_B, respectively, for GNCs and N-GNCs. Our μ_{eff} calculation shows that, for a cluster of 1000 carbon atoms, there are ~ 200 interacting spins for GNCs. The μ_{eff}, for N-GNCs, shows that this number is reduced to ~ 50 spins for the same number of carbon atoms. This shows the number of electrons participating in spin interaction is reduced after nitrogen doping. The computed μ_{eff} and χ are correlated as: $\mu_{eff} = 2.83 \cdot \left[\chi_{pauli} \cdot T \right]^{1/2} \mu_B$ [19]. At room temperature, it was found that the magnitude of $\chi_{GNCs} = 1.63 \times 10^{-5}$ and $\chi_{N-GNCs} = 1.17 \times 10^{-6}$. The values obtained for χ_{palui} using the VSM technique show a slight variation in the magnitude of χ_{pauli} computed using ESR. For GNCs, the variation is around two orders, and it is one order for N-GNCs. The observed variations could be attributed to the accuracy and precision of the systems used. The other parameters, obtained using the VSM technique, are listed in Table 7.2 [1,6,22–26].

The analysis of ESR and VSM revealed that spin transport and exchange interactions occurring in GNCs are modified after incorporating nitrogen. Specifically, the spin–orbit interactions couple electronic states with opposite spin projections in π and σ bands. This is indeed reflected in variation in magnitudes of L_i and ρ obtained for GNCs and N-GNCs. The L_i is connected to the inter-σ–π band separation via μ (chemical potential). In contrast, the physical entity, ρ, is associated with the density of states of carriers at the Fermi level. The questions to be asked are the following: (a) How to quantify σ–π variations? (b) Does it have implications on the electrical transport of the system? Using chemical analysis, one may shed light on the first issue, whereas the second could be addressed on the basis of electrical measurements [25–27].

7.1.5 ELECTRON SPECTROSCOPIC CHEMICAL STUDIES

For the analysis of ESCA spectrum, in terms of contribution from individual components representing various species, each obtained peak was fitted by a combination of components by minimizing the total squared error (least squared error) of the fit. Individual components were represented by a convolution of a Lorentzian function representing the life time broadening and a Gaussian function to account for the instrumental resolution [27]. The Gaussian broadening was kept same for different components. A Shirley background function is considered to account for the inelastic background in the ESCA spectrum.

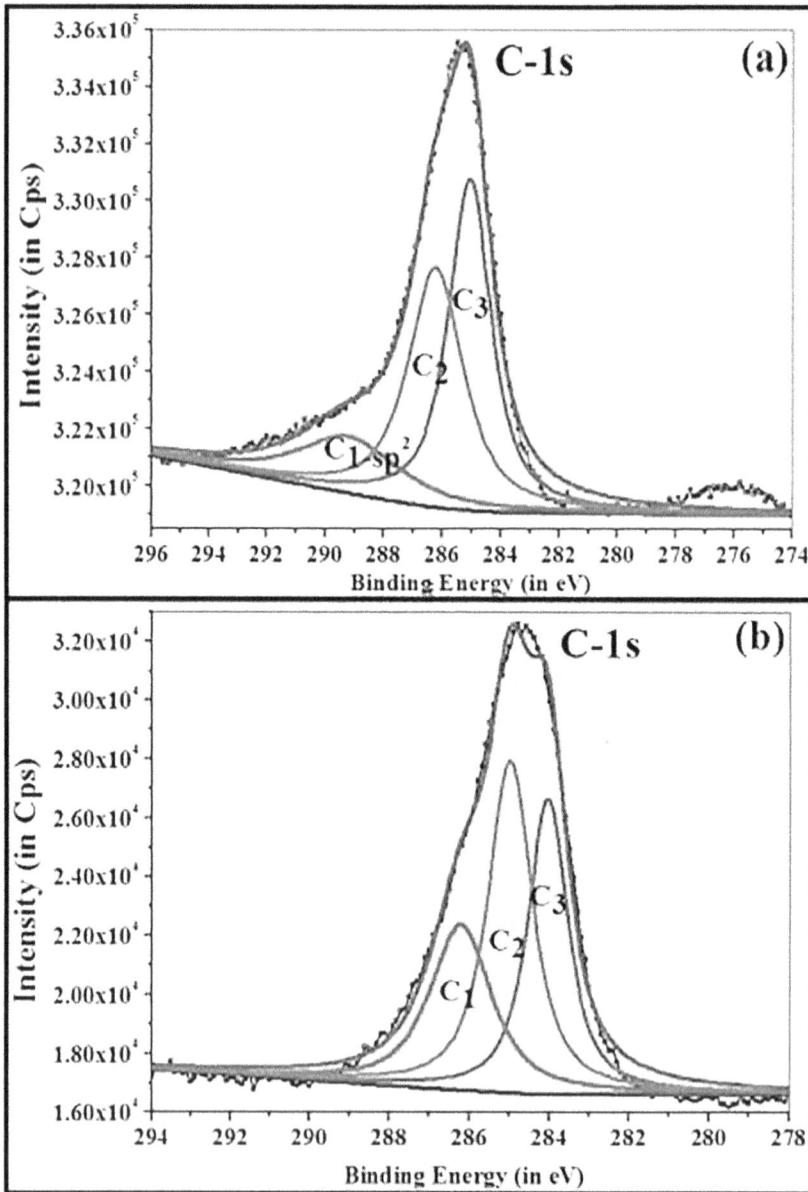

FIGURE 7.4 The recorded C-1s spectra for (a) GNCs and (b) N-GNCs, using electron spectroscopy for chemical analysis.

Figure 7.4 shows the recorded ESCA peak of C-1s recorded for (a) GNCs and (b) N-GNCs. The C-1s core spectrum of GNCs located at 285.33 eV is shifted to lower binding energy at 284.66 eV after incorporating nitrogen in GNCs. The full width at half maximum (FWHM) of C-1s is 2.50 eV for GNCs and 2.33 eV for N-GNCs. The peak intensity is 3.35×10^5 counts per second (Cps) and 3.20×10^4 Cps, respectively, for GNCs and N-GNCs [26,27].

The observed decrease in FWHM and intensity reconfirms successful doing of nitrogen in GNCs [28]. The amount of TDAE in GNCs suspension is maximum 0.5 mg/mL. The addition of TDAE in GNCs seems to be negligibly small because atomic % (at %) of carbon estimated in both samples is 93.00%. The C-1s peak is deconvoluted into three peaks. For GNCs, the C_1 peak located at 289.20 eV is attributed to O=C–OH, i.e. for sp^2 carbon. The composition of O=C–OH moiety is 9.32 at%. The C_2 at 286.23 eV is assigned to sp^3-based –C–O–C– group. And C_3 appeared at 285.07 eV corresponds to C–C of sp^3 origin. The compositions of C_2 and C_3 are 40.45 and 43.23 at%, respectively. These two peaks are characteristic peaks for GNCs phase. For N-GNCs, the C_1 peak located at 286.21 eV is attributed to C–NH_2, which is an amine-based moiety, having state of hybridization sp^3 [29]. The composition of C–NH_2 is 31.32 at%. The peaks C_2 at 284.99 eV and C_3 at 284.04 eV are assigned to C–C. The compositions of C_2 and C_3 are 40.21 and 32.09 at%, respectively. The features of these peaks are having similar characteristic to that of C-1s peak. The analysis of N-1s and O-1s is also carried out. The details are provided in Appendix C, Figure C.7.4.1. To summarize, N-1s appeared at 400.45 eV (GNCs) is shifted down at 399.89 eV (N-GNCs). The chemical shift is 0.45 eV and negative. The satellite feature present at 405.06 eV, for GNCs, disappeared after nitrogen doping [30]. The overall peak intensity of N-1s peak recorded for GNCs is lowered. This indicates that p-type dopant is received by nitrogen moiety. The p-type dopant is available in GNCs in the form of bound Dirac hole. For O-1s, the prominent observation is the decrease in at % oxygen moieties in N-GNCs. The GNCs contain native oxygen ~3.25 at%, which is reduced to 1.45 at%. The reduced oxygen is utilized for radicalization/fragmentation of TDAE. This is consistent with our FTIR and UV–visible analysis presented in Figures C.7.4.2 and C.7.4.3. Thus, the ESCA analysis indicates that chemical reduction takes place after treating GNCs with TDAE. The % decrease in oxygen content indicates that the reaction takes place at O=C–OH (carboxyl) and –C–O–C (epoxy) sites. The nitrogen (–NH_2, amine) is introduced in the sp^2 network. Nitrogen gets bonded by accepting p-type dopant from GNCs. The exchange interaction is dominant at oxygen site that exchanges couple hole, replaces oxygen, and modifies state of hybridization from sp^2 to sp^3. As a result, population of sp^3 content is increased appreciably. The modification could bring change in the *makeup* of exchange correlations and spins transport in GNCs [25–27].

7.1.5.1 Reduced Exchange Correlations in Nitrogenated GNCs

Magnetism, fundamentally, requires an unpaired electron with lattice atom that carries a net non-zero magnetic moment. These moments can be coupled via exchange interaction with other electrons. The observed decrease in μ_{eff} shows that the moments, coupled via exchange interactions, are reduced. Electron–electron and lattice–electron interactions play an important role in magnetization of honeycomb carbon [31]. In such a network, every sp^2 unit contains a half-empty p_z orbital and a half-filled localized π-electron. The π-electron is symmetrically bound to a hole. They form a bound pair. Further, TDAE is an organic material that shows a strong tendency to donate electron via charge transfer mechanism, due to presence of nitrogen [32]. Nitrogen plays a crucial role by donating one of the electrons from its loan pair. The donation is exchange based, and the exchange takes place with half-empty p_z orbital. The itinerant electron gets bound and spin coordinated with localized π-electron at carbon lattice. The coupling takes place in the Screen–columbic repulsive environment. This introduces distortion in laterally oriented σ bonds. The proposal is based on calculation of % variation (%$_{var}$) of T_{ss} and T_{sl} computed for GNCs and N-GNCs samples. The %$_{var}$ is given by relation: %$_{var}$ = $|T (GNCs) - T(N - GNCs)| \times 100$. The T_{ss} %$_{var}$ is ~46% and for T_{sl} is ~74% computed across GNCs to N-GNCs. This indicates that %$_{var}$ for spin–lattice interaction is dominant over spin–spin interaction. This is possible; in fact, in graphene, each Dirac cone is positioned, longitudinally, at a distance of C–C, i.e. 1.421 Å apart. The lateral σ band, connecting nearest two sp^2 carbons, is an immediate neighbour of the bound pairs of Dirac cone. During exchange coupling, the perturbation offered by an uncoordinated itinerant electron, to empty p_z orbital, could distort σ bond sensitively. The schematic is shown in Figure 7.5. Figure 7.5a shows the attack of itinerant electron on hole and (b) indicates σ bond deformation. As a result, the strength of spin–orbit coupling and carrier concentration at Fermi level could be affected.

The clue obtained by ESR, VSM, and ESCA is sufficient to address the second issue raised above.

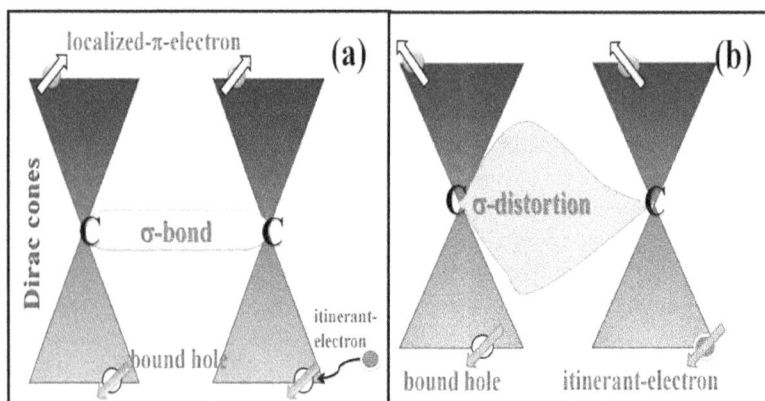

FIGURE 7.5 Schematic representation of exchange interaction between itinerant electron and half-filled localized π-electron. (a) The coupling takes place by replacing bound hole. The interaction is short range, and coulombically repulsive, which could distort lateral σ bands. (b) The itinerant electron is donated by the nitrogen of TDAE to sp^2 carbon network of GNCs [25,26].

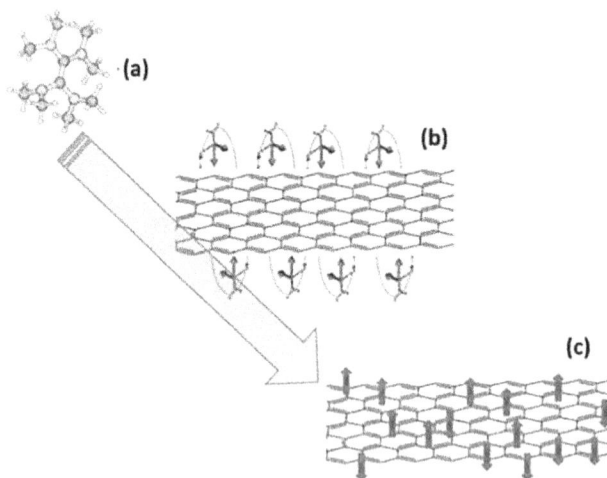

FIGURE 7.6 Schematics of magnetic ordering in TDAE-treated grapheme in which (a) TDAE molecule, (b) TDAE bonded graphene and (c) exchange coupled interactions.

7.1.5.2 RKKY Interactions

Figure 7.6 shows the possible mechanism of magnetic ordering in the TDAE-treated graphene. Figure 7.6a shows the structure of TDAE molecule, whereas Figure 7.6b shows graphene network with fragments of TDAE. These fragments could be responsible for magnetic ordering in graphene, as shown in Figure 7.6c. The fragments of TDAE contain itinerant electrons. On attaching to the graphene network, the transfer of itinerant electron takes place. However, the host graphene site already consists of localized electrons. Now, the questions to be addressed are as follows: (a) How itinerant electrons generate spin ordering in localized electrons? (b) What is the nature of coupling between itinerant and localized electrons? Magnetism, fundamentally, requires an unpaired electron with lattice atom that carries a net non-zero magnetic moment. These moments can be coupled via exchange interaction with other electrons. In a neutral graphene network, each sp^2 carbon atom contributes one p_z orbital and one π-electron, where π-electron system is half-filled and symmetrically bound to a hole. Electron–electron interactions play an important role in magnetization of sp^2 carbon network [23,28–30]. Thus, one could address this problem on the basis of RKKY interaction [28–30]. The RKKY interaction is an indirect exchange interaction between two localized spins induced by itinerant electrons in a host material. These interactions are short range and repulsive in nature, where magnetic coupling is mediated by mobile charge carriers [31–33]. The itinerant and localized electrons interact only if they occupy p_z atomic orbital in the host lattice site, cloaking the coulombic repulsion. The condition for finite temperature magnetic ordering is the magnitude of on-site coulombic repulsive energy, U, between itinerant and localized electrons and must be less than hopping length, t (expressed in hoping energy), of itinerant electrons. In other words, one can say that magnetic interaction between itinerant and localized electrons takes place with cloaking of coulombic repulsion. This assumption is valid for local as well as for total magnetization in TDAE-treated graphene and mainly depends upon the ratio U/t [33]. Thus, the probability of finding random localized electron spin at host site is modulated by charge-transferred itinerant electron spin. As a result, spin polarization of conduction electron in host graphene matrix takes place and could be expressed by a spin wave. The modulated spin wave gets transmitted as a damping oscillation to the next lattice point where itinerant electron spin resides with the separation of lattice constant. If the phase of the spin wave matches at each lattice point, then all the localized electron spin align to the same direction achieving a ferromagnetic spin ordering in TDAE-treated graphene.

7.2 SPIN DYNAMICS IN GNCs VS GRAPHENE: ROLE OF ADATOMS

The focus of this section is to study electron spin dynamics in graphene, GNCs, and their nitrogen (N)-doped derivatives. The spin transport data were obtained using ESR spectroscopy, carried out at 123–473 K. Transport parameters were derived by deconvoluting peak-to-peak line width (ΔH_{pp}) and estimating anisotropy in Lande's g-factor (Δg). The obtained parameters were examined for their temperature as well as interdependence. The study revealed that there is a significant difference in spin transport properties of graphene, and GNCs. The spin bath behaves in the opposite fashion after doping N, for both systems. Our calculations showed that the number of non-interacting spins for graphene and GNCs is reduced, significantly, after N doping, particularly for GNCs. Although the overall magnetization is reduced, the presence of N in GNCs seems to be useful to provide desired spin degrees of freedom. Understanding the spin transport in GNCs may provide a clue to construct solid-state quantum bits for future computational devices [34,35].

7.2.1 SPIN TRANSPORT IN CARBON: THE BACKGROUND

Broadly speaking, harnessing quantum laws in nanocarbon system is very promising for spintronic applications such as processing information in the form of strings of quantum bits [36], for storage and communication. Such quantum systems *communicate* with their environment through up or down orientations of electron spin [37,38]. In general, the spin degree of freedom of electron endowed to its lattice is on the *talking term* and could be exploited, primarily, using spin–orbit (SO) and exchange interactions [39]. Nonetheless, the crucial requirements on spin bath are quantum entanglement [40], coupling spin to lattice vibration [41], weak decoherence [42], and hyperfine interaction (HFI) with surrounding nuclear spin [43]. In graphene, the SO interaction has intrinsic [44], Bychkov–Rashba (BR) [45], ripple, and extrinsic contributions [46]. These interactions are supposed to be weak, in μeV regime [47], due to low atomic number of carbon. The exchange interactions that originate due to spin-flip and spin splitting are proposed to be identical for disordered and ordered sp^2 graphene network. In fact, these interactions are predicted to be independent of SO coupling. The spin–lattice relaxation time (T_{sl}) is theoretically estimated in the 10–100 ns range [25]. The HFI are supposed to be absent or low, due to zero-spin ^{12}C nuclei (abundance

≈99.0%–98.9%) [7,35,34]. However, in recent experiments, the T_{sl} is measured as short as 60–150 ps [47,48]. In another study, Klein tunnelling was observed, indicating *leaky* spin density wave propagation due to difficulty in creating gap-tuneable graphene quantum dots [49]. The tunnel couple dots had found hindrance in Heisenberg–spin exchange coupling. This is due to valley degeneracy that exists in the vicinity of Dirac point in graphene band [26]. The role of strength and contribution of SO components is still a subject of discussion [19]. The emphasis on the study of pseudo-spin [27,34] has neglected the fourfold spin-degenerated state that exists in graphene band structure [23]. Thus, less attention has been paid to spin dynamics in realistic graphene structure which contains inversion, broken inversion symmetry, and heterostructure (intrinsic/extrinsic) lattice environment. The present scenario demands experimental validation for studying transport of spin in realistic graphene super lattice. Principally, spin–spin relaxation time (T_{ss}) is inseparable from SO interactions. It has direct dependence on spin transport parameters such as momentum relaxation rate (Γ), spin relaxation rate (Γ_{spin}), spin-flip factor (|b|), SO coupling constant (L_i), pseudo-chemical potential ($\tilde{\mu}$), effective magnetic moment (μ_{eff}), spin (S_c), defect concentration, and Pauli spin susceptibility (χ_{spin}) of the carbon system [12]. On analysing them, a fundamental insight may develop in designing tuneable spin degrees of freedom, crucial for optimal SO and exchange interactions.

In following discussions, temperature-averaged magnitude of measured and estimated physical quantities are quoted (with standard deviation) using the symbol $<...>_T$. As a first step towards computing, temperature-dependent spin transport, ΔH_{pp}, and shape of the ESR spectra were investigated and are shown in Figure 7.7 for three temperatures 123, 300, and 473 K [1,6].

7.2.1.1 Temperature-Dependent Electron Spin Resonance

The average line width is narrowed inhomogeneously for N-GNCs. For an inhomogeneously narrowed resonant line, the number of individual resonance processes that are occurring after nitrogen loading does not commensurate under one resonance envelope. This could be due to the fact that the magnitude of applied field and modulation amplitude exceeds the strength of ΔH_{pp}.

The inhomogeneous narrowing is also indicative of two more sources: (a) the T_{sl} in N-free system that gets modified and (b) charge inhomogeneities (the so-called puddles). The puddles over the volume of the sample exceed the natural line width, $\dfrac{1}{\gamma T_{sl}}$, where γ is the gyromagnetic ratio. These puddles prevent relaxing electron from reaching the Dirac point, and they have the average minimal charge density ~10^9/cm^2. The spins associated with the puddles in various parts of the sample find themselves in different field strengths, and the resonance is narrowed in an inhomogeneous manner, after doping [28]. This indicates that there are more than two T_{sl} involved in GNCs that are merged into one overall line. However, after doping with N, there is an inhomogeneous narrowing trend in both the systems. For spin resonance in metallic system, Dysonian line shape analysis is given in addition to two common line shapes: Gaussian and Lorentzian. However, for GNCs, the line shapes are found to be Gaussian, perfectly, and for N-GNCs, they are mixed Lorentzian–Gaussian. The mixed feature is indicative of weak magnetic interactions. All measurements were conducted at constant microwave power level, which was kept sufficiently low in order to avoid saturation in samples. As a result, the intensity of the recorded spectrum is proportional to the number of spins and amount of moments. It can be seen that, at room temperature, the number of spins that are participating in resonance process are reduced after the incorporation of nitrogen. The same trend is valid over the entire temperature regime; however, variations are marginal for GNCs in contrast to their nitrogen counterparts. This indicates that spin system with short relaxation time is observed dominantly for N-GNCs [1].

7.2.1.2 Anisotropy in Effective g-Factor

If the angular momentum of a system is solely due to spin angular momentum, the tensor *g*-factor should be isotropic, with the value 2.00232, i.e. the spin moment of typical non-degenerated Pauli gas. Any anisotropy or deviation from this value involves contributions of: (a) orbital angular moment

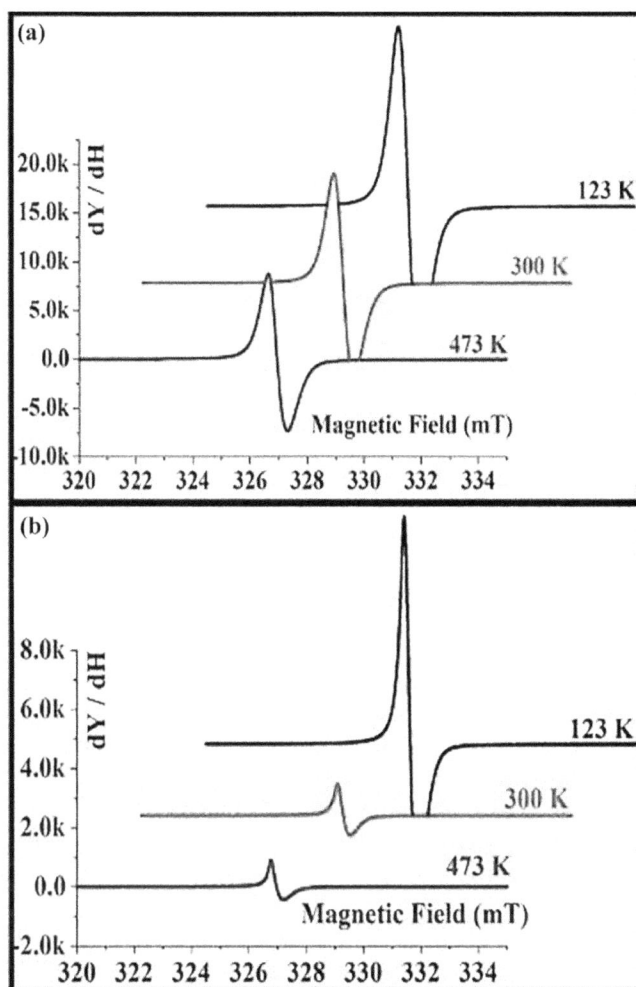

FIGURE 7.7 Recorded ESR spectra for (a) GNCs and (b) N-GNCs. Three typical profiles are shown recorded at temperatures 123, 300, and 473 K.

from the excited state and (b) spin–spin interaction [1]. Thus, it results in the effective g-factor. The magnitude of effective g-factor is estimated for the systems. The effective g-factor is observed to be less than 2.00232. The difference Δg between 2.00232 and effective g-factor for the system is computed. The small values of line widths (analysed in previous section) and the smaller deviation in effective g-factor suggest that spin does not originate from any transition metal impurities, but only from carbon-inherited spin species [29]. The values also indicate that the local magnetic environment in the carbon system is distinctly different than that exists in a conventional magnetic material. For GNCs, $<\Delta g>_T$ is $4.6 \times 10^{-3} \pm 7.9 \times 10^{-5}$ and reduced marginally to $4.3 \times 10^{-3} \pm 9.5 \times 10^{-5}$ for N-GNCs. For grapheme, anisotropy is comparably high, i.e. $5.2 \times 10^{-3} \pm 2.1 \times 10^{-4}$, and increased to $5.6 \times 10^{-3} \pm 4.0 \times 10^{-4}$, after incorporating N. The behaviour underlines following facts: (a) The contribution of orbital moments and spin–spin interactions are increased in graphene, after doping, (b) the overall strength of this component seems to be less in GNCs, compared to graphene, and reduced further for N-GNCs, (c) orbital moments of electrons of nitrogen contributed, vectorially, to N-doped systems, and (d) the response of spin of extrinsic adatom to inversion (graphene) and broken inversion symmetry (GNCs) lattice environment seems to be different [24,25].

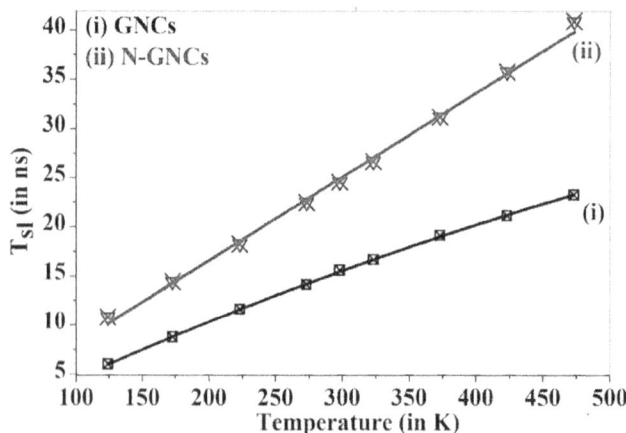

FIGURE 7.8 Variation in T_{sl} for (i) GNCs and (ii) N-GNCs.

7.2.1.3 Spin–Spin (T_{ss}) and Spin–Lattice (T_{sl}) Relaxation

The characteristic time required to relax spin $<T_{ss}>_T$ is ~ 8.0 ps for GNCs and increased significantly to ~13.0 ps after loading N. This indicates that the inter-spin correlations are modified after incorporating nitrogen in GNCs. However, the magnitude of $<T_{ss}>_T$ is found almost in the same range, i.e. ~8.0 ps for graphene, and its nitrogen derivative to that of GNCs.

The principal parameter governing spintronic usability is T_{sl} which quantifies variation of non-thermal spin state around the lattice. From the perspective of spintronic applications, the theoretical estimate of T_{sl}, for graphene, is 10–100 ns, whereas experimental spin–transport measurements, surprisingly, showed that it is as short as 60–150 psec. Spin–lattice relaxation in graphene is associated with low-energy excitation around K–K' Brillouin zone corner point of triangular sublattice. The magnitude of T_{sl} is computed using the relation: $\frac{1}{T_{sl}} = \frac{28.0 \text{ GHz}}{T \text{ (in K)}} \times \Delta H_{pp}$. The variation in T_{sl} as a function of temperature for the measured carbon systems is shown in Figure 7.8. The value of T_{sl} is found to be varied, linearly, from 6.0 to 24.2 ns, for GNCs. After incorporating nitrogen in GNCs, the overall magnitude is increased. And for N-GNCs, T_{sl} is found to be varied, exponentially, from 10.0 to 41.0 ns. Over the temperature regime, both the systems showed a fourfold change in T_{sl}. However, for graphene, N-graphene (results not shown here), and GNCs, the variations in T_{sl} are quite closer and almost identical. Basically, electron spin relax by transferring energy selectively to those lattice modes with which they resonate. The resonant modes are on *talking term* with the spins. And one can modify their *crosstalk* by introducing break in the symmetry inversion (i.e. disorder) or adatom in the sp^2 carbon network. However, at a higher temperature, the resonant levels seem to be somewhat broadened and hence a saturation trend is observed for GNCs [1,26,27].

7.2.1.4 Estimation of Spin (Γ_{spin}) and Momentum Relaxation Rates (Γ)

The spin relaxation rate, Γ_{spin}, is related to T_{sl} by a relation: $\Gamma_{spin} = \hbar / T_{sl}$, where $\hbar = 6.59 \times 10^{-16}$ eV–s and is listed in Table C.7.1.2. Specifically, Γ_{spin} depends on impurity in graphene. The Γ_{spin} can be significantly modified by resonant scattering at the impurity site. If the amount of impurity is high, the modification in resonant scattering is in that proportion. The relation between Γ_{spin} and Γ is given by: $\Gamma_{spin} = \propto \cdot \left(\frac{L_i}{\Delta}\right)^2 \cdot \Gamma$. Figure 7.9 shows variations in Γ as a function of temperature for the systems. Our observations on the obtained result are as follows: The value of   decreases gradually as the temperature increases. For graphene, Γ is between 3.94 and 1.25 meV range, with $<\Gamma>_T = 1.94 \pm 0.93$ meV. For N-graphene, GNCs and N-GNCs, the range of Γ is, respectively, 4.06–0.82 meV ($<\Gamma>_T = 1.59 \pm 1.09$ meV), 5.22–1.36 meV ($<\Gamma>_T = 2.25 \pm 1.20$ meV), and 5.87–1.66 meV

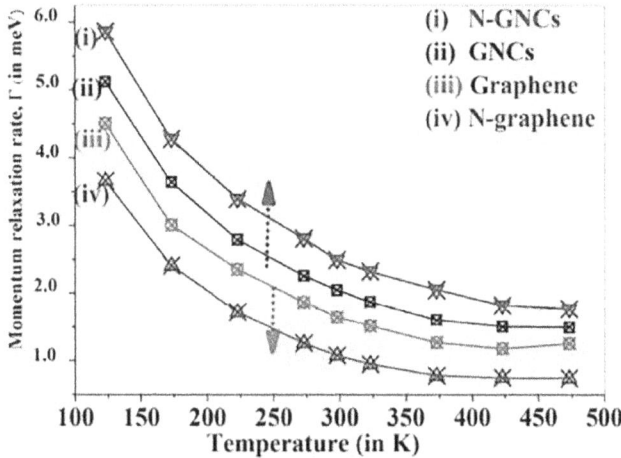

FIGURE 7.9   Variations in Γ as a function of temperature for (i) N-GNCs, (ii) GNCs, (iii) graphene, and (iv) N-graphene. Arrows indicate the opposite behaviour for graphene and GNCs, over the temperature range.

($<\Gamma>_T = 2.83 \pm 1.39$ meV). This indicates that the magnitude of Γ for GNCs is greater than that for graphene. After doping N, Γ is increased for GNCs, whereas it is reduced for graphene. At room temperature, the values of Γ are almost the same for GNCs and N-GNCs. This indicates that N loading has no effect on the change in momentum relaxation rate in the GNC system; however, the scenario is opposite for graphene. At the lowest measured temperature, somewhat opposite behaviour is observed. The open ellipse indicates their behaviour. The presence of N adatom adsorbed and its position in the sp^2 environment has a profound effect on the spin configurations of these systems and is consequently reflected in the variations in Γ [27].

7.2.1.5   Estimation of SPIN Susceptibility (χ_{spin}), and Concentration (S_c)

Figure 7.10a shows the variation in Γ as a function of $\tilde{\mu}$. The quantity Γ has L_i^2 (spin–orbit coupling constant) dependence. Over the observed variations, the computed curves are for total contribution of L_i, i.e. intrinsic, Bychkov–Rashba (BR) type, ripple, and extrinsic (in case of doped systems). The extrinsic contribution is added in an opposite way for both the systems. The ripple relaxation contribution depends on Γ only when $\mu \ll \Gamma$, where it resembles an Elliott–Yafet relation $\Gamma \propto L_{ri}^2 \Gamma \ln (D/\Gamma)$ [36].

In our case, Γ ranges in meV, whereas the computed $\tilde{\mu}$ is in the range of eV. Thus, one can neglect ripple contribution for all the systems. The performance of ESR is given by the limit of detection of signal, i.e. the number concentration of non-degenerated spin $\frac{1}{2}$ particles measured over the temperature. The limit of detection gives a signal-to-noise (S/N) ratio, and for our system, the S/N ratio is 20 at 0.05 mT line width. The value of limit of detection for our system is 5×10^9 spins/mT. For the ESR system, the gain of the band amplifier and phase-sensitive detector was tuned much smaller than the measured line width. As a result, the recorded line shape transformed into the first derivative (dY/dH) of the absorption line (Y). By measuring the area (A) under the obtained dY/dH, one can compute the number density of unpaired spins in the samples. The A, usually, is proportional to the S_c in the samples. From Figure 7.7, we estimated S_c for the systems [37]: $Y_j = \left(H_j - H_{j-1}\right)\sum\limits_{i=j}^{m} Y_i'$, where Y_j, Y_i', and H_j, are the components of absorption line width, its first derivative, and the corresponding field, respectively. Figure 7.10b shows the variations in S_c as a function of temperature. Figure 7.10d is the complementary data, for S_c, i.e. defect concentration, obtained from the χ_{spin} calculations. One can see that the concentration of spins for GNCs is varied from 8.77×10^{25} to 8.79×10^{25} s/m^3 with mean $<S_c>_T = 8.69 \times 10^{25} \pm 9.49 \times 10^{25}$ s/m^3, whereas, for

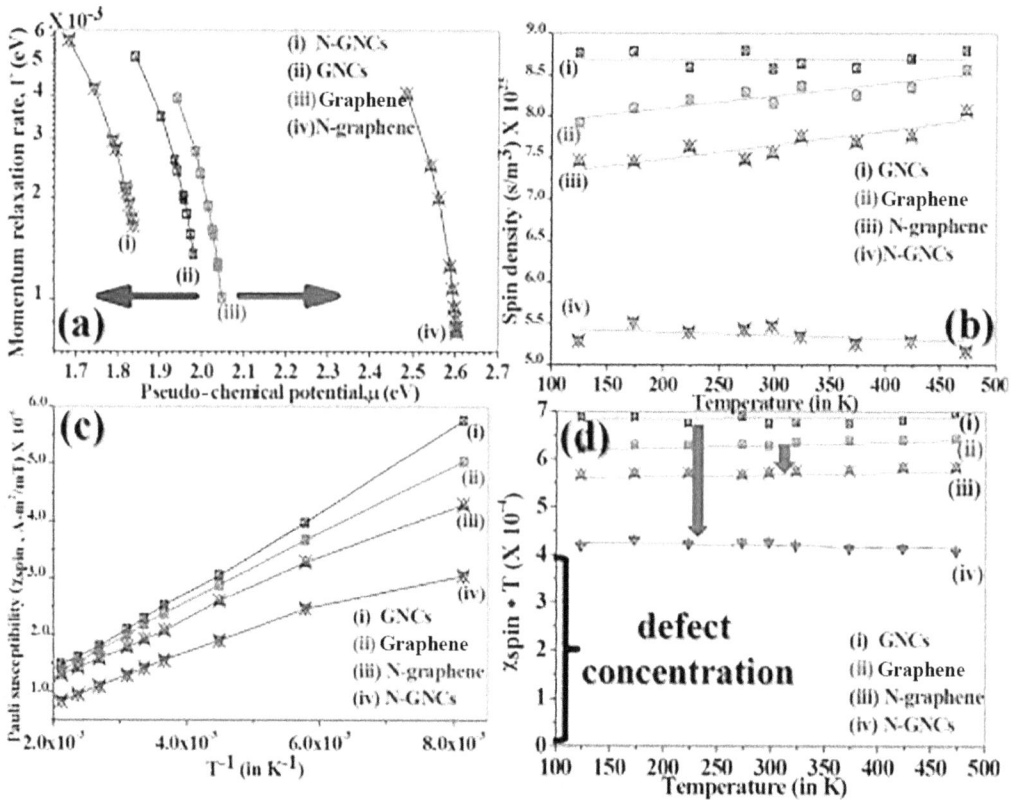

FIGURE 7.10 (a) Change in Γ with $\tilde{\mu}$ (arrows indicate how the two systems behave after nitrogen loading), (b) behaviour of S_c (measured in s/m³) over the temperature, (c) χ_{spin} as a function of T^{-1}, and (d) variations in $\chi_{spin}*T$ as a function of T; arrows indicate the magnitude of change is large for GNCs compared to graphene.

graphene, the range is 7.93×10^{25}–8.57×10^{25} s/m³ with $<S_c>_T = 8.25 \times 10^{25} \pm 1.80 \times 10^{24}$ s/m³. After N doping, the magnitude of $<S_c>_T$ is decreased drastically to $5.36 \times 10^{25} \pm 1.15 \times 10^{24}$ s/m³, for GNCs. For N-graphene, the change, $<S_c>_T$, is marginal, i.e. $7.66 \times 10^{25} \pm 1.98 \times 10^{24}$ s/m³. Usually, the localized spins could be assigned either to unpaired spins of the free radical molecules or to defects such as dangling bonds, trapped carriers, and vacancies. The exact number could be estimated in the form of χ_{spin} of such spins and is given by the following relation: $\chi_{spin} = \mu_O \, \mu_B^2 \, \dfrac{N}{V} \cdot (k_B T)^{-1}$, where μ_O is the permeability of vacuum, μ_B is the Bohr magneton, N/V is the spin concentration, k_B is Boltzmann's constant, and T is the temperature. The variations in χ_{spin} as a function of T^{-1} are shown in Figure 7.10c. The temperature dependence of total χ_{spin} has two components: one χ_{orb}, i.e. orbital diamagnetic response, and χ_n due to itinerant nature of electrons and the Pauli paramagnetism due to their spin. At a low temperature, χ_{spin} originates from ρ. The value of χ_{spin}, for N-GNCs, varies from 3.05×10^{-6} to 8.42×10^{-7} A–m²/mT, over the measured temperature range. For N-GNCs, a large deviation has been observed from the linearity, which could be attributed to antiferromagnetic ordering in GNCs, after doping N. The graphene honeycomb superlattice consists of two complementary sublattices A and B and segregation of conduction electron spin density, n^c, over the sublattices generates (a) ferromagnetic ($n_A^c = n_B^c \, 0$), (b) ferrimagnetic ($n_B^c = 0$, $n_A^c \, 0$), and (c) antiferromagnetic ($n_A^c = - \, n_B^c \, 0$) ordering [38]. The observed behaviour of GNCs indicates antiferromagnetic ordering. So on each sublattice, holes and electrons could be segregated. For graphene,

χ_{spin} is 5.05–1.39×10⁻⁶ A–m²/mT, whereas, for N–graphene and GNCs, its value is, respectively, 4.29–1.36×10⁻⁶ A–m²/mT and 2.74–1.38×10⁻⁶ A–m²/mT.

Frequently, the Curie law is also plotted in modified form χ^*T as a function of T. Figure 7.10d shows the variations in χ^*T as a function of T. The extrapolated intersection of each profile in Figure 7.10d indicates its proportionality to the number of magnetically ordered species. For GNCs, the value of the intersection is highest, whereas for N-GNCs, it has the lowest magnitude. The μ_{eff} is computed using values of the obtained χ_{spin} for the samples. The μ_B is used as a basic physical constant for magnetic moment of electrons per single atom. The μ_B expresses an intrinsic electron magnetic dipole moment, which is ~1 Bohr magneton for electron [39,40]. The study of μ_B of a single carbon atom is critical because it verifies how many μ_B a single carbon atom contains. The μ_{eff} is computed using the relation [41]: $\mu_{eff} = 2.83 \left[\chi_{spin} T \right]^{1/2} \mu_B$. By and large, our μ_{eff} calculations showed that, for a cluster of 1000 carbon atoms, there are ~72 non-interacting spins for graphene and, after N doping, the value remains almost the same, i.e. ~69. For GNCs, this number is somewhat high ~74 and reduced substantially to ~58 spins for N-GNCs, for the same number of carbon atoms. This shows the number of electrons participating in spin interaction is reduced significantly, for GNCs, after N doping. This could provide enhanced spin degrees of freedom to spin bath, for N-GNCs.

7.2.2 N-GNCs Qubit

What follows in, from the analyses of spin dynamics? The graphene and GNCs systems are fundamentally working in opposite ways. Another question is: Which system is well suited for solid-state qubit construction? It seems to be N-GNCs. From the perspective of spin degrees of freedom offered to electron, pure graphene seems to be a tight system. But, one could expect more interesting and diversified lattice environment for electron spin degrees of freedom in GNCs. Although GNCs form two-dimensional array of carbon rings, in contrast to graphene, it consists of conformational and vacancy disorder. Thus, nano-engineered electron spin degrees of freedom, in carbon, may provide optimum SO interaction with desired exchange coupling, essential for spintronic applications. The present analysis underlined the following facts: (a) By and large, properties of spin bath differ, significantly, for graphene and GNCs, (b) spin transport behaves in an opposite fashion after doping, and (c) the net magnetic moment is reduced in the presence of N. The local bond configuration of GNCs, after doping, seems to be favourable to construct solid-state qubits.

7.3 MOLECULAR AND SPIN INTERACTIONS OF TELLURIUM ADATOMS IN REDUCED GRAPHENE OXIDE

In this section, the work is presented on rGO doped with tellurium (Te) that was prepared by the modified Wolff–Kishner method and Te atom (1–4 atomic %) added, in situ, during the reduction process. Both the systems were subjected to electron spectroscopy for chemical analysis, vibration spectroscopy, conductometry, electron spin resonance spectroscopy, electron microscopy, and cyclic voltammetry. The presence of Te was one over three carbon rings and has a decisive impact on transport of electrons, force constant of sp^2 lattice, photo-excited vibration states, electric conductivity, and charging–discharging characteristics. At such dilute level, incorporation of Te modified degree of stacking order, stability of two dimensional networks in terms of strain and frustration that, consequently, altered both longitudinal and transverse optical and acoustic mode in carbon sublattices. It has an impact on the electrical conductivity that is departed, strongly, from the standard model. From the multifunctional aspects such as spintronics and energy storage, probing, exploiting, and architecture of physical properties of Te–rGO system are crucial.

7.4 MOLECULAR SPINTRONICS IN 2D CARBON WITH ADATOMS

Functionalization makes graphene challenging due to chemical modification and alteration in its idealized chemical structure. Substitutional doping of heteroatoms in the graphene lattice has proven as an efficient and robust route for tuneable properties. Such heteroatom doping can also facilitate catalytic activity at graphene surface, while creating local charge centres in the delocalized electron system. The substitution of adatom can either donate or withdraw free electrons from sp^2 network and drastically alter band structure and is demonstrated successfully in many studies. Such systems show enhanced chemical activity, which is potentially useful for various applications such as fuel cells, environmental remediation, and energy storage/conversion. Due to a small mismatch in atomic sizes, boron (B) and nitrogen (N) are popular doping elements in comparison with phosphorus (P) and silicon (Si) [11,25]. Moreover, in chalcogenide groups, oxygen (O) is the most electronegative element, but the substitutional doping of an O atom is bit difficult because of its strong electro-negativity and large size compared to carbon in graphene. Due to the similar electro-negativity of S (2.58 eV) and C (2.55 eV), negligible polarization (or charge transfer) exists in carbon (C)–sulphur (S) bond. In contrast to the zero spin density of pristine graphene, there is a mismatch of the outermost orbital of S and C that induces a non-uniform spin density distribution in S-doped graphene, which consequently endows graphene with catalytic properties useful for many applications [12,26]. In this scenario, the effect of dilute substitution of other chalcogenide such as tellurium (Te) on physico-chemical properties of reduced graphene oxide has not been reported so far and thus is the focus of our present communication.

7.4.1 Chemical State Analysis of Te in rGO Using Electron Spectroscopy

In Figure 7.11, C-1s consists of five components in both the systems. For rGO, carbon associated with C–C ring that appeared at 284.59 eV is shifted to the lower binding energy side by 90 meV and full width at half maximum (FWHM) reduced negligibly by 2.00 meV, by the presence of Te. The major change it faces is the reduction in peak area by 30%. This indicates that ring content is reduced significantly [17,18]. For C–N, neither peak position (285.90 eV) nor its area is changed, except peak broadening 290 meV. For C–O, at 286.50 eV, is though unshifted, however, has a dramatic reduction in peak area 60%. In contrast, for carbonyl group (287.8 eV), a significant amount of broadening of 600 meV is observed with almost two times enhancement in area. The carboxylate appeared at 289.09 eV that contains one double and one single bond character, shifted as well to higher energy side by 10 meV. The higher side shift is suggestive of the increase in sp^2 character, but peak area is reduced by 30%. The % proportion of sp^3:sp^2 is computed for C–O:C=C. For rGO, the proportion is 7% and increased to 22%.

7.4.2 Bond Molecular Environment of Te in rGO: Raman Studies

The trivial requirement in vibration spectroscopy of carbon is to identify all members of the carbon family. This tool is simple, ultrafast, and non-destructive and provides a large structural/electronic insight with high resolution and precision. But, the price for its simplicity is paid while interpreting the obtained spectra. FTIR spectra for rGO and Te–rGO are shown in Figure 7.12a. It represents the identical features for both systems except oxygen-related sites. The skeletal vibrations, appeared, at 664 and 1525, 2312 cm⁻¹, respectively, associated with C–H (in-plane bending modes) and C=C remained unchanged. The peaks related to C–O–C (epoxides), C=O (carbonyl), and C=O (carboxylate), respectively, at 1210, 1370, and 1730 cm⁻¹ are reduced in Te–rGO [35].

7.4.2.1 Structure of rGO and Te–rGO: Molecular Parameters

As per the formulation presented before (Chapter 6, Section 6.1.3.1), L_D^2 was computed to be 82.25 and 58.33 nm², whereas L_a, 10.3 and 22.4 nm, respectively, for rGO and Te–rGO. In addition, the defect density, $n_D \left(cm^{-2} \right)$, for rGO was $1.22 \times 10^{-15}/cm^2$ and $1.72 \times 10^{-16}/cm^2$, for Te–rGO. For

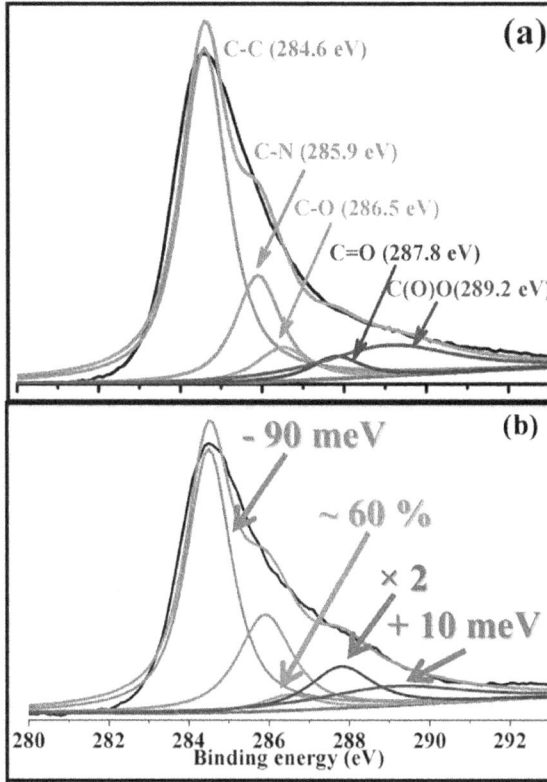

FIGURE 7.11 Chemical analysis of C-1s for (a) rGO and (b) Te–rGO by electron spectroscopy. Top panel indicates a considerable degree of oxidation with five components of functional groups: non-oxygenated C–C ring (284.6 eV), C–N (285.9 eV), C in C–O (286.0 eV), carbonyl group (287.8 eV), and carboxylate carbon (O–C= O, 289.0 eV). Bottom panel: The number shows changes in the corresponding peaks, C–C ring, and carboxylate carbon chemical shift, respectively, by −90 and +10 meV, 60% reduction in C–O content, and increment in C=O by two times. Atomic % of Te is 4.

rGO, the Fermi velocity (in m/s), v_F, is 3×10^6 m/s and reduced by a factor of two, to 1.5×10^6 m/s, by adding Te atoms. Moreover, one can calculate contribution to the electric resistivity, σ, using $\sigma = \dfrac{1}{144} \dfrac{m v_F}{n e^2} b^2 N_D$, where m is the mass of carbon expressed in atomic mass unit, n, the number of carbon atoms/cm^2 (3.81×10^{15}), e, the electronic charge (1.62×10^{-15}), b, Burger's vector at dislocation, having magnitude equal to a_0, and N_D is computed as above. This simplifies to $\sigma \sim \dfrac{4 \times 10^{-21} N_D}{n}$ in Ω-cm. For rGO, σ is 3.28 and 0.23 $\mu\Omega$-cm, respectively, for rGO and Te–rGO [25].

The enrichment in π environment offered an opportunity to compute bond frustration and stretching, modelled through valence force field [24]. These forces are usually short range for σ bonds and extended more than the fifth nearest neighbour. In π-bonded scenario, the forces are proportional to bond orders and inter-bond polarizabilities. The estimates for force constant, k, are approximately given by the relation: $\omega_q = \sqrt{k_q / \mu}$, where k_q is the dynamic variable of force constant and μ is the reduced atomic mass of carbon [23]. Figure 7.13 shows the change in force constant, k, for rGO and Te–rGO with photon energy. Figure 7.13A is the estimation of k, using D dispersion, whereas Figure 7.13B using G-peak. For rGO, the value of $\dfrac{\partial \omega_D}{\partial E}$ is 53.29 and 41.18

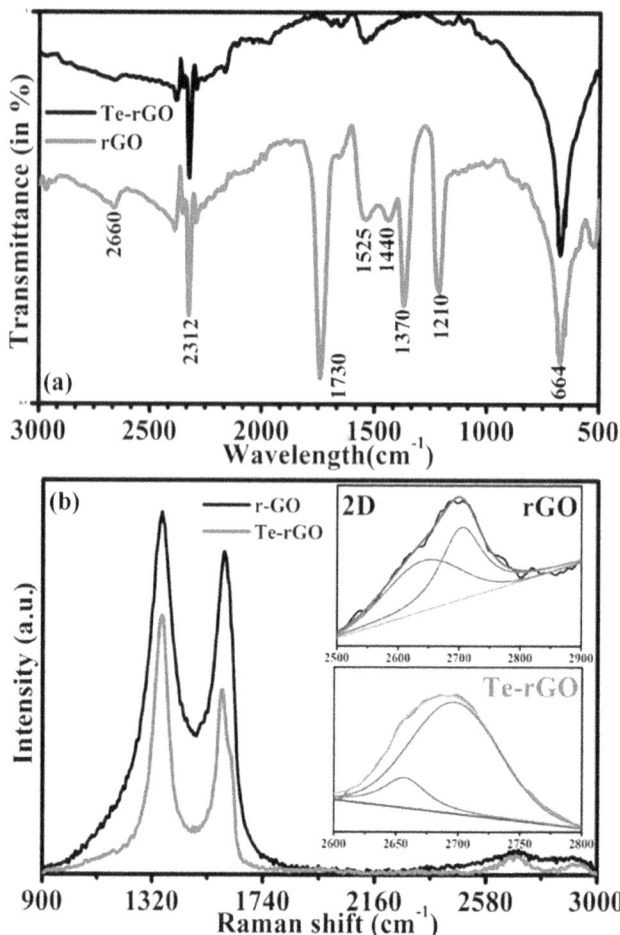

FIGURE 7.12 (a) FTIR for rGO and Te–rGO; (b) Raman spectra at 532 nm. Inset shows 2D deconvolutions indicating number of layers less than 10.

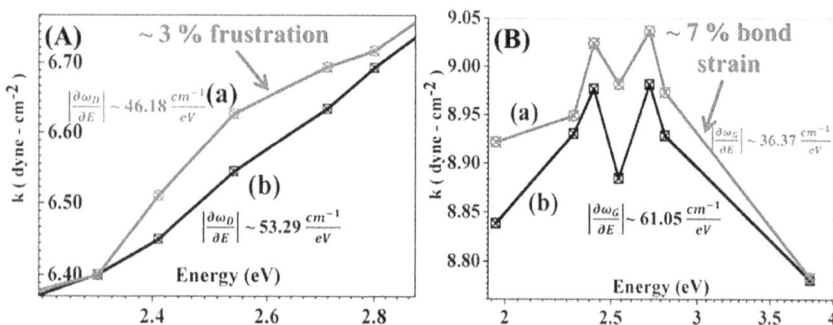

FIGURE 7.13 Dynamical force constant, k_q, estimated by dispersion in D- and G-peaks. (A) Change in k due to $\dfrac{\partial \omega_D}{\partial E}$ and (B) by $\dfrac{\partial \omega_G}{\partial E}$. Profiles (a) rGO and (b) Te–rGO samples. The reduced atomic mass, μ, has been approximated using $\dfrac{1}{\mu} = \dfrac{1}{\sum_1^n m_i}$ in g. The effective $\dfrac{\partial \omega_G}{\partial E}$ and $\dfrac{\partial \omega_D}{\partial E}$ are found to be reduced for rGO, after doping Te.

$\dfrac{cm^{-1}}{eV}$ for Te–rGO. A similar trend is seen for G, in which $\dfrac{\partial \omega_G}{\partial E}$ modifies from 61.05 to 36.37 $\dfrac{cm^{-1}}{eV}$.

This dynamical variable is directly related to electronic density of states, which we have computed experimentally in Section D. The nature of hybridization changes the spatial arrangements of C-lattices and consequently surrounding electrons as Te gets added. The electron density redistribution between C-Te modifies the dynamical variable of force constant, k_q. In our case, what we have found is that Te generated a bond frustration of 3% with respect to the undoped counterpart. This is at C(O)O site. This phenomenon, consequently, causes around 7% strain in bonds, locally, around neighbouring sp² sites. Thus, the usual strong covalent σ bond character between neighbouring carbon atoms is locally modified by the presence of Te atoms [35].

7.4.2.2 Electrolytic Conductance of rGO and Te–rGO

Figure 7.14, upper panel, indicates the prepared suspensions and experimental set-up. Figure 7.14B indicates the change in specific conductance, L_s, of (a) rGO and (b) Te–rGO with filler concentration, whereas Figure 7.14C shows the profiles of variations in equivalent conductance, Λ. Two distinctly different regions of L_s can be seen in rGO. The conductance, L_s, enhances from 21.28 $\mu\Omega^{-1}$ to a peak value of 35.83 $\mu\Omega^{-1}$. In the presence of external electrostatic field, the charge centres in the form of π-electron puddles of sp² carbon, oxygen di-vacancies, and donor loan pairs of native nitrogen participate in forming net charge of equal and opposite magnitude around the central conducting sheet of rGO [25].

For mobile charge carriers, these centres provide the path for the conduction, under applied field. This is valid up to 50 wt% (shown in top left inset) and L_s started sliding down, to 19.41 $\mu\Omega^{-1}$ (at 87 wt%). As wt% increases, the number density of rGO sheet per unit volume increases, bringing more rGO together. The proximal presence results in the neutralization of fractional net charge on

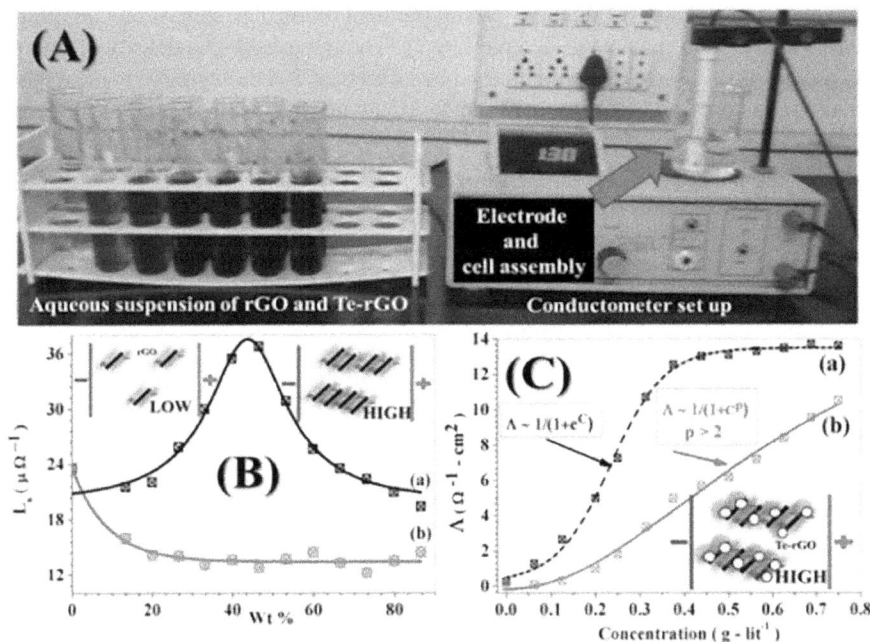

FIGURE 7.14 (A) Digi-cam photograph of experimental set-up. Left portion indicates suspension of rGO and Te–rGO in aqueous medium. Right portion indicates instrumental set-up including conductometer electrode and cell assembly. (B) Variations in conductance, L_s, in $\mu\Omega^{-1}$ as a function of filler wt%; (C) computed equivalent conductance, Λ, measured in $\mu\Omega^{-1}$-cm² at variable concentrations in g/L.

rGO lowering L_s at higher regime. The trend is departed, severely, for Te–rGO. The overall conductance is observed to be low for this system. The well-behaved exponential decrease in L_s indicates Te–rGO as a weak electrolyte solid. Te modified volume concentration of mobile charge carriers, distinctly, when compared to its counterpart. This has led to the change in the inherent degree of ionization of rGO by Te. To investigate this, another fundamental parameter equivalent conductance, Λ, has been evaluated as a function of g/L. Juxtaposed to L_s, Λ is observed to be increased with an increase in g/L. The reason is the decrease in L_s compensated by the increase in the value of reciprocal concentration function. For rGO, at lowest filler fraction, the value of Λ is estimated at 0.51 Ω^{-1}–cm^2, which gradually increased up to a saturation state 12.96 Ω^{-1}–cm^2 at 0.45 g/L. The nature of, Λ, is correlated to concentration, C, by taking the help of a mathematical function that reads as: $\dfrac{1}{1+e^c}$. This indicates that, for larger C values, Λ yields to saturation. For such an incomplete dissociation state, all columbic inter-carrier effects pop up. Moreover, the carrier migration heavily depends upon co-carriers and contributes to total equivalent conductance with a definitive share that decisively depends on the nature of dopant as well as its associated carriers. From this, ionization energy has been estimated to be 0.9 meV for mobile charge carriers in rGO. For Te–rGO, the scenario is different. No definitive saturation regime existed. Perhaps, fitting of mathematical function reads to be: $\dfrac{1}{1+C^p}$ with $p>2$ [25,35].

7.4.2.3   Spin Dynamics: rGO vs Te–rGO

In general, analysis of ESR provides insights into the spin dynamics of mobile charge carriers in both the systems. Spin dynamic parameters were estimated for both the systems and are displayed in Table 7.3.

By and large, an inhomogeneous ESR broadening was indicative of modification in spin–lattice relaxation time by Te. Due to the increase in the volume of charge puddles, they prevented to relax electron from reaching the Dirac point. The contribution of orbital moments, and spin–spin interactions are increased in rGO, by Te due to p-orbital moment of Te. The spin–orbit coupling constant for rGO was 445 meV and reduced to 225 meV by Te. The integrated spin density of states per unit energy per atom was 0.16 for rGO and increased by a factor of 3. The segregation of oxygen around Te-substituted sights confined electron between Te-O and modified spin dynamics [35,39].

TABLE 7.3

Dynamic Molecular Parameters Estimated Using Raman Spectroscopy for rGO and Te–rGO

Parameters (Unit)	Samples	
	rGO	Te–rGO
ΔH_{pp} (mT)	58.963 ± 3.3	61.223 ± 2.27
T_{ss} (ps)	10.38	11.00
T_{sl} (ns)	05.50	05.71
Δg	$93.30 \times 10^{-3} \pm 7.9 \times 10^{-5}$	$61.70 \times 10^{-3} \pm 9.5 \times 10^{-5}$
Δ (eV)	4.77	3.65
L_i (meV)	445.04 ± 7.32	225.20 ± 3.7
Γ_{spin} (μeV)	0.12	0.12
Γ (μeV)	0.138	0.315
$\tilde{\mu}$ (eV)	2.3876	1.8249
ρ (states per eV-atom)	0.169	0.515

7.5 TETRAKIS(DIMETHYLAMINO)ETHYLENE-INDUCED MAGNETISM

In our previous sections, we have discussed the room-temperature magnetic interactions in graphene-like nanocarbons (GNCs) [14]. In continuation, we have now studied magnetic interactions using SQUID magnetometer at temperatures from 2 to 300 K. In order to induce magnetic interactions, we have selectively chosen tetrakis(dimethylamino)ethylene (TDAE), a high-temperature organic ferromagnet [15] and a strong spin polarized donor to functionalize graphene and GNCs. The effective interactions are increased in TDAE-incorporated graphene compared to GNCs. The focus of the current work is to reveal variations in the magnetic behaviour of graphene and GNCs incorporated with TDAE to explore their prospects in spintronics applications.

7.5.1 NITROGENATED NANOCARBONS: THE INTRIGUING SYSTEMS

Among all the dopants, nitrogen (N) is one of the most popular and well-studied dopants for graphene due to similarity in its atomic size with carbon atom [8–10]. With one additional electron as compared to a carbon atom, nitrogen acts as an *n*-type (electron-donating) dopant. Nitrogen doping therefore affects the electric conductivity and the electronic density of states; moreover, graphitic nitrogen is able to provide π-electrons close to the Fermi level of graphene. Various doping methods have been used to form nitrogen-doped graphene such as gas mixture of nitrogen, liquid organic precursors (acetonitrile, pyridine, and aniline) [11,12]. It has been proposed theoretically as well as experimentally that depending on the concentration and packing geometry of doped N atoms, it is possible to induce a magnetic response in graphene [13]. The nature of dopant plays a crucial role in predicting the magnetic properties.

7.5.1.1 Thermo-Magnetic Behaviour

Figure 7.15 shows the room-temperature (300 K) and low-temperature (2 K, inset) behaviours of magnetization M, as a function of applied field H/T for (a) GNCs and N-GNCs and (b) graphene and N-graphene. The hysteresis curves recorded exhibit saturation behaviour, for all samples. The observed behaviour is described by the *Brillouin function* for magnetization [26,34], given by:

$$M = N \cdot g \cdot J \cdot \mu_B \left\{ \frac{2J+1}{2J} ctnh\left[\frac{(2J+1)x}{2J}\right] - \frac{1}{2J} ctnh\left(\frac{x}{2J}\right) \right\}, \text{ where } N \text{ is the number of interact-}$$

ing spins at saturation level (per g), g is Lande's g-factor, J is the angular momentum number defined by the initial slope of M vs (H/T) curve, and $x = \dfrac{g\ J\ \mu_B\ H}{k_B\ T}$, where k_B is the Boltzmann

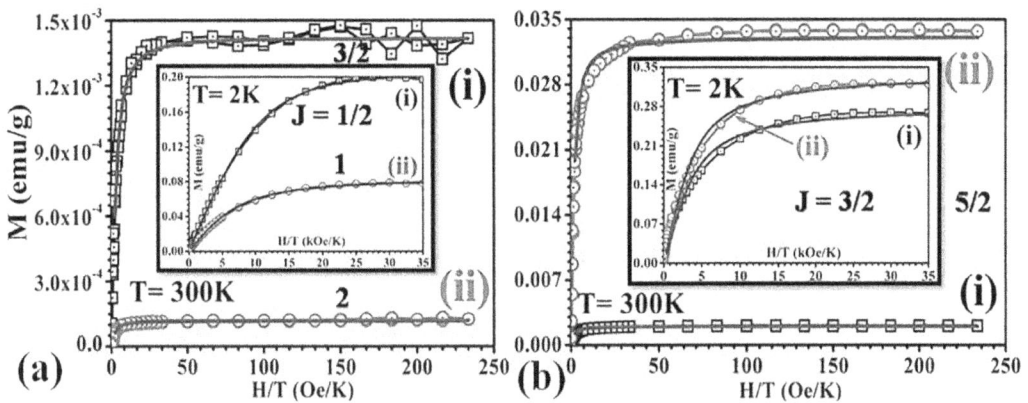

FIGURE 7.15 Recorded M vs H/T response for (a) GNCs and (b) graphene systems at 300 K (inset: 2K). In plot (a), curve (i) is for GNCs and curve (ii) is for N-GNCs. In a similar fashion, in (b), curve (i) is for graphene and curve (ii) is for N-graphene.

constant. The x is the mean molecular magnetization field, and the magnitude of μ_{eff} is given by: $\mu_{eff} = N \cdot g \cdot J \cdot \mu_B$. For given g, the *Brillouin function* provides an excellent fit for J as shown in Figure 7.15. For GNCs, at 2 K, $J=\frac{1}{2}$ is associated with free electron paramagnetic contribution and is not seen in other classes of samples studied herein.

The quantitative information can be obtained by analysing N, J, and g in terms of saturation magnetization (M_s). It indicates that magnitude of magnetization for GNCs is 1.3×10^{-3} emu/g, which is further reduced to 1.5×10^{-4} emu/g in N-GNCs. The value of magnetization is 0.002 emu/g in graphene, and it is increased to 0.0032 emu/g in N-graphene. The response is opposite for the two systems because of different behaviours of nitrogen in the two systems.

In addition, the values of electron spin moment μ_{eff} have been estimated using values of J and N obtained by simulating equation for M for the curves. The estimates are made for moments on area of 40×40 nm^2 for room temperature and on area of 4×4 nm^2 for 2K. For GNCs, at 300 K the itinerant electron spin moment contribution is 10, which is further reduced to 1 in N-GNCs. In graphene, the contribution is ~100 and is further enhanced to 200 in N-graphene. At 2 K, spin moment is 1 for GNCs and is reduced to 0.5 in N-GNCs. A somewhat reverse trend is seen for graphene, the itinerant electron spin moment contribution is 25 for graphene and is enhanced to 50 for N-graphene.

In Figure 7.16, a comparison of FC and ZFC thermo-magnetic curves is shown for (a) GNCs, (b) N-GNCs, (c) graphene, and (d) N-graphene. For all systems, a marked irreversibility in magnetization is observed by the bifurcation in the FC–ZFC curve right from 300 K.

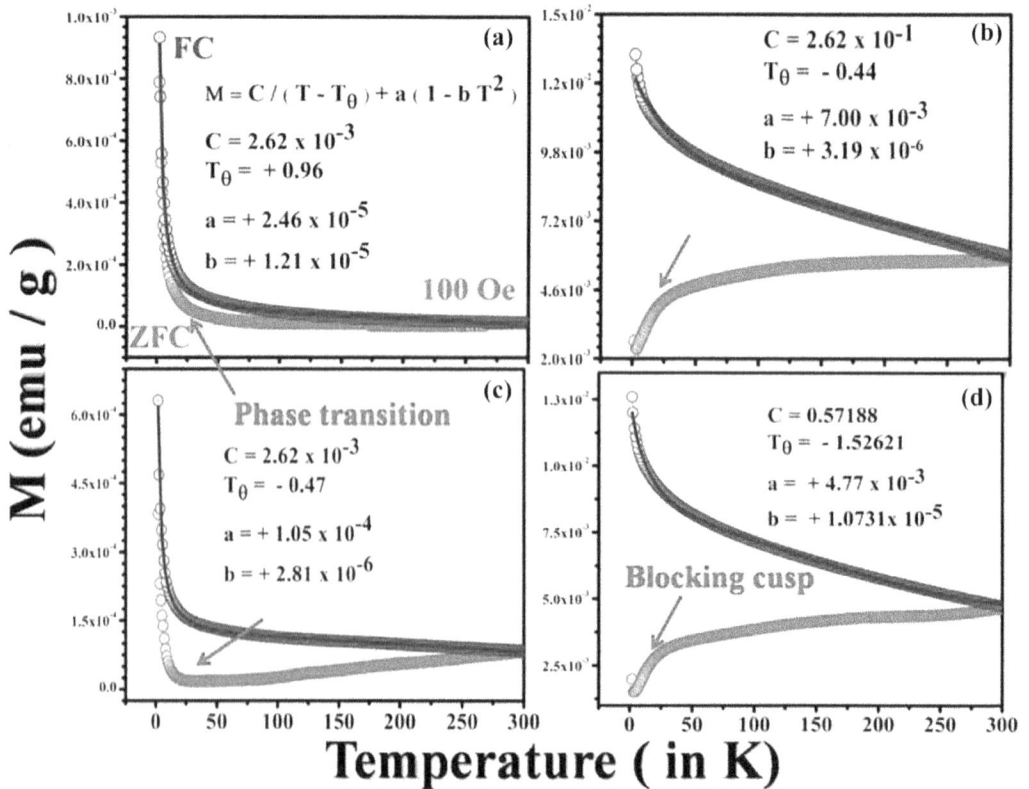

FIGURE 7.16 Recorded FC (blue) and ZFC (red) curves for (a) GNCs, (b) N-GNCs, (c) graphene, and (d) N-graphene showing thermo-magnetic behaviour at 100 Oe. Arrows in (a) and (b) indicate magnetic phase transition, and in (c) and (d), the blocking temperature.

At 2K, GNCs are perfect paramagnetic contributors with $J=\frac{1}{2}$. At this temperature, N enhances contribution by 50%, and at room temperature, the same trend is observed. In ZFC curve of N-GNCs, a shallow and broad thermal trap ~25 K (denoted by arrow in Figure 7.16b) has been observed and is associated with dominant paramagnetic spin interactions. In GNCs, the temperature is estimated to be 1 K and is reduced to 0.5 K in N-GNCs. However, the overall strength of magnetization is less as compared to graphene.

For graphene/N-graphene, a blocking cusp is observed at 25–30 K (denoted by arrows in Figure 7.16c and d). In addition, ZFC curves show magnetic phase transition for the existing ordering. For graphene, magnetic phase transition temperature is ~0.5 K and is increased to 1.5 K in N-graphene. For both the systems, FC curves show distinctly different behaviour of magnetization. It has been recognized that there is a striking resemblance between $\chi(T)$ and FC curves. All thermo-magnetic curves are simulated for the Curie–Weiss law: $M = \dfrac{C}{T + T_\Theta}$, with an additional Pauli-enhanced term [$a\left(1 - b\,T^2\right)$] [26,33]. The total expression reads as: $M = \dfrac{C}{T + T_\Theta} + a\left(1 - bT^2\right)$, where C is the Curie constant, T_θ is the Curie temperature, a is a constant, and b is a parameter associated with DOS of electrons on radical carbon. The estimated values of C for all systems are correspondingly shown in Figure 7.16 [26]. This is indicative of dramatic variations in exchange interaction in two systems. The parameter b also changes in both the systems, after loading TDAE.

The analysis carried out using SQUID technique indicates that GNCs and graphene have diamagnetic character; however, after functionalization, they show paramagnetic and antiferromagnetic contributions, respectively.

For GNCs, the added nitrogen containing free radicals migrate to the defect site and form bonds that saturate the unpaired electron. They do not show magnetic interactions. It seems that concentration of isolated spin decreases with increase in the attached functional groups at edge states. As a result, no free charge carriers are available to respond to external magnetic field. In graphene, increase in magnetization is observed on attaching the TDAE molecule, which is due to fragments of TDAE lying on the basal plane by changing the nature of bonding from sp^2 to sp^3 to some extent. Unlike GNCs, graphene is *structurally clean* having no defects. In addition, initially TDAE fragments and gets attached at the edges of graphene. The already existing free spin edge states interact with the spins of the TDAE fragments. They do not add to Curie spin concentration, but kill them. On increasing the number of radicals, they attach at the edges (physisorption) within the basal planes and create linear defects, where carbon radicals are responsible for changing state of hybridization from sp^2 (graphitic) to sp^3 (diamond-like), which is consistent with the discussions presented before, in Raman analysis. The exchange interactions are through itinerant lone pair of electrons from nitrogen radicals formed by TDAE. These radicals seem to be following the zigzag directional attachment. This results in the creation of imaginary peculiar edge state within graphene whose electronic structure is analogous to that of zigzag graphene mechanism. Such interactions are antiferromagnetically ordered.

The experimental graphene and GNCs are modelled theoretically as pristine graphene nano-ribbon (GNR) and GNR with monovacancy defect, respectively.

7.5.1.2   DFT Calculations

7.5.1.2.1   *Computational Details*

Theoretical studies have been carried out for pristine graphene nano-ribbons (GNR) and GNR with monovacancy defect. Both armchair (AGNR) and zigzag GNR (ZGNR) have been considered. The calculations have been performed using the Quantumwise Virtual Nanolab (VNL) package [18]. A supercell approach has been used for calculating the minimum energy geometry and subsequently the electronic structure. In this cell, the nano-ribbon is placed along the z-direction. We have considered 10×10 atom cell for armchair and 14×7 atom cell for zigzag nano-ribbons so that the number of atoms is comparable in the two cases. We have considered vacuum of 15 Å on each side in the y-direction and 10 Å in the x-direction to prevent interaction between the periodic images of the system. Two positions –CN$_2$ molecule at the middle and at the edge of the AGNR have been considered. The binding energy is calculated using the Slater–Koster tight-binding method [19] as implemented in the software [20]. The preferred position for the molecule on the AGNR has been accordingly decided to be at the edge. This observation also agrees with the inputs given from the experiments. For the ZGNR, we have accordingly considered only the edge position. The preferred geometries are then relaxed without any constraints using spin-polarized density functional theory (DFT) [21–23]. The Fritz Haber Institute (FHI) pseudopotential [24] with the Perdew–Burke-Ernzerhof (PBE) exchange correlation energy functional [25] has been used in the calculation. The basis set used is double zeta polarized (DZP) [26]. Optimization is performed till all the force components are reduced to less than 0.01 eV/Å. A $1 \times 1 \times 17$ Monkhorst-Pack **k**-point mesh is used in the calculations as per the underlying geometry of the graphene nano-ribbon and for convergence of the total energy. Calculations have also been carried out for GNRs passivated with hydrogen atoms. The binding energy (BE) is calculated using the following formula [26]:

$$BE = \left(\text{Energy of the molecule adsorbed GNR}\right) - \left(\text{Energy of GNR} + \text{energy of molecule}\right).$$

Larger negative values of the BE indicate a more stable configuration.

The tight-binding calculations show that the CN$_2$ molecule, from TDAE, prefers to be at the edge of the ribbon in case of the pristine armchair graphene nano-ribbon (AGNR). In the case of AGNR with a monovacancy, –CN$_2$ prefers to be at the defect site. The tight-binding results show that the binding energy (BE) for AGNR-CN$_2$ is −642.45 eV when –CN$_2$ is at the edge, while it is 641.94 eV when –CN$_2$ is at the centre. In the case of AGNR with monovacancy, the BE is −1083.66 eV at the edge and −1093.47 eV at the defect site [26].

In case of AGNR with a monovacancy, the vacancy defect leads to breaking of bonds which the missing C atom makes with three neighbouring atoms. On relaxing the geometry, two of the atoms, marked in orange, come closer forming a weak bond as shown in Figure 7.17a. The distance between these two atoms changes from 2.48 to 1.87 Å. This leads to quenching of magnetic moment at these two sites. There is a magnetic moment of 0.91 μ_B at the site of the third atom. Our calculations show that on relaxing the geometry of adsorbed CN$_2$ molecule on AGNR with monovacancy, the C atom of CN$_2$ molecule goes at the vacancy site, healing the defect and restoring the pristine AGNR structure. This quenches the local magnetic moment near the defect site. The two N atoms are detached and form a N$_2$ molecule as shown in Figure 7.17b and c.

From the unconstrained optimization of the geometry, using DFT, for the passivated pristine nano-ribbon, we find that –CN$_2$ occupies the bridging position between the atoms on the edge of the ribbon. The molecule has been displaced by 0.32 Å from the line joining the two C atoms, in the outward direction from the ribbon. The two N atoms lie on a line perpendicular to the line joining the two C atoms. The difference in the heights of N atoms is 0.27 Å, and they are 1.26 Å apart. Both the N atoms are at a distance of 1.45 Å from the C atoms and make an angle of ~51° at the C atom [26,36].

The experimental observations suggest that the TDAE molecule splits at the carbon double bond site on its interaction with GNCs or graphene. We have thus considered geometry of half of the TDAE molecule with unpassivated pristine AGNR. The molecule was initially placed at the bridging position of two edge atoms, perpendicular to the nano-ribbon. However, on unconstrained geometry optimization, we find that the molecule shifts outwards with the C atom lying in the plane of the ribbon, as shown in Figure 7.18. The two N atoms lie on a line perpendicular to the ribbon at a distance of 1.44 Å from the C atom on either side of the GNR and make an angle of ~110° at the C atom. The binding energy for the system is −4.83 eV. No magnetic moment is seen in pristine AGNR before and after attaching the half TDAE molecule.

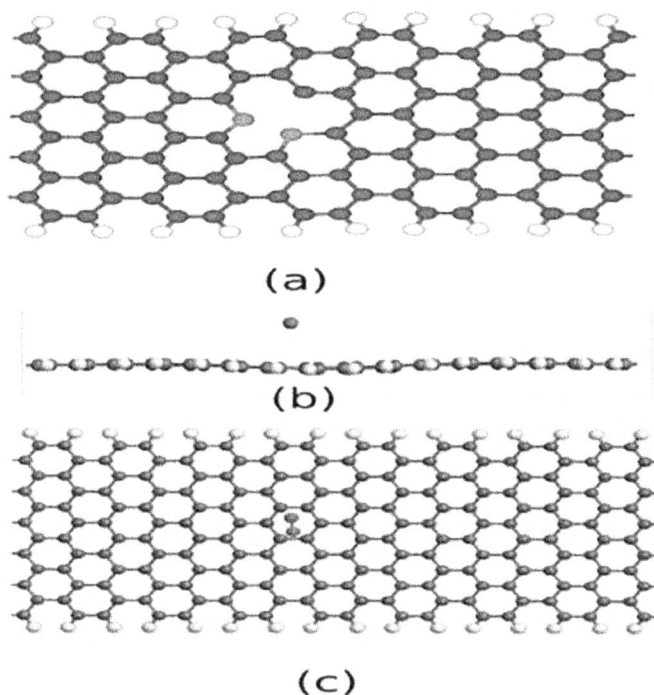

FIGURE 7.17 (a) AGNR with a monovacancy (AGNR-def); the paired atoms are shown in orange. The relaxed geometry of AGNR-def-CN_2: (b) side view and (c) top view. Hydrogen atoms are used at the edges for passivation.

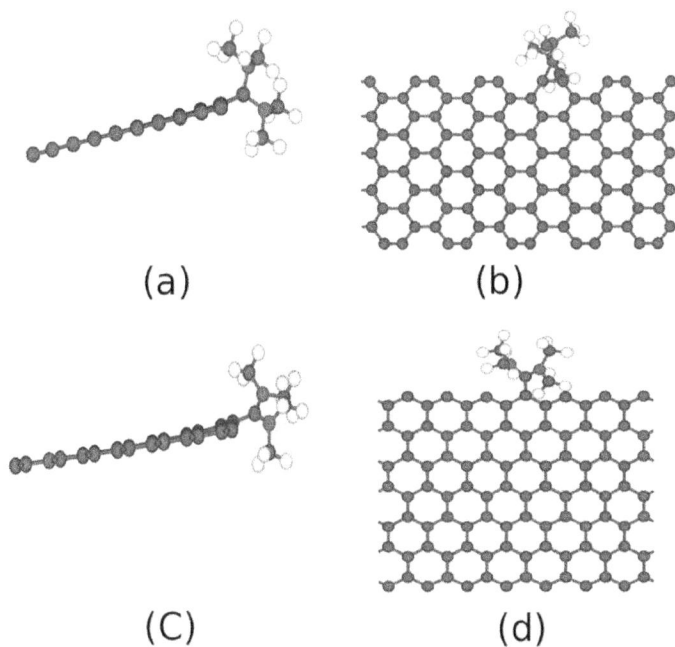

FIGURE 7.18 Geometry of TDAE on pristine unpassivated AGNR: (a) side view and (b) top view. Geometry of TDAE on pristine unpassivated ZGNR: (c) side view and (d) top view.

In case of half TDAE molecule on unpassivated zigzag nano-ribbons (ZGNR), we find that the central C atom of the half TDAE molecule is attached to an edge atom. It lies in the plane of the ribbon as in the case of the AGNR; however, the two N atoms lie on a line making an angle of ~26° with the line perpendicular to the ribbon (i.e. the x-axis) as shown in Figure 7.8c and d. The two N atoms subtend an angle of 116° at the C atom. The pristine ZGNR has a local magnetic moment of 1.16 μ_B at each of the edge atoms. The corresponding spin density plot is shown in Figure 7.9a. The Mullikan population analysis of the TDAE-ZGNR shows that the magnetic moment at the site where TDAE attaches is quenched. The spin density plot (shown in Figure 7.19b) shows that all the edge atoms in the ZGNR have a non-zero spin density; however, the spin density at the atom where the TDAE atom attaches in TDAE-ZGNR has zero spin density, which agrees with the Mullikan population results. The binding energy for the system is −6.89 eV, which is more negative than that of the TDAE-AGNR case indicating higher stability of this system. It may be noted that in AGNR the edge atoms dimerize removing the dangling bonds at the edges. This does not happen in case of ZGNR. This leads to ZGNR being more chemically active at the edges, and therefore, the half TDAE molecule has stronger binding [26].

This chapter, by and large, illustrated the magnetism scenarios in graphene and GNCs when functionalized with TDAE. GNCs consist of $sp^2 + sp^3$ heterostructure carbon, and graphene has a crystalline sp^2 network. The scope for electron spin to interact successively is variable due to the nature of inversion symmetry of GNCs and graphene. Such systems are useful magnetic media for spintronic applications. In SQUID, the magnitude of total angular momentum, J, is increased from 3/2 to 2 in N-GNCs, whereas J is 5/2 and remains unaltered in N-graphene. The total magnetization is, however, enhanced in graphene, while it is reduced in GNCs. The thermo-magnetic characteristic showed a reduction in magnetic phase transition temperature from 1 (GNCs) to 0.5 K (N-GNCs) and a rise from 0.5 (graphene) to 1.5 K (N-graphene). The system possesses high exchange anisotropy that is assigned to Pauli-enhanced antiferromagnetic order. At molecular levels, probed by FTIR, asymmetric $N_2C=CN_2$ (1628 cm^{-1}) and C-N stretching + bending mode (1260 cm^{-1}) emerges in N-GNCs. On the other hand, the degree of symmetry of central $N_2C=CN_2$ is retained with loss

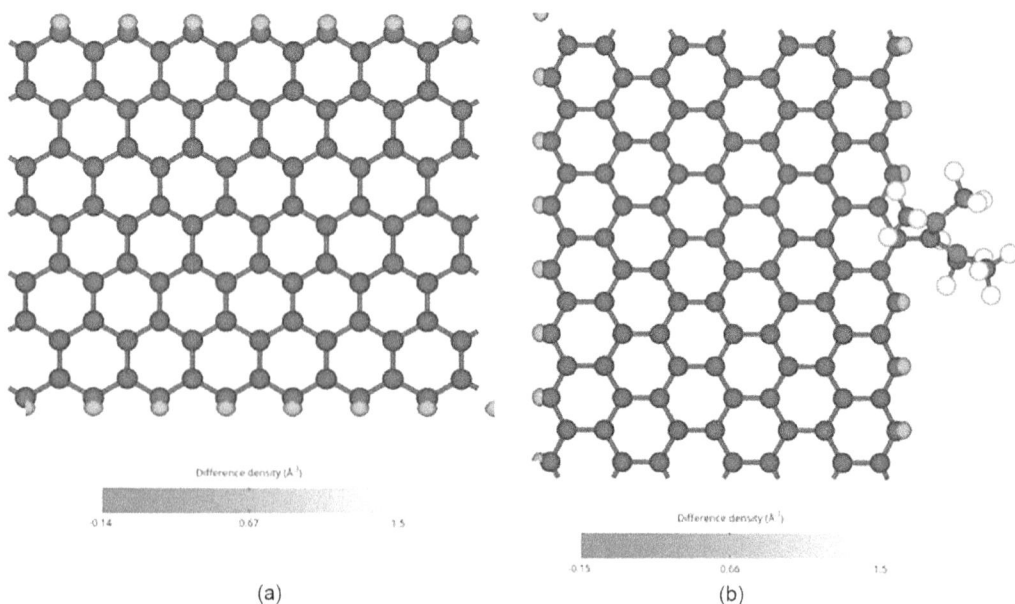

Difference density (Å^{-1})

0.14   0.67   1.5

(a)

Difference density (Å^{-1})

0.15   0.66   1.5

(b)

FIGURE 7.19 Spin density isosurfaces for (a) pristine unpassivated ZGNR and (b) TDAE molecule on pristine unpassivated ZGNR. Spin density is seen at all the edge atoms in (a), while it is quenched at the atoms where TDAE gets attached as seen in (b).

of bending character in C–N, in N-graphene. This occurs due to repulsion offered by $-(CH_3)_2N-$ fragment to π-electron puddles. Consequently, graphene and GNCs acquired, respectively, di- and mono-ionic states. In Raman, electron–phonon coupling is reduced in N-GNCs, whereas coupling is increased in N-graphene. This has resemblances with spin–orbit coupling estimated by ESR (reported elsewhere). In ESCA, the blueshift in N-graphene (1.82 eV) is four times than that in N-GNCs (0.45 eV). Nitrogen acts as a p-type dopant in GNCs, while it acts as an n-type dopant in graphene, selectively reacting with carboxylic (284.60 eV) and epoxide (286.60 eV) sites, in both. Our spectro-microscopic analysis has sketched the scheme of molecular fragmentation, the emergence of photo-excited mid-gap states, and surface modifications in two systems [26]. In GNCs, TDAE radicals are present in the form of tetramethylurea and dimethylamine, while in graphene, tetramethylamines are present, exhibiting a mid-gap of 5.26, 4.23, and 3.52 eV, respectively. Graphene and GNCs are modelled as pristine GNR and GNR with monovacancy defect, respectively. Theoretical results predict the position of the TDAE molecule at the edge of the AGNR, and the results corroborate with the experimental observations. DFT-based calculations, carried out on pristine unpassivated and passivated armchair and zigzag GNRs, indicate magnetic moments at the edge atoms of only unpassivated zigzag GNR. Adsorption of TDAE molecule on this GNR quenches the moment at the adsorption site. Adsorption of $-CN_2$ molecule on the GNR with monovacancy heals the vacancy and frees the N_2 molecule. Magnetism scenarios seen experimentally in graphene and GNCs functionalized with TDAE molecule could not be asserted theoretically, and we therefore feel that simple models used in the calculations may not describe the experiments judiciously [50].

REFERENCES

[1] Alegaonkar, Ashwini P., Arvind Kumar, Satish K. Pardeshi, and Prashant S. Alegaonkar. "Spin dynamics in graphene and graphene like nanocarbon doped with nitrogen the esr analysis." *arXiv preprint arXiv:1308.5291* (2013).

[2] Owens, Frank J., Zafar Iqbal, Lioubov Belova, and K. V. Rao. "Evidence for high-temperature ferromagnetism in photolyzed C_{60}." *Physical Review B* 69, no. 3 (2004): 033403.

[3] Esquinazi, P., A. Setzer, R. Höhne, C. Semmelhack, Y. Kopelevich, D. Spemann, T. Butz, B. Kohlstrunk, and M. Lösche. "Ferromagnetism in oriented graphite samples." *Physical Review B* 66, no. 2 (2002): 024429.

[4] Li, Shandong, Guangbin Ji, and Liya Lü. "Magnetic carbon nanofoams." *Journal of Nanoscience and Nanotechnology* 9, no. 2 (2009): 1133–1136.

[5] Sasaki, Munetaka, and Olivier C. Martin. "Temperature chaos, rejuvenation, and memory in Migdal-Kadanoff spin glasses." *Physical Review Letters* 91, no. 9 (2003): 097201.

[6] Alegaonkar, Ashwini P., Arvind Kumar, Sagar H. Patil, Kashinath R. Patil, Satish K. Pardeshi, and Prashant S. Alegaonkar. "Spin transport and magnetic correlation parameters for graphene-like nanocarbon sheets doped with nitrogen." *The Journal of Physical Chemistry C* 117, no. 51 (2013): 27105–27113.

[7] Nair, R. R., M. Sepioni, I-Ling Tsai, O. Lehtinen, J. Keinonen, A. V. Krasheninnikov, T. Thomson, A. K. Geim, and I. V. Grigorieva. "Spin-half paramagnetism in graphene induced by point defects." *Nature Physics* 8, no. 3 (2012): 199–202.

[8] Enoki, Toshiaki, and Kazuyuki Takai. "The edge state of nanographene and the magnetism of the edge-state spins." *Solid State Communications* 149, no. 27–28 (2009): 1144–1150.

[9] Kiguchi, Manabu, Kazuyuki Takai, V. L. Joseph Joly, Toshiaki Enoki, Ryohei Sumii, and Kenta Amemiya. "Magnetic edge state and dangling bond state of nanographene in activated carbon fibers." *Physical Review B* 84, no. 4 (2011): 045421.

[10] Fernández-Rossier, Joaquín, and Juan José Palacios. "Magnetism in graphene nanoislands." *Physical Review Letters* 99, no. 17 (2007): 177204.

[11] Tao, Chenggang, Liying Jiao, Oleg V. Yazyev, Yen-Chia Chen, Juanjuan Feng, Xiaowei Zhang, Rodrigo B. Capaz et al. "Spatially resolving edge states of chiral graphene nanoribbons." *Nature Physics* 7, no. 8 (2011): 616–620.

[12] Sheng, W., Z. Y. Ning, Z. Q. Yang, and H. Guo. "Magnetism and perfect spin filtering effect in graphene nanoflakes." *Nanotechnology* 21, no. 38 (2010): 385201.

[13] Yazyev, Oleg V., Wei L. Wang, Sheng Meng, and Efthimios Kaxiras. "Comment on graphene nano-flakes with large spin: broken-symmetry states." *Nano Letters* 8, no. 2 (2008): 766–766.

[14] Khveshchenko, D. V. "Magnetic-field-induced insulating behavior in highly oriented pyrolitic graphite." *Physical Review Letters* 87, no. 20 (2001): 206401.

[15] Andriotis, A. N., and M. Menon. *Clusters and nano-assemblies, physical and biological systems*, pp, 199–200. World Scientific, Singapore, 2005.

[16] Lehtinen, Petri O., Adam S. Foster, Andres Ayuela, Arkady Krasheninnikov, Kai Nordlund, and Risto M. Nieminen. "Magnetic properties and diffusion of adatoms on a graphene sheet." *Physical Review Letters* 91, no. 1 (2003): 017202.

[17] Yazyev, Oleg V., and Lothar Helm. "Defect-induced magnetism in graphene." *Physical Review B* 75, no. 12 (2007): 125408.

[18] Elliott, R. J. "Theory of the effect of spin-orbit coupling on magnetic resonance in some semiconductors." *Physical Review* 96, no. 2 (1954): 266.

[19] Yafet, Y. "g Factors and spin-lattice relaxation of conduction electrons." In *Solid state physics*, vol. 14, pp. 1–98. Academic Press, UK 1963.

[20] Ochoa, H., A. H. Castro Neto, and F. Guinea. "Elliot-Yafet mechanism in graphene." *Physical Review Letters* 108, no. 20 (2012): 206808.

[21] Dyakonov, M. I., and V. I. Perel. "Spin relaxation of conduction electrons in noncentrosymmetric semiconductors." *Soviet Physics Solid State* 13, no. 12 (1972): 3023–3026.

[22] Kumar, Arvind, Sumati Patil, Anupama Joshi, Vasant Bhoraskar, Suwarna Datar, and Prashant Alegaonkar. "Mixed phase, sp^2–sp^3 bonded, and disordered few layer graphene-like nanocarbon: synthesis and characterizations." *Applied Surface Science* 271 (2013): 86–92.

[23] Alegaonkar, Ashwini P., Arvind Kumar, Prashant S. Alegaonkar, Shobha A. Waghmode, and Satish K. Pardeshi. "Exchange interaction of itinerant electron donors of tetrakis (dimethylamino) ethylene with localized electrons in graphene." *Synthesis and Reactivity in Inorganic, Metal-Organic, and Nano-Metal Chemistry* 44, no. 10 (2014): 1477–1482.

[24] Alegaonkar, Ashwini P., Satish K. Pardeshi, and Prashant S. Alegaonkar. "Spin dynamics in graphene-like nanocarbon, graphene and their nitrogen adatom derivatives." *Applied Physics A* 124, no. 7 (2018): 1–10.

[25] Alegaonkar, Ashwini, Prashant Alegaonkar, and Satish Pardeshi. "Exploring molecular and spin interactions of Tellurium adatom in reduced graphene oxide." *Materials Chemistry and Physics* 195 (2017): 82–87.

[26] Alegaonkar, Ashwini P., Aniruddha S. Kibey, Prashant S. Alegaonkar, Anjali Kshirsagar, and Satish K. Pardeshi. "Experimental and theoretical study of tetrakis (dimethylamino) ethylene induced magnetism in otherwise nonmagnetic graphene derivatives." *Materials Chemistry and Physics* 222 (2019): 132–138.

[27] Alegaonkar, Ashwini, Prashant Alegaonkar, and Satish Pardeshi. "Magneto chemistry and spin dynamics in graphene and graphene derivatives." In *Spintronics: a review and direction for research*, vol. 1, pp. 35–55, Switzerland, 2019.

[28] Ruderman, Melvin A., and Charles Kittel. "Indirect exchange coupling of nuclear magnetic moments by conduction electrons." *Physical Review* 96, no. 1 (1954): 99.

[29] Kasuya, Tadao. "A theory of metallic ferro-and antiferromagnetism on Zener's model." *Progress of Theoretical Physics* 16, no. 1 (1956): 45–57.

[30] Yosida, Kei. "Magnetic properties of Cu-Mn alloys." *Physical Review* 106, no. 5 (1957): 893.

[31] McCulloch, D. G., J. L. Peng, D. R. McKenzie, Shu Ping Lau, D. Sheeja, and B. K. Tay. "Mechanisms for the behavior of carbon films during annealing." *Physical Review B* 70, no. 8 (2004): 085406.

[32] Winberg, H. E., J. R. Downing, and D. D. Coffman. "The chemiluminescence of tetrakis (dimethyl-amino) ethylene[1]." *Journal of the American Chemical Society* 87, no. 9 (1965): 2054–2055.

[33] Pisani, L., J. A. Chan, B. Montanari, and N. M. Harrison. "Electronic structure and magnetic properties of graphitic ribbons." *Physical Review B* 75, no. 6 (2007): 064418.

[34] Kumar, Arvind, and Prashant S. Alegaonkar. "Properties of spin bath of graphene–like nanocarbon." *International Journal of Innovative Research in Science, Engineering and Technology (IJIRSET)* 3 (2014): 14049–14055.

[35] Alegaonkar, Ashwini P., Prashant S. Alegaonkar, and Satish K. Pardeshi. "Electrochemical performance of a self-assembled two-dimensional heterostructure of rGO/MoS_2/h-BN." *Nanoscale Advances* 2, no. 4 (2020): 1531–1541.

[36] Neumann, P., R. Kolesov, B. Naydenov, J. Beck, F. Rempp, M. Steiner, V. Jacques et al. "Quantum register based on coupled electron spins in a room-temperature solid." *Nature Physics* 6, no. 4 (2010): 249–253.

[37] Merali, Zeeya. "The power of discord: physicists have always thought quantum computing is hard because quantum states are incredibly fragile. But could noise and messiness actually help things along?" *Nature* 474, no. 7349 (2011): 24–27.

[38] Carter, Samuel G., Timothy M. Sweeney, Mijin Kim, Chul Soo Kim, Dmitry Solenov, Sophia E. Economou, Thomas L. Reinecke, Lily Yang, Allan S. Bracker, and Daniel Gammon. "Quantum control of a spin qubit coupled to a photonic crystal cavity." *Nature Photonics* 7, no. 4 (2013): 329–334.

[39] Huertas-Hernando, Daniel, Francisco Guinea, and Arne Brataas. "Spin-orbit coupling in curved graphene, fullerenes, nanotubes, and nanotube caps." *Physical Review B* 74, no. 15 (2006): 155426.

[40] Monz, Thomas, Philipp Schindler, Julio T. Barreiro, Michael Chwalla, Daniel Nigg, William A. Coish, Maximilian Harlander, Wolfgang Hänsel, Markus Hennrich, and Rainer Blatt. "14-qubit entanglement: creation and coherence." *Physical Review Letters* 106, no. 13 (2011): 130506.

[41] Golovach, Vitaly N., Alexander Khaetskii, and Daniel Loss. "Phonon-induced decay of the electron spin in quantum dots." *Physical Review Letters* 93, no. 1 (2004): 016601.

[42] Childress, L., M. V. Gurudev Dutt, J. M. Taylor, A. S. Zibrov, F. Jelezko, J. Wrachtrup, P. R. Hemmer, and M. D. Lukin. "Coherent dynamics of coupled electron and nuclear spin qubits in diamond." *Science* 314, no. 5797 (2006): 281–285.

[43] Kane, Charles L., and Eugene J. Mele. "Quantum spin Hall effect in graphene." *Physical Review Letters* 95, no. 22 (2005): 226801.

[44] Bychkov, Yu A., and Emmanuel I. Rashba. "Oscillatory effects and the magnetic susceptibility of carriers in inversion layers." *Journal of Physics C: Solid State Physics* 17, no. 33 (1984): 6039.

[45] Dóra, Balázs, Ferenc Murányi, and Ferenc Simon. "Electron spin dynamics and electron spin resonance in graphene." *EPL (Europhysics Letters)* 92, no. 1 (2010): 17002.

[46] Gmitra, Martin, Sergej Konschuh, Christian Ertler, ClaudiaAmbrosch-Draxl, and Jaroslav Fabian. "Band-structure topologies of graphene: spin-orbit coupling effects from first principles." *Physical Review B* 80, no. 23 (2009): 235431.

[47] Zhao, You Xiang, and Ian L. Spain. "X-ray diffraction data for graphite to 20 GPa." *Physical Review B* 40, no. 2 (1989): 993.

[48] Cançado, Luiz Gustavo, Ado Jorio, Erlon Henrique Martins Ferreira, F. Stavale, Carlos Alberto Achete, Rodrigo Barbosa Capaz, Marcus Vinicius de Oliveira Moutinho, Antonio Lombardo, Tero S. Kulmala, and Andrea Carlo Ferrari. "Quantifying defects in graphene via Raman spectroscopy at different excitation energies." *Nano Letters* 11, no. 8 (2011): 3190–3196.

[49] Józsa, C., T. Maassen, M. Popinciuc, P. J. Zomer, A. Veligura, H. T. Jonkman, and B. J. Van Wees. "Linear scaling between momentum and spin scattering in graphene." *Physical Review B* 80, no. 24 (2009): 241403.

[50] Niyogi, Sandip, Elena Bekyarova, Mikhail E. Itkis, Jared L. McWilliams, Mark A. Hamon, and Robert C. Haddon. "Solution properties of graphite and graphene." *Journal of the American Chemical Society* 128, no. 24 (2006): 7720–7721.

8 Multi-Functional Nano-Carbons

From Meta-Materials to Non-Liner Optics and Gas Sensing to Mechanically Tough Fibre Mat Application

8.1 MULTIFUNCTIONAL ASPECTS

The search for multifunctional material, particularly for gas sensing and electromagnetic interference (EMI) shielding, is of great importance in military and civil domains. In recent years, nanocarbons such as carbon nanotubes, reduced graphene oxide, and graphene have become most popular sensing probes for gas detection [1], particularly for ammonia (NH_3). It has been reported that nanocarbon showed good NH_3 sensitivity due to unique structure, e.g. small size, large specific surface area [2], and outstanding electronic properties such as high electron mobility [2], sensitivity to electrical perturbations through NH_3 molecules. Hu et al. demonstrated NH_3 gas sensor based on reduced graphene oxide, at room temperature [3]. Cui et al. developed silver nanocrystal incorporated carbon nanotubes for NH_3 sensors at room temperature [4]. Ghosh et al. found NH_3 sensitivity in the range of 200–2800 ppm, for reduced graphene oxide sensor probe [5]. Mao et al. reviewed sensor study, focused on challenges and opportunities of nanocarbon-based materials for gas detection [6]. Further, the importance of EMI shielding has increased greatly as today's fast developing society is more dependent on electronics and growth of radio frequency radiation sources. The electromagnetic radiation, particularly, that at microwave frequencies tend to interfere with radars and electronics. Such interference is assumed to be strategically adverse in defence and, moreover, is hazardous in civil sector [7]. Cao et al. has studied temperature dependence of permittivity of ultrathin graphene composites [8], and multi-region microwave absorption of nano-needle-like ZnO [9]. In another study, He et al. reported tuneable electromagnetic attenuation capability of magnetic nanoparticles decorated reduced graphene oxides in microwave region [10]. Wen et al. found the thinnest and most lightweight materials with highly efficient microwave attenuation performance of reduced graphene oxide [11]. In most of these studies, the emphasis is on graphene based nanocarbon material correlating dielectric, EMI shielding, and microwave-absorption performance [12,13]. Another class of carbon is the amorphous nanocarbon, the broad name used for soot of the carbonaceous element. It is primarily composed of nanocarbon in the form of agglomerated nanoparticles with diameter of about 10–50 nm. They have neither graphite nor diamond-like phase of carbon [14] and lead to interesting physical properties. Their degree of sp^2 graphitic ordering ranges from nanocrystalline graphite to glassy carbon. One can define diamond-like carbon as amorphous carbon which can have mixture of sp^3 and sp^2 sites with composition depending upon formation of amorphous phase. The basic element which makes this material interesting is their bond molecular environment that consists of σ and π states. They have significantly different behaviour and properties changes dramatically with composition of sp^2 and sp^3. Such nanocarbon is light weight, easy to synthesize, cost effective, stable at high temperature, and eco-friendly.

DOI: 10.1201/9781003317258-8

The focus of this section is to evaluate performance of such nanocarbon, obtained by pyrolysis of camphor at low temperature, for NH_3 gas sensing and EMI shielding applications. The fabricated nanoparticles consisted of sp^2+sp^3 carbon having bond disorder in terms of unsaturated dangling bonds, vacancy mediated defects, etc. The nanocarbons were integrated with optical fibre-based gas sensor. The performance characteristic of the designed sensor was investigated for full scale detection, response function and limit of detection. The evaluation of nanocarbon for EMI shielding was demonstrated by analysing dielectric function, DC and AC conductivity, skin depth and % reflection [15,16].

8.1.1 OPTICAL GAS SENSING

8.1.1.1 Optical Band Structure: The Sensor Characteristics

In the present work, we have demonstrated applications of the obtained nanocarbon for optical based gas sensing and EMI shielding. To utilize nanocarbon for optical applications, one need to know optical band gap and available photo-excited state of the material. UV–visible spectroscopy along with PL provides information about band gap and excited states. Figure 8.1 shows recorded UV–visible spectrum of fabricated nanocarbon in non-aqueous medium. Figure 8.1a shows variation in normalized absorbance (α) as a function of wavelength (in nm). It showed a sharp peak centred ~207 nm and a broad shoulder extended from 225 to 400 nm.

The sharp peak was finger print of π-σ transition and broad feature was associated with π–π^* transitions in nanocarbon. Figure 8.1b is Tauc graph showing $(\alpha h\nu)^2$ vs energy. The extrapolation of straight line to $(\alpha h\nu)^2=0$ axis gives the value of band gap [17]. The constructed tangents (dotted lines) to the curve indicated respective optical band gap for π–σ and π–π^* electronic environments. The bandgap of the nanostructure was 4.5 eV having mid-gap of 2.8 eV. Interestingly, the optical charge carriers have mid-gap state which were non-radiative in nature and having photo carrier excited states [18].

Figure 8.2a shows PL emission having two excited states one at 2.27 and other 1.6 eV. It showed that nanocarbon consisted of two photo-excited states lower than mid-gap state. Thus, nanocarbon also possesses non-radiative fluoro-excited states. The fluoro-excited states were contributions from the dangling edges of sp^2+sp^3 hetero-structures. In nanocarbon case the lower unoccupied molecular orbital (LUMO) band consisted of both s and p orbitals, contain a total number of eight electrons per Brillion zone [19]. The band scheme is shown in Figure 8.2b. By and large analysis revealed that several states existed within HOMO-LUMO gap. These states have the wide dispersion of oscillator strength especially active in infrared and mid infrared region. This makes nanocarbon useful for optical-based applications [20].

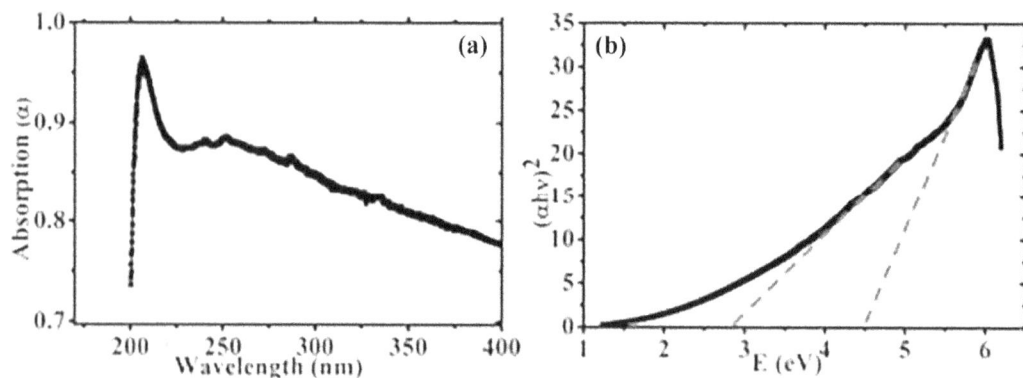

FIGURE 8.1 Recorded UV–Vis absorption spectra of nanocarbon colloidal. Plot (a) absorbance (α) vs wavelength, and (b) $(\alpha h\nu)^2$ vs energy. The dotted lines indicate tangent to Tauc curve.

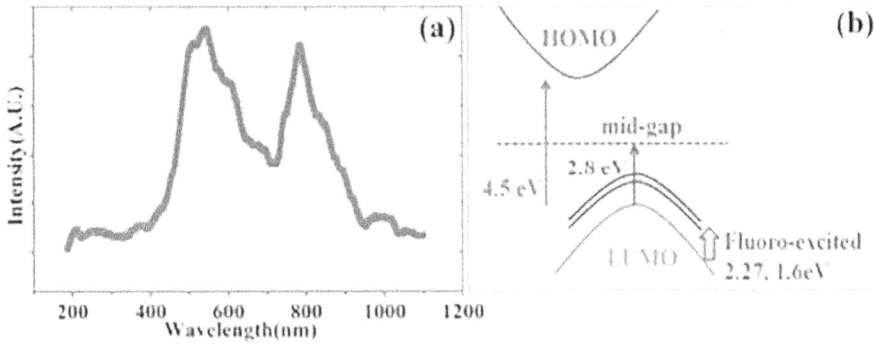

FIGURE 8.2 (a) PL spectrum recorded for nanocarbon at excitation wavelength of $\lambda \sim 300\,nm$. (b) Optical band scheme of nanocarbon.

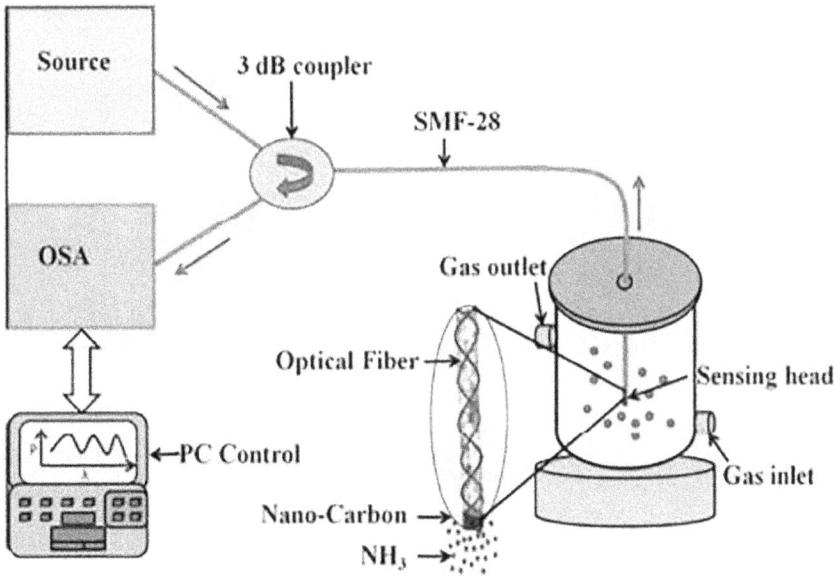

FIGURE 8.3 Elements of Fabry–Perot interferometer optical fibre gas sensor set-up that consisted of SMF-28, 3 dB coupler, OSA, gas sensing chamber, and PC control unit. The photograph of zoomed in portion of sensor probe is shown with schematics of NH_3 molecule sensing.

8.1.2 MEASUREMENTS OF SENSING PARAMETERS: NH_3 GAS A CASE STUDY

For utility, nanocarbon specimen was deposited on optical fibre used for gas sensing application. The schematic representation of Fabry–Perot interferometer optical fibre NH_3 gas sensor is shown in Figure 8.3.

In this set-up, the incident mid-infrared radiation (wavelength 1510–1590 nm) from the source gets bifurcated from 3 dB coupler into the ratio of 50:50. The 50% of light travels through SMF-28 whose one end was perfectly cleaved normal to the fibre axis. The light from the fibre end reflects back giving rise to interference pattern. Thus, the cleaved fibre end was extremely sensitive to the reflection of light. In Figure 8.4, zoomed portion of the fibre tip decorated with nanocarbon is shown in schematics. On interaction with NH_3 molecules, the optical band gap of the tip coating changes with change in the interference pattern. The remaining 50% radiation passes through OSA and was

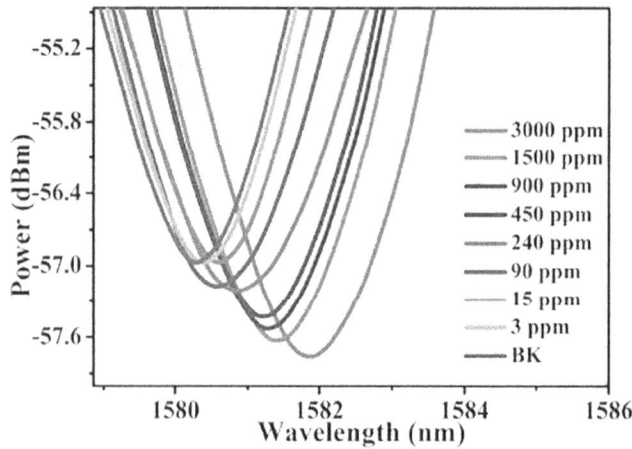

FIGURE 8.4 Sensing response of nanocarbon tip to NH_3 molecules, for a period of 180 seconds, at molecular concentrations ranging from 3 to 3000 ppm.

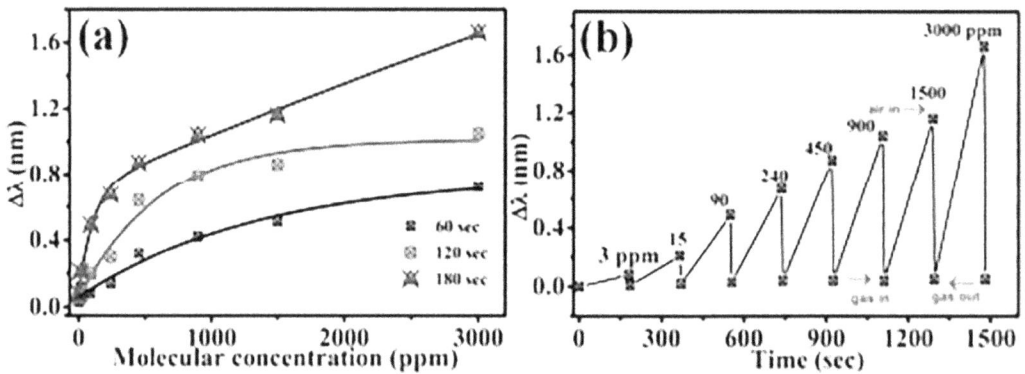

FIGURE 8.5 (a) Response function characteristics of wavelength shift ($\Delta\lambda$) with molar concentration (C in ppm) for different time periods, and (b) functional response of the sensor in terms of ($\Delta\lambda$) and time at different ppm levels.

coupled to PC control unit that displayed sensing output for power and wavelength shift, as seen in Figure 8.4.

Initially, the spectrum was recorded for the tip in absence of NH_3. Following this, 3 ppm of NH_3 insufflate into gas chamber and onset, spectrum was recorded. The process iterated from 3 to 3000 ppm and at each iteration the response/recovery was monitored and recorded. It has been observed that, with subsequent increment in ppm level, there was a shift in the wavelength. The sensing measurements were carried at room temperature and reliability as well as reproducibility in the sensor response has been noted well.

In Figure 8.5a wavelength shift ($\Delta\lambda$) as function of molar concentration (C in ppm) of NH_3 is plotted for various probe/gas interaction time, 60–180 seconds. At 3000 ppm, the sensor showed the wavelength shift of 1.66 nm for 180 seconds. For 3 ppm of gas, the wavelength shift of 80 pm was observed for 180 seconds. It has been observed that, as the NH_3 concentration increases from 0 to 3000 ppm, the sensor showed prominent wavelength shift. All measurements were perfectly reproducible [15].

8.1.2.1 Sensor Transfer Function

Transfer function of a sensor system quantifies response of stimuli as a function of input. The nature of $\Delta\lambda - C$ has been simulated for the measured time interval in the form of transfer function characteristics. At smaller time intervals, i.e. up to 60 seconds, the function took simple form $\Delta\lambda = e^{C/M}$, where M is interaction volume, K, is sensitivity constant expressed as the product of molecular interaction length and molar concentration (C). In this time domain, the simulated curve parameters, indicated that the change in wavelength was of the order of 0.8–1.0 nm affecting 1000–1500 carbon molecules. The sensitivity for molecular NH_3 was ~7–8 atomic distances of carbon. For higher temporal domain, the statistical interaction scenario between NH_3/nanocarbon system became more complex. This was particularly due to other physical component that started dominating the interaction such as diffusion of NH_3 in sub-surface region of individual nanocarbon, mutual chemical potential (μ) and physisorption potential (ε) experienced by NH_3/carbon ensemble, onset, within interaction volume, mismatch in the molecular vibration, etc. More diffusivity of NH_3 increased the number of carbon atom that were participating in interaction, in this time domain, which was estimated ~5000–6000. This had implication on both sensitivity and molar length in which sensitivity improved marginally to 10–15 molar length. For still higher sensing time, the $\Delta\lambda - C$ transfer function took the form: $\Delta\lambda = \prod_{i,\,j=1}^{3} K\, e^{C/m_j}$, in which molar interaction went in the order of 103–104 and beyond. The transfer function characteristic discussed here is, strictly, at 300 K. Any departure from this value introduces adiabatic perturbative term of higher order which has potential dependence on thermodynamics of interacting system [21]. Figure 8.5b shows variation in $\Delta\lambda$ as function of time for both response and recovery. The empirical analysis of the curve indicated that it was composed of linear and impulse response function which is express as,

$$\Delta\lambda = \prod_{i,\,j=1}^{3}\left\{ K_i T + \left[1 - \exp\left(\frac{\dot{\mu}\dot{\varepsilon}^{-1} + \dot{\varepsilon}\dot{\mu}^{-1}}{K_j} \right) \right] \right\}$$ where T is interaction time, B is diffusivity (cm²/s) of

NH_3 molecule, ε and μ, respectively, rate of change of physisorption and chemisorption potential. The response and recovery time of about 5 and 8 seconds were noted, respectively. Table 8.1 indicate performance characteristics of our sensor [15].

TABLE 8.1

Performance Sheet of Nanocarbon Sensor System

Parameters	Performance and Range
Gas selectivity	NH_3 (assessed and evaluated)
Full-scale detection (instrumental)	3–3000 ppm
Pulse rise time	5 seconds
Rise response	Linear
Pulse decay time	8 seconds
Decay response	Impulse
Limit of detection (intrinsic)	1–5 ppb per ten cycles (irreversible)
Time	0–180 seconds
Molar transfer function	$\Delta\lambda = \prod_{i,j=1}^{3} K\, e^{C/m_j}$
Temporal transfer function	$\Delta\lambda = \prod_{i,j=1}^{3}\left\{ K_i T + \frac{K_j B}{T^3}\left[1 - \exp\frac{\left(\dot{\mu}\,\dot{\varepsilon}^{-1} + \dot{\varepsilon}\,\dot{\mu}^{-1} \right)}{K_j} \right] \right\}$
Temperature range	300 ± 10 K

8.1.2.2 Sensing Mechanism

The nanocarbon consisted of sp^2+sp^3 phase of carbon in which sp^3 phase was distributed in-homogeneously and isotropically. At the sp^2+sp^3 interface the coordination number of carbon atom connecting two phases is in disproportionate and lead to one under coordinated electron. In addition, there were $\pi-\pi^*$ conjugated electrons in sp^2 zone. Moreover, Raman analysis revealed vacancy mediated disorder in nanocarbon system. At the vacancy site, there was non-uniform electron sharing between three uncoordinated carbon atoms. The two-charge deficient carbon played role of acceptor atoms, wherein, the third carbon as the donor. The under coordinated electrons, $\pi-\pi^*$ conjugated electrons, and vacancy sites acted as the photo-carriers that participated in optical excitation process. This results into the behaviour of nanocarbon as an n-type photoconductor. The molecular NH_3 on interaction with nanocarbon, transiently, changed the mid-gap, fluoro-excited gap and, consequently, optical gap. In NH_3, nitrogen that carries lone pair of electrons get attracted to the charge deficient carbon atom sites by van-der-Waal interactions. The hydrogen atom in NH_3 physically adsorb at the interface of sp^2 and sp^3. Since NH_3/nanocarbon interaction was statistical in physico-chemical nature, depending upon the stereo-regular configuration of sp^2+sp^3, there could be possibility of complete charge transfer. This will have indentation on molecular vibrations of nanocarbon, analysed before. Interestingly, variations have been observed in the post-NH_3 sensed nanocarbon revealed by the Raman spectroscopy [15].

8.1.2.3 Molecular Imprint of NH₃

In Figure 8.6b there are significant changes in the post NH_3 treated nanocarbon sensing probe when compared to free standing nanocarbon (Figure 8.6a). Broadly, G-band was decreased and D-band was increased and broaden. The broadening of the band indicated reduction in crystallite size of

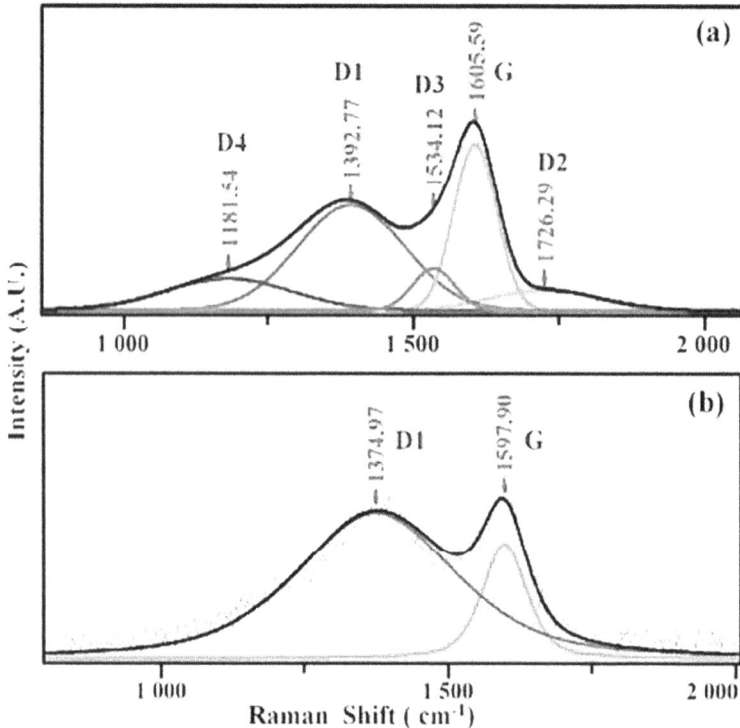

FIGURE 8.6 Recorded Raman spectrum of (a) nanocarbon decorated on optical fibre tip; (b) post-NH_3 sensing ($\lambda \sim 457\,nm$).

nanocarbon. Moreover, G-peak was shifted from 1605.59 to 1597.9 cm^{-1}, after NH$_3$ sensing. The downshifting of G line position was attributed to enhancement in bond-angle disorder at G sites [22]. The peaks D_4 and D_3 disappeared, whereas D_2 practically vanished. This indicated that sp^2 +sp^3 sites and zones related to amorphous carbon were altered. The submerging of D_3 and D_4 into D_1 was indicative of increased in unsaturated dangling bond character in nanocarbon. Although the sensing action of nanocarbon probe seems to be completely reversible, macroscopically, at molecular level, the imprint of NH$_3$/nanocarbon interaction is identified, clearly, by Raman spectroscopy. The sum of molecular contribution of amorphous carbon moieties, i.e. D_4+D_3, was of the order of 7×10^3 which got sacrificed during sensing action. On the contrary, from sp^2 environment $\sim 3 \times 10^3$ dangling carbon bonds were modified in sensing 3–3000 ppm of NH$_3$ molecules, at the end of several cycles. The rupture at sp^3 was nearly twice than at sp^2. Though the cumulative data is shown at the end of the cycles, one can investigate the effect after individual cycle. The result suggested that the disordered carbon in nano-shell forms interstitial defects leading to molecular imprints [15,16]. Further, the absence of D_2 showed degradation in crystallinity which form non-graphitic phase of nanosized dimension having polycrystalline/amorphous carbon nature. This was reflected in reduction of L_a from 3 (free-standing nanocarbon) to 1 nm (post NH$_3$ treated). The change corresponds to a second maximum in the graphite vibrational density of states near the M point of the Brillion zone boundary, which became prominent in small graphite crystallites, after NH$_3$ treatment. This resulted into lack of a long-range translation symmetry which led to a breakdown of the k-momentum conservation rule [18,19].

8.1.3 EMI Shielding: The Added Feature

Shielding of an object from incoming electromagnetic radiation is important in both military and civil sectors. EMI shielding is achieved by reflection as well as absorption of electromagnetic radiation by a material, which thereby acts as a shield against penetration of radiation [23].

8.1.3.1 Coating Characteristics

Figure 8.7a shows recorded SEM cross-sectional view of nanocarbon film. The image showed thickness of nanocarbon ~ 60 μm on ~ 25 μm copper foil. The film was uniform, however, having columnar deposition inhomogeneities at certain places. In general, most of the film was continuous laterally and longitudinally. Also, at some places the peeling effect was observed. A typical micrograph is shown and others are provided in Appendix C, Figure C.8.7.1. In order to see the elemental composition of deposited nanocarbon, energy-dispersive X-ray analysis (EDAX) was carried out. Figure 8.7b indicates peaks associated with oxygen, carbon, and copper. The inset table shows elemental composition. The high amount of oxygen ~ 12 atomic % was attributed to the presence of native oxide layer on copper [15].

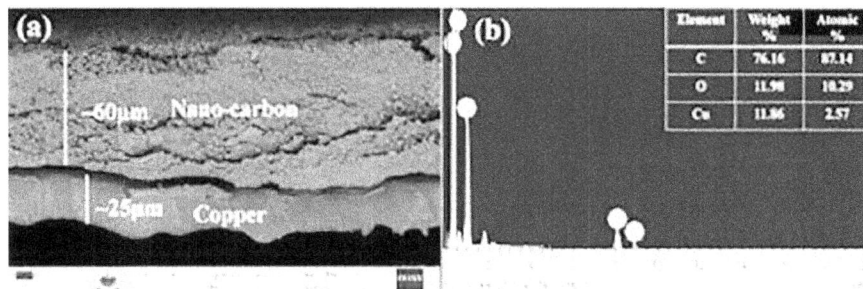

FIGURE 8.7   (a) FESEM cross-sectional view of nanocarbon film deposited onto copper; (b) the recorded EDAX spectrum. Inset in (b) shows elemental composition.

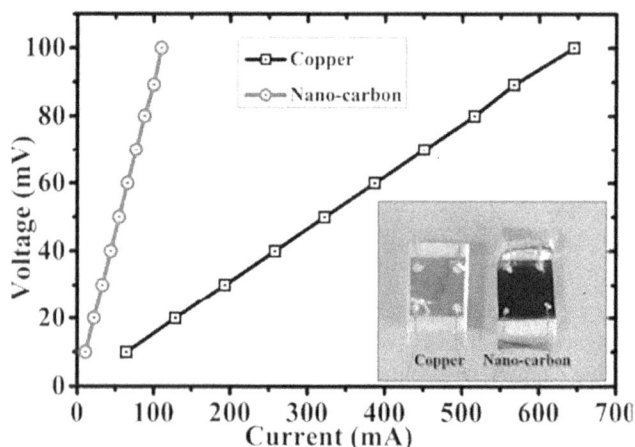

FIGURE 8.8 Measured current–voltage (*I–V*) characteristic of nanocarbon and copper foil. Inset shows typical photograph of samples with electrical connections and contact developed.

TABLE 8.2
***I–V* Measurement Parameters for Copper and Nanocarbon**

Parameters	Copper	Nanocarbon
Slope (*m*)	0.16013	0.90712
Thick (*h*)	25 µm	60 µm
Width (*w*)	2 cm	2 cm
Length (*l*)	2 cm	2 cm
DC conductivity (σ_{dc})	2.5×10^5 s/m	1.8×10^4 s/m
Skin depth (δ)	11 nm	42 nm

8.1.3.2   DC Conductivity (σ_{dc})

The electromagnetic rays in X-band, particularly, interact with nanocarbon material in the form of various losses such as electric conduction, hysteresis and electron spin resonance. Since nanocarbon is non-magnetic in nature, eddy current and electron spin resonance are dominant effect in microwave absorption. In this work, we have not carried out any study on Reflection is the primary mechanism of EMI shielding which depends on conductivity, both AC and DC, of shielding material. EMI shielding depends on the interaction between mobile charge carriers (electrons or holes) of the material and the electromagnetic field. The shielding material with good electrical conductivity is necessary requirement in shield technology. Specifically, reflection loss is the function of ratio of σ_r/μ_r, where σ_r electrical conductivity relative to copper and, μ_r, relative magnetic permeability [24,25].

Figure 8.8 shows measured *I–V* characteristic of nanocarbon compared with copper. The conductivity was estimated using: $\sigma_{dc} = \dfrac{l}{m \times h \times w}$, where *l* is length of active channel, *m*, slope, *h*, thickness and, *w*, width of conducting channel. Table 8.2 shows *I–V* measurement parameters estimated for both the systems.

8.1.3.3   % Reflection Loss

Figure 8.9 shows % reflection data obtained for copper and nanocarbon, in X-band regime. The amount of reflection from nanocarbon was recorded to be ~85%, comparable to copper (95%). For EMI shielding material, the reflection, including reflection from surface and interface scattering, will increase with increasing conductivity of the materials [25].

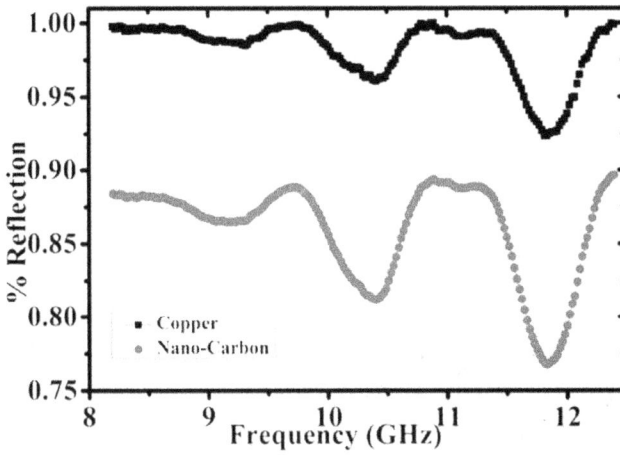

FIGURE 8.9   Recorded % reflection for both the systems, in X-band.

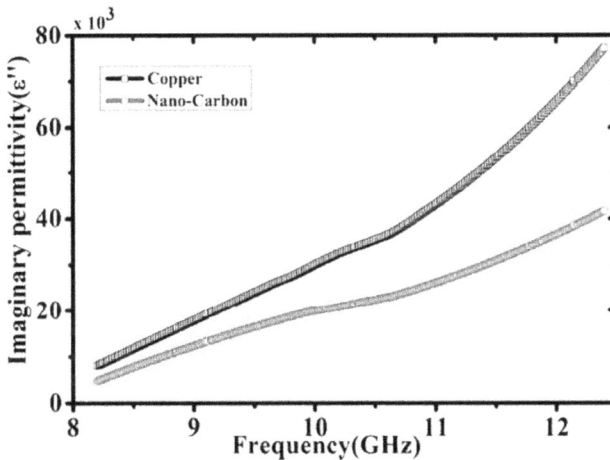

FIGURE 8.10   Recorded imaginary permittivity, ε'', as a function of frequencies, exhibiting monotonic relationship between frequency and ε''.

Further, from Table 8.2, the magnitude of σ_{dc} for copper was ~ten times high compared to nanocarbon. Another important estimated parameter is skin depth, δ, the extent to which incident radiation interacted with material and given by: $\delta = \sqrt{\dfrac{1}{\pi \cdot f \cdot \mu \cdot \sigma_{dc}}}$, where f, applied frequency, $\mu = 1$ (for copper). For 10 GHz, the value of δ was ~10 and 40 nm for copper and nanocarbon, respectively. In both, copper and nanocarbon, reflection loss caused due to eddy current was more due to skin effect [26].

Figure 8.10 shows imaginary permittivity, ε'' of copper and nanocarbon. Reflection loss of microwave radiation is mainly due to impedance matching condition and the change of electromagnetic parameters of complex permittivity [17]. Free electrons plays important role in imaginary part of complex permittivity due to good electrical conductivity [17,27,28]. According to the free electric theory [29], ε'' could therefore be obtained by using: $\varepsilon'' = \dfrac{\sigma_{dc}}{2\pi \cdot \varepsilon_o \cdot f}$. The relation showed that $>P@$ plays the dominating role in ε''. It seems that the σ_{dc} of copper was marginally high compared to nanocarbon. This change was due to bond environment of copper and nanocarbon medium [26].

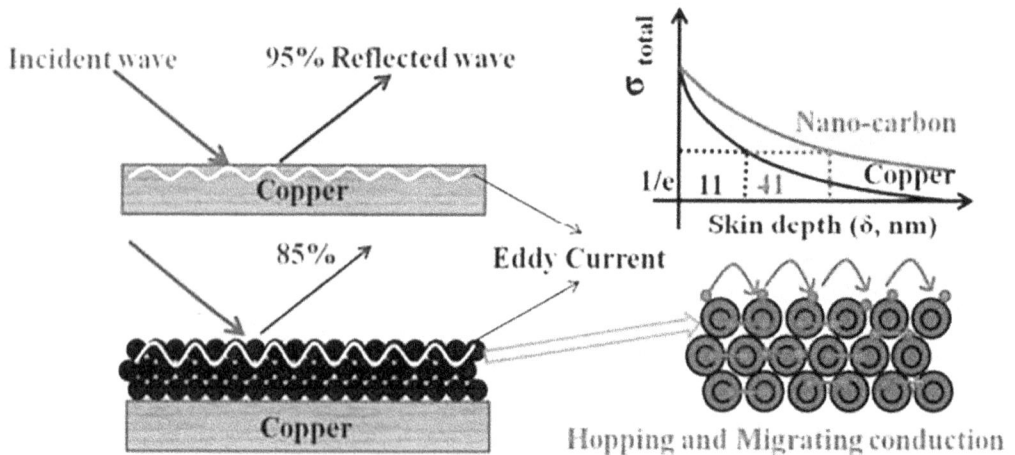

FIGURE 8.11 Schematic illustration showing electromagnetic wave interaction of nanocarbon and copper with differing skin depth level, in X-band.

8.1.4 SHIELDING MECHANISM

The scheme of interaction mechanism of incident electromagnetic radiation with nanocarbon and copper medium is shown in Figure 8.11. In copper, the conduction mechanism is due to free mobile charge carriers. While in nanocarbon there exist both migrating conduction and hopping conduction due to hetero-structure sp^2+sp^3 carbon zones [17,21,22,28,30]. Thus, increasing total conductivity in disordered spherical amorphous nanocarbon concentric shells enhances imaginary permittivity ε'' resulting in more reflection, comparable to copper. However, incident radiation of ~15% is attenuated within the nanocarbon layer due to larger penetration depth of electromagnetic radiation in nanocarbon compared to copper [15,16].

On interaction the incident electric field component gets coupled strongly with the molecular electric field of sp^2 σ-bond. This resulted into larger dissipation of field into heat in nanocarbon. Thus, comparative microwave reflection property of nanocarbon with copper, particularly in X band region, is advantageous. Copper being metal is high density, corrosion prone, and expensive material for shielding technology. On the other hand, nanocarbon is light in weight, eco-friendly, stable at high temperature, easy to synthesize and cost effective. Thus, nanocarbon is well suited candidate for EMI shielding over copper.

8.2 SPLIT-RING RESONATORS: FERRO-NANOCARBON METAMATERIALS

In this section, we have revealed split-ring resonators (SRRs) properties of ferro-nanocarbon (FNC) as a bi-anisotropic left-handed medium (LHM) for its application as an electromagnetic cloak operational at 8.5–10 GHz (X-band) regime. The FNC, prepared by the pyrolysis of camphor with variable wt % of Cobalt (Co), [31] was studied for its structure-property relationship using electron microscopy, Raman, dielectric relaxation spectroscopy, and VSM. In analysis, formation of Co-sp^3 phase, emergence of low frequency polarization modes, and enhanced internal magnetic fields were found to be advantageous to implement FNC as SRRs. They offered peculiar characteristic of constitutive parameters extracted using Nicolson–Ross–Weir and retrieval technique revealing bi-anisotropic LHM behaviour of SRRs. The computational electromagnetic work indicated concealing of radiated field at SRRs generating a cloak-like response over the sub-wavelength (8.5–10 GHz) regime. The theoretical result resembles with the experimentally obtained scalar microwave scattering parameters to a certain extent [16,18–20,23–26].

8.2.1 ORIGIN OF LEFT-HANDED MATERIAL SYSTEMS: THE VESELAGO MEDIUM

The threat spectrum of a tracking radar comprises of surveillance, guidance, and homing mode. In general, the tracking trail of a target due to its interaction with microwave (radar) radiations is assumed to be, strategically, hostile activity. There are several penetration aids such as radar absorbing material, booster fragmentation, jammers, chaff, decoys, etc. to clutter or shield the signature of such object [31]. In recent years, cloaking techniques have been emerged as a promising countermeasure to make the target invisible [32].

In this, the paths of electromagnetic waves are controlled within a material by introducing a specific spatial variations in (constitutive parameters) permeability (μ) and permittivity (ε) [20,33]. The cloak architecture involves coordinate transformation of constitutive parameters that squeezes incident field space volume into a shell surrounding the concealing volume; achieved via a cellular pattern. The pattern obeys a scattering condition: $a \ll \lambda \left(= \dfrac{2\pi c}{\omega} \right)$, where a is periodic dimension of the pattern, λ, incident wavelength, ω, frequency of radiation, and c, velocity of light [16]. Physically, incoming wave field gets coupled to the cellular pattern to pop-up magnetic or electric dipole, at the cell, depending upon charge dynamics within the pattern. This causes asymmetric reflection, or asymmetric transmission, or complete concealing of the incident field. Thus, cloaking effect comes under the category of metamaterial, which is exotic having negative constitutive parameters, not necessary for cloaking, however, useful for wave propagation in backward direction, zero-point phase velocity, and offers negative refractive index to the medium. Formally, Veselago [16] proposed metamaterial, also popularly, known as negative index material (NIM), or LHM [16] and reference cited therein. For normal medium, achieving metamaterial properties is challenging due to restrictive values of constitutive parameters because of myopic response of atoms and molecules to the incident radiation. For cellular architecture, the field behaviour could be studied by analysing tensors of permittivity, permeability, chiral parameter, non-reciprocity conditions, forward/backward reflections, scattering (S)-parameters (or impedances), scattering power, reflection coefficient, magnetoelectric coupling factors, refractive index, etc., to classify them as an anisotropic, bi-anisotropic, and/or chiral metamaterials.

The first theoretical realization of LHM occurred in studying a wired metallic medium whose permittivity was estimated to be negative due to artificial electric plasma properties of the medium [16]. By constructing metal split-ring resonators (SRRs), the magnetic plasma properties of cellular medium were manipulated to achieve negative permeability. The first artificial LHM was demonstrated by Smith et al. (2001) by combining metal wires and SRRs in which phenomenon of the negative refraction was confirmed [33]. In recent years, LHM has been the focus of both theoretical exploration and experimental study [20] including the discovery of perfect as well as superlenses [33]. The elements of LHM could, artificially, be fabricated by a macroscopic composite of periodic or aperiodic structure, whose function is due to both the cellular architecture and the chemical composition. Nevertheless, LHM has unavoidable disadvantages such as large loss, narrow bandwidth, and metal composition. Particularly, metal based SRRs are corrosive, and inflexible in nature. In so far, the electromagnetic properties of non-metal-based SRRs such as nanocarbon composition is not revealed in the literature. They are simple to fabricate, and pattern, moreover, relatively cost effective compared to lithographically obtained metal SRRs.

8.2.2 FERRO-NANOCARBON AS LHM: PREPARATION AND ASSESSMENTS

The NC/FNC specimen, synthesized as per the descriptions in Chapter 2, Section 2.1.2 and deposited collected by gently scrubbing the substrate using a razor blade. FNC samples were subjected to electron microscopy, Raman, dielectric studies, VSM, and VNA measurements. Other characterization techniques, except dielectric and VSM, were performed in same fashion as mentioned in the preceding chapters. Details of the undescribed two are given below:

8.2.2.1 Dielectric Measurements

The dielectric relaxation studies were performed to measure the permittivity and conductivity of the samples, over 10–30 MHz. The dielectric spectrometer (Novocontrol broadband) equipped with an analyser (Alpha-A) interfaced to the sample cell was used along with Win Fit software.

8.2.2.2 VSM Studies

The magnetization measurements were performed using 16 T PPMS-VSM, Quantum Design, magnetometer having DC sensitivity 10–5 emu at 1 T with temperature range 2–300 K with field sweep 100 Oe/sec. The thermo-magnetic measurements (cooling/heating) were carried out at a rate of 1.5 K/min.

8.2.2.3 Fabrication of Ferro-Nanocarbon Split-Ring Resonators (FNC-SRRs)

Initially, FR4 dielectric substrate was cut into 2.286 (a) × 1.016 (b) × 0.1 (d) cm^3 dimension (Figure 8.2). The composite of NC/FNC was made with the help of readily available, standard colloidal silver liquid for imprinting SRRs onto the substrate. The fabricated SRRs consisted of a planar set of two concentric conducting rings with inner ring diameter 5 and outer 8 mm (thickness ~ 100 μm) having a gap (split, g = 1mm) on each ring opposite to each other [16].

Figure 8.12a shows co-ordinate variables along with the geometry and the sample axis of FNC SRR unit cell. It was assumed that three co-ordinate axes $\widehat{e_1}$, $\widehat{e_2}$, $\widehat{e_3}$, of the unit cell oriented in the z-, x-, and y-directions, respectively. Figure 8.12b and c shows corresponding, in-plane view and implemented structure, respectively.

8.2.2.4 Computational Electromagnetics

To simulate field scattered from SRRs, RF module of COMSOL Multiphysics 5.2 was used by employing harmonic propagation analysis mode and parametric solver in X-band regime. In our analysis, the longitudinal configuration was assumed for the propagation of the wave. The configuration was homogenized and completely fills the cross section of the waveguide. The fundamental mode (TE_{10}^z) was propagating in the z direction from air $\left(\dfrac{\partial}{\partial y} = 0\right)$ only TE_{m0}^z modes was supported

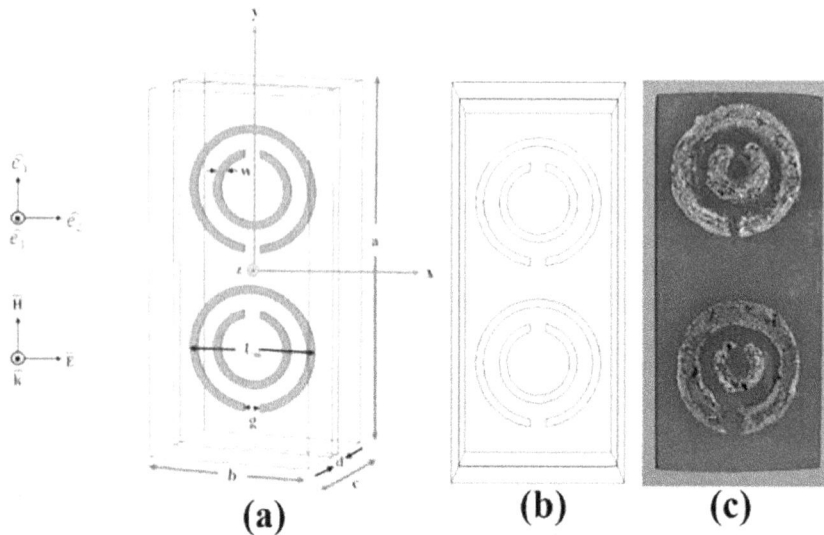

(a)   **(b)**   **(c)**

FIGURE 8.12 (a) Configurations of principal axis and geometry of the FNC SRR unit cell. The field directions and propagation vector. L_m = 8 mm, w = 0.5 mm, g = 1 mm, d = 1 mm (thickness of the FR4 substrate), a = 22.86 mm (height), b = 10.16 mm (width), c = 9.78 mm (distance from the input and output port), (b) in-plane view, (c) implemented FNC SRRs.

by the longitudinal configuration changing with x and y to simulate radiated field around SRRs and microwave (scattering) S-parameters [16].

8.2.2.5 Microwave Measurements on FNC-SRRs

Experimental X-Band (8–12 GHz) measurements were performed using PNA network analyser (Agilent, N5222A), equipped with waveguide (dim. $2.3 \times 1.1\,cm^2$), for measuring S-parameters of NC/FNC SRRs. Prior to measurements, microwave source was started for 2 hours for stabilization. Full two port calibration was performed on the test specimen to avoid errors due to directivity, isolation, source, load match, etc. S-parameters were determined from two port measured scattering data with the help of commercially available Agilent software module 85071, based on the procedure given in Agilent product note. The cellular structures were mounted into a quarter wave plate slot to perform measurements. The set-up is shown in Figure 8.13 [16].

From measured S-parameters, values of permittivity and permeability were obtained using Nicolson–Ross–Weir method. Moreover, using self-consistent approach, other parameters such as n, refractive index, z_{ws}^+, z_{ws}^-, forward and backward normalized wave impedances, permeability, permittivity, S_{11} and S_{21} were extracted by estimating Γ_1, Γ_2 reflection coefficients, T propagation factor, β_{0z} phase constant in z direction, using more advanced retrieval technique [16].

8.2.3 MOLECULAR CHARACTERISTICS OF FNC-SRRs

Figure 8.14a shows recorded Raman spectrum of NC, which consisted of two peaks, D and G. They were, respectively, assigned to sp^3 and sp^2 phases of NC. The emergence of these phases is a peculiar characteristic of amorphous carbon indicative of a large amount of disorder. To quantify this, a curve fitting was carried out in terms of spectroscopic parameters such as peak position, peak width, line shape (i.e. Gaussian, Lorentzian or a mixture of both), and band intensity using Labspecs 5.0 software. The result of the line decomposition is indicated in Figure 8.14. Several fits were tried leaving all spectroscopic parameters free to progress and the best fitting was invariably obtained for all recorded spectra [16].

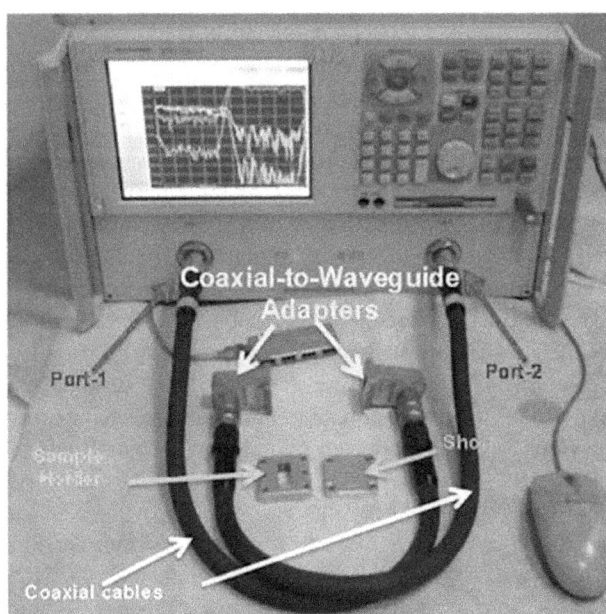

FIGURE 8.13   X-band measurement set-up.

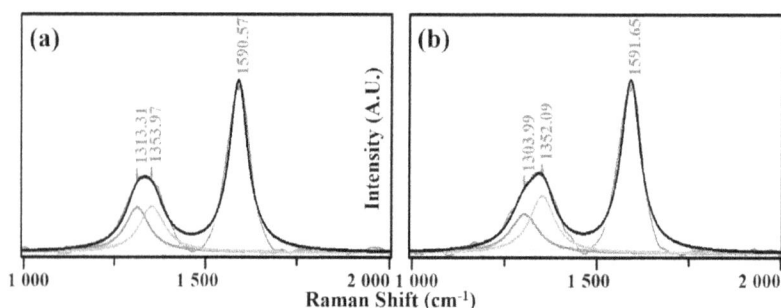

FIGURE 8.14 Raman spectra recorded for (a) NC and (b) FNC 20% at 532 nm excitation wavelength.

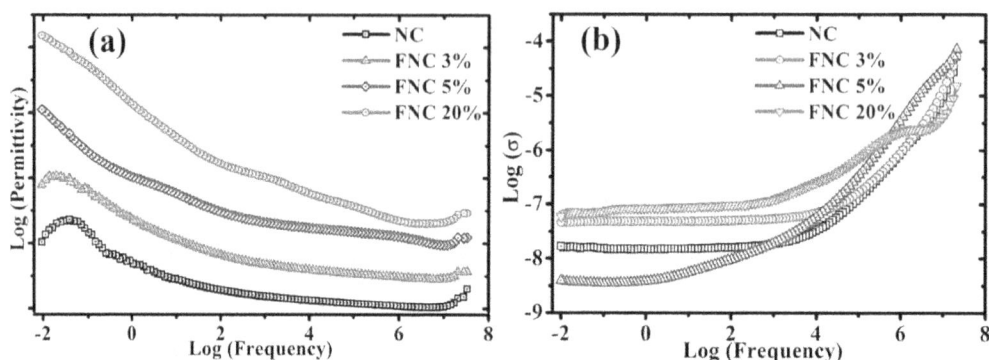

FIGURE 8.15 Recorded log–log plots for (a) permittivity and (b) AC conductivity (σ_{ac}) as a function of frequency in the range 10–30 MHz.

For NC, D-peak was deconvoluted for two components at 1313.31 and 1353.97 cm^{-1} having effective peak width ~110 cm^{-1}. In the case of FNC, the major change that has been observed in recorded spectra was variations in line shape and width of D-peak, as seen in Figure 8.14b. For FNC, the effective width was increased to ~150 cm^{-1}. It consisted of D-peak doublet appearing at 1303.99 and 1352.09 cm^{-1}. However, the G-peak appearing in NC and FNC was found to be invariant in peak position and width, respectively, recorded to be ~1590 and 60 cm^{-1}. Broadly, it seems that Co incorporated in the sp^3 zone of NC. The packing fraction of sp^3 is low compared to the sp^2 network. Due to this, the incorporated Co gets an opportunity to migrate into the interstitial space of the sp^3 zone, available within NC. This has implication on molecular and, consequently, dielectric characteristics of NC and FNC. The dielectric characteristic is associated with relaxation behaviour of NC, in terms of motion of sp^2 and sp^3 molecules which subsequently may get modify by adding Co. Since most of the carbonaceous medium exhibit more than one dielectric molecular relaxation region, generally, no single molecular model is adequate to describe the behaviour over a wide frequency and temperature range. Here, we have conducted frequency relaxation study mainly.

8.2.3.1 Dielectric Relaxation and Anisotropy

To study molecular relaxation in NC and FNC, the equal amount powder was mixed with a polymer whose relaxation behaviour is known and non-overlapping with our NC. Figure 8.15a shows variation in permittivity (in arbitrary units) as a function of frequency. It indicates very prominent low-frequency orientational polarization appeared between mHz and Hz. The observed peak could be attributed to, mainly, chair-to-chair motion of sp^2 segments of NC. The incorporation of Co

(3%) showed a broadening of the low-frequency peak and the trend was quite systematic with subsequent increase of Co % in NC, showing the emergence of broader shoulder peak for FNC 20%. Figure 8.15b shows variations in σ_{ac} as a function of frequency which corresponds to dielectric loss component. At higher frequency, ranging from kHz to MHz, no prominent change has been observed for NC and FNCs. This indicates that the electronic polarization is not affected as significantly as that of orientation polarization. This is consistent with the Raman data analysed before indicating interstitial encaging of Co in NC, mainly affecting molecular environment associated with the sp^3 carbon. Co is magnetic in nature. The encaged Co in sp^3 clusters of NC though show peculiar dielectric properties, one needs to investigate the magnetic parameters of FNC, as well. This is important from the viewpoint of effective permeability that depends on the magnetization of the medium [16].

8.2.3.2 Magneto-Molecular Assessments

Figure 8.16a shows $M–H$ curves recorded for NC and FNC. The hysteresis curve recorded for NC exhibit no saturation behaviour. To estimate saturation level of magnetic moment, M_s, the geometrical projection of $M–H$ curve was taken on the magnetic moment axis. The value of M_s was estimated to be ~3.1×10^{-3} emu/g for NC. As the wt % of Co increases in NC, the value of M_s was observed to be increased, subsequently. For FNC 20%, the M_s was obtained to be 2.03 emu/g which is three orders magnitude high compared to NC. The effective permeability, μ_{eff}, was computed using the values of M_s obtained for all the samples. The μ_{eff} was found to be 0.0015 μ_B and 10 μ_B, respectively, for NC and FNC 20%. There is about four orders of magnitude difference in effective permeability of FNC compared to NC. This showed that the extent of internal magnetic field was more than thousand atoms in FNC, whereas in NC the field dies by a factor of (1/6)th within the carbon sub-lattice. The schematic shown below in Figure 8.17 indicate variation in effective permeability as a function of lattice distance [16].

The thermo-magnetic analysis, carried out using FC-ZFC measurements, on NC (inset) and FNC 20% is shown in Figure 8.16b. From the inset, one can see that the exchange anisotropy involved in NC medium was smaller compared to FNC 20%. This indicates that the inherent structural defects (vacancies), topological disorder, and impurities influenced exchange anisotropy. Mostly, they exist in the form of radical spin moment available on the carbon atom. Since such structural inhomogeneities are randomly spaced with smaller concentration, the interaction between them is indirect and assisted by a carbon sub-lattice mediator. The dipole moment induced by encaged Co in FNC play the crucial role as a mediator. The Coulomb interaction between itinerant sp^3 electrons and magnetic dipole ensures a means for the dominant exchange interaction between the magnetic moments.

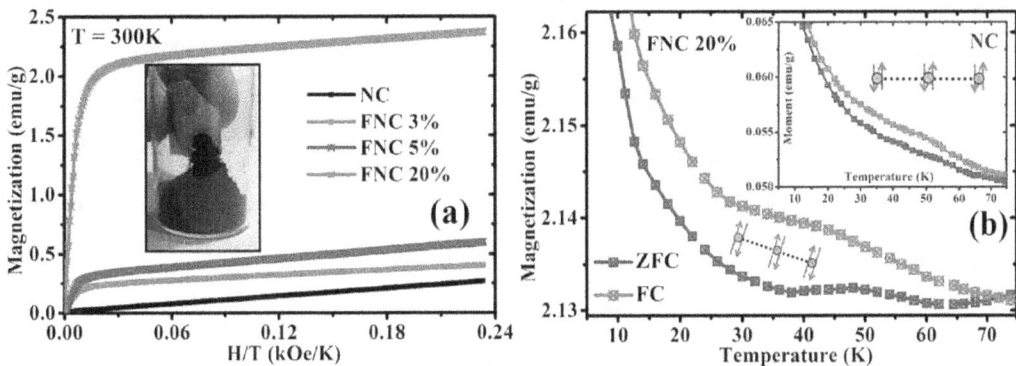

FIGURE 8.16 (a) Magnetization hysteresis curves ($M–H$) measured for NC and FNC at 300 K. Inset photograph shows the response of FNC 20% to permanent magnet, (b) $M–T$ curve recorded for NC (inset) and FNC 20%.

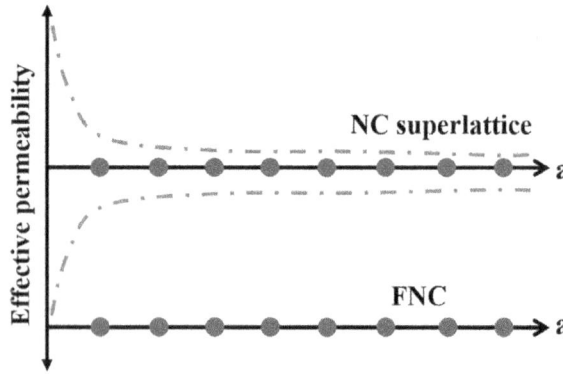

FIGURE 8.17 Schematic representation of decay and rise in effective permeability with superlattice distance for NC and FNC.

Thus, significant irreversibility has been observed causing large exchange anisotropy in FNC 20%. It would be of interest to investigate the nature of magnetization, range of ordering and correlation of ordering, however, it is in purview of the present discussions [15,16].

By and large, the Co-sp^3 molecular magnetic environment is responsible for generating a low-frequency dipolar field which is highly anisotropic, inhomogeneous and distributed non-uniformly within sp^2 nanocarbon framework. Though the atomic and molecular character of FNC seems to be advantageous, however, provide rather restrictive values of constitutive parameters to build a material with peculiar electric and magnetic properties, especially, at frequencies in the gigahertz range. To exploit naturally unavailable properties of FNC, they are implemented in the form of SRRs, as described in experimental section.

8.2.4 Modelling and Simulation: Computational Electromagnetics

The transformation of FNC into SRRs is a small step to replace the atoms of the original concept with the structure on a larger scale. The periodic structure was having a unit cell of a characteristic dimension a, which obeyed the scattering wave relation $a \ll \lambda = 2\pi C/\omega$. The electromagnetic resonance response of FNC SRRs has been investigated, numerically. Port boundary conditions were placed on the input and the output boundaries of the waveguide. For the input, incident transverse electric field (TE$_{10}$ mode, $\bar{E}$) obeys Maxwell's wave equation [34]:

$$\frac{1}{\mu_r}\nabla^2\bar{E} - \bar{\nabla}\left(\bar{\nabla}\cdot\bar{E}\right) - k^2\left(\epsilon_r - \frac{j_T}{\omega\sigma}\right)\bar{E} = 0, \tag{8.1}$$

where μ_r, relative permeability, ϵ_r relative permittivity, j_T, total current, σ, AC conductivity.

The dispersion of incident wave vector, k, is given by,

$$k^2 = \mu_r\varepsilon_r\omega^2 + \mu_r\,\omega\sigma, \tag{8.2}$$

The total time reversal electric field solution is given by,

$$\bar{E}_{i,r} = \overline{E_1}\,exp\left\{i\omega\left(\frac{a}{C}-t\right)\right\} + \overline{E_2}\,exp\left\{-i\omega\left(\frac{a}{C}+t\right)\right\}. \tag{8.3}$$

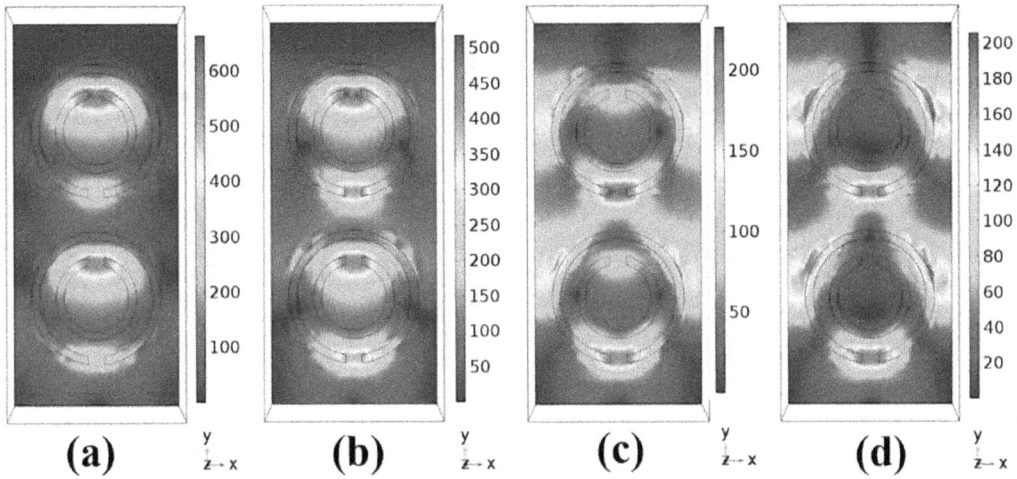

FIGURE 8.18 Simulated electric field, in x–y plane, by FNC 20% SRRs unit cell indicating variations in the topology of the field in which (a), (b), (c), and (d) are, respectively, for 8, 8.5, 10, and 12 GHz response frequency.

where $\overline{E}_{i,r}$ is total electric field, $\overline{E}_1$ and $\overline{E}_2$ corresponding incident and reflected components, respectively. The related boundary condition used in terms of Poynting vector (dimension: energy/area×time) is given by [20,34],

$$\frac{dS_T}{d\Omega} = \frac{\overline{E}_{i,r} \cdot \overline{E}_1 - \left|\overline{E}_1\right|^2}{\left|\overline{E}_1\right|^2} + \frac{\overline{E}_{i,r} \cdot \overline{E}_2}{\left|\overline{E}_2\right|^2},$$

(8.4)

where $d\Omega$ is the volume element through which scattering occurs at SRRs. In above equation, the first term is associated with the input and the second term with the output port. The port boundary automatically determined reflection and transmission characteristics in terms of S-parameters. In Figure 9.63, the electric field distribution, in x–y plane, is simulated for a typical FNC 20% SRRs unit cell, that indicated squeezing and localization of incident field around the cell. It has frequency dependence which comes through the capacitive action that was generated by SRRs. On interaction with incident field, the structure acts as an infinitely conducting cylinder in the high-frequency limit that generates oppositely directed alternating currents, due to split, moving within the ring structure. The curl of currents was responsible for producing a net dipole moment vector orthogonal to the plane of the ring. Thus, the imprinted SRRs together with its split acted as an LC circuit. Over the bandwidth (8–12 GHz), the resonance response due to dipole moment was observed to be varied as seen in Figure 8.18.

8.2.4.1 Scalar S-Matrix Parameter and Constitutive Analysis

Further, the simulated scalar S-parameters for FNC 20% SRRs is shown in Figure 8.19a. The magnitude S-parameters in terms of shielding effectiveness (SE) measured in dB is given by [20,34]:

$$SE\ (dB) = 10\log\left[\frac{P_T}{P_1}\right],$$

(8.19)

From computational techniques, S_{11} parameter was well below −20 dB at ~8.5 GHz, whereas at the same frequency, the S_{21} was nearing to 0 dB. This indicated that, at this frequency, the SRRs was

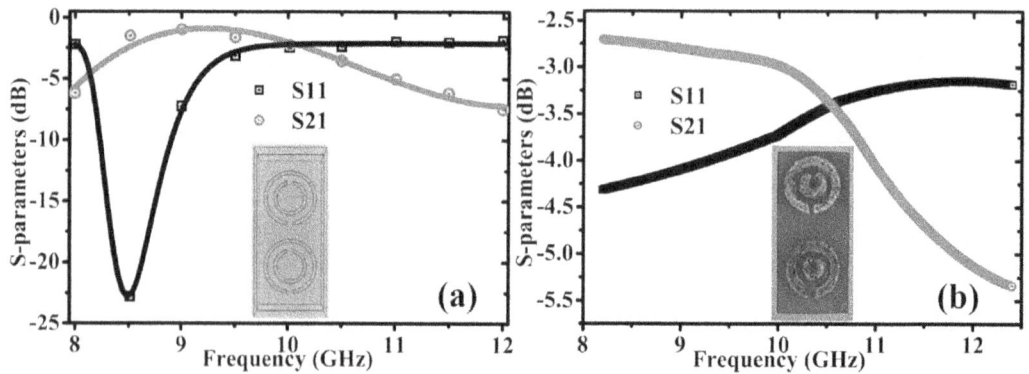

FIGURE 8.19 (a) Simulated and (b) experimental S-parameters for FNC 20% SRRs for X-band region. Inset indicates the corresponding simulated and experimentally fabricated SRRs unit cell.

fully transparent to the incident radiation by squeezing the field within the unit cell. Figure 8.19b shows, average of the experimental microwave scattering S-parameters obtained statistically for FNC 20% SRRs. There is a marked difference between measured and simulated S-parameters of FNC SRRs which could be attributed to factors such as design of unit cell, field received and interacted within the cell, response of boundaries and interfaces. In simulated cell, these factors could be operative at the optimum level in contrast to practical one. Qualitatively, the field received in simulated cell would experience no corner reflection and losses within annular structures of SRRs. Further, simulated cell is strictly a homogeneous medium and having infinite mismatch at the edges and the split region. In the practical cell the condition of homogeneity of medium would be somewhat graded due to presence of magnetic impurity in inherent dielectric carbon network. Such locally inhomogeneous electromagnetic medium cannot be simulated that effectively. As a result, one can see the variations in simulated and experimental S-parameters, however, similarity in S_{11} and S_{21} cut-off region appeared at 10.5 GHz.

8.2.4.2 Nicolson–Ross–Weir Formulism

In the current study, we have used two port rectangular waveguide method in which S_{11} and S_{22} parameters were, invariably, symmetric. This is due to the fact that incident magnetic field perpendicular to the plane containing SSRs ring will induce magnetic excitation in the ring along the z axis producing electric dipole along the x axis. Further, the dipole field perpendicular to the slit axis, i.e. x-axis will cause charges of opposite polarities to accumulate over the ring yielding a magnetic dipole symmetric along z axis. Thus, for the TE_{10}^z mode, the scattering S_{11} and S_{22} will be symmetric. In order to calculate total loss of incident radiation in FNC SRRs, we have plotted $|S_{11}|^2 + |S_{21}|^2$ as function of frequency and provided below. The medium is less lossy over 8–10 GHz compared with high frequency regime [16,20,34].

Figure 8.20a and b shows estimated values of constitutive parameters by Nicolson–Ross–Weir model for NC and FNC, over X-band region. The overall response of permittivity is negative with cut-off at high frequency for all the systems and almost similar trend is observed for recorded permeability. In general, the response of dipole, atom, and electron to the harmonically oscillating electromagnetic field is such that the charge segregation is in the same direction as that of oscillating field below resonance frequency. Above resonance, the charge lag occurs due to the mass involved in segregation of polar moieties harmonically bound within the medium. The observed negative response suggests that FNC is LHM in nature in SRRs configuration. Further, Nicolson–Ross–Weir formalism is based on estimation of constitutive parameters for a homogeneous, isotropic materials, directly, by examining the measured S-parameters. However, this approach has some challenges for studying metamaterial cellular elements due to limitation on homogenization of scattered electromagnetic field.

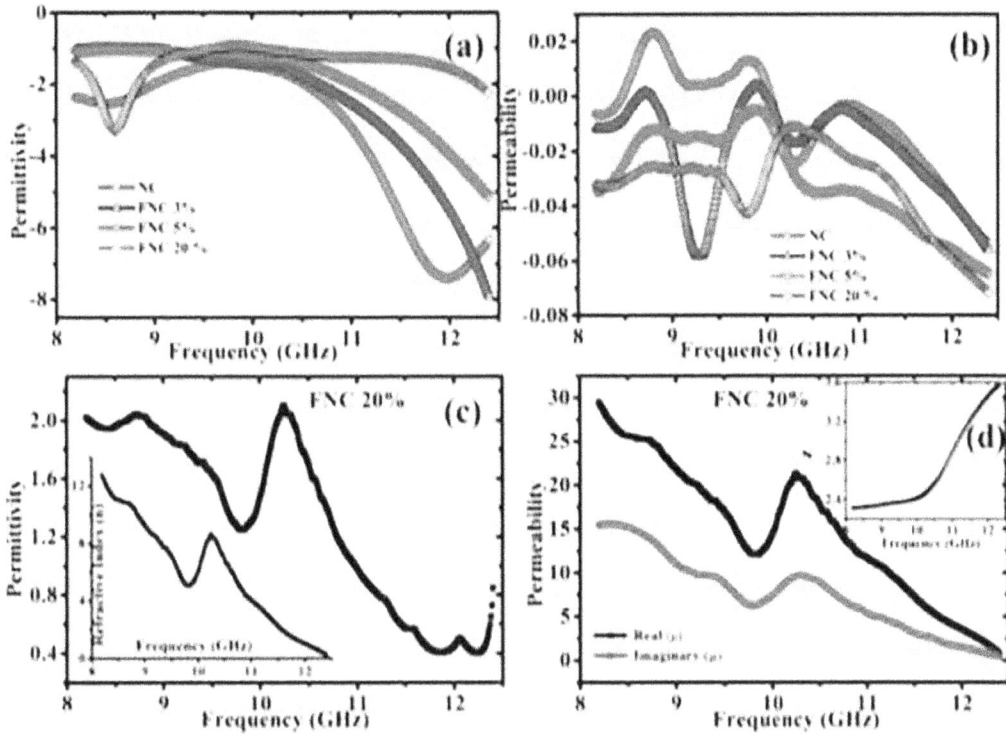

FIGURE 8.20 Extracted (a) permittivity (ε), (b) permeability (μ), for NC and FNC by Nicolson–Ross–Weir model, (c) permittivity, refractive index (n) inset, and (d) permeability real and imaginary parts, total normalized wave impedance (z), for FNC 20% SRRs obtained by retrieval technique.

8.2.4.3 Retrieval Methodology

In a cellular metamaterial structure, the electromagnetic response depends on three factors: (a) production of magnetic dipole by generation of circulating surface current due to magnetic field interaction of incident wave with the cell, (b) generation of electric dipoles by induction of charge densities with opposite polarities due to accumulation of charges at corners/edges of the cell, and (c) magnetoelectric field coupling. The interaction transforms the metamaterial property from anisotropic to bi-anisotropic or to chiral behaviour resulting asymmetric reflection and transmission, respectively. Mostly, metamaterials exhibits bi-anisotropic property when transformed in to SRRs [35].

The bi-anisotropic medium can be specified by analysing constitutive, chirality, and magnetoelectric parameters. They have different forward and backward scattering parameters/impedances, a wider stopband transmission spectrum, they differ in forward and backward powers, having variable reflection coefficient and magnetoelectric coupling factors. It is difficult to estimate all of them merely using S_{11}, S_{22} and S_{21}, however, in the present communication we have estimated a few of them by retrieval technique [32] using following equations:

$$S_{11}^b = \frac{\Gamma_1\left(1 - T^2\right)}{1 - \Gamma_1\,\Gamma_2\,T^2} \quad S_{22}^b = \frac{\Gamma_2\left(1 - T^2\right)}{1 - \Gamma_1\,\Gamma_2\,T^2} \tag{8.20}$$

$$S_{21}^b = S_{12}^b = \frac{T\left(1 - \Gamma_1\Gamma_2\right)}{1 - \Gamma_1\Gamma_2\,T^2} \tag{8.21}$$

$$\Gamma_1 = \frac{z_{ws}^+ - 1}{z_{ws}^+ + 1}; \Gamma_2 = \frac{z_{ws}^- - 1}{z_{ws}^- + 1}; \quad z_{ws}^+ = \frac{\mu_1 \ \beta_{0z}}{\beta_{sz} + ik_0\xi_0} \tag{8.22}$$

$$z_{ws}^- = \frac{\mu_1 \ \beta_{0z}}{\beta_{sz} - ik_0\xi_0}; T = e^{+i\beta_{sz}L} \ ; \ \beta_{sz} = n\beta_{0z} \tag{8.23}$$

where b denotes bi-anisotropic feature of FNC SRRs, in equations (8.20) and (8.21). The intermediate variables, Γ_1, Γ_2 reflection coefficients, T propagation factor, β_{0z} phase constant in z direction, z_{ws}^+, z_{ws}^- forward and backward normalized wave impedances, n refractive index were computed for longitudinal wave propagation configuration with $L_m = 8$mm, for SRRs. From equations (8.20)–(8.23), it has been noted that, for two-port rectangular waveguide measurements magnetoelectric coupling, $\xi_0 = 0$, and $\Gamma_1 = \Gamma_2$, $S_{11}^b = S_{22}^b$. In order to extract ε, μ_r, μ_i, one has to determine z_{ws}^+, z_{ws}^- from above equations.

$$z_{ws}^+ = \frac{-\wedge_2 \pm \sqrt{\wedge_2^2 - 4 \wedge_1 \wedge_3}}{2 \wedge_1}; z_{ws}^- = \frac{z_{ws}^+ + \wedge_4}{1 + z_{ws}^+ \wedge_4} \tag{8.24}$$

$$\wedge_1 = \left(S_{21}^b\right)^2 - \left(1 - S_{11}^b\right)\left(1 - S_{22}^b\right); \wedge_2 = 2\left(S_{11}^b - S_{22}^b\right) \tag{8.25}$$

$$\wedge_3 = \left(1 + S_{11}^b\right)\left(1 + S_{22}^b\right) - \left(S_{21}^b\right)^2; \wedge_4 = \frac{S_{11}^b - S_{22}^b}{S_{11}^b + S_{22}^b} \tag{8.26}$$

$$\xi_0 = \frac{in_s \beta_{0z}}{k_0}\left(\frac{z_{ws}^+ - z_{ws}^-}{z_{ws}^+ + z_{ws}^-}\right); \mu_r = z_{ws}^+\left(n_s + \frac{ik_0\xi_0}{\beta_{0z}}\right) \tag{8.27}$$

$$\mu_i = zn; z = \pm\sqrt{\frac{\left(1 + S_{11}\right)^2 - S_{21}^{\ 2}}{\left(1 - S_{11}\right)^2 - S_{21}^{\ 2}}} \tag{8.28}$$

$$\varepsilon = \frac{1}{\mu_r}\left(\xi_0^{\ 2} + \frac{\beta_{0z}^{\ 2}n_s^{\ 2} + \frac{\mu_r}{\mu_i}\beta_{0x}^{\ 2}}{k_0^{\ 2}}\right) \tag{8.29}$$

Using equations (8.25)–(8.29), the variations in extracted parameters can be studied as a function of frequency. In addition, values of S_{11} and S_{21} obtained from (8.20) to (8.21) as function of frequency are provided shown in Figure 8.22. Figure 8.20c and d, typically, shows extracted ε, μ_r, μ_i, n, z parameters for FNC 20%. One can see that there is a variation in the nature of profiles obtained by both the methods. The study of constitutive parameters of FNC SRRs by retrieval technique is in its infancy stage [35].

The results of S- and constitutive parameters, broadly, suggests that the higher anisotropy and out of phase response of Co-sp³ dipolar field within supra-molecular carbon domains segregate the charge in opposite direction to the oscillating incident field. Further, computationally, FNC SRRs created a resonance condition at ~8.5 GHz, and a trade-off ~10 GHz (experimentally resembling as well). In this sub-band regime, the average dipolar field gets coupled strongly to the incident field, in

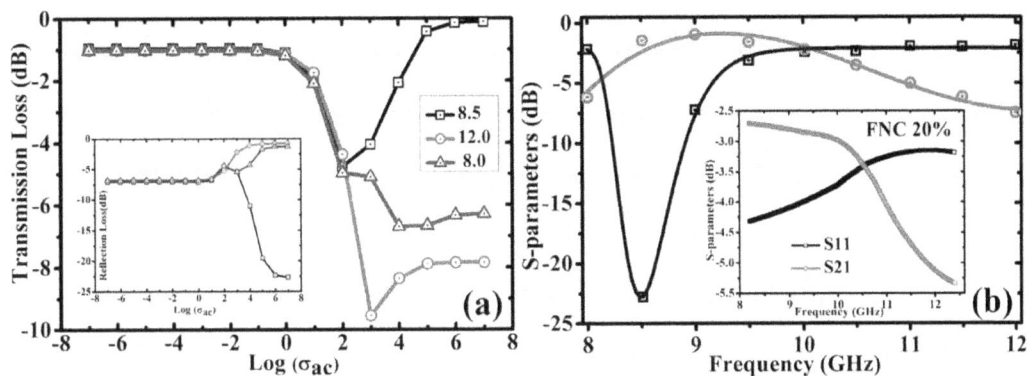

FIGURE 8.21 Comparison of losses (both transmission and reflection) and S-parameter by Nicolson–Ross–Weir and retrieval models.

somewhat reverse fashion. This sets a gradient in constitutive parameters, that made FNC as a LHM to generate cloak-like response, particularly between 8.5 and 10 GHz, when imprinted as SRRs. The field distribution incident on the surface of SRRs was transported uniformly across the dielectric slab and generated a cloak-like effect in this frequency regime, evidently seen in Figure 9.66.

8.2.5 CELLULAR ARCHITECTURE AND BI-ANISOTROPY

Our discussions revealed that heterostructure molecular environment present in FNC is responsible to behave like an electromagnetic cloak in narrow X-band regime that squeezes the incident field effectively and bring out the possibility to make the object invisible. The transparency of an object to the radar threat spectrum at guidance and tracking range is a strategically important application.

In summary we have studied electromagnetic character of ferro-nanocarbon (FNC) split-ring resonators (SRRs) in X-band (8–12 GHz) region. The FNC SRRs acted as a bi-anisotropic left-handed material (LHM) generating a cloak-like response between 8.5 and 10 GHz, as revealed by analysis of the constitutive parameters. Initially, FNC was synthesized using camphor precursor (1,7,7-trimethyl-bicycloheptan, $C_{10}H_{16}O$), by incorporating variable wt % (3–20) of $Co(C_2H_3O_2)_2$. To understand the structure-property correlations, the obtained FNC was subjected to electron microscopy, Raman, dielectric relaxation spectroscopy, and vibrating sample magnetometry. In analysis, FNCs were 40–50 nm spherical, self-assembled, interconnected three-dimensional nanocarbon network with sp^2/sp^3 heterostructure molecular environment in which Co was encaged within sp^3 phases. The molecular relaxation studies on NC and FNC clearly indicated a prominent low frequency (mHz–Hz) orientation polarization attributed to chair to chair motion of sp^2 segment for NC which disappeared systematically with increasing Co content. The amount of magnetization was increased from 3.1×10^{-3} (NC) to 2.03 emu/g (FNC 20%) with dramatic enhancement in effective magnetic permeability by ~7000 times. The exchange anisotropy of the medium was high due to random volume distribution of Co substitutional impurities that interacted indirectly with carbon lattice network. This has implication on microwave scattering properties of FNC when transformed into SRRs, as studied experimentally. The numerous parameters were extracted such as scattering (S), permittivity, permeability, forward/backward scattering impedances, magnetoelectric coupling factors, refractive index, phase constant, etc., using Nicolson–Ross–Weir and retrieval methods. In addition, S- parameters and field profiles were simulated, indicating concealing of radiated field at FNC SRRs with S11 around −20 and S21 ~ 0 dB, at 8.5 GHz. In mechanism, the incident electromagnetic wave, especially, magnetic field component, extended along the y axis, interacted with SRR structures oriented in xy plane, in waveguide configuration. It generated the circulating surface currents in structures producing a magnetic dipole due to ferro carbon phase in FNC. Moreover, rich π-electron characteristic associated with sp^2 bonding induced charge densities with opposite polarities at corners and edges of the ring structure, developing an electrical dipole, which got coupled with the incident field. This made FNC SRRs a bi-anisotropic LHM generating a cloak-like response at 8.5–10 GHz. The obtained SRRs were nanocarbon based material, non-corrosive, flexible, environmental friendly, and inexpensive. The FNC SRRs showed the emerging possibility to generate basic building block for narrow bandwidth X-band cloak [16,35].

8.3 MECHANICAL PROPERTIES OF GNC NANOCOMPOSITES

Investigation of mechanical properties is probably one of the most studied phenomena in epoxy composites due to their wide range of applications from aerospace to wind-mill. In this context, there are consistent efforts to reduce the amount of filler content in epoxy matrix in order to minimize the nanoparticle agglomeration, and to achieve better dispersion and improvements in mechanical properties [36–38]. The overall performance of the composites is the manifestation of the combination of various mechanical parameters such as tensile strength and modulus, flexural strength and modulus and fracture toughness, etc. Fracture toughness is the most crucial parameter particularly for structural applications. Several research efforts have been attempted to improve the fracture toughness and to study the fracture behaviour of epoxy composites [36] and reference cited there in. The improvements in mechanical properties of composite material are highly influenced by the physical and chemical properties of fillers, such as surface area, geometry, surface morphology [37], chemical functionality [38], interfacial chemistry [39], and agglomeration tendency [17]. Rafiee et al. [40] compared the mechanical properties of composites using graphene platelets (GPL), single-walled carbon nanotubes and multiwalled carbon nanotubes (MWCNT) as fillers in epoxy matrix.

They found that the highest improvement in mechanical properties was obtained using GPL amongst the fillers used. The Young modulus, tensile strength, fracture toughness (K_{IC}), and critical strain energy release rate (G_{IC}) of 0.1 wt% GPL composites were found to increase by ~31%, ~40%, ~53%, and ~126%, respectively, as compared to pure epoxy. Bortz et al. [29] reported the effect of incorporation of GO on mechanical properties of epoxy. The flexural strength and modulus, K_{IC} and G_{IC} showed a monotonic increase with weight fractions between 0.1 and 1 of GO; whereas the tensile properties showed improvements, but not invariably [29]. At 1 wt% loading of GO, the flexural strength and modulus, K_{IC} and G_{IC} were found to improve by 23%, 12%, 63%, and 111%, respectively. Tang et al. [17] investigated the effect of filler dispersion on mechanical properties of graphene/epoxy composites. In comparison with poorly dispersed reduced GO composites, highly dispersed composites showed higher improvement (~52%) in KIC with 0.2 wt% filler. Both the composites showed marginal improvements in tensile and flexural modulus as compared to neat epoxy [17].

In most of the above studies, graphene and its derivatives are shown to be superior nanofillers for improving the mechanical properties of polymers compared to other carbon based fillers. Further the filler contents used in these composites were 0.05 wt% or more [36–38]. However, the mechanical properties and fracture mechanisms in epoxy composites at a very low content (<0.05 wt%) of graphene derivatives have not been explored and not fully understood. Further, GNC being a new derivative of graphene, their nanocomposites have not been reported in the literature. GNC contains polar hydroxyl and epoxide groups with highly disordered graphitic backbone and with mixed hybridization (sp^2 and sp^3) [27]. Moreover, GNC has several advantages over GO; such as the former is prepared by a simpler method and from a less expensive precursor material. Similarly, GNCs have advantages over carbon nanotubes (CNTs); such as: (a) GNCs are sheet-like structure, hence offers higher surface area to interact with matrix as compared to CNTs, (b) GNCs have polar functional groups which probably reduce the interlayer van der Waals forces between their sheets and thereby facilitate their dispersion in polymer solution by gentle sonication.

In this section, GNCs were used as reinforcement in epoxy matrix at weight fractions between 0.005 and 2 wt%. The dispersion of GNCs in epoxy matrix is investigated by means of various characterization techniques. The bulk dispersion of GNCs in nanocomposites was investigated by interplay between dispersion agglomeration phenomena using optical microscope. The mechanical and thermal properties were investigated in detail. Further, the toughening mechanisms and energy absorption through crack propagation were studied using fractography.

8.3.1 Dispersion of GNCs in Epoxy Matrix

Mechanical properties of nanocomposites are mainly controlled by (a) homogenous dispersion and agglomeration tendency of nanofillers, (b) their stiffness and (c) interfacial adhesion between the matrix and filler [27,36–38].

Dispersion of nanofiller was studied spectral Raman mapping. As both GNC and epoxy are carbon based materials, we found that it was difficult to identify GNCs in epoxy due to poor contrast difference in low wt%. The spectral Raman mapping, a promising technique, has been successfully used previously to evaluate the nanofiller dispersion in polymer matrix [34]. We adopted the same technique in the present study to assess the GNC dispersion. In this technique, the spectral features of Raman spectra collected from each specific spatial coordinates within the selected scan area in the specimen are used as pixels for final imaging. Therefore, thousands of Raman spectra are used to create an image. For the Raman mapping, the nanocomposite samples were cut using a diamond cutter to obtain ultrafine surface. The typical Raman spectra were recorded in the range of 1000–1800 cm^{-1}. However the spectral variation over the range 1250–1570 cm^{-1} (the spectral range covering D and G peaks of GNCs) were used to map the dispersion of the GNCs in the nanocomposites. The obtained images consist of spectral dispersion of epoxy, GNCs and both, which are shown in Figure 8.21. In these images, the red and green regions indicate epoxy and GNC spectral responses, respectively. The boundary between GNC and epoxy is indicated by dark colour. The spectral image were taken in large areas (40×40 µm^2) for better understanding of the GNC dispersion. Figure 8.22a shows the Raman image of 0.005 wt% nanocomposite, in which epoxy region is dominant with very small islands of GNCs and they are dispersed uniformly in the epoxy matrix. In the case of 0.01 wt% nanocomposites, the GNCs (green) are clearly seen to form a network-like structure (Figure 8.22b), indicating a homogeneous dispersion on the surface. This may be due to the exfoliation of GNCs to individual sheets in epoxy matrix.

Further at higher content of GNCs (0.5 wt%) in nanocomposites, GNCs occupied a large area of the image as shown in Figure 8.22c, which may be due to aggregate formation as shown in optical micrograph (discussed below) owing to high aspect ratio of GNCs [31]. Optical micrography was

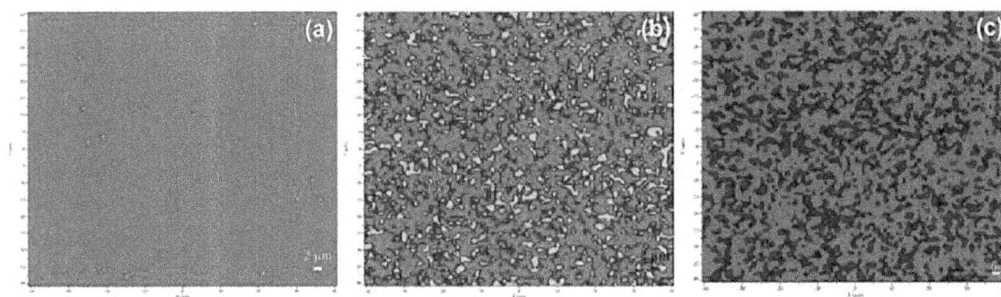

FIGURE 8.22 The spectral Raman image of (a) 0.005 wt% and (b) 0.01 wt%, and (c) 0.5 wt%. (Scale bar: 2 mm, area 40×40 µm^2). The typical spectral dispersion of GNCs (green) in the epoxy matrix (red) obtained for specimens with GNCs concentration (for the interpretation of the references to colour in this figure legend, the reader is referred to the web version of the article [27,36–38]).

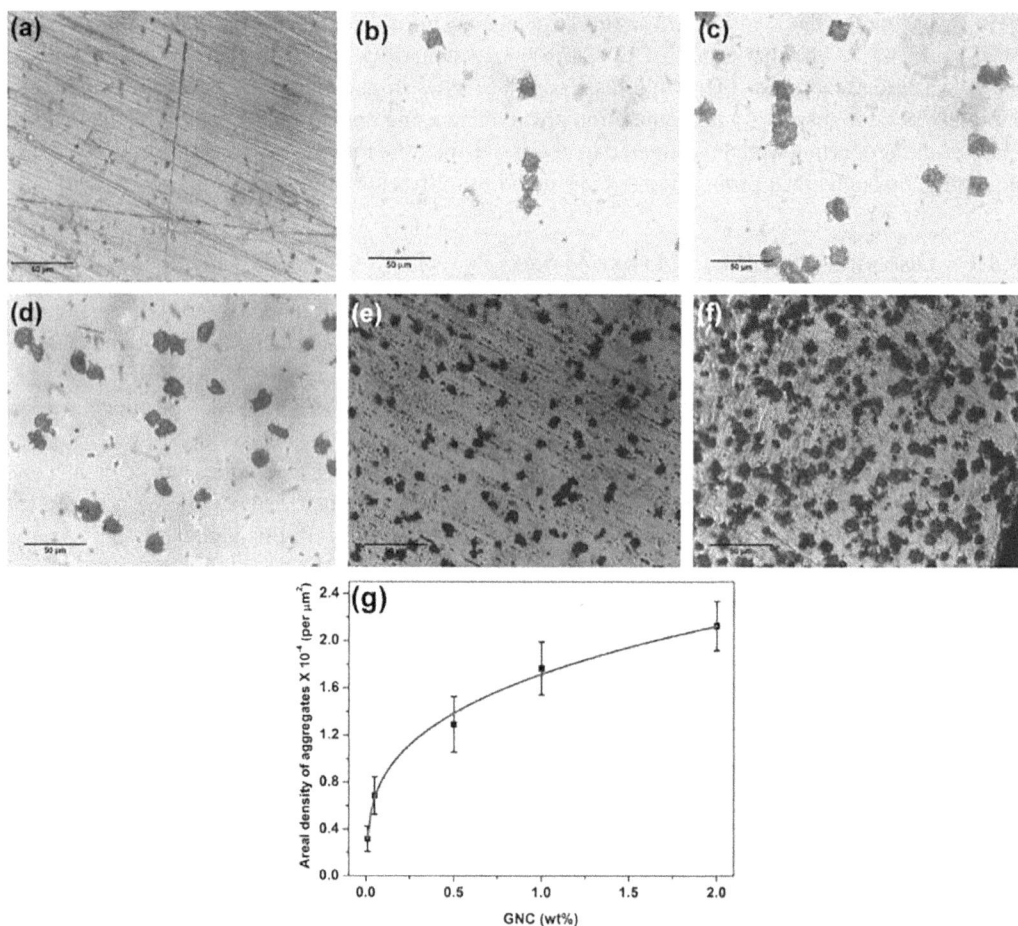

FIGURE 8.23 Optical micrographs on the surface of the as-prepared specimens of (a) neat epoxy and different weight fractions of GNC nanocomposites: (b) 0.01 wt%, (c) 0.05 wt%, (d) 0.5 wt%, (e) 1.0 wt%, and (f) 2.0 wt% taken at 20× magnification (scale bar: 50 mm), (g) areal density of aggregates (measured in number of aggregates/mm^2) as a function of GNC wt%. The areal density of aggregates obeyed the power law: ADA α $(W{-}W_c)^\gamma$, where W=GNCs wt%, W_c critical GNCs wt%, and γ is a critical exponent. The curve parameters with the best fit are: $W_c \sim 0.006$ and $\gamma \sim 0.3$ (correlation factor = 0.025).

used to obtain more insights into the state of dispersion. Figure 8.23 presents optical micrographs of epoxy and nanocomposites with different concentrations of GNCs. The optical micrograph of pristine epoxy does not show any distinct features (Figure 8.22a). The micrographs of nanocomposites are shown in Figure 8.22b–f. In these images several dark island-like structures are clearly observed and can be attributed to agglomerates of GNCs in the epoxy matrix. The number of these dark islands is found to increase with weight fraction of GNCs (Figure 8.23b–f). The following parameters are determined using the micrographs: (a) size of the aggregates and (b) areal density of the aggregates (ADA) (measured in per mm^2). Images were taken at least on three sheet-like specimens of dimension ~1 cm^2 to generate aggregate statistics. About 300–350 images were taken at different sites and the size as well as the number density of agglomerates was determined. It was observed that the colour intensity (black) of the aggregates increases with increasing the weight fraction of GNCs (Figure 8.23b–f). This indicates that with increase in the GNC concentration, the nanoparticle agglomeration increases in composites. Aggregates with irregular geometrical shapes are well-distributed in the matrix. The average aggregate size is found to be ~13±2 mm for 0.01 wt% and

TABLE 8.3

Average Size of GNC Aggregates with Standard Deviation in Epoxy Matrix

Sample	Size of Aggregates in μm
0.005 wt%	-
0.01 wt%	13 ± 2
0.05 wt%	20 ± 4
0.5 wt	20 ± 5
1 wt%	09 ± 3
2 wt%	11 ± 4

~20±4 mm for 0.05 wt% nanocomposites. For higher weight fractions, the aggregate size decreases marginally with increase in GNC content up to 2 wt%.

The average sizes of GNC aggregates in the composites are given in Table 8.3. The variation in the areal density of aggregates (ADA) with weight fractions of GNCs is shown in Figure 8.23g. For 0.01 wt% nanocomposites, ADA is found to be $(3.16 \pm 1.09) \times 10^{-5}$/mm^2 and it increases with weight fraction of GNCs (Figure 8.22g) and obeyed a power law curve. From this study, it can be concluded that the nano-dispersion of GNCs with aggregate formation is a challenging task, even at such a low GNC content. One might improve the properties further if the formation of agglomeration can be prevented or minimized [27,36–38].

8.3.2 MECHANICAL PROPERTIES OF NANOCOMPOSITES

8.3.2.1 Tensile Properties

The specific tensile modulus (E_S) (defined as elastic modulus divided by density) and the ultimate tensile strength (UTS) of neat epoxy and the nanocomposites are presented in Figure 8.23a [27,38].

It is interesting to note that with only 0.005 wt% of GNCs the composites showed improvements in UTS (~2%) and E_S (~8%) as compared to pristine epoxy, though marginally (Figure 8.24a). Further, with 0.01 wt% the composites exhibited a moderate increase in UTS (~4%) and a significant increase in ES (~15%) as compared to pristine epoxy (Figure 8.24a), which is accompanied by a marginal decrease in the tensile toughness (K) (area under the stress-strain curve): from ~10 MN/m^2 for pure epoxy to ~8 MN/m^2 for 0.01 wt% nanocomposites. The significant increase in modulus with such a low weight fraction of GNCs is likely due to effective GNC reinforcement owing to the chemical interaction between functional groups of GNCs with that of epoxy. Further, the nano-level dispersion, barring the agglomeration formation, of GNCs in epoxy matrix provides a huge surface area to interact with the matrix. This leads to efficient stress-transfer from matrix to filler. To the best of authors' knowledge, the enhancement in tensile properties; particularly the elastic modulus, with 0.01 wt% GNCs has not been reported previously. The minimum weight fraction of nanofiller (graphene) used so far in composite systems is 0.05 wt% as reported by Tang et al. [17]. The advantage of composites containing very low weight fraction of filler, as in the present case, is that the agglomeration of nanofillers in matrix could be minimized. To further investigate the reinforcement and agglomeration effect of GNCs in epoxy, tensile properties of the composites with higher weight fractions (up to 2 wt%) are also studied; where the tensile properties showed marginal improvement. For example, the composites between 0.05 and 1 wt% showed an increase of 3%–10% in E_S as compared to pure epoxy (Figure 8.24a); whereas the UTS showed a moderate increase, which is associated with a marginal decrease in K for all the cases.

However, in the case of nanocomposites with 2 wt% GNCs, the tensile properties do not show much improvement; rather the strength decreased slightly and the modulus is comparable to that of pure epoxy. In the present case maximum increase in tensile properties are exhibited by 0.01 wt%

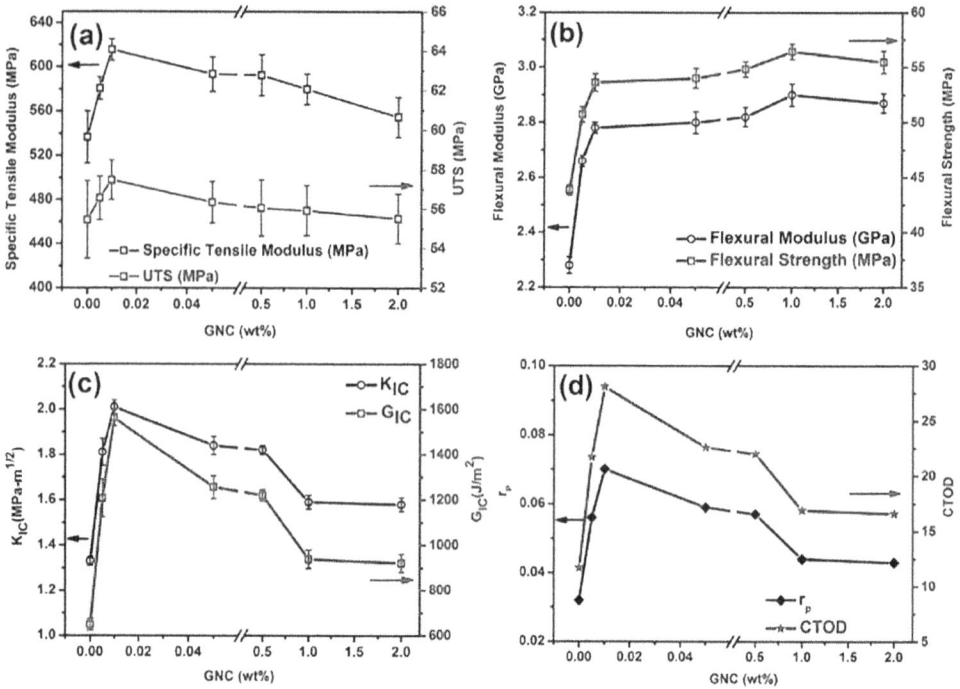

FIGURE 8.24 Mechanical properties of epoxy and nanocomposites with various GNCs wt%. (a) specific tensile modulus and UTS vs. GNCs weight fraction, (b) flexural modulus and strength, (c) fracture toughness (K_{IC}) and critical strain energy release rate (G_{IC}) as a function of GNCs weight fraction, (d) theoretical values of CTOD (δ) and plastic zone size (r_p) using experimental results for neat epoxy and nanocomposites.

nanocomposites. This is apparently due to uniform dispersion and good chemical interaction of GNC sheets with epoxy. Moreover, the GNCs possess large surface area due to their high aspect ratio [31]. Above this concentration the graphene platelets probably stalk with each other to form aggregates, which resulted in reduced reinforcement effect. Further, in 2 wt% composites the number of stalked GNC layers increases to an extent that they form clusters of aggregates (Figure 8.24). These aggregates seem to affect adversely on the tensile properties, since tensile strength is sensitive to the defects produced due to aggregates in nanocomposites [27,36–38].

8.3.2.2 Flexural Properties

The effect of GNCs on flexural properties of epoxy has been examined by means of three point bending test. Figure 8.24b presents the flexural strength and modulus of nanocomposites with different weight fractions of GNCs. Interestingly; we have observed significant improvements in the flexural strength (~16%) and modulus (~17%) with only 0.005 wt% of GNCs in matrix as compared to that of pure epoxy. Similarly, in the case of 0.01 wt% nanocomposites, the flexural strength and modulus are increased by ~22% and ~23%, respectively. The flexural strength and modulus further increased gradually up to 1 wt% of GNC composites (Figure 8.24b). The 1 wt% nanocomposite showed the highest value of flexural strength and modulus, which are ~29% higher than that of pure epoxy. These results indicate that the optimum flexural properties can be obtained with GNCs up to 1 wt%; presumably due to stiffening effect of GNCs in epoxy, in spite of forming large number of aggregates in the matrix. Hence, unlike tensile properties, the flexural or bending properties appear to be less sensitive to the defects in the specimen in terms of GNC aggregates, which is consistent with the literature for other nanofillers [27,36–38]. However, above a critical weight fraction of filler (in the present case ~1 wt%), the flexural properties are also adversely affected; though marginally.

8.3.2.3   Fracture Toughness

The K_{IC} and G_{IC} of pure epoxy matrix and that of nanocomposites are determined using the linear elastic fracture mechanics (LEFM) approach. The K_{IC} and G_{IC} values are plotted with the GNC weight fraction in Figure 8.23c. The K_{IC} and G_{IC} for epoxy are found to be 1.33 (±0.34) MPa.m$^{1/2}$ and 649 (±25) J/m^2, respectively. The K_{IC} and G_{IC} showed significant improvements: ~36% and ~86% for 0.005 wt% composites, and ~51% and ~140%, respectively, for 0.01 wt% composites as compared to that of neat epoxy. The percentage increase in K_{IC} and G_{IC} are much higher in the present case for 0.01 wt% nanocomposites as compared to results reported by Bortz et al. [29] for 0.1 wt% graphene oxide composites and the increase is comparable to that of Rafiee et al. [40] for 0.1 wt% graphene platelets composites. The improvement of notch toughness properties also gives an indirect indication of nano-level dispersion of GNCs in epoxy matrix. These sheets when dispersed uniformly in the matrix render obstacle to crack propagation leading to higher energy absorption. The K_{IC} and G_{IC} showed marginal decreasing trends in the weight fractions between 0.05 and 2 wt% (Figure 8.24c); though the values are higher than that of neat epoxy. The increases in K_{IC} and G_{IC} for the weight fractions between 0.05 and 2 wt% compared to neat epoxy are in the ranges of 19@40% and 42@96%, respectively. The nanocomposites with 2 wt% GNCs showed the least increase in the K_{IC} (~19%) and G_{IC} (~42%) as compared to neat epoxy. This is likely due to the formation of large number of GNC aggregates as shown in optical microscopy studies. These aggregates are likely to act as defect sites in polymer matrix.

Both the tensile and K_{IC} showed similar trends with respect to filler content (Figure 8.24a and c) (increase in properties till 0.01 wt% GNC followed by decrease); however, flexural properties showed invariable increase with GNC content. It is important to mention here that in tensile and single-edge-notch-bend (SENB) specimens, fracture occurs through crack initiation and propagation. However in case of flexural test there is no crack formation; hence no complete fracture; although it is a destructive testing method. The test was stopped at 5% deflection. Therefore probably flexural test (bending in the absence of pre-crack) is less defect-sensitive. Further, on account of the layered morphology of GNCs and its ability to improve the mechanical properties remarkably, GNCs could be promising material for preparation of bio-inspired (e.g. seashell [35]) layer-by-layer composites [36,37]. The crack tip opening displacement (CTOD) (δ) and plastic zone size (r_p) for plane strain condition are calculated from G_{IC} and K_{IC} using the equations: $\delta = G_{IC}/\sigma_y$ and $r_p = 1/6\pi \, (K_{IC}/\sigma_y)^2$, where σ_y is the yield strength of epoxy. The values of CTOD and r_p for epoxy and GNC composites are illustrated in Figure 8.24d. It is observed that the CTOD and r_p values increase significantly up to 0.01 wt% of GNCs compared to pure epoxy. These values show decreasing trends above 0.01 wt%, though the values are higher than that of pure epoxy. These results suggest that the composites show larger plastic deformation at the crack tip as compared to pure epoxy [27,36–38].

8.4   MECHANICAL PROPERTIES OF ELECTROSPUN PVA/CNT COMPOSITE NANOFIBRES

Several recent investigations have suggested that CNTs could be a well suited candidate for novel reinforced composite nanomaterials. The bending stiffness of an individual nanotube is in the range of 1–5 TPa [28], and techniques are being devised to produce thin films comprising aligned carbon nanotubes. This suggests that nanotubes could be used as reinforcing fibres in high toughness nanocomposites [30,41]. However, the chemical parameters, such as the filler dispersion, orientation and interfacial bonding could predominantly affect the synthesis of nanotube–polymer composites [21]. Among the various methods of producing nanotube–polymer composites, electrospinning is a novel and efficient tool for their fabrication. The composites fabricated by electrospinning can be efficiently utilized to assemble fibrous polymer sheets with fibre diameters ranging from the μm to the nm range. The electrospinning set-up used to obtain randomly orientated or aligned nanofibres consists of a bipolar high voltage source (the high voltage can be applied between the negatively charged composite fluid and collector), a syringe injector coupled to a needle (to carry the polymer fluid from the syringe to the spinneret) and a conducting collector. Various collector configurations can be used, such as a flat plate or a rotating drum, in order to provide greater control over the fibre alignment [22,42].

The electrospun nanofibres can be aligned to construct unique functional nanostructures such as nanotubes [43], nano-wires, thin non-woven mats and aligned arrays of nanocomposites [44]. Furthermore, depending on the polymer type, polymer nanofibres with a wide range of properties, such as electrical conductivity [45], high strength, low weight and low porosity, can be achieved, along with tailored surface functionalities [46]. Such properties make the polymer nanofibres a well suited candidate for many diverse applications. It was previously reported that finely dispersed nanotubes could be wrapped by water soluble polymers [47]. The wrapping of nanotubes in water soluble polymers is a well-known phenomenon, which is driven largely by the thermodynamics of the composite system causing it to eliminate the hydrophobic interface between the tubes and their aqueous medium [13]. Nevertheless, to obtain a fine dispersion, a stable and uniform suspension of nanotubes in the polymer is required, which can be achieved by using various surfactants, such a sodium dodecyl sulphate and Gum Arabic. However, this could be inconvenient from the standpoint of the removal and separation of the surfactants and may lead to undesirable effects. In the present work, we avoided using a surfactant by employing a water soluble polymeric system, and studied the effect of the nanotube concentration on the optical colouration and mechanical properties of the PVA/MWCNTs composites [48].

8.4.1 Fibre Electrospinning

In the present work, PVA obtained from the DC Chemical Co., Ltd. (polymerization ~1700 and degree of hydrolysis ~85.5%–88.5%) was used as a matrix material for the fibre and multi-walled nanotubes (MWCNTs, Iljin Nanotech Co., Ltd., diameter ~15–25 nm) grown by chemical vapour deposition (CVD) were used as a filler. To obtain a uniform dispersion, first the raw MWCNTs were functionalized [48] using sulphuric and nitric acid solution (95% H_2SO_4: 65% HNO_3= 3:1). The solution was sonicated for ~4–7 hours. After the acid treatment, the functionalized MWCNTs were collected on a PTFE (polytetrafluoroethylene) membrane (pore size ~1 μm). To neutralize the acid treated MWCNTs, the collected nanotubes were thoroughly rinsed four to five times with deionized water, followed by drying at ~90°C under atmospheric conditions. The functionalized nanotubes were mixed with distilled water and PVA was added. The weight fraction of PVA was equal to that of the MWCNTs.

Subsequently, more PVA was added until the viscosity of the solution attained its optimum level. The concentration of PVA in the aqueous solution was ~9 wt% and the corresponding viscosity was maintained at ~400–500 centipoises. Following this process, a well dispersed MWCNTs/ PVA solution was obtained, with a variation in the weight of the nanotubes of ~1–7.5 wt%. The MWCNTs/ PVA solution was spun into fibres using an electrospinning system (NanoNC Co., Korea). The experimental set-up used for the electrospinning process is shown in Figure 8.1. A syringe was filled with the MWCNTs/PVA solution. The solution was transferred through a syringe needle at a flow rate of ~0.1–5 mL/h. The electrospun fibres were collected on a paper wrapped rotating metal drum. The drum diameter was ~9 cm and the speed of rotation was ~25 rpm. The inner diameter of the syringe needle was ~0.2 mm and the distance between the syringe needle and the rotating metal drum was maintained at ~10–15 cm. The potential difference between the MWCNTs/PVA solution loaded needle and metal drum was ~10–20 kV. The obtained MWCNTs/PVA nanofibres were coated with a conducting platinum coating and their textural morphology was studied by environment scanning electron microscopy (ESEM: XL30 ESEM-FEG, Philips) and transmission electron microscopy (TEM: JEM 3010, JEOL). However, for the ESEM studies, the nanofibre samples were prepared on an aluminium plate, whereas for the TEM samples they were spun directly on a TEM-copper-grid. The mechanical properties of the PVA/MWCNTs non-woven sheets were analysed using a universal

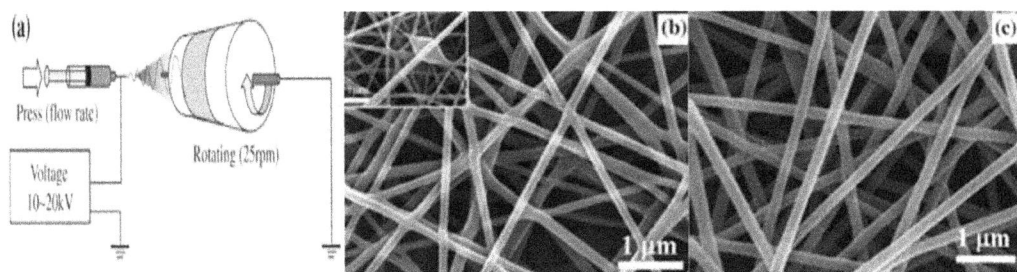

FIGURE 8.25 (a) Schematic depiction of electrospinning process, (b) recorded SEM image of PVA nanofibres sheets; inset shows the formation of beads and notches and the deformation of the fibre during spinning and (c) SEM image of PVA/MWCNTs composite nanofibre sheets (filler concentration 1 wt%).

testing machine (UTM, LLOYD Instrument). All of the UTM measurements were carried out at room temperature. The extension rate was kept at ~50 mm/min, the load cell used was ~250 N with a gauge length of ~30 mm. The dimensions of the sheet used were ~2 cm×5 cm ×~100–150 µm (width×length ×thickness).

8.4.2 MECHANICAL PROPERTIES

Figure 8.25a shows a schematic depiction of the electrospinning process, Figure 8.25b shows a recorded SEM image for the PVA spun fibres (the inset shows details of the morphology of the PVA nanofibres), and Figure 8.25c shows a SEM image of the PVA/ MWCNTs spun fibre with a nanotube composition of ~1 wt%. Figure 8.25b and c demonstrate that the PVA/MWCNTs nanofibres were successfully spun by electrospinning at ~15 kV. The diameter of the nanofibres varies in the range of ~100–200 nm. However, Figure 8.24b shows that there are some undesirable forms of fibres, such as beads (inset) on fibre strings, notches and local non-uniformity in the thickness of the nanofibres, due to the non-optimization of the experimental conditions. Moreover, in Figure 8.25c, the textural morphology of the PVA/MWCNTs nanofibres (nanotube concentration ~1 wt%) can be seen to be distinctly different from that of the polymer spun fibres. The diameter and textural morphology of the spun nanofibres depends on various conditions [48] such as the viscosity, charge density and flow rate of the fluid, the distance from the needle to the collector, the voltage, and the dispersion uniformity of the fillers during the electrospinning process, as well as the properties of the solution. As the concentration of the PVA solution was increased, the morphology of the nanofibres changed from beads on a fibre string to uniform fibre structures with a consequent increase in the fibre diameter. Thus, it is thought that the addition of the nanotubes to the polymer altered the effective fluid charge density, which modified the effect of the electric force used for whipping and spinning the fibres. Nevertheless, even though the fluid charge density was changed, the variation in the fibre diameter was found to be insignificant. Figure 8.25 shows photographic images of the non-woven nanofibres sheets for (a) PVA and (b)–(e) the PVA/MWCNTs composites with various concentrations of nanotubes ranging from ~1 to 7.5 wt%. In order to obtain non-woven thick sheets, which could be easily handled, the nanofibres were spun for ~5 hours. The dimensions of the sheets were ~10 cm×20 cm (l×b) and their thickness was found to vary in the range of ~100–150 µm.

Furthermore, it can be seen that the PVA nanofibre sheets are colourless and transparent. As the wt% of the nanotubes increases in the nanofibre composites, the colour of the nanofibre sheets changes. The wt% of the nanotubes shown in Figure 8.26b is ~1 wt% and the colour of the nanofibre sheets is ash/grey, whereas the concentration of the nanotubes was increased, their colour became dark blue. These results suggest that colour centres of distinct scattering wavelengths are formed after the incorporation of different amounts of nanotubes into the PVA nanofibres [46–48]. Moreover, because the cross sections of the electrospun nanofibres are small (~100–200 nm), the composites fabricated using this method can maintain their optical transparency. This property

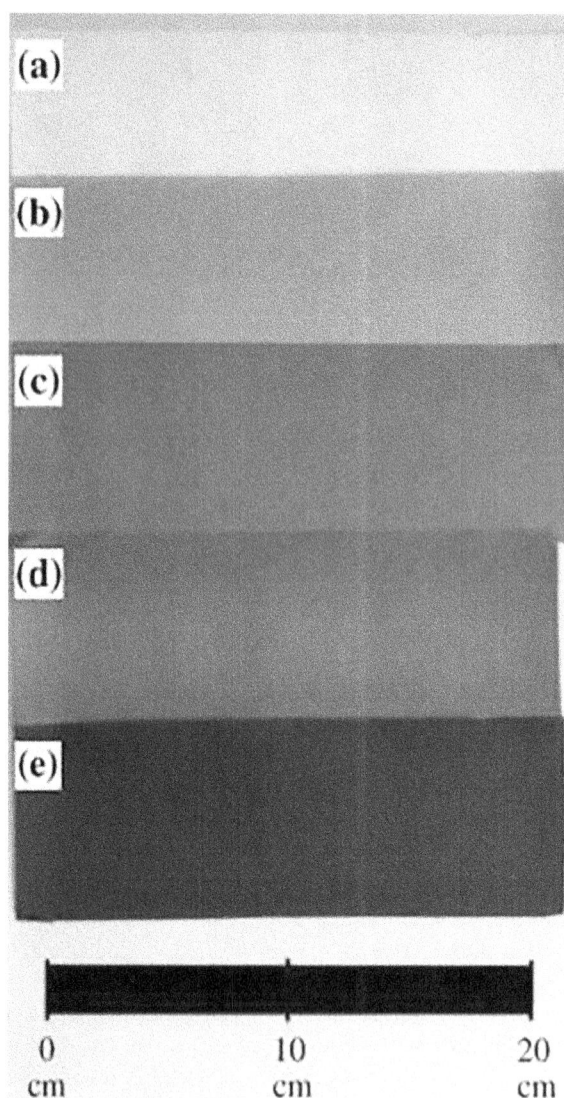

FIGURE 8.26 Recorded photographs of non-woven nanofibre sheets for (a) PVA and (b)–(e) PVA/MWCNTs composite, with increasing concentration of fillers from ~1 to 7.5 wt%. (a) Colourless, (b) ash/grey, (c) dirty grey, (d) deep brown and (e) deep blue fibre sheets. (For interpretation of the references to colour in this Figure legend, the reader is referred to the web version of this article.)

can make them suitable candidates for various novel optical-electronic applications, such as optical waveguides, non-linear optical materials and optical filters for the near-visible–ultraviolet–visible wavelength region [48].

Figure 8.27a shows the variation in the tensile stress (MPa) as a function of strain (%) for the (a) PVA and (b)–(e) 1–7.5 wt% PVA/MWCNTs non-woven nanofibres and Figure 8.27b shows the plot of the tensile modulus (MPa) as a function of the wt% of nanotubes in the composite nanofibres. Figure 8.27c shows the plot of the tensile strength as a function of the wt% of nanotubes (in the composite). Figure 8.27d shows the variation in the elongation at break as a function of the wt% of fillers. Compared to the PVA, the tensile yield stress of the PVA/MWCNTs sheets containing only 1.0 wt% of nanotubes is increased by ~15% from 105 to 133 MPa, and the tensile modulus is increased by ~2% from 175 to 178 MPa.

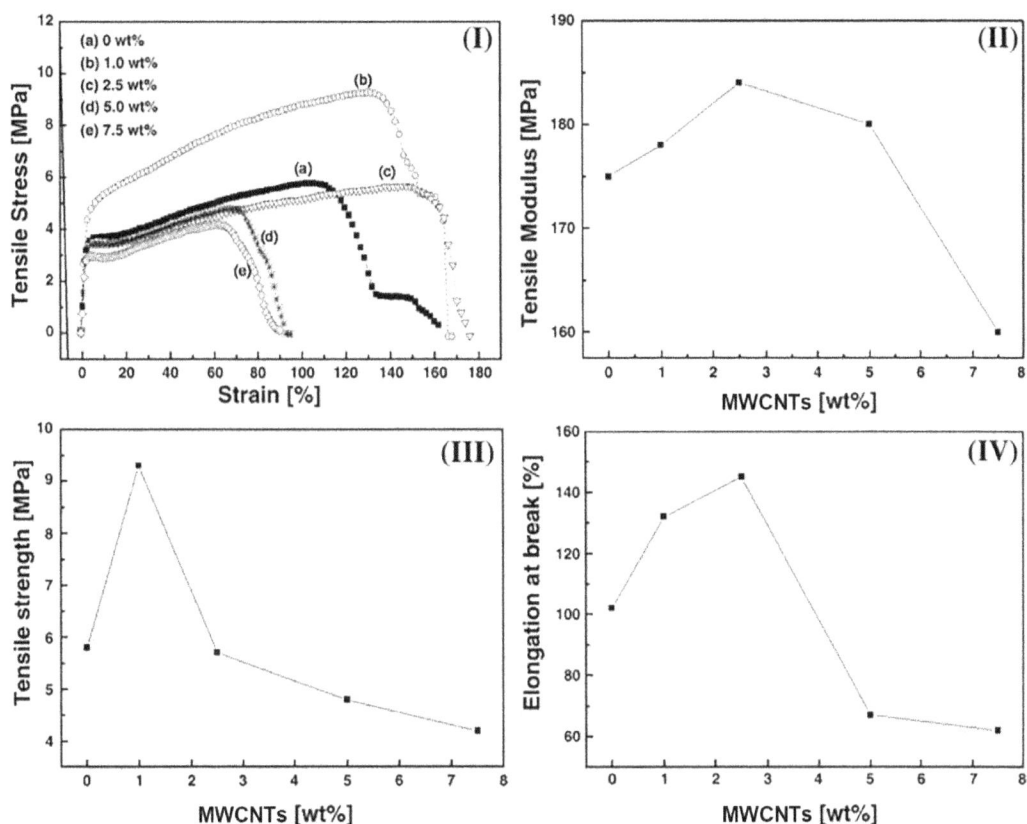

FIGURE 8.27 (a) Shows the variation in the tensile stress (MPa) as a function of strain (%) for the (a) PVA and (b)–(e) 1–7.5 wt% PVA/MWCNTs non-woven nanofibres and (b) shows the plot of the tensile modulus (MPa) as a function of the wt% of nanotubes in the composite nanofibres. (c) Shows the plot of the tensile strength as a function of the wt%. (d) Shows the variation in the elongation to break (in %) as a function of the wt% of fillers.

The tensile strength of the PVA/MWCNTs sheets with a concentration of nanotubes of 1 wt% is found to be much higher than that of the PVA fibre films. Moreover, as the wt% of nanotubes is increased from 1 to 2.5 wt%, the tensile modulus increases by ~5% from 175 to 184 MPa. All of the films were tested after being conditioned in a laboratory environment (23°C and 35% relative humidity) for 24 hours under identical conditions. However, as the wt% of nanotubes was further increased, both the tensile stress and tensile modulus decreased. The tensile strength of the composites containing 5 wt% or more of the nanotubes is almost the same as that of the polymer fibres, ~5.7 MPa. However, the decrease in the tensile modulus is marginal (~3% from 184 to 180 MPa), as compared to that of the 2.5 wt% nanofibre sheets. In general, the transfer of the load depends on the interfacial shear stress between the filler nanotubes and the wrapped polymer matrix. A high interfacial shear stress will transfer the applied load to the nanotubes over a short distance, whereas a low interfacial shear stress will require a long distance [47,48]. At a lower filler wt%, the load transfer to the inner nanotubes is low, which enhances the tensile properties. There are three main mechanisms of load transfer from a wrapped polymer matrix to the nanotube filler. The first is micromechanical interlocking, which is likely to be difficult in the nanotube composites due to their atomically smooth surface.

The second is chemical bonding between the nanotubes and the matrix. This is not guaranteed, but a recent study indicated that the interfacial shear stress due to bonding could be as high as

~0.05 GPa [48]. The third mechanism is a weak van der Waals bonding between the filler nano-tubes and the wrapped matrix. In order to assess the aptitude of the nanotubes to act as fillers, the issue of load transfer needs to be addressed. In addition, the effect of the extent of the nanotube orientation on the modulus of the composite has been studied theoretically [47,48], by computing the deformation induced orientation of the nanotubes and their Herman's orientation factor [48]. The initial increase in the tensile strength and tensile modulus (i.e. for the composites containing 1 and 2.5 wt% of filler) is attributed to the high degree of orientation of the filler nanotubes in the wrapped nanofibres. As the wt% of filler in the nanofibres is further increased (i.e. 5 and 7.5 wt%), agglomeration of the nanotubes takes place, which subsequently decreases their degree of anisot-ropy. Moreover, the agglomeration of the nanotubes increases the weak van der Waals interaction between the nanotube bundles and the walls of the wrapped nanofibres. However, these interactions are negligible at a lower wt% of nanotubes, due to the directional anisotropy. It is worth noting that the tensile strength of the PVA nanofibres (non-woven sheet) was found to be ~130 times lower than that reported in previous studies [48].

However, this difference in strength is attributed to the process adopted to cast the fibre sheets used for characterizing the tensile strength. Shaffer et al. found that the tensile properties of polyac-rylonitrile (PAN) non-woven sheets on a stainless steel substrate could be improved by compressing them mechanically at ~100°C for 3 hours. Thus, a crucial question is: Will nanotube-filled wrapped polymers be the materials with high modulus and strength? There is some doubt about this, since there is only a weak coupling between the outer polymer and inner layers of the nanotubes. This is especially true under tensile loading when the load transfer occurs from the matrix only to the outer nanotube layer that is bonded to the polymer. The situation in the nanotube composites may be similar since, in the agglomerated nanotube fibres [48], only the peripheral tubes bonded to the polymer will be stressed. This situation could be changed, however, if the polymer infiltrates into the interstices of the nanotube ropes and creates strong interlocking [35].

REFERENCES

[1] Kong, Jing, Nathan R. Franklin, Chongwu Zhou, Michael G. Chapline, Shu Peng, Kyeongjae Cho, and Hongjie Dai. "Nanotube molecular wires as chemical sensors." *Science* 287 (2000): 622.

[2] Javey, Ali, Jing Guo, Qian Wang, Mark Lundstrom, and Hongjie Dai. "Ballistic carbon nanotube field-effect transistors." *Nature* 424 (2003): 654.

[3] Hu, Nantao, Zhi Yang, Yanyan Wang, Liling Zhang, Ying Wang, Xiaolu Huang, Hao Wei, Liangmin Wei, and Yafei Zhang. "Ultrafast and sensitive room temperature NH_3 gas sensors based on chemically reduced graphene oxide." *Nanotechnology* 25, no. 2 (2013): 025502.

[4] Cui, Shumao, Haihui Pu, Ganhua Lu, Zhenhai Wen, Eric C. Mattson, Carol Hirschmugl, Marija Gajdardziska-Josifovska, Michael Weinert, and Junhong Chen, "Fast and selective room-temperature ammonia sensors using silver nanocrystal-functionalized carbon nanotubes." *ACS Applied Materials & Interfaces* 4 (2012): 4898–4904.

[5] Ghosh, Ruma, Anupam Midya, Sumita Santra, Samit K. Ray, and Prasanta K. Guha. "Chemically reduced graphene oxide for ammonia detection at room temperature." *ACS Applied Materials & Interfaces* 5, no. 15 (2013): 7599–7603.

[6] Mao, Shun, Ganhua Lu, and Junhong Chen. "Nanocarbon-based gas sensors: progress and challenges." *Journal of Materials Chemistry A* 2, no. 16 (2014): 5573–5579.

[7] Li, Ning, Yi Huang, Feng Du, Xiaobo He, Xiao Lin, Hongjun Gao, Yanfeng Ma, Feifei Li, Yongsheng Chen, and Peter C. Eklund. "Electromagnetic interference (EMI) shielding of single-walled carbon nanotube epoxy composites." *Nano Letters* 6, no. 6 (2006): 1141–1145.

[8] Cao, Wen-Qiang, Xi-Xi Wang, Jie Yuan, Wen-Zhong Wang, and Mao-Sheng Cao. "Temperature depen-dent microwave absorption of ultrathin graphene composites." *Journal of Materials Chemistry C* 3, no. 38 (2015): 10017–10022.

[9] Liu, Jia, Wen-Qiang Cao, Hai-Bo Jin, Jie Yuan, De-Qing Zhang, and Mao-Sheng Cao. "Enhanced permittivity and multi-region microwave absorption of nanoneedle-like ZnO in the X-band at elevated temperature." *Journal of Materials Chemistry C* 3, no. 18 (2015): 4670–4677.

[10] He, Jun-Zhe, Xi-Xi Wang, Yan-Lan Zhang, and Mao-Sheng Cao. "Small magnetic nanoparticles decorating reduced graphene oxides to tune the electromagnetic attenuation capacity." *Journal of Materials Chemistry C* 4, no. 29 (2016): 7130–7140.

[11] Cao, Mao-Sheng, Xi-Xi Wang, Wen-Qiang Cao, and Jie Yuan. "Ultrathin graphene: electrical properties and highly efficient electromagnetic interference shielding." *Journal of Materials Chemistry C* 3, no. 26 (2015): 6589–6599.

[12] Hu, Qingmei, Rongliang Yang, Zichao Mo, Dongwei Lu, Leilei Yang, Zhongfu He, Hai Zhu, Zikang Tang, and Xuchun Gui. "Nitrogen-doped and Fe-filled CNTs/NiCo$_2$O$_4$ porous sponge with tunable microwave absorption performance." *Carbon* 153 (2019): 737–744.

[13] Wen, Bo, Maosheng Cao, Mingming Lu, Wenqiang Cao, Honglong Shi, Jia Liu, Xixi Wang et al. "Reduced graphene oxides: light-weight and high-efficiency electromagnetic interference shielding at elevated temperatures." *Advanced Materials* 26, no. 21 (2014): 3484–3489.

[14] Darmstadt, Hans, Christian Roy, Serge Kaliaguine, Guoying Xu, Michele Auger, Alain Tuel, and Veda Ramaswamy. "Solid state 13C-NMR spectroscopy and XRD studies of commercial and pyrolytic carbon blacks." *Carbon* 38, no. 9 (2000): 1279–1287.

[15] Ugale, Ashok D., Resham V. Jagtap, Dnyandeo Pawar, Suwarna Datar, Sangeeta N. Kale, and Prashant S. Alegaonkar. "Nano-carbon: Preparation, assessment, and applications for NH$_3$ gas sensor and electromagnetic interference shielding." *RSC Advances* 6, no. 99 (2016): 97266–97275.

[16] Jagtap, Resham V., Ashok D. Ugale, and Prashant S. Alegaonkar. "Ferro-nano-carbon split ring resonators a bianisotropic metamaterial in X-band: Constitutive parameters analysis." *Materials Chemistry and Physics* 205 (2018): 366–375.

[17] Tang, Long-Cheng, Yan-Jun Wan, Dong Yan, Yong-Bing Pei, Li Zhao, Yi-Bao Li, Lian-Bin Wu, Jian-Xiong Jiang, and Guo-Qiao Lai. "The effect of graphene dispersion on the mechanical properties of graphene/epoxy composites." *Carbon* 60 (2013): 16–27.

[18] Tauc, J. "Optical properties of amorphous semiconductors." In *Amorphous and liquid semiconductors*, edited by J. Tauc, pp. 159–220. Springer, Boston, MA, 1974.

[19] Harrison, Walter A. "Bond-orbital model and the properties of tetrahedrally coordinated solids." *Physical Review B* 8, no. 10 (1973): 4487.

[20] Pendry, John B., A. J. Holden, D. J. Robbins, and W. J. Stewart. "Low frequency plasmons in thin-wire structures." *Journal of Physics: Condensed Matter* 10, no. 22 (1998): 4785.

[21] Park, Cheol, Zoubeida Ounaies, Kent A. Watson, Roy E. Crooks, Joseph Smith Jr, Sharon E. Lowther, John W. Connell, Emilie J. Siochi, Joycelyn S. Harrison, and Terry L. St Clair. "Dispersion of single wall carbon nanotubes by in situ polymerization under sonication." *Chemical Physics Letters* 364, no. 3–4 (2002): 303–308.

[22] Theron, A., Eyal Zussman, and Alexander L. Yarin. "Electrostatic field-assisted alignment of electrospun nanofibres." *Nanotechnology* 12, no. 3 (2001): 384.

[23] Ramo, Simon, John R. Whinnery, and Theodore Van Duzer. *Fields and waves in communication electronics*. John Wiley & Sons, 1994.

[24] Lu, Ming-Ming, Wen-Qiang Cao, Hong-Long Shi, Xiao-Yong Fang, Jian Yang, Zhi-Ling Hou, Hai-Bo Jin, Wen-Zhong Wang, Jie Yuan, and Mao-Sheng Cao. "Multi-wall carbon nanotubes decorated with ZnO nanocrystals: mild solution-process synthesis and highly efficient microwave absorption properties at elevated temperature." *Journal of Materials Chemistry A* 2, no. 27 (2014): 10540–10547.

[25] Chung, D. D. L. "Electromagnetic interference shielding effectiveness of carbon materials." *Carbon* 39, no. 2 (2001): 279–285.

[26] Feng, Dong, Dawei Xu, Qingqing Wang, and Pengju Liu. "Highly stretchable electromagnetic interference (EMI) shielding segregated polyurethane/carbon nanotube composites fabricated by microwave selective sintering." *Journal of Materials Chemistry C* 7, no. 26 (2019): 7938–7946.

[27] Kumar, Arvind, Sumati Patil, Anupama Joshi, Vasant Bhoraskar, Suwarna Datar, and Prashant Alegaonkar. "Mixed phase, sp^2–sp^3 bonded, and disordered few layer graphene-like nanocarbon: Synthesis and characterizations." *Applied Surface Science* 271 (2013): 86–92.

[28] Treacy, M. M. Jebbessen, Thomas W. Ebbesen, and John M. Gibson. "Exceptionally high Young's modulus observed for individual carbon nanotubes." *Nature* 381, no. 6584 (1996): 678–680.

[29] Bortz, Daniel R., Erika Garcia Heras, and Ignacio Martin-Gullon. "Impressive fatigue life and fracture toughness improvements in graphene oxide/epoxy composites." *Macromolecules* 45, no. 1 (2012): 238–245.

[30] Bower, Chris, Robert. Rosen, Li. Jin, Jong-Seob Han, and Otto. Zhou. "Deformation of carbon nanotubes in nanotube–polymer composites." *Applied Physics Letters* 74, no. 22 (1999): 3317–3319.

[31] Fetter, Steve, Andrew M. Sessler, John M. Cornwall, Bob Dietz, Sherman Frankel, Richard L. Garwin, Kurt Gottfried et al. "Countermeasures: a technical evaluation of the operational effectiveness of the planned US national missile defense system." (2000).

[32] Pendry, John B., David Schurig, and David R. Smith. "Controlling electromagnetic fields." *Science* 312, no. 5781 (2006): 1780–1782.

[33] Frank, Michael C., and Noah D. Goodman. "Predicting pragmatic reasoning in language games." *Science* 336, no. 6084 (2012): 998–998.

[34] Veselago, Victor Georgievich. "The electrodynamics of substances with simultaneously negative values of ϵ and μ." *Soviet Physics Uspekhi.* 10, no. 4 (1968): 509.

[35] Hasar, Ugur C., Yunus Kaya, Joaquim José Barroso, and Mehmet Ertugrul. "Determination of reference-plane invariant, thickness-independent, and broadband constitutive parameters of thin materials." *IEEE Transactions on Microwave Theory and Techniques* 63, no. 7 (2015): 2313–2321.

[36] Kumar, Arvind, Devesh Kumar Chouhan, Prashant S. Alegaonkar, and T. Umasankar Patro. "Graphene-like nanocarbon: an effective nanofiller for improving the mechanical and thermal properties of polymer at low weight fractions." *Composites Science and Technology* 127 (2016): 79–87.

[37] Chouhan, Devesh K., Sangram K. Rath, Arvind Kumar, P. S. Alegaonkar, Sanjay Kumar, G. Harikrishnan, and T. Umasankar Patro. "Structure-reinforcement correlation and chain dynamics in graphene oxide and Laponite-filled epoxy nanocomposites." *Journal of Materials Science* 50, no. 22 (2015): 7458–7472.

[38] Chouhan, Devesh K., Arvind Kumar, Sangram K. Rath, Sanjay Kumar, Prasant S. Alegaonkar, G. Harikrishnan, and T. Umasankar Patro. "Laponite-graphene oxide hybrid particulate filler enhances mechanical properties of cross-linked epoxy." *Journal of Polymer Research* 25, no. 2 (2018): 1–12.

[39] Starr, Francis W., Thomas B. Schrøder, and Sharon C. Glotzer. "Molecular dynamics simulation of a polymer melt with a nanoscopic particle." *Macromolecules* 35, no. 11 (2002): 4481–4492.

[40] Rafiee, Mohammad A., Javad Rafiee, Zhou Wang, Huaihe Song, Zhong-Zhen Yu, and Nikhil Koratkar. "Enhanced mechanical properties of nanocomposites at low graphene content." *ACS Nano* 3, no. 12 (2009): 3884–3890.

[41] Wagner, Hanoch Daniel, O. Lourie, Y. Feldman, and Robert Tenne. "Stress-induced fragmentation of multiwall carbon nanotubes in a polymer matrix." *Applied Physics Letters* 72, no. 2 (1998): 188–190.

[42] Katta, P., Massimiliano Alessandro, R. D. Ramsier, and G. G. Chase. "Continuous electrospinning of aligned polymer nanofibers onto a wire drum collector." *Nano Letters* 4, no. 11 (2004): 2215–2218.

[43] Li, Dan, and Younan Xia. "Direct fabrication of composite and ceramic hollow nanofibers by electrospinning." *Nano Letters* 4, no. 5 (2004): 933–938.

[44] Li, Dan, Yuliang Wang, and Younan Xia. "Electrospinning of polymeric and ceramic nanofibers as uniaxially aligned arrays." *Nano Letters* 3, no. 8 (2003): 1167–1171.

[45] Ra, Eun Ju, Kay Hyeok An, Ki Kang Kim, Seung Yol Jeong, and Young Hee Lee. "Anisotropic electrical conductivity of MWCNT/PAN nanofiber paper." *Chemical Physics Letters* 413, no. 1–3 (2005): 188–193.

[46] Cha, Dong Il, Hak Yong Kim, Keun Hyung Lee, Yong Chae Jung, Jae Whan Cho, and Byung Chul Chun. "Electrospun nonwovens of shape-memory polyurethane block copolymers." *Journal of Applied Polymer Science* 96, no. 2 (2005): 460–465.

[47] Bergshoef, Michel M., and G. Julius Vancso. "Transparent nanocomposites with ultrathin, electrospun nylon-4, 6 fiber reinforcement." *Advanced Materials* 11, no. 16 (1999): 1362–1365.

[48] Jeong, Jin Su, Jin-San Moon, Sung Yun Jeon, Jae Hong Park, Prashant Sudhir Alegaonkar, and Ji Beom Yoo. "Mechanical properties of electrospun PVA/MWNTs composite nanofibers." *Thin Solid Films* 515, no. 12 (2007): 5136–5141.

9 Application Engineering of Nanocarbon-Reinforced Composites

9.1 FIELD ELECTRON EMISSION ASPECTS OF CNTs: THE PASTE APPROACH

Field emission is a quantum mechanical phenomenon with no classical analogue [1]. When a sufficiently high electric field is applied to the surface of a metal or semiconductor, the surface potential barrier is deformed so as to provide a finite length through which an electron can "tunnel" and emit to the vacuum. Based on this principle, the prototype field emission (FE) displays (FEDs) and other vacuum electronic devices were produced as a proof of concept by several major companies [2]. Most of these devices were based on the Spindt-type structures [3] or tip of the carbon nanotubes (CNTs) as the FE source. However, FE properties of CNT emitters depends upon number of parameters like functionalization of tubes, their dispersion in paste, inter tube distance, quality, and texture of the host matrix, inter-electrode distance, effective emission area, mean, and effective field enhancement factor, including vacuum and packaging conditions.

Particularly, at an industrial level, CNT is implemented as the cathodes layers in FED technology with following important modifications: (a) integrating of CNTs in the form of the paste (or composite), (b) their coating by the screen printing process, and (c) activation of the emitters using adhesive tape. Such process is in turn cost, and time effective including a long shelf life for FED applications [4–12].

In FE, the mean field or field enhancement factor, γ, which, fundamentally, estimates the electric field intensity near the tip of an emitter, is one of the most important parameters and depends upon the geometry and tip-anode spacing [4]. The easiest way to quantify the enhancement factor is from the plot of the $\ln(I/V^2)$ as function of $1/V$ [4–8].

9.1.1 FOWLER–NORDHEIM THEORY: BASIC FORMULATION OF FIELD ENHANCEMENT FACTOR (γ)

Figure 9.1 shows the geometry involved in FE. It is important issue to understand how the field enhancement factor appears in the Fowler–Nordheim equation? In order to do this let us consider current density according to the Fowler–Nordheim theory (FNT) [13]. The analysis of density of the emission current was shown in detail by Diamond [14], Forbes [15] and Bonard [16].

According to the FN theory, for 1D case with plane electrode configuration, the emission current density, J, can be expressed as: $J = I/A = (B/\phi t^2(y)) F_0^2 \exp[-C \phi^{3/2} v(y)/F_0]$, where I, emission current, A, effective emission area, ϕ, work function of emitter [in eV]; F_0 (= V/d), V, d, are, respectively, inter-electrode field, voltage and distance, $t(y)$ and $v(y)$, elliptical dimensionless Nordheim functions, B and C are constant; with $B \equiv q_e^3 / 8\pi h = 1.54 \times 10^{-6}$ A-eV/V^2, $C \equiv 8\pi/3q_e h (2m_e)^{1/2} = 6.83 \times 10^9$ eV$^{-3/2}$-V/m, mass of free electron, $m_e = 9.11 \times 10^{-31}$ kg, elementary charge, $q_e = 1.60 \times 10^{-19}$ C, Planck constant, $h = 6.63 \times 10^{-34}$ J-s $= 4.14 \times 10^{-15}$ eV-s. To find the current density from above equation it is necessary to evaluate the elliptical functions $t(y)$ and $v(y)$, whose values are computed in [17,18]. The argument of the functions $t(y)$; $v(y)$ is the parameter of the Nordheim [13]: $y = \left[q_e^3 F_0 / 4\pi\varepsilon_0\phi^2 \right]^{1/2} = 3.79 \times 10^{-5}$ $\sqrt{F_0} \phi$, where, ϕ in eV; permittivity of free space $\varepsilon_0 = 8.85 \times 10^{-12}$ F/m. The values of y are belongs to the interval 0–1. Since, $t(y)$ is a weak function of y, then it is approximated as: $t^2(y) = 1.1$ [13]. For the real field strength at the tip of the emitter $\sim(2–5)10^9$ V/m the following approximation of $v(y)$ is

DOI: 10.1201/9781003317258-9

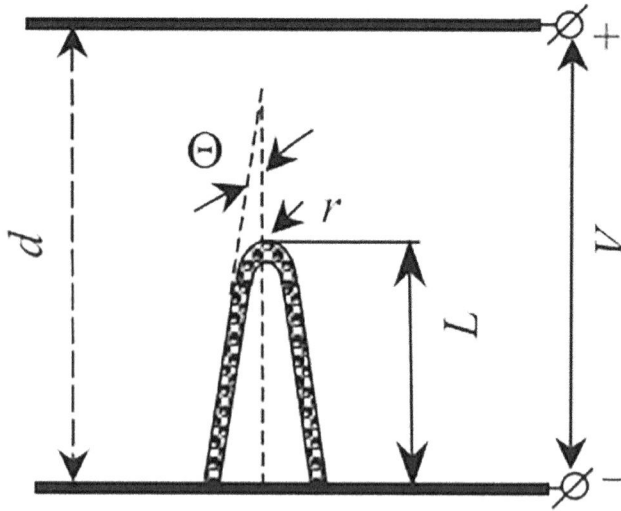

FIGURE 9.1 Geometry of an individual CNT emitter in diode system. L – height of emitter; r – radius of curvature; Θ – tip semi-angle for cone form ($\Theta = 0$ for cylindrical emitter).

used usually [8,13]: $v(y) = 0.956 - 1.062\, y^2$. Using expressions of y and $v(y)$ in current density, J, equation: $J = I/A = B/(1.1\ \phi)\exp[10.4461/\sqrt{\phi}]\, F_0^2\exp[-0.956\, C\,\phi^{3/2}] = B/(1.1\ \phi)\exp[10.4461/\sqrt{\phi}]\,(v/d)^2\exp[-0.956\, C\,\phi^{3/2}]$. If the array of the protrusions or micro tips emitters located on the substrate surface serves as an (nano)-emitters and contribute to the electrical field, then above could be modified. As a result, the local field at the tip of an individual emitter, F_{loc}, is appreciably more than for a field for plane electrode configuration, thus, $F_{\text{loc}} \gg F_0$.

Furthermore, in order to find F_{loc} two parameters could be used in the modern literature: the geometrical field factor, β, and the field enhancement factor, γ. The functional dependences of the β and the γ on the geometry of the carbon emitter were considered by Edgcombe and Valdre [19,20]. They had suggested the followings simulated expressions for enhancement factor of a cylindrical emitter and a flat anode: $\gamma \approx 1.2\,[2.15 + L/r\,]^{0.9}$. In this case, the value $L/2r$ could be considered as an aspect ratio for the cylindrical emitter. The computational model of Edgcombe and Valdre [19,20] was based on the evaluation of the electric field intensity near the emitter tip and did not contain any detail information about the enhancement factor as a function of the CA distance.

The geometrical field factor, β, was introduced by Dyke et al. [21], and later by Gomer [22], who suggested that field at a distance ~ 1 nm from the tip of the emitter $F_{\text{loc}}^{(1)}$ is defined by the radius of curvature of the emitter tip: $F_{\text{loc}}^{(1)} = V/k\, r$, where k is the coefficient depending on the length of the emitter ($k \approx 5$ for $L \sim 100\, r$, $k \approx 100$ for $L \approx 100r$ in according to [5]). As was shown by Dyke et al. [21], one can express the local field as: $F_{\text{loc}}^{(2)} = \beta V$, where β, geometrical field factor, and depends on geometry of emitter and substrate where emitter is located. Comparing $F_{\text{loc}}^{(1)}$ and $F_{\text{loc}}^{(2)}$ one can see that: $\beta = 1/k\, r$. Except for the geometrical field factor, β, the field enhancement factor, γ, exists [15], and is defined as: $\gamma = F_{\text{loc}}^{(2)}/F_0$; i.e. $F_{\text{loc}}^{(2)} = \gamma\, F_0$. Taking into account of γ, the current density, J, for an individual emitter could be expressed as: $J = I/A = (B/1.1\ \phi)\exp[10.4461/\sqrt{\phi}]\,\gamma^2\,(V^2/d^2)\exp[(-0.956\, C\,\phi^{3/2}d)/\gamma\, V]$. As the current density is a function of the inter-electrode distance, it makes sense to find this dependence, which is important for the other applications. One can see that, $\gamma = F_{\text{loc}}^{(2)}/F_0 = \beta V/\,(V/d) = \beta d$. Hence, for the given geometry of the emitter (if, the geometrical enhancement factor, $\beta = $const, see expressions for $F_{\text{loc}}^{(1)}$ and $F_{\text{loc}}^{(2)}$, the field enhancement factor is a linearly increasing function of the inter-electrode distance. This result is not consistent with the experimental and computed data of Bonard, Figures 2 and 3 of the reference. For an individual tube, one can see that $\gamma(d)$ is a decreasing function, especially, in the area close to the tip. The same result followed from the reference [6].

This leads to a conclusion that, the geometrical field factor, β, cannot be considered as a constant. As was shown by Dyke et al. [21] for a conical metal emitter $\beta_0 \cong 1/d^n$, where β_0 – the geometrical field factor at the tip apex, $n - a$ non-integer order of the Legendre function $P_n \cos(\Theta)$, Θ-angle in the plane polar coordinates. Also, the same type dependence, β, by tip-anode distance for the conical tip metal emitter was shown by Charbonnier [23]: $\beta \cong 1.4/[r^{2/3} (d-h)^{1/3} (\tan \Theta)^{0.2} (d/h)^{0.5}]$, where d-tip-anode distance, h-height of the emitter cone, Θ half-angle at the cone apex. In the last expression, if $d \gg h$, then: $\beta \cong 1.4 h^{0.5}/[r^{2/3} (\tan \Theta)^{0.2} d^{0.83}] \propto 1/d^{0.83}$. From this equation one can see that, the geometrical field factor is a non-linear decreasing function of d: $\beta \propto 1/d^c$, where c is a parameter, which depends on the geometrical configuration of the emitter. Since $c < 1$, then $\gamma \propto d^{1-c}$ is a non-linear increasing function of inter-electrode distance.

9.1.1.1 Array FE Configuration of CNT Paste

Initially, the photosensitive CNT paste was prepared by add-mixing the MWCNTs, spin on glass (SOG), organic vehicle, photosensitive monomers, photosensitive oligomers, and the photo initiators. The CVD grown MWCNT powder and SOG were used as an electron emission source and inorganic binder, respectively. The mixture of the CNT powder, organic vehicle and inorganic binder was premixed through solder paste softener for 15 minutes. After that, the three-roll mill process was carried out for mixing and dispersion of the CNT powders in the organic vehicle as the polymer matrix. The details of the CNT paste preparation was described in our previous report [4–6]. Mechanically, well-dispersed CNT paste was printed onto an indium tin oxide (ITO)-coated soda lime glass. The uniform thickness of the CNT paste film was obtained by the backside exposure and development process after printing and drying of the CNT paste in a forced convection oven for a period of ~15 minutes at a temperature ~90°C. The residue of the organic vehicle leads to such problems as out gassing and arcing during the field emission measurements. The CNT paste film was annealed at a temperature 400°C–450°C in the air or in the nitrogen ambient in order to degas the organic materials from the paste to obtain a stable electron emission characteristic. The CNT paste with the SOG easily formed a uniform thick film and the annealing was carried out at relatively high temperature ~450°C in air or nitrogen. In accordance with our previous results, annealing at a temperature ~450°C under the nitrogen atmosphere is a suitable condition for the photosensitive paste with the CVD-MWCNTs [7,8]. The field emission scanning electron microscopy (FESEM) was employed for the characterization of the CNT paste. Figure 9.2 shows the SEM images of the screen-printed CNT paste with contain of 10 wt% of the SOG. It is known that the screen-printed CNT-FEA generally needs a special type of surface treatment, such as laser irradiation, ion irradiation, and surface rubbing with adhesive tape for high emission current density and uniform emission site through protrusion of the CNT [4–12] and reference cited there in. In the

(a) (b)

FIGURE 9.2 SEM images of the acryl-based photosensitive CNT paste films with SOG content 10 wt. %. (a) Top view and (b) cross-sectional view.

present work, initially the activation treatment was carried out by using an adhesive tape followed by the measurements of field emission of the CNT paste. All the field emission measurements were carried out with DC voltage. The emission characteristics of the CNT paste were measured in the vacuum chamber at a pressure level of ~5×10^{-6} Torr.

To compare the enhancement factor $\gamma(d)$ for an individual CNT and the effective enhancement factor $\gamma(d)$ for a randomly arrayed tubes in the CNT paste, Figure 9.2b, we carried out the simulation for the electric field intensity in the diode system (cathode – individual CNT, anode – infinitive conducting plate). The geometry of the CNT was taken from the real size of the CNTs, which are applied in the FED and observed in the experiments. The CNT had a typical length, $L \sim 10$ μm, and radius, $r \sim 10$ nm and the gap between the cathode and the anode could be varied from ~250 to 1500 μm. For the simplification the geometry of the CNT under investigation was imagined as a conducting rod, kept at a potential level, $V = 1000$ V (see Figure 9.3a). For such a physical situation, the function of $\gamma(d)$ was simulated by using the finite element method (FEM) with ANSYS 9.0 software. The inter-electrode (cathode and anode) potential level distribution was evaluated by using Laplace's equation, $\nabla^2 \phi = 0$, with the corresponding boundary conditions $\phi = 0$- at the cathode; and $\phi = \phi_1$-at the anode in the body centred angle Ω. In addition, the electric field distribution was computed and found to be axially symmetric relative to the longer axis of the CNT, and the simulation was modelled for the half CNT structure. It can be seen from Figure 9.3b that a large amount of change in the electric field intensity near the tip of the CNT; however, the detail analysis depends on the correlation between ratios L/r and d/r. In general, the ratios $d/r \gg L/r$, e.g. $d/r = 10^5$ L/r, hence, it is important to estimate the distance of an anode from tip of the CNT, where the changes of electric field is more predominant. Figure 9.3c shows the simulation of the enhancement factor, $\gamma(d)$, as a function of the distance of the anode from the CNT tip (i.e. $d-L$), and in the inset is the electrical field intensity as function of the distance of the anode from the CNT tip. Furthermore, it can be seen from inset of Figure 9.3c that the most visible changes in the field distribution occur near the tip of the CNT at a distance ~100 μm. As a result, two different possibilities emerge: (a) anode located near the CNT tip ($d < 100$ μm) and (b) anode located far from the CNT tip at the $d > 100$ μm. In case of the anode located far from the CNT tip, in a uniform electric field, it will be the weak dependence of the enhancement factor on inter-electrode distance. This yields the classical electrostatic law, where $F(d) \propto 1/d^2$ [24], with $\gamma(d)$ as a weak function of the inter-electrode distance at large d. Furthermore, for the inter-electrode distance, $d > 100$ μm, the enhancement factor is a weak function of inter-electrode distance could be defined through aspect ratio of the CNTs, i.e. $L/2r$. The values of the enhancement factor in Figure 9.3c correspond to the experimental data for an individual nanotube with $30 \leq \gamma \leq 260$ and are consistent with reference [18]. The result shown in Figure 9.5

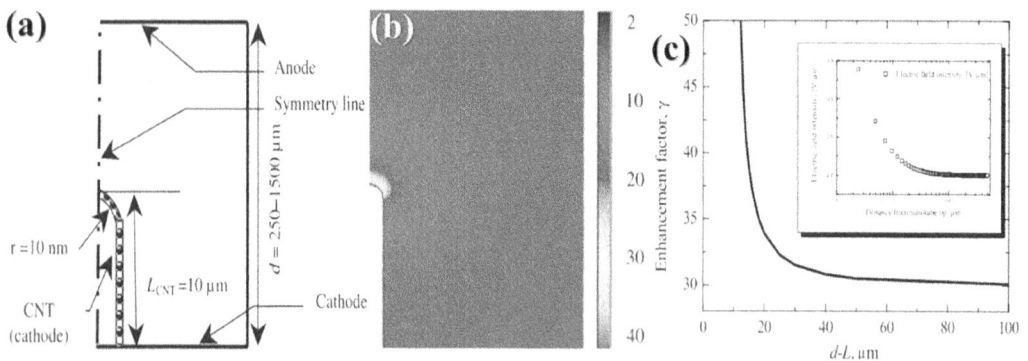

FIGURE 9.3 (a) The model for simulation and geometry of a CNT, (b) the distribution of the enhancement factor, γ, near the tip of an individual CNT, and (c) the field enhancement factor, γ, as a function of distance between the tip of a CNT and an anode. The inset shows the distribution of the electrical field as a function of distance from the tip of a CNT.

agrees qualitatively with the result of Bonard [18] and shows that the enhancement factor changed marginally near the tip of the emitter, but if d/L, then the enhancement factor is slightly depends on the inter-electrode distance.

In the design of FED and field emission experiments, the measured inter-electrode, d, distance is a distance between flat substrate, where an array of CNT emitters are located, and flat anode. Usually d belongs to interval as: $100\ \mu m \leq d \leq 7000\ \mu m$, then $V\ 10^{-2}\ V/\mu m \leq F_0 \leq V\ 0.143 \times 10^{-3}\ V/\mu m$. The typical values of length of CNTs on the flat substrate are: $1\ \mu m \leq L \leq 50\ \mu m$, therefore the condition of d/L is mostly applied to practice and enhancement factor, γ, is weakly decreasing function of the CA distance at constant voltage, V. If d approaches to L, then one should take into account the strong dependence of the enhancement factor on the distance. In contrast to the enhancement factor for an individual emitter, when an array of CNTs on the substrate is considered, one could introduce the effective enhancement factor, γ. Moreover, the estimation the effective enhancement factor could be carried out by analysing the FN region in the plot of $\ln(I/V^2)$ as a function of $1/V$. The effective enhancement factor is an additive function of the individual emitter tips and the total emission current, I, is given by, $I = I_1 + I_2 + \ldots + I_N = \sum_{i=1}^{N} I_i$, where $i = 1 \ldots N$-the number of emitter or emitting sites. With the assumption, $\phi_i = \phi$, $A_i \approx A$ and $d_i = d$ (since $d \gg L_i$), the expression for the total emission current, I, is given by the following equation: $I = (A \times B\ /\ 1.1\ \phi) \exp[10.4461/\sqrt{\phi}]\ (V^2/d^2) \sum_{i=1}^{N} \gamma^2 \exp$

$[(-C\ \phi^{3/2} d)/\gamma_i\ V]$. It can be seen from this equation that the effective enhancement factor, $\bar{\gamma}$ is an additive function of the individual emitter sites, then $\bar{\gamma} > \gamma_i$.

Furthermore, it was shown [4–12] that for an array of the CNTs fabricated on a metal tip the effective enhancement factor, $\bar{\gamma}$, could be increased with increasing the inter-electrode distance, which was also reported in [24] for the emission of the multi-walled CNT film. It has to be noted that the dependences of the enhancement factor as a function of inter-electrode distance is entirely different for an individual nanotube and for the array of the CNTs. However, this could be explained by the definition regime of the enhancement factor, i.e. constant voltage or constant current. Since the effective enhancement factor, $\bar{\gamma}(d) \propto d^{1-c}$ then in case of the FEA, $\gamma(d)$ is an increasing function of the inter-electrode distance. This leads to a conclusion that in an array of the CNTs the geometry of the nanotube could be slightly changed from a cylindrical to the conical form with a hemispherical cap; hence, $\gamma(d)$ varies with d^{1-c} where the value of the exponent, $c \cong 0.83$. For an array of CNT the effective enhancement factor, $\bar{\gamma}(d)$ will have the typical value, especially in the case of the random distribution of CNT arrays. Furthermore, the effective enhancement factor, $\bar{\gamma}(d)$, could be defined from the experimental findings, at know I–V values, by using expression for J (i.e. $J = I/A = (B/1.1\ \phi) \exp[10.4461/\sqrt{\phi}]\ \gamma^2\ (V^2/d^2) \exp[(-0.956\ C\ \phi^{3/2} d)/\gamma\ V]$), we get: $\bar{\gamma}^2 \exp[(-0.956\ C\ \phi^{3/2} d)/\bar{\gamma}\ V]) = I \times 1.1\ \phi/A \times B \exp[-10.4461/\sqrt{\phi}]\ (d^2/V^2)$.

The evaluation of the effective enhancement factor $\bar{\gamma}(d)$ for an array of the nanotubes from the CNT paste was carried out by performing the field emission experiments at the different inter-electrode spacing. The inter-electrode distance, d, was varied at a 500; 1100; and 1400 μm. Figure 9.6a and b show the current voltage (I–V) characteristics and corresponding FN plots for the CNT paste emitter with an area 1×1 cm as a function of inter-electrode distance, d. To find the effective enhancement factor $\bar{\gamma}(d)$ a FN area was selected in every FN plot, see Figure 9.6b. In the present study, we have used a two-region-field-emission (TRFE) model which was proposed in [4–8]. According to this model, the effective enhancement factor, γ, is given by equation: $\frac{1}{\bar{\gamma}} = \frac{d_2}{d} + \frac{1}{\gamma_0}$, where d_2-width of the enhancement region, and γ_0-absolute enhancement factor. According to $(J = I/A = (B/1.1\ \phi) \exp[10.4461/\sqrt{\phi}]\ \gamma^2\ (V^2/d^2) \exp[(-0.956\ C\ \phi^{3/2} d)/\gamma\ V])$:

$\ln\left[\dfrac{I}{F_0^2}\right] = \ln A \times B \exp\left[-10.4461/\sqrt{\phi}\right]\ \bar{\gamma}^2) - 0.956\ C\ \phi^{3/2}/\bar{\gamma}\ F_0 = b - \dfrac{m}{F_0}$, where m, is the slope in the FN area, see Figures 9.4b and 9.5a.

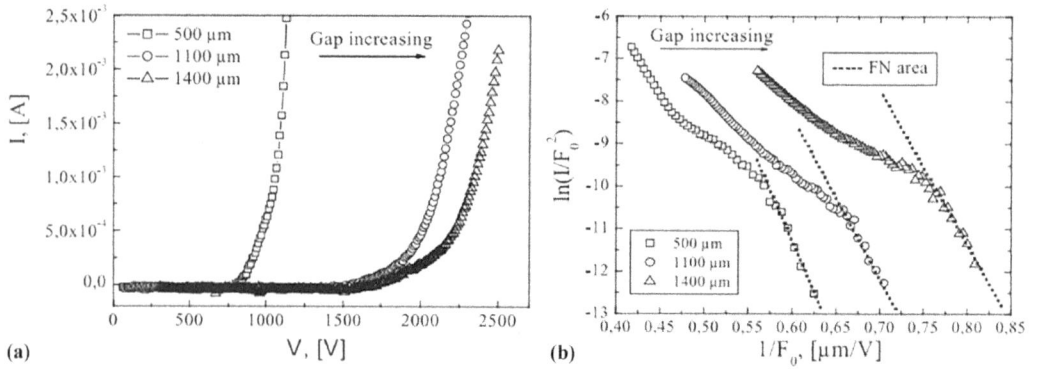

FIGURE 9.4 (a) Current voltage (*I–V*) characteristics and (b) FN plots for CNT paste emitter as a function of different cathode–anode spacing.

TABLE 9.1

Field Emission Parameters for the Carbon Nanotube Composite Synthesized by Various Methods

	Turn-on Field $\left(V/\mu m\right)$ (at 10 $\mu A/cm^2$)	Current Density $(\mu A/cm^2)$ (at 3 $\left(V/\mu m\right)$)	Mean Field Enhancement Factor	Area (m^2)
(a)	1.698	467.8	6881.7	5.9×10^{-18}
(b)	1.649	613.7	6760.3	8.7×10^{-18}
(c)	1.574	824.4	7123.9	8.5×10^{-18}
(d)	1.374	2030.2	7851.4	1.2×10^{-17}

The work function for the CNTs $\phi = 5\,eV$ [8]. As a result, the enhancement factor, γ, from the slope is defined as: $\gamma = 6.53 \times 10^3 \phi^{3/2}/|m|$, with the effective emission area, $A = 1.67 \times 10^{-14} m^2\ \phi^2$ exp $[b - 10.44\sqrt{\phi}\ [m^2] = [7.13 \times 10^5\ \phi/(\overline{\gamma})^2\ ]$ exp $[b - 10.44/\sqrt{\phi}]$ in $[\mu m^2]$. One can get the same values of effective enhancement factor, $\overline{\gamma}$, by solving the non-linear equation $(\overline{\gamma}^2$ exp $[(-0.956\,C\ \phi^{3/2}d)/\overline{\gamma}\ V]) = I \times 1.1\ \phi/A \times B$ exp $[-10.4461/\sqrt{\phi}]\ (d^2/V^2))$ numerically. The values of the absolute slope $|m|$, effective enhancement factor, γ and b, and the effective emission area A are enlisted in Table 9.1. The values of, $\overline{\gamma}$, which were defined from slope of FN plot and from equation above, are equal.

According to the TRFE model, the computed values of $\overline{\gamma}_0 = 3135.3$ and $d_2 = 205\,nm$, refer to Figure 9.5b. The effective enhancement factor, $\overline{\gamma}$ (d) plot is given in the inset of Figure 9.5b. $\overline{\gamma}$ $(d) = A_0 + A_1 d + A_2 d^2$, where $A_0 = 447.0$, $A_1 = 2.25\ \mu m^{-1}$, $A_2 = -7.38 \times 10^{-4}\ \mu m^{-2}$. Since $\overline{\gamma} = F_{loc}^{(2)}/F_0 = \overline{\beta}$ $V/V/d = \overline{\beta}d$, then $\overline{\beta} = \overline{\gamma}d$. The average geometrical factor $\overline{\beta}$ is a decreasing function of inter-electrode distance, d as shown in Figure 9.6a, $\overline{\beta}$ $(d) = \overline{\beta}_0\ 1/d^c\ \mu m^{-1}$, where $\beta_0 = 85.9$; and $c = 0.55$.

Further, it is important to understand why the enhancement factor has shown an opposite dependence in case of an individual CNT (Figure 9.3c) and a randomly arrayed CNT paste (inset in Figure 9.5b), as a function of inter-electrode distance, *d*. We suppose that it is necessary to distinguish two ways for the determination of the enhancement factor: first is at *V*=constant; *d*=variable; and secondly, at *I*=constant; *d*=variable. The first case could be realized for the calculation of the enhancement factor for an individual CNT at a constant voltage; hence, the local field $F_{loc}^{(2)}$ and the enhancement factor, $\overline{\gamma}$, are found to be decreasing functions of the inter-electrode distance. In another case, with constant current, and variables values of voltage and inter–electrode distance, the ratio should necessarily satisfy the condition: $\overline{\gamma}\ V/d$=constant, then the enhancement factor, $\gamma(d)$ has to be an increasing function of the inter-electrode distance. Moreover, it is illustrated in Figure 9.5b,

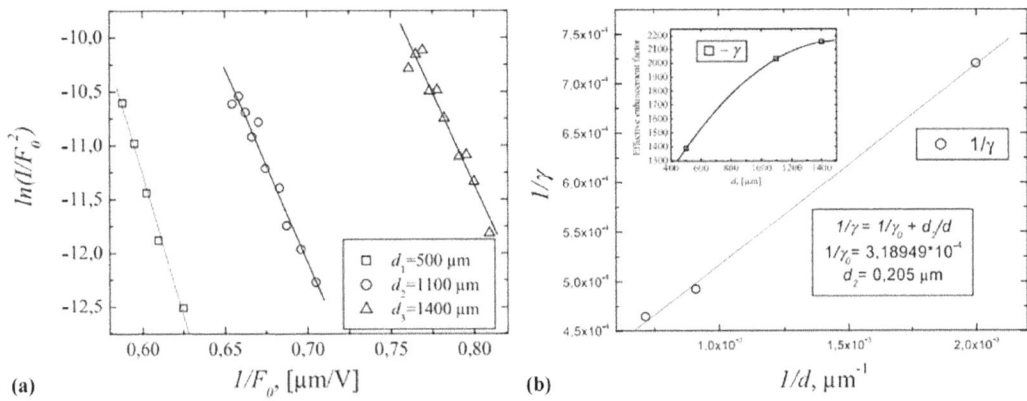

FIGURE 9.5 (a) FN plots for different CA spacing and (b) relationship of the field enhancement factor, $\bar{\gamma}$, and cathode–anode spacing, d. In the inset: effective enhancement factor, $\bar{\gamma}$ (d).

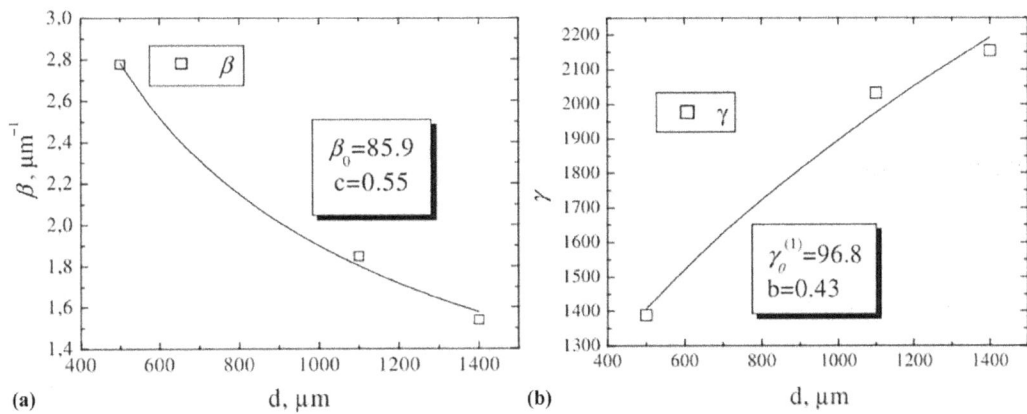

FIGURE 9.6 (a) Dependence geometrical enhancement factor, β, as a function of CA gap. (b) Dependence enhancement factor as a function of CA gap from formula $\bar{\gamma} = \bar{\beta}\, d = \bar{\gamma}_0^{(1)}\, d^s$.

where enhancement factor, $\bar{\gamma} = \bar{\beta}\, d = \bar{\gamma}_0^{(1)}\, d^s$, $\bar{\gamma}_0^{(1)}$ corresponds to $\bar{\beta}_0$ with an accuracy of ±13%, with $s \approx 1-c$ and had accuracy of ±5%.

9.1.1.2 Screen-Printed Triode-CNT FE Arrays for Flat Lamps

The CNT emitters are one of the best candidates for field emission devices such as, field-emission displays (FEDs) [4–12], X-ray tubes [25] and backlight units (BLUs) [9] for liquid crystal displays (LCDs). In general, three known methods could be adopted, to use CNTs as the field emitters; screen printing [4–8], spray deposition and chemical vapour deposition (CVD) [8,9]. Among them, screen printing method is the most cost effective method because of its easy appliance to a large screen size area. And hence, it has become a more matured technology in other display manufacturing areas. In general, backside expose [4] method was used for the patterned CNT emitters. Thus, expensive indium tin oxide (ITO) which has good transparent property was commonly prepared as the cathode materials. Also, to use cathode electrode at triode structure, ITO should been patterned by photolithography and etching technology [8]. The use of ITO material as cathode electrode is relatively expensive and it needs complicated sub-processes. And hence, research and development of other type of transparent electrode is required to fabricate simplified triode structure.

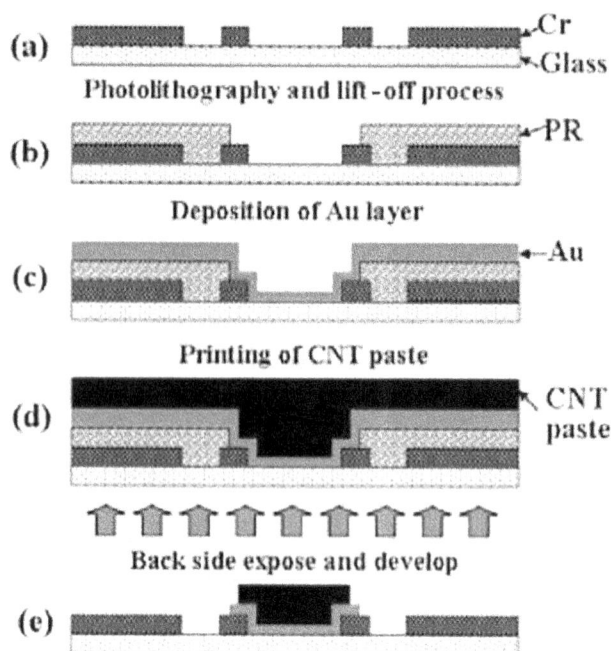

FIGURE 9.7 Schematic depiction of fabrication processes for triode structure with CNT emitters.

In this section, we have described a simple fabrication process. The patterned ITO film was replaced by thin Au film with thickness ~5–20 nm to serve as a transparent electrode material. Therefore, the ITO patterning process was eliminated. Moreover, coated Au film on photoresist (PR) layer was acted as an Ultraviolet (UV) block. The details of the fabrication process, conducting and transmission properties of the thin Au layer are discussed in this section.

The schematic cross section of parallel triode type CNT field emitter arrays is showed in Figure 9.7.

The soda lime glass substrates were sonicated in acetone bath for a period of ~10 minutes. After rinsing the substrates in methanol and DI water, respectively, they were baked in a furnace at a temperature ~110°C for 5 minutes. The substrates were coated by photoresist (PR, AZ1500) using spin coating method (2500 rpm, 35 seconds), and then a opened-glass line was formed by the developer after UV exposure. To make cathode and gate electrode, thin Cr layer of thickness ~3000 Å was deposited on the substrate by the e-beam evaporation technique. The lift-off method (Figure 9.7a) was adopted to fabricate cathode and gate electrode. To fabricate thin film of Au in cathode electrode and field emitter with CNT paste, PR was coated on the substrate (Figure 9.7b). Thin layers of Au and Ti with thickness ~150 and 20 Å, respectively, were deposited by e-beam evaporation technique (shown in Figure 9.7c and d). And photosensitive CNT paste was printed by screen printing method. The photosensitive CNT paste was composed of CNT powders, organic vehicles, photo initiator, and inorganic binder. To obtain a uniform thickness of CNT paste, the backside exposure and development process were carried out after printing process [6–10].

To obtain thickness layer of ~0.9 μm the CNT paste was exposed to UV for a period of ~285 seconds. The drying of CNT paste was carried out in forced convection oven for 20 minutes at 90°C. And, the PR layer was removed by soaking in acetone. Generally, the residue of organic vehicle in CNT paste leads to problems such as out-gassing and arc discharging during the field emission process. Organic materials in paste were removed in order to obtain stable emission characteristics. Therefore, the CNT paste was fired at 450°C for 10 minutes in nitrogen (N_2) ambient. The schematic diagram of fabricated triode structure with CNT emitter is shown in Figure 9.7e. After the surface treatment using adhesive tape, the field emission characteristics of parallel gated type CNT

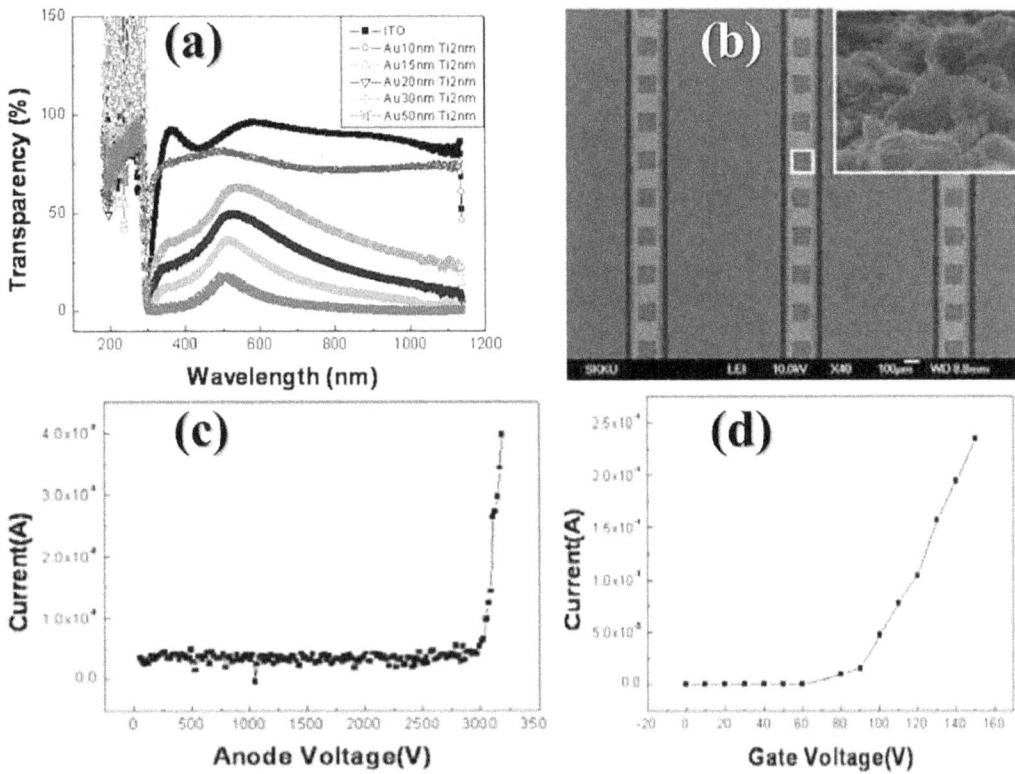

FIGURE 9.8 (a) Plot of variation in transmission (in %) over the measured wavelength range as a function of thickness of Au layer, (b) recorded SEM image for parallel triode structure and CNT emitters on a cathode electrode after surface activation by adhesive tape, and (c) recorded *I–V* characteristics in triode and diode configuration with CNT emitters (anode voltage is 2700 V).

field emitters were measured. A super used stainless steel (SUS) was used as an anode plate. The distance between the cathode and the anode plate was kept constant ~1000 μm. DC voltage was supplied between the anode and cathode electrodes using a high-voltage power supply (Fug, HCN 700–3500MOD) with different gate voltage for investigation of the triode characteristics. The fabricated triode structure and surface morphologies were observed by Field emission scanning electron microscopy (JEOL, JSM 6700F) [4–8].

Figure 9.8a is recorded UV–Vis spectra (Scinco, UV S-2100) for different thickness of Au film and compared with conventionally used ITO-coated glass (SAMSUNG CORNING). In the present work, the Ti layer between Au and glass was used as an adhesion layer. The result revealed that, transmittance of ITO-coated glass is more than 90% measured at I-line wavelength (365 nm). Normally, ITO is coated by various methods such as, chemical vapour deposition (CVD), physical vapour deposition (PVD) and spin coating process. These methods need high temperature processing for make better electrical and optical properties. However, in our fabrication sequence, high temperature processing is avoided because of inclusion of the PR processing. Therefore, the thin metal layer (Au) as transparent electrode is used. The thin metal layer which is located between PR and CNT paste can also function as a block layer because the patterned PR layer can be dissolved by a solvent in CNT paste. And when back side expose, thin metal layer on PR is used as UV block [8–12].

The thickness of the Au layer was varied from 5 to 50 nm. The sheet resistance and transmittance of Au layer as a function of thickness of layer are shown in Figure 9.8b. The sheet resistance was measured by four probe method (CHANG MIN, CMT-SR 1000N) and the thickness of Au layer was measured by atomic force microscope (AFM) (THERO-MICROSCOPES CP Research). The resistance of the Au film with thickness more than 10 nm was found to be always lower than that of the resistance (8.198 $\Omega/\square$) of ITO film (film thickness: 185 nm). It was found that the transmittance was decreased as the thickness of the Au layer was increased. When pattern of photosensitive CNT paste was formed, the transmittance of electrode and substrate is very important factor. Lower transmittance of Au layer was solved with the increased energy of radiation. The expose time of CNT paste at back side of glass was decided with this numerical formula [2 J/(20 mW/cm^2×transparency)=expose time]. The thin metal layer of Au (15 nm) and Ti (2 nm) has characteristic of sheet resistance (4.99 $\Omega/\square$) and transparency (35%) at a wavelength of 365 nm. The Au thin film was introduced to make simple triode structure [4–8].

The fabricated triode structures are shown in Figure 9.8b. The width of gate and cathode were formed with 720 and 120 μm, respectively. The distance between the gate and cathode electrodes was 80 μm. After surface activation using adhesive tape thickness and morphologies of CNT paste on a cathode electrode is shown in Figure 9.8b. The cathode electrodes with CNTs of thickness of 0.9 μm were formed by screen printing, back exposure techniques and firing process. The emission characteristics of CNT field emitters as triode structure is shown in Figure 9.8c. The cathode size of triode structure was 1×1 cm^2. The distance between cathode and anode was about 1000 μm. The variation of anode current was measured when anode voltage was changed from 0 to 3200 V. The turn on voltage in diode was approximately 3000 V. In the triode mode, the anode voltage was set at 2700 V, which was slightly lower than the turn-on voltage for the anode in the diode configuration. And the gate voltage was increased from 0 to 150 V at a step of 10 V. Turn on voltage of triode emitter was 80 V. The maximum current density 234 μA/cm^2 was obtained at the gate voltage 150 V as seen in Figure 9.8d. Based on these results, it may be concluded that the simple triode structure with CNT emitters using pattered substrate could be used for the fabrication of light source for the flat lamp application [4,6].

9.1.1.3 Enhanced FE Properties of CNTs: Role of SiO$_x$ Coating

This section describes the synthesis, characterization, and FE properties of SiO$_x$-coated t-MWCNTs. A routine chemical method is adopted for coating SiO$_x$ onto the t-MWCNTs, and the samples are characterized using the high resolution transmission electron microscopy (HRTEM), Fourier transform infrared (FTIR), and thermogravimetric analysis (TGA) techniques. The field emission properties of the thin SiO$_x$-coated t-MWCNTs are compared with those of the raw and functionalized t-MWCNTs. The analysis reveals that the turn-on fields, current densities, and current stabilities are improved for the coated nanotubes [10].

We used t-MWCNTs, which were obtained from Iljin Nanotech Co., Ltd., as the raw material. The diameter of the t-MWCNTs was found to vary in the range from ~5 to 8 nm and their length was (75–80 μm). The t-MWCNTs/SiO$_x$ composite was fabricated as follows: initially, acid treatment of the t-MWCNTs was carried out in a mixture of concentrated sulfuric and nitric acids (95% H$_2$SO$_4$: 65% HNO$_3$=3:1 volume ratio) to induce the formation of various functional groups such as –COOH, –OH, vCO, and C–O on the side walls 15 of the nanotubes. The acid treatment was carried out under mild ultrasonic conditions for a period of 3–7 hours. The acid treatment time and ultrasonic power conditions were optimized to minimize the damage caused by the acid treatment and/or sonication. After the acid treatment, the solution was filtered by an evacuation technique using a polytetrafluoroethylene (PTFE) membrane the pore size of the membrane was ~0.2 μm. The nanotubes collected on the membrane were extensively washed four to five times in deionized (DI) water in order to neutralize the walls of the nanotubes. In the final stage, the nanotubes were washed with acetone to remove the water and dried at a temperature of ~90°C under atmospheric conditions for ~1 hour. This process was carried out to avoid the agglomeration of the nanotubes. The procedure used for coating a thin layer of SiO$_x$ onto the t-MWCNTs is as follows: The acid-treated t-MWCNTs (0.1 wt%) were redispersed in ethanol solution using an ultrasonicator (200 W) for ~15 minutes, forming a stable and homogeneous CNT suspension. A spin on glass (SOG) solution (SOG, Honeywell Electronic Materials Accuglass 512B, with 15 wt% methyl groups bonded to Si atoms in the Si–O backbone, 10 wt%) was added to the nanotube dispersed suspension and the resulting mixture was sonicated for a period of ~5 hours to prevent the phase separation of the SOG and t-MWCNTs/ethanol solution. The resultant suspension was filtered through a PTFE membrane filter and washed two times to remove the excess unsuspended t-MWCNTs and SOG particulates. Finally, the SiO_x-coated t-MWCNTs collected on the membrane filter were dried at a temperature of ~90°C under atmospheric conditions for ~1 hour. The morphology of the thin SiO_x-coated t-MWCNTs was characterized by field emission scanning electron microscopy (FE-SEM; JEOL JSM90), HRTEM (JEOL JEM2100F) with an electron beam energy of 300 kV, and energy dispersive spectrometry (EDS; Oxford INCA energy). The nature of the molecular bonding of the coated nanotubes was investigated by FTIR (Bruker IFS-66-S). TGA was carried out at ~1000°C with a constant ramp rate of ~10°C/min. All TGA measurements were carried out under atmospheric conditions. For the field emission measurements, the raw, functionalized, and thin SiO_x-coated t-MWCNTs were dispersed in isopropyl alcohol (IPA) solution (concentration ~1–3 mg/mL) for the quick evaporation of the solvent. The solution was sonicated for ~10 minutes. The well-dispersed nanotube suspension was directly sprayed on indium tin oxide (ITO)-coated soda lime glass (substrate size ~1 × 1 cm^2) using an air brush pistol. The sprayed nanotubes were activated by activation treatment. The emission characteristics of all of the samples were measured in a high vacuum chamber with a parallel diode-type configuration, [10] at a pressure of ~5 × 10^{-6} Torr. The inter-electrode distance was kept fixed at ~200 μm, while the field emission current was measured using a Keithley 237 source measurement unit. The analysis of coating including stability is discussed in [11].

Further, SiO_x is a wide-band-gap material (~2.26 eV), with a small positive electron affinity of ~0.6–0.8 eV [25,26]. The thin coating of SiO_x material on the t-MWCNTs deposited by the chemical synthesis route is useful in decreasing the effective work function of the carbon nanotube emitters, which increases their electron emissivity. The thin SiO_x coating acts as a protective layer for the sharp conductive core, under the stable SiO_x/C interface conditions, and the high saturated drift velocity of ~2.7 × 10^7 cm/s for SiO_x also allows for its operation at high temperatures. Moreover, the silicon carbide in the SiO_x coating can withstand a high breakdown electric field (~2.2 × 10^6 V/cm), thus allowing for high voltage and high power operation without any danger of avalanche breakdown and for the emitters with Si–C and Si–O–C coatings to be closely packed together. Hence, it is important to study the field emission characteristics of thin SiO_x-coated t-MWCNTs, which could be useful for FED devices. It is noteworthy that the samples used for this field emission study were fabricated by a spray coating technique; it is possible that the morphology and density of the nanotubes may vary from sample to sample. Hence, the effective field emission site in each sample could be affected by its individual morphology and emitter density. To compare the field emission properties of the raw, functionalized, and coated samples, three samples from each category were prepared and subjected to the field emission characterization study. The averaged results are compared. In a similar fashion, the analysis of the field emission lifetime was carried out for the different categories of samples

Figure 9.9A shows the plot of the field emission current density (A/cm^2) as a function of the applied electric field (V/μm) recorded for the (a) raw t-MWCNTs, (b) functionalized t-MWCNTs, and (c) thin SiO_x-coated t-MWCNTs. Before conducting the measurements, all of the samples were subjected to a multi scanning process to obtain more stable and reproducible *I–V* profiles without arcing. The bias voltage was applied for a period of ~1 hour to reach stable emission conditions. Each current density in Figure 9.9A is the average value of four/five measurements from each category of samples. It can be seen that for the raw t-MWCNTs, the turn-on field is (1.37 V/μm at a current of ~1 μA, and the recorded current density is ~1.75 mA/cm^2 at a field of ~2.5 V/μm. For the same magnitude of field strength, the emission current density is found to be lowered to ~116 μA/cm^2, with an increase in the magnitude of the turn-on field value to ~1.7 V/μm (current ~1 μA) for the functionalized t-MWCNTs. However, for the thin SiO_x-coated t-MWCNTs, the value of the turn-on field is decreased to ~1.08 V/μm and the current density, measured at a field of ~2.5 V/μm, is found to be markedly enhanced to ~5.8 mA/cm^2. Thus, for the coated nanotubes, the magnitude of

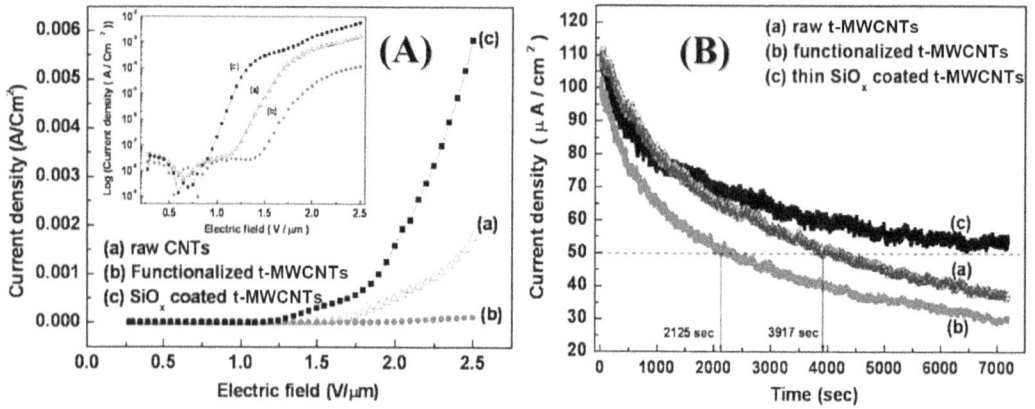

FIGURE 9.9 (A) Recorded field emission current density (A/cm²) as a function of applied electric field (V/μm) for (a) raw t-MWCNTs, (b) functionalized t-MWCNTs, and (c) thin SiO$_x$-coated t-MWCNTs, in a diode-type configuration. The inter-electrode distance is ~200 μm. Inset is the log (current density) as a function of applied electric field and (B) recorded field emission lifetime characteristics for (a) raw t-MWCNTs (lifetime ~1 hour), (b) functionalized t-MWCNTs (lifetime ≤1 hour), and (c) thin SiO$_x$-coated t-MWCNTs (lifetime >2.5 hour).

the emission current density is enhanced by approximately three times and the value of the turn-on field is lowered by a factor of ~1.3, as compared to the emission characteristics of the raw nanotubes [10]. The observed increase in the emission current density for the coated nanotubes is attributed to the effect of conduction band bending in the space charge layers, which depends on the dielectric constant K and the conductivity of the SiO$_x$ coating [27]. As mentioned above, the coating on the nanotubes is composed of several nanophases such as Si–C, Si–O–C, and SiO$_x$. In the case where an electric field, E_{vac}, is applied to the (C/SiO$_x$) coating-vacuum interface on the vacuum side, the strength of the electric field inside the C/SiO$_x$ interface, E_s, is smaller than E_{vac} by a factor of K, the dielectric constant of the C/SiO$_x$ interface layer. Under the applied electric field, E_{vac} the relation between the Fermi band ε_F, the work function of the C/SiO$_x$ interface φ_o, the amount of band bend-

ing φ_B, and the dielectric constant K of the interface state is given by: $\varepsilon_F + \varphi_o + \varphi_s = \left(\dfrac{K-1}{K+1} E_{vac} \right)^{\frac{1}{2}}$ [10,25,26].

From this equation one can see that the work function of the emitter (i.e. the coated nanotube assembly), φ_o is independent of the amount of band bending of the emitted electrons, but determines the tunnelling probability via E_{vac}, and thus the field emission current. This leads us to speculate that the density of states for the coated nanotubes may be localized close to the Fermi energy and contribute to their efficient field emission behaviour. Moreover, the conduction of electrons in the thin layers of Si–C and Si–O–C has a high saturated drift velocity value of ~2.7×10^7 cm/s. In addition, the thin interface layers possess a low dielectric constant $K \sim 2.5$ at 10 Hz) and could act as Schottky barriers [28]. The excess tunnelling current, J_{st}, through such Schottky barriers depends on the local temperature of the emitter. During the field emission process, the temperature of an individual emitter may rise to ~1500–2000 K. As a result, a large saturation current, J_{st}, flows through the Si–C and Si–O–C layers, which is given by: $J_{st} = S \times A \times T^2 \exp(-q\, \phi_o / k_B T)$. In this equation, S is the effective area of the thin layer coating, A, is a constant ~ 146 A m^{-2}/K^2 for SiO$_x$, and T is the local temperature of the emitter. Thus, for the thin SiO$_x$-coated t-MWCNTs, the current density is found to be increased with a consequent downshift in the turn-on electric field. However, in the present work, the field emission properties were obtained from the bulk samples and not from the individual SiO$_x$-coated MWCNTs [10,28].

Figure 9.9B shows the recorded field emission lifetime spectra for the (a) raw t-MWCNTs, (b) functionalized MWCNTs, and (c) thin SiO$_x$-coated t-MWCNTs. The field emission life-time is defined as the time period during which the magnitude of the current density decays to one-half of its initial value. Before the lifetime measurements, the current density (~1 mA/cm^2) was normalized with respect to time (~1 hour) and all of the lifetime spectra were recorded for ~2 hours (~7000 seconds). It can be seen from Figure 9.9B that, initially, the current density decreases significantly, which is attributed to the poor adhesion between the nanotube emitter and the substrate. From the lifetime characteristics plot, the estimated life-time for the (a) raw t-MWCNTs is ~3917 seconds (~1 hour), that of the (b) functionalized t-MWCNTs is ~2125 seconds (<1 hour), and that of the (c) thin SiO$_x$-coated t-MWCNTs is more than 7000 seconds. From the slope of the curve, the lifetime of the coated nanotubes is found to be ~9000 seconds (~2.5 hours) [10].

During the field emission process, in the ensemble of the chamber, gas could be ionized by the electrons ejected from the CNT emitters. These ionized species form various bonds with the carbon nanotubes, such as C–O, C–N, and C–Hx bonds. The gradual decrease of the current density with time for the raw t-MWCNTs could be attributed to the etching effect, which consequently leads to a decrease in height, degradation of the tip, and local erosion of the nanotubes. However, the thin SiO$_x$ coating can protect the nanotubes from the reactive sputter etching, since the coating is robust and inert. Thus, the outer surface of the coated nanotubes does not experience any structural degradation during field emission. Hence, the thin layer SiO$_x$ coating on the t-MWCNTs plays an eminent role as a protective layer against the reactive plasma etching. The analysis reveals that the thin SiO$_x$-coated t-MWCNTs with an appropriate layer thickness possess high field emission characteristics and a more favourable lifetime, which is ~2.5 times greater than that of the raw t-MWCNTs [4,6,10].

9.1.1.4 FE Properties of Plasma-Treated CNT Cathode Layers

In this section we present work on the effect of plasma treatment on the MWCNT paste cathode layers in an attempt to improve the field emission properties. We realize that the plasma treatment acts as a form of "conditioning" which, in effect, gives the cathode layers an initial "burn-in" so as to remove the unstable, high-aspect-ratio MWCNTs. Thus, longer and more energetic plasma treatment leads to better field emission behaviour. This section centres on the effect of plasma treatment on the MWCNT paste cathode layers in an attempt to obtain improved field emission properties [11,29].

In experimental preparations, CNT paste was synthesized by mixing multiwalled carbon nanotubes (MWCNTs, diameter <10 nm) in an organic solution and glass frit. The prepared MWCNT paste was screen-printed on the indium tin oxide (ITO) glass to obtain the cathode layers. These cathode layers were subjected to surface activation treatment [19] to obtain vertically aligned MWCNTs. The surface activated cathode layers were subjected to Xe/Ne plasma conditioning. During the plasma treatment, the distance between the cathode layers and anode was kept constant at ~1.8 mm. The plasma treatment was carried out for periods of either ~1 or 3 minutes at two different biasing voltages, viz. ~250 and 300 V. The cathode layers which were treated at a biasing voltage of ~250 V for ~1 and 3 minutes are designated as S_1 and S_2, respectively. The cathode layers which were treated at a biasing voltage of ~300 V for ~1 and ~3 minutes are designated as S_3 and S_4, respectively, and the untreated virgin cathode layers are designated as S_V. A large number of S_V and S_{1-4} samples were prepared. The sample area was kept constant at ~2×2 cm^2. A few samples (S_V and S_{1-4}) were studied with the scanning electron microscopy technique. A few virgin S_V and all the plasma-treated samples, S_{1-4}, were subjected to a field emission characteristics and lifetime study. Moreover, the lifetime measurements were carried out under a vacuum of ~10^{-6} Torr. The measurements were carried out at a constant value of current density of ~1 mA/cm^2, for a period of ~48 hours. In another study, the lifetime measurements were carried out for a period of ~220 hours, under a high vacuum condition of ~10^{-8} Torr. All measurements were carried out at a fix cathode-to-anode distance of ~200 µm and constant voltage value of ~1200 V. Moreover, the measurements were carried out in a parallel diode configuration coupled to a pulse generator with a duty ratio of 1/100 [11].

During the plasma treatment, the cathode layer film is bombarded by plasma ions and electrons. The metastable ionic species, such as Xe$^+$, Xe^{2+}, and Ne^{2+}, and energetic electrons could rupture the graphene shells and dissociate the CNTs. During their interaction with the cathode layer film, these

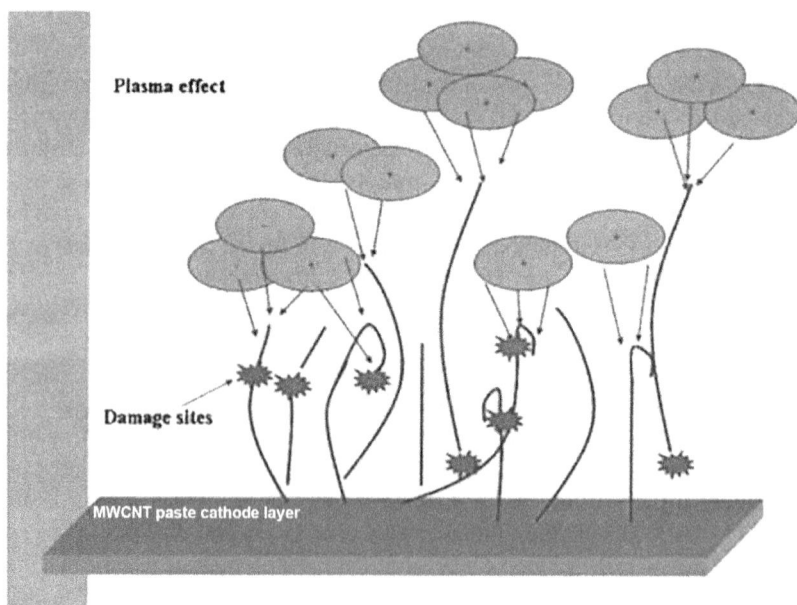

FIGURE 9.10 Schematic depiction of the plasma conditioning of the MWCNT paste cathode layers.

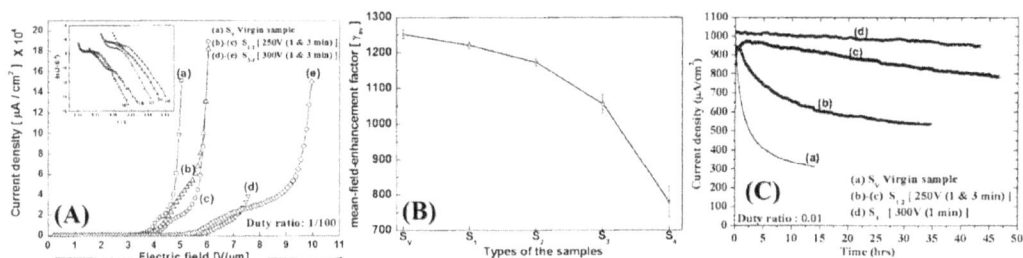

FIGURE 9.11 (A) Plot of the variation in the measured current density (μA/cm^2) as a function of the macroscopic electric field (V/μm) for (a) the virgin sample S_V and [(b)–(e)] the plasma-treated samples S_{1-4}. The inset is the plot of ln (J/E^2) as a function of (1/E). (B) Variations in the estimated value of the mean field-enhancement-factor γ_{av} for the virgin sample S_V and the plasma-treated samples S_{1-4} values are estimated form Figure 9.11 (A) inset, (C) Plot of the lifetime behaviour for (a) the virgin sample S_V and [(b)–(d)] the plasma-treated samples S_{1-3}. [No data are provided for the S_4 sample, due to the arcing and discharging which occurred during the measurements (vacuum condition of ~10^{-6} Torr).]

species may transfer their initial mobile energy [20] to the carbon atoms associated with the CNTs. This energy transfer could dissociate the bonding and lead to the ejection of carbon atoms, physically, from the CNTs. These charge particle interaction events may take place anywhere at the local sites of the CNTs (shown in Figure 9.10). The loss of carbon atoms from the CNTs has several possible effects, such as the erosion of the nanotubes from their local sites, the fracturing of the tubule structure, the knee bending of the tip, or the loss of the shape of the nanotubes. Figure 9.10 describes the effect of the plasma treatment on the cathode layer samples, schematically [29].

Figure 9.11 shows the plot of the variation in the measured current density (μA/cm^2) as a function of the macroscopic electric field (V/μm) for (a) the virgin sample S_V and [(b)–(e)] the plasma-treated samples S_{1-4}. The inset is the plot of ln(J/E^2) as a function of 1/E. From Figure 9.11, one can see that the value of the onset macroscopic field is smaller for the virgin samples, S_V, than for the plasma-treated, S_{1-4}, samples. For the S_V samples the estimated value of the onset macroscopic field was

~4.65 V/μm. For the plasma-treated, S_{1-4}, samples, however, the onset macroscopic field was found to increase gradually from 5.45 ± 0.05 to 0.05 ± 0.05 V/μm. It is important to note that, during the field emission measurements, the applied voltage (~1200 V) and cathode-to-anode distance (~200 μm) were kept constant. For the virgin samples, S_V, it is possible that most of the emission comes from a relatively a small number of nanotubes with a high aspect ratio and that the performance of these emitters is particularly degraded due to the plasma treatment [29]. It is also reasonable to consider that the longer and more energetic the bombardment, the greater will be the loss of performance. Thus, in Figure 9.11, one can observe a tendency for the onset macroscopic field to increase with increasing plasma biasing voltage and processing time.

Figure 9.11B is the variation in the estimated value of the mean field-enhancement factor γ_{av} for the virgin sample S_V and the plasma-treated samples S_{1-4}. From Figure 9.11B, for the virgin sample S_V the value of the mean field-enhancement factor γ_{av} is estimated to be 1251 ± 13. Moreover, the mean field-enhancement factor γ_{av} decreases gradually from 1210 ± 10 to 779 ± 43 for the plasma-treated samples S_{1-4}. Thus, the longer and more energetic the plasma-ion bombardment, the smaller the values of the mean field-enhancement factor γ_{av}. This is consistent with the relationship "onset macroscopic field = onset-barrier field/γ," under the assumption that field emitters of a specified work function Φ always turn on at the same value of the barrier field. Thus, the onset macroscopic field gradually increases for the plasma-treated samples S_{1-4} with a subsequent decrease in the mean field-enhancement factor γ_{av}. This implies that the ion bombardment effectively reduces the density of the active field emitter sites in the MWCNT paste cathode layers. Thus, the reduction in the mean field-enhancement factor γ_{av} decreases the areal density of the active field emitter sites, which in turn reduces the inter-tube screening effect [25]. Therefore, electrostatic screening could affect the onset macroscopic field, emission stability, and lifetime of the cathode layers [29]. Figure 9.11C shows the plot of the lifetime behaviour for (a) the virgin samples S_V and [(b)–(d)] the plasma-treated samples S_{1-3}. These measurements were carried out at a vacuum condition of ~10^{-6} Torr with duty ratio of 1/100. The field emission half-lifetime $t_{1/2}$ is defined as the time period during which the magnitude of the measured current density decreases to one-half of its initial value. From half-lifetime $t_{1/2}$ one can estimate the lifetime of the field emitter samples. Figure 9.11C(a) shows that, for the virgin sample S_V, the value of the current density decreases drastically, from its initial value of ~1000 μA/cm^2, over the measured time period. However, from Figure 9.11C(b)–(d) one can see that, for the plasma-treated samples S_{1-3}, the nature of the lifetime curve is changed dramatically. The decay in the current density with time is reduced and the stability is markedly enhanced [11].

9.1.1.5 CNT Composite: Dispersion Routes and FE Parameters

By and large, CNT nanocomposites could be synthesized by: (a) surface modification (functionalization) of the CNTs [12]; (b) their homogeneous dispersion; and (c) integration to desired epoxy/polymer matrix. However, the dispersion of the CNTs in the nanocomposites is a major problem to be solved for nanotube-reinforced polymers. Due to their large surface area (>1000 m^2/g), which is several orders of magnitude larger than the surface of conventional fillers, CNTs exhibit a strong tendency to form agglomerates. The functionalization enhances dispersibility and interfacial bonding of the nanotubes in the hybrid material system [29]. An effective exploitation of the desired properties of the CNT nanocomposites is, therefore, related to the homogeneous dispersion of CNTs in the polymer matrix or an untanglement of the agglomerate. A considerable amount of work related to the functionalization of CNTs and synthesis of CNT nanocomposites has been carried out [12,29,30]. The synthesis and characterization of graphitic carbon nanofibres functionalized by reactive linker molecules derived from diamines and tri-amines has been reported [30]. An alternative approach, to strong acid surface modification of the nanotubes, has been suggested in which the CNTs were treated with triethylenetetramine (TETA) solution in UV/O$_3$ environment [31]. The effect of plasma surface modification and ultrasonication time on the mechanical properties of multiwall carbon nanofibres–polycarbonate has been investigated [32]. Etching effects on the multiwall CNTs by treating them with ethanol, methanol and water vapours has been studied [27]. In most cases, such studies were carried out to explore the mechanical properties of the CNT nanocomposites. In general, not much attention has been paid to studies of the field emission parameters of the CNT nanocomposites synthesized by dispersing the CNTs via various dispersion routes [33].

In this section, a correlation has been presented between the dispersibility of the t-MWCNTs in the composite to their FE parameters. The dispersion has been achieved via the chemical and mechanical dispersion routes. In chemical dispersion route, the t-MWCNTs has been ACOOH functionalized by the H_2O_2 treatment, whereas, in the mechanical dispersion route, the shear mixing of the t-MWCNTs has been carried out to fabricate the CNT composite cathode layers. The field emission analysis revealed that, better field emission parameters have been achieved for the CNT composite synthesized by the combination of the chemical and mechanical dispersion routes.

The thin multiwall carbon nanotubes (t-MWCNTs (Iljin-Nanotech, Korea), purity 95%, diameter: 5–7 nm, length 10 lm) have been used as the starting material. The functionalization of the pristine t-MWCNTs has been carried out by soaking the nanotubes in the H_2O_2 solution followed by the sonication of the solution for about 12 hours. These t-MWCNTs have been filtered and characterized further using Fourier transform infrared (FTIR) technique to confirm the functionalization of the t-MWCNTs [12,29].

Figure 9.12 is the synthesis scheme for the t-MWCNT composites. One can see that four different approaches have been adopted to synthesize the t-MWCNT composite. For the method (a) the pristine t-MWCNT has been admixed with the a-terpineol ($C_{10}H_{18}O$, (R)-2-(4-Methyl-3-cyclohexenyl)-2-propenol) solution and sonicated for ~12 hours, followed by the stirring with the ethyl cellulose for about 1 hour. These samples were designated as the raw-composites. In method (b), the mechanical dispersion route (MDR) has been adopted, in which the pristine t-MWCNTs along with the ethyl cellulose and terpineol has been premixed for a period of 10–15 minutes [12].

Following this, the composite has been subjected to the three-roll milling (calendering) process (Kyong Yong Machinery Co. Ltd. Korea) for a short dwell (residence) time 15 minutes to disperse the pristine t-MWCNTs in the polymer composite under the enormous shear forces. The viscosity of the composite has been monitored constantly. In method (c), the chemical dispersion route (CDR) has been adopted, in which the pristine t-MWCNTs have been functionalized by refluxing them in the hydrogen peroxide (H_2O_2) solution for 12 hours and filtered. The filtered nanotubes followed

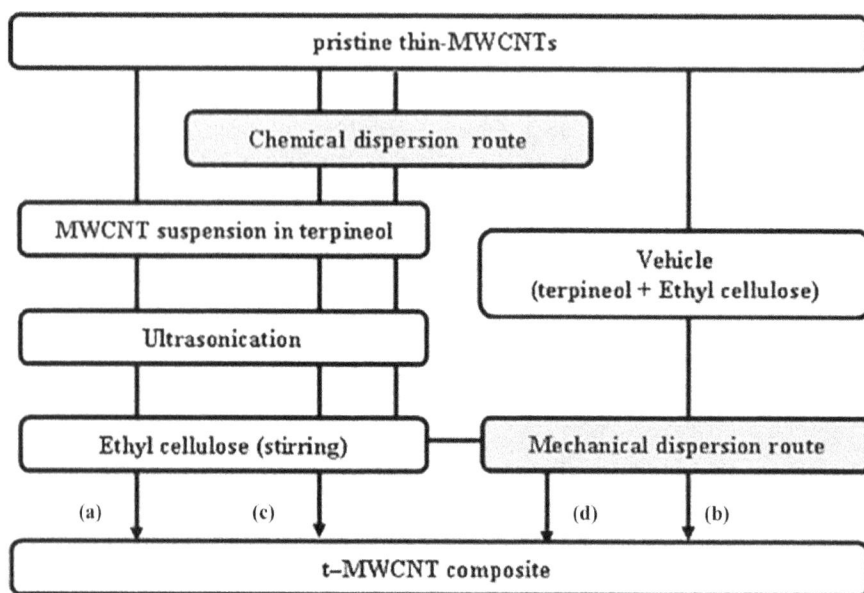

FIGURE 9.12 Synthesis scheme for the thin multiwalled carbon nanotube composite.

the same process sequence as described in the method (a). These samples were designated as the CDR-composites. In method (d), initially, the chemical functionalization of the nanotubes has been carried out followed by the three-roll milling of the composite at the end of process sequence. Since, the method (d) was the combination of the method (b) and (c), the samples synthesized by this method were designated as the CMDR-composites. Thus, the composite have been synthesized using the pristine, functionalized, shear mixed and combination of functionalized/shear mixed t-MWCNTs [29]. In all the cases, the weight content of the t-MWCNTs has been kept constant, i.e. 0.1 wt%. The obtained composite has been screen-printed (printing area: 1×1 cm^2) on the indium tin oxide (ITO)-coated (thickness 200 nm) glass substrate and kept in the oven for 30 minutes for the drying. The volatile organic components in the t-MWCNT composite cathode layers have been removed by curing them at ~450°C for 10 minutes. A large amount of samples has been prepared and kept ready for the optical microscopy studies. The aggregation of t-MWCNTs in the composite has been studied by using an optical microscope (OM, Olympus BH-2) equipped with an image analyser. The photograph has been snapped at a constant magnification of 50 X. For the field emission measurements, initially, the t-MWCNT composite cathode layers have been subjected to the tape activation treatment to protrude the nanotube emitters perpendicular to the surface of the substrate. The field emission measurements have been carried out in the diode configuration with a fixed inter-electrode distance ~200 µm. The applied voltage sweep of 50–700 V has been employed and all measurements have been carried out at a vacuum level of ~10^{-7} Torr [12,29].

9.1.1.5.1 Dispersibility of Pristine and H$_2$O$_2$-t-MWCNTs: Chemical Dispersion Route

Figure 9.13a is the recorded digi-cam photograph for: (i) pristine t-MWCNTs; and (ii) H$_2$O$_2$-treated t-MWCNTs suspended in the terpineol solution. The photograph has been snapped after about 20 days. From Figure 9.13a, a considerable amount of difference has been observed for the suspension of the pristine and H$_2$O$_2$-treated t-MWCNTs in the solution. The nanotubes have observed to be gradually sedimented from upper part of the solution within 30 minutes, after dispersing the raw t-MWCNTs in the terpineol solution. However, the suspension of the H$_2$O$_2$-treated t-MWCNTs in terpineol has observed to be dispersed and stabilized better, even more than 20 days. The effect of H$_2$O$_2$ treatment on the t-MWCNTs has been studied using the FTIR technique. In general, the characteristic frequencies of the vibrational oscillations for the various molecular groups in a functional composite lie in the infrared region and can easily be determined by this method. Figure 9.13b shows the recorded FTIR spectra for the (i) pristine t-MWCNTs and (ii) H$_2$O$_2$-treated t-MWCNTs [12].

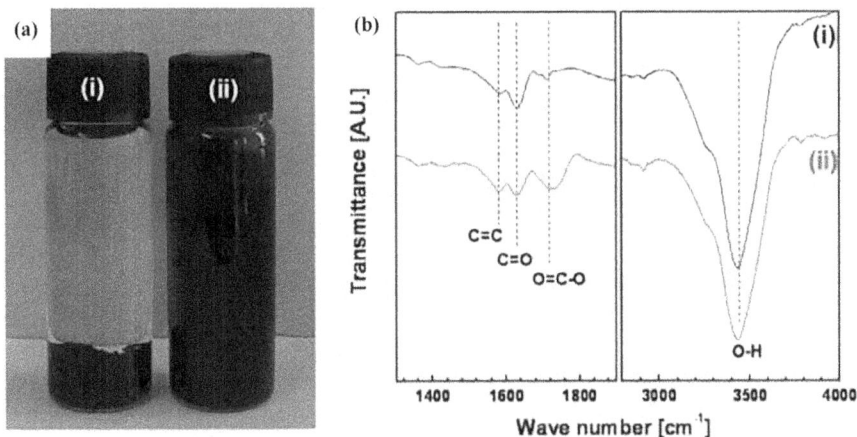

FIGURE 9.13 (a) Recorded digi-cam photograph for (i) pristine t-MWCNTs and (ii) H$_2$O$_2$-treated t-MWCNTs dispersed in terpineol (t-MWCNTs, 0.1 wt%). The photograph has been recorded after 20 days. (b) Recorded Fourier transform infrared (FTIR) spectra for: (i) pristine t-MWCNTs and (ii) H$_2$O$_2$-treated t-MWCNTs.

The spectra have been recorded in the transmittance (%) mode over the wave number range of 500–4000 cm^{-1}. The FTIR spectrum in: (i) shows characteristics bands at 1580, 1630, 1725 and 3433 cm^{-1}, which corresponds to the C=C, -C=O, -O=C=O and -OH vibrations, respectively [10]. It is important to note that, t-MWCNTs (Iljin-Nanotech, Korea) samples used in this study were synthesized by the chemical vapour deposition (CVD) synthesis. Such nanotubes, generally, contained amorphous carbon and (metal) catalyst which was removed by the thermal oxidation and acid treatment. As a result, the bands corresponding to –C=O, -O=C=O and -OH vibrations, in case of the pristine t-MWCNTs, has been observed. However, in spectrum (ii) the overall transmittance and the nature of bands at 1580 and 1725 cm^{-1} has been modified due the H_2O_2 treatment. The observed modifications, in the spectrum indicate the molecular vibrations associated with the -COOH functional group attached to the side walls of the nanotubes. The functionalization of nanotubes could positively contribute to the dispersion of the t-MWCNTs [12,29].

From Figure 9.13a, one can see that the colloidal property of -COOH functionalized t-MWCNT suspension in terpineol solution is improved significantly as compared to the pristine t-MWCNT suspension in terpineol. The improved dispersibility and stability of the t-MWCNTs in the terpineol could be attributed to the high viscosity of the terpineol. The terpineol has a ring structure attached to the branched -OH group. However, the polar -COOH functional present on the t-MWCNTs may interact with the terpineol via π–π* stacking mechanism. The π–π* interaction between –COOH-terpineol and presence the bulky -OH group in both structures may hinders the hydrogen bonding tendency of the t-MWCNTs with the medium. As a result, the Lewis basicity (i.e. availability of a free electron pair) of terpineol medium is stimulated towards the -COOH functionalized t-MWCNTs [7,16,29]. However, this channel is absent for the pristine t-MWCNTs. Thus, surface-functionalized tubes showed good degree of dispersion and stabilization in the terpineol medium.

Moreover, it is also crucial to functionalize t-MWCNTs without significantly affecting their properties. In the present study, the functionalization of t-MWCNTs has been carried out using the H_2O_2 reagent which is "mild" as compared with the HNO_3/H_2SO_4. The nanotube etching (shortening) rate for the H_2O_2 functionalization process is 130 nm/h, whereas, for the HNO_3/H_2SO_4 acid treatment, the nanotube etching rate is 200 nm/h [16]. Retaining a high aspect ratio of the t-MWCNTs is of special interest towards an improvement of the field emission performance of a CNT. In general, the mean field enhancement factor, cm and turn-on-field depend on the aspect ratio of the CNTs, for grater aspect ratio CNTs one could achieve higher magnitude of mean field enhancement factor, γ_m and lower turn-on-field [7]. Thus, the precaution has been taken not to cut short t-MWCNTs by the functionalization process.

9.1.1.5.2 Three-Roll Milling (Calendering) Process: Mechanical Dispersion Route

The mechanical interlacing of the nanotubes presents significant processing challenges. A high degree of nanoscale dispersion is desired but large aspect ratios are preferred for property enhancements. For t-MWCNTs, adequate shear stresses must be applied so that the nanotubes untangle and disperse uniformly in the ethyl cellulose matrix but not damage the nanotubes by substantially reducing their length. The three-roll milling approach offers nearly pure shearing as compared with the other types of milling process which relies on compressive impact as well as shear stress [18]. Figure 9.14, schematically, shows the general configuration of a three-roll mill. The intense shear mixing occurs in the region between the feed and centre rolls. The inter roll narrow gap and mismatch in angular velocity of the adjacent rolls results in high shear rates. Due to surface tension, the composite material flows under and over the adjacent rollers and is collected by placing a collector blade in contact with the apron roll. Within a short residence time 15 minutes, the three-roll mill yield in locally very high shear forces. In the present work, the calendering process, for dispersing t-MWCNTs, has been carried out by utilizing a commercially available laboratory scale three-roll mill, consisting of three stainless steel rolls that are 80 mm in diameter. The mismatch between the angular velocity of adjacent rolls has been fixed where the centre roll rotates three times faster than the feed roll ($\omega_2 = 3\ \omega_1$) and the apron roll rotates three times faster than

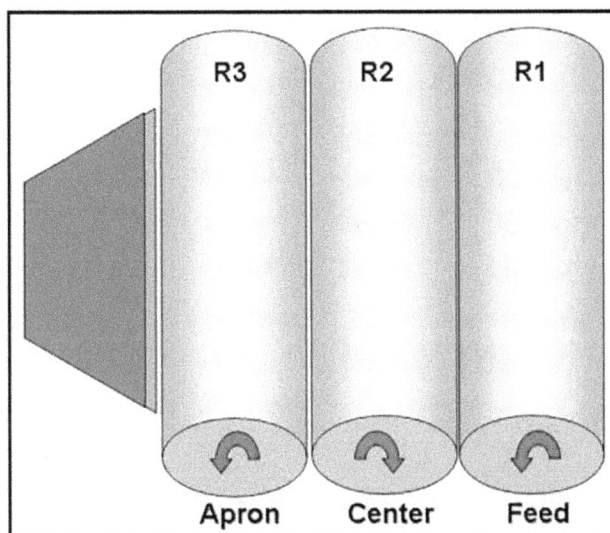

FIGURE 9.14　Schematic diagram showing the general configuration of a three-roll mill.

the centre roll ($\omega_3 = 3\ \omega_2 = 9\ \omega_1$). The mill setting has been controlled electronically and enables precise monitoring of the process. Comprehensive studies have been made of the effects of roller gap setting in the calendaring process on the structure and multi-functional properties of CNT nanocomposites [29]. The gap setting is adjusted accurately by the pressure gauge control system to maintain a specified value at 20 μm.

Figure 9.4a–d shows the four typical optical microscopy images recorded for the composite samples prepared by the methods a–d. The dark spots have been observed for the raw-composites, as seen in the photograph (a). The spot size and density is observed to be varied locally. This indicates that, a large amount of t-MWCNT aggregation has been occurred in the raw-composite. For the MDR-composite, the spot size as well as the areal density of the spots is reduced significantly (photograph (b)). This indicates that, dispersion of the t-MWCNTs is uniform in the composite. The achieved dispersion reveals that, the three-roll milling could effectively disperse and untangle the nanotubes in the ethyl cellulose matrix. However, one can still observe a small amount of t-MWCNT aggregation spots as seen in the photograph (b). Moreover, photograph (c), recorded for the CDR-composite shows the aggregation of nanotubes in the composite as well. In contrast, the photograph (d) shows uniform dispersibility of the t-MWCNTs in the CMDR-composite. The observed reduction in the t-MWCNT agglomeration indicates that, the nanotubes maintain their dispersibility even after mixing them in the ethyl cellulose polymer. It seems that, the presence of O- and EtO-legend in the backbone of the ethyl cellulose offers stearic hindrance to the bulky -COOH functional present on the side wall of the t-MWCNTs. In addition to primary chemical dispersion, further agglomeration of the nanotubes might have hindered, during the mechanical dispersion process [7,16,29] (Figure 9.15).

The composite experiences the knead-vortex between the rolls in which the final untanglement and dispersion of the nanotubes occurs in the area between the rolls. Thus, the shear mixing could provide good dispersion of the nanotube in the composite due to untanglement from the shear force. In this process, the degree of disentanglement depends on the type of polymer composite and the surface modifications of the nanotube fillers [19]. Thus, the -COOH surface modified CNTs positively participates in the mechanical dispersion process. As result, the composite obtained by the method (d) has a high dispersibility of the t-MWCNTs as compared with the composites synthesized by other routes. The field emission parameters have been measured for the synthesized composites [16,29].

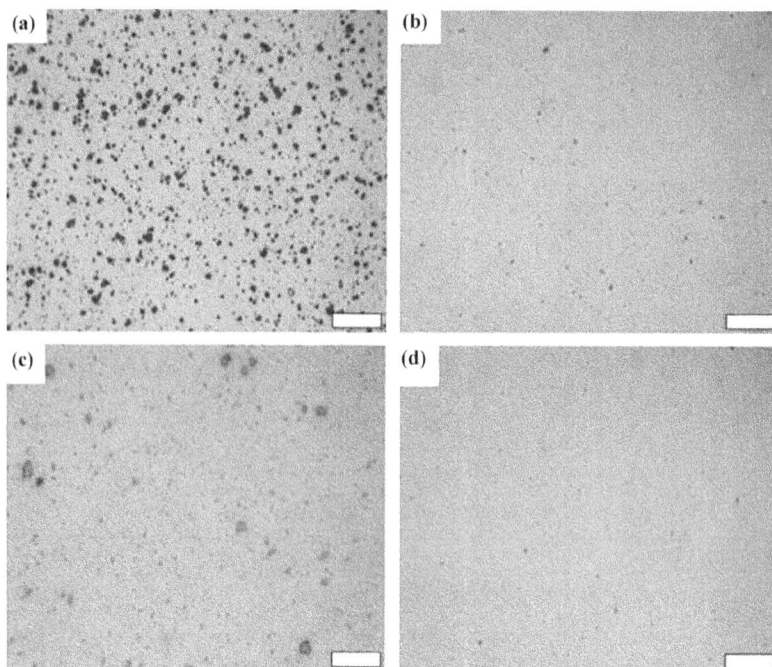

FIGURE 9.15 Recorded optical micrographs for (a) raw-composite, (b) MDR-composite, (c) CDR-composite and (d) CMDR-composite (scale bar: 20 μm).

9.1.1.5.3   FE Parameters for the t-MWCNT Composite

Figure 9.16 is the plot of variation in the measured current density (mA/cm²) as a function of applied electric field (V/μm) for the composites prepared by the methods (a)–(d). Each plot is the average of five recorded profiles. Inset shows the corresponding Fowler–Nordheim (F–N) plot for log $(I–V^{-2})$ as a function of V^{-1}. From Figure 9.16 one can see that, the magnitude of the current density monotonically increases with increase in the applied electric field, however, no major variations in the current density has been observed up to 2 V/μm, for the plots (a)–(d). Moreover, the J–E behaviour typically follows the F–N tunnelling mechanism at low electric field, from 2 to 2.5 V/μm, and current densities. With subsequent increase in the field, from 2.5 to 3 V/μm, and increased current density (>0.5 mA/cm²), the plot (d) exhibit a marked current density variation that sharply deviates from the plots (a)–(c). With further increase in the applied electric field, from 3 to 3.5 V/μm, and current density, the J–E characteristics yield to the electron tunnelling behaviour governed by the F–N theory. For each plot, the magnitude of the turn-on-electric field (measured in V/μm) has been estimated at a current density ~10 V/μm, whereas the magnitude of measured current density, J, is coated at 3 V/μm. Table 9.1 shows the field emission parameters enlisted for the composite synthesized by various methods. One can see that, for the raw-composite, the turn-on-field is 1.698 V/μm and decreased up to 1.649 V/μm for the MDR-composite [16,29].

The magnitude of the turn-on-field lowered down further, 1.574 V/μm, for the CDR-composite. And the turn-on-field has achieved the lowest value 1.374 V/μm for the composite synthesized by the combination of the chemical and mechanical dispersion routes, i.e. the CMDR-composite. Thus, significant improvement in the turn-on-field parameter has been achieved for the CMDR-composite. Furthermore, the magnitude of current density, J, (measured at 3 V/μm) is 467.8 μA/cm² for the raw-composite and gradually observed to be increased from 613.7 to 824.4 μA/cm² to finally 2030.2 μA/cm² for the MDR-, CDR- and CMDR-composites, respectively. As compared with the current density, J, of the raw-composite, the magnitude of current density is almost twice and 4.5 times for the

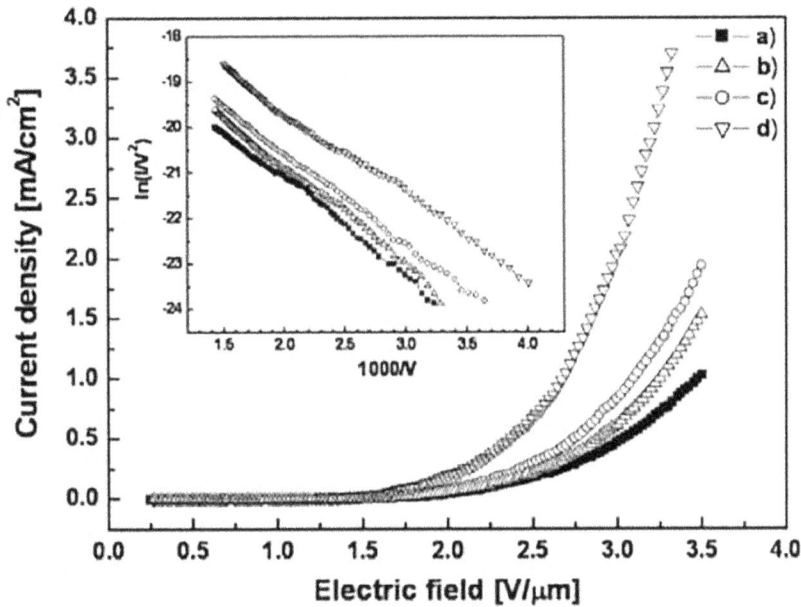

FIGURE 9.16 Variation in field emission current density (measured in mA/cm²) as a function of applied electric field (V/μm) for (a) raw-, (b) MDR-, (c) CDR- and (d) CMDR-composites. Inset is the corresponding Fowler–Nordheim plot.

CDR- and CMDR-composite, respectively. In general, it is difficult to extract physically meaningful parameters from a fit of the F–N equation to the I–V plots [17]. This is especially true for I–V plots obtained from randomly distributed multiple CNT emitting tips such as those on the t-MWCNTs-composite cathode layers. However, a qualitative estimation can be made by computing the mean field enhancement factor, cm, for the CNT field emitters, using the generally accepted F–N theory [13] in terms of the voltage-based notion.

From Figure 9.16, a F–N plot has been determined (shown as inset) and its slope has been measured in order to estimate the mean field enhancement factor, cm, for the raw – as well as the other composites. The detailed process is described elsewhere [22]. The slope correction factor, s, has been taken as ~1 and the local work function, U, of the nanotubes was assumed to be a constant ~5 eV. It can be seen that, the FN equation itself is degenerate, because it describes two mathematical parameters (offset a and slope b) and three canonical physical quantities such as, curvature of the emitter tip, d, conversion factor, b, and work function, U. Thus, only a few comments could be made on the details of the emission physics from the plot of variations in current density with the applied electric field, since the electronic properties of the tips cannot be easily separated from the geometric field enhancement factor. Nevertheless, one could extract an average of the distribution of the field emission parameters. The mean field enhancement factor, cm, computed for the raw-composite is 6881.7 and decreased down to 6760.3 for the MDR-composite and thereafter, a gradual increase has been observed from 7123.9 to 7851.4 for the CDR- and CMDR-composite, respectively. Thus, the computed mean field enhancement factor, cm, increases by 14% for the CMDR-composite as compared to the raw cathode layers. For the raw composite, the computed geometric area participating in the electron field emission process is $5.9 \times 10^{-18} \, m^2$ and increases gradually from 8.5 to $8.7 \times 10^{-18} \, m^2$ and up to $1.2 \times 10^{-18} \, m^2$ for the MDR-, CDR-, and CMDR-composites, respectively. The active area of the field emission is increased approximately two times for the CMDR-composites as compared with the raw-composite. In general, field emission parameters, measured for array of emitters, depend on the position of the individual emitter on the cathode surface. One can find that superior field emission properties have been obtained for the CNT composite synthesized by the combination of the chemical and mechanical dispersion routes. This indicates that, CNTs that were dispersed well in the nanocomposite when printed onto the cathode layers have an optimum position to give favourable field emission properties. It is interesting to note that, the weight content of the t-MWCNTs in the composite was 0.1 wt% which is order of magnitude smaller as compared with the other reports [11] presented before [29].

9.2 TIO$_2$-COATED CNTs: DYE-SENSITIZED SOLAR CELLS

Dye-sensitized solar cells (DSCs) have been attracting considerable attention because of their high efficiency, simple fabrication process and low production cost. Cost effectiveness is an important parameter for producing DSCs as compared to the widely used conventional Si-solar cells [34]. Moreover, enhanced dye-sensitized solar cell efficiency would provide enormous economic advantages [35] and reference cited therein. Recently, TiO$_2$ nanoparticles have been used as a working electrode for DSCs due to their higher value of efficiency than any other metal oxide semiconductor. However, the highest conversion efficiency so far reported for this device is ~10% under air mass (AM) 1.5 (100 mW/cm^2) irradiation when liquid electrolytes containing I^-/I_3^- redox couples was used as conjunction [36,37]. Because, photo-generated charge recombination should be prevented for enhanced efficiency, solely enlarging the oxide electrode surface area is not sufficient. Strategies to enhance efficiency include the promotion of electron transfer through film electrodes and the blockage of interface states lying below the edge of conduction band.

Interface states facilitate recombination of injected conduction band electrons with I_3^- ions. The efforts have been made to improve the conversion efficiency by modifying TiO$_2$ film. CNTs are remarkable materials, which are being widely studied because of their extraordinary electronic and mechanical properties. Polymer composites with CNTs have recently been investigated for improved electrical conducting layer, optical devices and high strength composites. A composite of poly(p-phenylene vinylene) with CNTs in a photovoltaic device showed good quantum efficiency, owing to the formation of a complex interpenetrating network with the polymer chains [35]. CNTs also conferred electrical conductivity to metal oxide nanocomposites [35]. However, only few reports have been found in the literature where CNTs were used in TiO$_2$ films of DSCs, despite of their expected potential to enhance solar energy conversion efficiency due to favourable electrical conductivity. Thus, we introduce CNTs in DSCs to improve the electrical conductivity of TiO$_2$ film. In this study, we incorporated TiO$_2$-CNTs in porous TiO$_2$ films. As a result, the value of the J_{sc} of DSSCs was increased. To prevent leakage current in device, thin passivated layer was prepared between the transparent conducting glass (FTO) substrate and porous TiO$_2$ film.

9.2.1 Preparation and Production

Multi-walled CNTs (MWCNTs, supplied by ILJIN Nanotech) synthesized by the thermal chemical vapour deposition (thermal CVD) method were used in the present study. The raw powder contains MWCNTs of diameter 25 nm, amorphous carbon, and carbon-encapsulated metal nanoparticles. MWCNTs were oxidized in a hydrogen peroxide (H$_2$O$_2$) solution under ultrasonication condition for 24 hours at the temperature 50°C to produce finely dispersed MWCNTs terminated with carboxylic acid groups. The resulting solution was filtered by a polytetrafluoroethylene (PTFE) membrane with pore size 1 µm. At this step, the carbonaceous impurities were removed from the as-grown MWCNTs. Raman spectrometer and Fourier transform infrared spectrometer (FTIR) were used to identify the formation of carboxylic acid groups on MWCNTs. The Sol–Gel solution (SGS) was prepared using titanium tetraisopropoxide Ti(OPri)$_4$, isopropanol (IPA), nitric acid (HNO$_3$) and distilled water (H$_2$O). The weight ratio for the SGS preparation is kept as 1:10:1:0.2 for Ti(OPri)$_4$:IPA:H$_2$O:HNO$_3$ [35]. The solution was reflux at the temperature 80°C for a period of 1 hour, using a magnetic stirrer. For each sample, 1 g of MWCNTs were mixed with 100 mL of SGS and stirred in close vials for 3 hours. The impregnated MWCNTs were separated from the solution by filtration process [35].

To obtain TiO_2-CNTs, the filtrated nanotubes were dried in an oven at 80°C for 1 hour under atmospheric conditions followed by thermal treatment at 450°C for 1 hour. The passivation layer was introduced between the fluorine-doped SnO_2 (FTO) substrate and porous TiO_2 layer. To obtain a uniform and flat surface, the Sol solution ($Ti(OPri)_4$:IPA: HNO_3= 1:10:0.2) was spin coated. After being dried in air, the passivation layer was annealed for 1 hour at 500°C, under atmospheric conditions. Scanning electron microscope (SEM) measurement revealed the thickness of the passivation layer 70 nm. This solution was also reflux at 80°C for 1 hour, using a magnetic stirrer before spin coating. Porous TiO_2 films were prepared by coating a passivated transparent conducting glass substrate (Solaronix; fluorine-doped SnO_2 over-layer; sheet resistance: 17 Ω/sq) with viscous slurry of TiO_2 powder and TiO_2-CNTs dispersed in an aqueous solution.

Initially, TiO_2-CNTs (0.1–0.3 wt.%) were added in IPA and sonicated during 1 hour to obtain well dispersed solution of TiO_2-CNTs in IPA. Commercially available TiO_2 powder (0.5 g, P25, Degussa) and IPA included TiO_2-CNTs (1 g) were ground in a mortar with distilled water (1 g), polyethylene glycol (0.1 g, Aldrich, MW 2000) and polyethylene oxide (0.1 g, Aldrich, MW 100,000) to break up the aggregate into a dispersed paste. Adhesive tape was placed on the edges of the conductive glass to form a guide for spreading the slurry using a glass plate. The film thickness was controlled by the amount of water in the slurry and by the thickness of adhesive tape. After being dried, the porous TiO_2 film mixed with TiO_2-CNTs was annealed for 1 hour at 500°C, under atmospheric conditions. The film thickness was 10–15 um and measured with a Tencor Alpha-Step profiler. Following this process, the resulting surface-modified TiO_2 films were immersed in absolute ethanol containing 0.3 mM $[RuL_2(NCS)_2]\cdot2H_2O$ (L=2,2'-bipyridine-4,4'-dicarboxylic acid; Solaronix) for 12 hours at room temperature. The dye-covered electrodes were then rinsed with absolute ethanol and dried. Pt counter electrodes were prepared by spreading a drop of 5 mM hexachloroplatinic acid (Fluka) in IPA on the FTO glass followed by heating at 400°C for 30 minutes in air.

The Pt electrode was placed over the dye-coated electrode, and the edges of the cell were sealed with 0.5-mm-wide strips of 100-μm-thick Surlyn (Dupont, grade 1702). The redox electrolyte consisted of 0.8 M lithium iodide (LiI), 40 mM iodine (I_2) and 0.2 M 4-tertbutylpyridine (TBP) in acetonitrile was introduced into the cell through one of the two small holes drilled in the counter electrode. The holes were then covered and sealed with small squares of microscope objective glass and Surlyn. To analyse the crystallinity of TiO_2-CNTs, X-ray diffraction (XRD) data was recorded. Investigations on film morphology were carried out by atomic force microscope (AFM) and field emission scanning electron microscope (FE-SEM). Current density–voltage (J–V) characteristics were recorded using Keithley (model 2400) as a source measure unit, which was connected between the FTO and Pt electrodes under an illumination of a 300 W Xe lamp (ILC technology Inc.). The voltage was scanned from −0.2 to 0.8 V in steps of 0.05 V. The incident light intensity (100 mW/cm^2) was calibrated using a Newport 818 UV photodiode detector [36].

9.2.2 Morphology and Solar Cell Performance

It is known that pristine MWCNTs have hydrophobic surface and poor dispersion stability. To avoid these problems the pre-treatment of MWCNTs is needed for many applications. Carboxylic acid groups could be generated easily by oxidation of MWCNTs, by H_2O_2 treatment. It is a less-destructive and mild oxidation method for removing impurities as well as forming carboxylic acid groups on nanotubes. H_2O_2 solution is a mild acid and easy to handle. Also, reaction gases such as CO_2 and H_2O are non-toxic and could be released safely during the oxidation processing [37,38].

H_2O_2-treated MWCNTs have a hydrophilic surface. The carboxylic acid groups on the surface of MWCNTs have a polar covalent bonding by the electronegativity difference. Thus, we could consider that H_2O_2-treated MWCNTs have a generally negatively charged surface. The negatively charged surface of MWCNTs enhances the stability of dispersion. These H_2O_2-treated MWCNTs were well dispersed in SGS (Sol–Gel solution). After reaction with SGS, morphology of TiO_2-coated MWCNTs, recorded by SEM, is shown in Figure 9.17A.

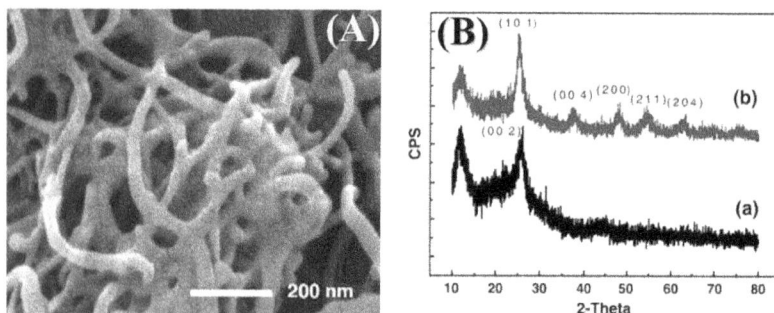

FIGURE 9.17 (A) SEM micrograph for TiO$_2$-coated MWCNTs (TiO$_2$-CNTs), (B) XRD spectra for (a) pristine MWCNTs and (b) TiO$_2$-CNTs annealed at ~450°C, under atmospheric conditions.

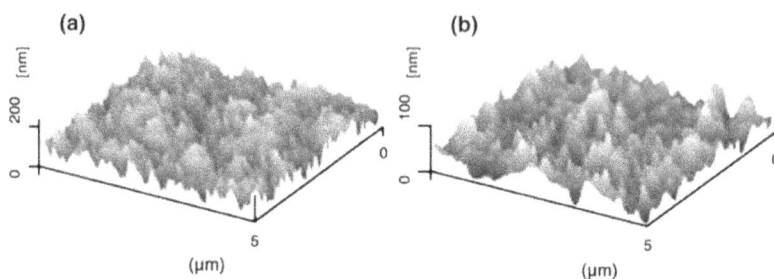

FIGURE 9.18 AFM images (tapping-mode) for (a) FTO glass and (b) TiO$_2$ passivated FTO glass.

The coated samples were thermally treated at 450°C in order to crystallize anatase on the nanotubes surface without any damage to MWCNTs. The recorded XRD spectra for (a) pristine MWCNTs and (b) thermally treated TiO$_2$-CNTs are shown in Figure 9.17B. The most intense peaks at (002) reflection (Profile (a)) corresponds to the MWCNTs and overlaps significantly with (101) band (Profile (b)), which corresponds to anatase TiO$_2$. The other peaks present in Profile (b) were attributed to the anatase form of TiO$_2$. It indicates that the surface of MWCNTs was covered with anatase form of TiO$_2$. Normally, the passivation layer consisted of materials of same or lower conduction band than that of porous TiO$_2$ film. The passivation layer made by spin coating of TiO$_2$ sol solution has a non-porous thin film. Thus, the passivation layer improves the property of interface surface and enhances the conversion efficiency by reducing the recombination of electron-hole pairs. Also, the passivation layer increases the adhesion property between the FTO glass and porous TiO$_2$ layer, reduces the leakage current by preventing direct contact of electrolyte and FTO glass [36–38].

Figure 9.18 shows the AFM images for (a) pristine FTO glass and (b) FTO glass passivated with TiO$_2$ thin film. For raw FTO glass, the surface is very rough with values of r.m.s. roughness (σ) of 32.1 nm. The passivated FTO glass shows σ ~16.1 nm, i.e. the morphology is found to be smoother than that of un-passivated FTO glass.

Figure 9.19a shows appearance of large amount of cracks on the surface of TiO$_2$ film. One can see that, these large sized islands consist of clusters of TiO$_2$ nanoparticles on there. The close circle along with an arrow in Figure 9.19a indicates the details of the cluster morphology as shown by close circle in Figure 9.4b. Thus, it can be seen from Figure 9.19b, that cluster consists of large amount of TiO$_2$ nanoparticles. Moreover, Figure 9.19b indicates that the clusters of TiO$_2$ particles entirely cover the aggregated TiO$_2$-CNTs. As a result, no TiO$_2$-CNTs are observed on the surface of porous TiO$_2$ film. However, it is noteworthy that, the amount of cracks developed on TiO$_2$ films, prepared by commercially available powder (P25) only, is found to be marginal. This phenomenon suggests that TiO$_2$-CNTs play the role of nucleation sites for clustering TiO$_2$ nanoparticles on the surface of the film.

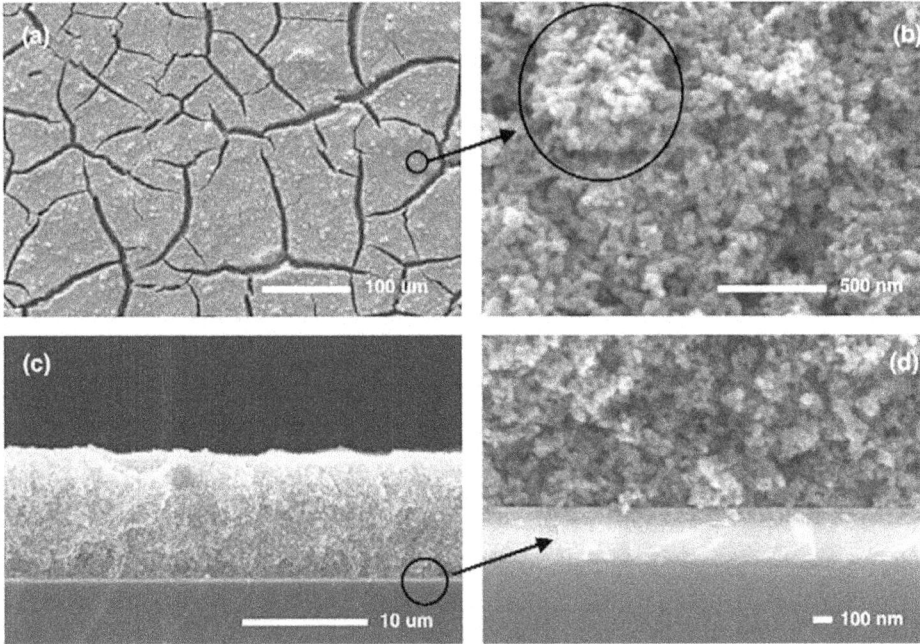

FIGURE 9.19 SEM images for (a) development of cracks on surface of porous TiO_2 electrode, (b) the details of TiO_2 cluster, which is marked by close circle in (a), indicates that TiO_2 particles entirely covers TiO_2-CNTs (content 0.1 wt%), (c) the thickness of the porous TiO_2 film and close circle with an arrow indicates details of TiO_2 passivation layer shown in (d).

Moreover, with increase in amount of TiO_2-CNTs, the number of cracks on the surface of the films is increased subsequently. It is thought that, the cracks generated on the surface could be reducing the number of adsorption sites on TiO_2 film as well as causing the discrimination in the conversion efficiency of DSCs. The thickness of porous TiO_2 film is shown in Figure 9.19c and close circle with an arrow indicate the details of the passivation layer as shown in Figure 9.19d. As described earlier, the thickness of porous TiO_2 film (10 μm) was controlled by an adhesive tape. In contrast, the thickness of passivation layer (70 nm) was controlled by the spin coating speed [36].

Figure 9.20 shows the $J–V$ characteristics for (a) as prepared TiO_2 films, (b) 0.1 wt% TiO_2-CNTs, (c) 0.2 wt% TiO_2-CNTs, and (d) 0.3 wt% TiO_2-CNTs, on TiO_2 films and Table 9.2 enlists other parameters of solar cells. The value of open circuit voltage, V_{oc}, is increased by ~6% from 0.60 to 0.63 V with subsequent increase in TiO_2-CNTs from 0 to 0.1 wt%. It has been observed that surface treatment usually increases the values of V_{oc} regardless of nature and characteristics of coated materials on electrode [36,39]. Furthermore, for 0.1 wt% TiO_2-CNTs, value of short circuit photocurrent density (J_{sc}) is found to be increased by ~60% from 8.49 to 13.5 mA/cm², when compared to as prepared TiO_2 electrode. However, J_{sc} decreases thereafter (from 13.5 to 9.69 mA/cm²) with subsequent increase in content of TiO_2-CNTs from 0.1 to 0.3 wt%. With increase in TiO_2-CNTs contents, the gradual decrease in J_{sc} is attributed to the increase in number of cracks on the surface of porous TiO_2 electrode. Consequently, the addition of TiO_2-CNTs enhances the electro-conductivity of porous TiO_2 electrode, but decrease the adsorption site for ruthenium dye to making a crack in TiO_2 film. Thus, TiO_2-CNTs (0.1 wt.%) contained TiO_2 electrodes have higher value of conversion efficiency ~50% higher compared with as prepared TiO2 electrode of cell (Table 9.1). However, subsequent increase in concentration of TiO_2-CNTs (from 0.1 to 0.3 wt%) does not help to increase the value of conversion efficiency further [35–38].

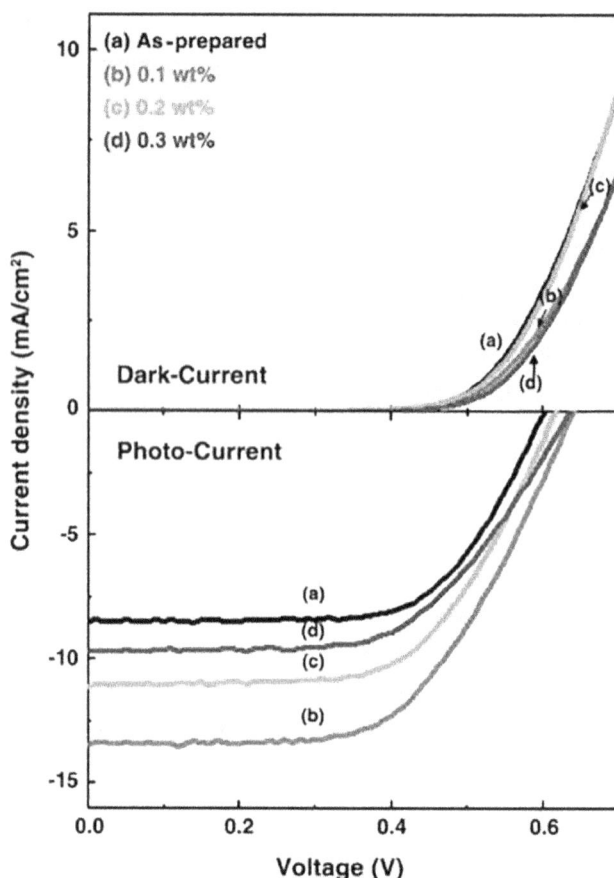

FIGURE 9.20 *J–V* characteristics for (a) as-prepared TiO$_2$ film, (b) 0.1 wt% TiO$_2$- CNTs, (c) 0.2 wt% TiO$_2$-CNTs, and (d) 0.3 wt% TiO$_2$-CNTs, on TiO$_2$ films.

TABLE 9.2

Photocurrent J–V Parameters for CNTs Incorporated TiO$_2$ Electrodes in DSCs

Composition of TiO$_2$ CNTs	J_{sc} (mA/cm²)	V_{oc}(V)	Fill factor	Efficiency (%)	Cell area (cm²)
0 wt.%	8.49	0.60	0.65	3.32	0.33
0.1 wt.%	13.5	0.63	0.59	4.97	0.36
0.2 wt.%	11.1	0.61	0.61	4.14	0.33
0.3 wt.%	9.69	0.63	0.59	3.60	0.33

9.3   CNT-EMBEDDED NYLON NANOFIBRE BUNDLES BY ELECTROSPINNING

Electrospinning is one of the effective methods to produce polymer composites and their nanofibres, with diameters ranging from nm to μm. In this technique, the electrostatic force is exerted on a polymer drop emerging from the jet. The drop experiences the electrostatically driven bending instability and transforms into the thin nanofibres. The electrospinning set-up consists of a bipolar high voltage source, a syringe injector coupled to a needle (to carry the polymer fluid from the syringe to the spinneret) and a conducting collector, to obtain nanofibres. The nanofibres obtained by this method are non-woven and randomly oriented. Such fibres have limited number of applications.

However, the traditional fibre and textile industry requires single, continuous nanofibres or uniaxial fibre bundles, which are useful in opto-electronic devices. Fabrication of individual aligned electrospun nanofibres could be difficult because the polymer jet trajectory is in a very complicated three-dimensional whipping way caused by the bending instability rather than in a straight line. Efforts are being made in various research groups all over the world [40] to control the alignment of the spun fibres. In this work, MWCNTs/nylon nanofibres were spun by routine electrospinning technique. The spun fibres were collected on water surface, in water bath. It is observed that, the fibres floating on water, during their transfer to the rotating drum were transformed into the aligned bundles. This issue is investigated. The MWCNTs are functionalized by amine treatment process and the effect of MWCNT concentration on the electrical property of the CNTs/nylon nanofibres is studied.

Initially, to obtain a uniform dispersion, first the raw MWCNTs were functionalized [40] using sulfuric and nitric acid solution (95% H_2SO_4: 65% HNO_3=3:1). The solution was sonicated for 7 hours. After the acid treatment, amine group can be attached to the functionalized MWCNTs by a synthetic route using thionyl chloride ($SOCl_2$) followed by ethylenediamine. The acid-treated MWCNTs of 0.5 g was sonicated in 200 mL of $SOCl_2$ for 5 minutes for the dispersion. The mixture was refluxed for 24 hours at 50°–70°, filtered, and washed with anhydrous tetrahydrofuran (THF) to remove any unreacted $SOCl_2$. The filtrate was stirred in 200 mL of ethylenediamine for 3 days at 50~70 and filtered using a 1 μm membrane. After filtration process, the residue was washed anhydrous ethyl alcohol to remove residual ethylenediamine [24,40].

The functionalized MWCNTs were then dried at ~70° under atmospheric conditions. FTIR spectrometer was used to characterize the functionalized MWCNTs. The functionalized MWCNTs were dispersed in formic acid by sonicating for 1–3 hours. Subsequently, nylon was added and heating for dissolution in solvent until the viscosity of the solution attained its optimum level. The concentration of nylon in the formic acid was 15 wt%. Following this process, dispersed MWCNTs/nylon solution was obtained, with a variation in the weight of the nanotubes of 6–20 wt%. The MWCNTs/nylon solution was spun into fibres using an electrospinning system (NanoNC Co, Korea). The experimental set-up of the electrospinning process is shown in Figure 9.22a. A syringe was filled with the MWCNTs/nylon solution. The solution was transferred through a syringe needle at a flow rate of 0.1~0.5 mL/hr. The electrospun fibres were collected on a metal collector or glass plate. The inner diameter of the syringe needle was 0.25 mm and the distance between the syringe needle and the rotating metal drum was maintained at 10 cm. The potential difference between the MWCNTs/nylon solution loaded needle and collector was 15~20 kV. The obtained MWCNTs/nylon nanofibres

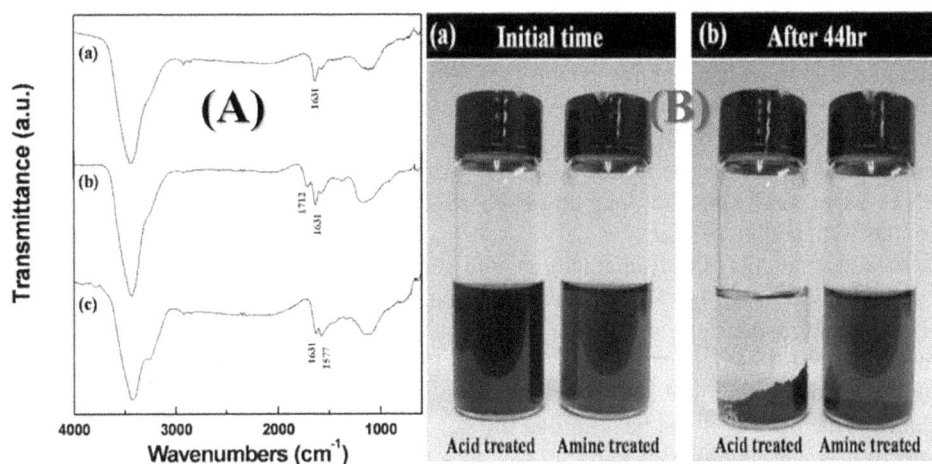

FIGURE 9.21 (A) FTIR spectra of MWCNTs. (a) purified MWCNTs, (b) acid-treated MWCNTs (carboxylated), and (c) amine functionalized MWCNTs, (B) the photographs of dispersed MWCNTs of 0.0002wt% in formic acid by sonication for 30 minutes. (a) Initial status and (b) after 44 hours.

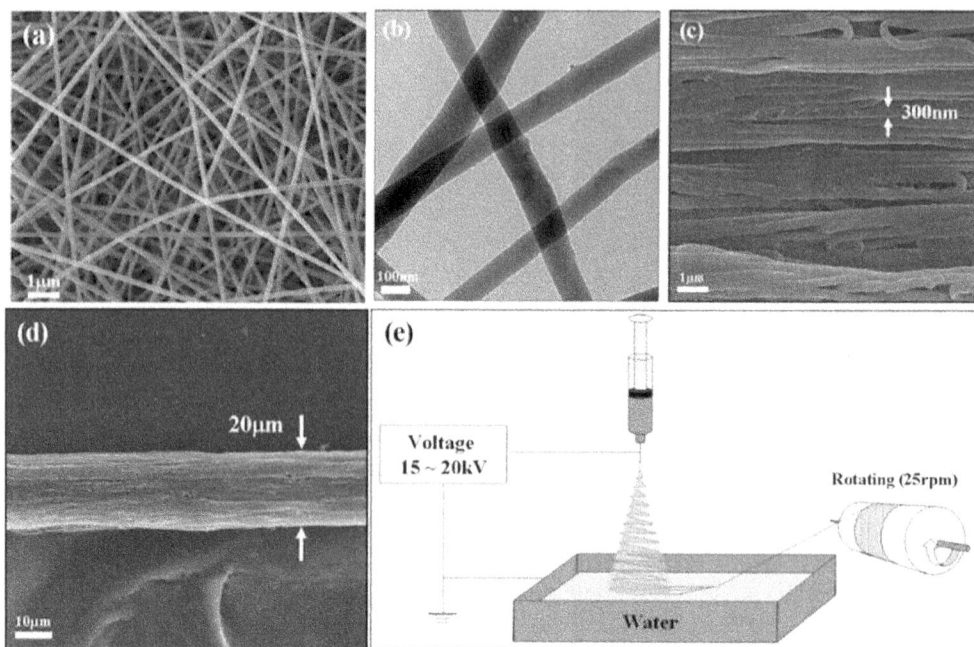

FIGURE 9.22 (a) and (b) are recorded SEM and TEM images of typical electrospun non-woven fibres, (c) and (d) shows aligned fibres bundle and its detail morphology, respectively, (e) schematic diagram for electrospinning process system to obtain aligned nanofibre bundles.

were observed by environment scanning electron microscopy (ESEM: XL30 ESEM-FEG, Philips) and transmission electron microscopy (TEM: JEM 3010, JEOL).

The amine functionalization of the MWCNTs was confirmed by FTIR. Figure 9.21A shows FTIR spectra result of functionalization of MWCNTs. The peak at $1631\,cm^{-1}$ is attributed to the MWCNTs C=C stretching mode associated with side wall attachment. After acid treatment of MWCNTs the peak was observed at $1712\,cm^{-1}$ which is assigned to the C=O stretch of the carboxylic (COOH) group. However, after amine treatment the $1712\,cm^{-1}$ peak was found to be disappeared and the peak at $1577\,cm^{-1}$ was observed instead of $1712\,cm^{-1}$ peak. The peak at $1577\,cm^{-1}$ indicates the N-H in-plane of amine groups [16]. This behaviour can be explained substitution of the -OH of carboxylic group for amine. Amine functionalized MWCNTs were more stable in the formic acid than other MWCNTs. Figure 9.21B shows dispersion of acid-treated MWCNTs and amine-treated MWCNTs in the formic acid after 44 hours in the same solvent. The amine-treated MWCNTs dispersion is stable for long times, whereas the acid-treated MWCNTs agglomerate and settle to the bottom after 44 hours.

The obtained nanofibres were characterized by SEM and TEM techniques. Figure 9.22a shows the recorded SEM image for MWCNTs/nylon nanofibres. It can be seen that these nanofibres are unwoven and randomly oriented in a plane. The diameter of the fibres was found ~200 nm. Figure 9.22b is the TEM image for the spun fibres. The concentration of the MWCNTs in fibre solution was 10 wt%. Figure 9.22c shows the aligned fibres in bundles and Figure 9.22d shows the details of the bundle. It can be seen that the bundle has diameter ~20μm approximately. It is noteworthy that, the diameter of fibre bundles was found to be dependent on the conditions of spinning process such as, RPM (velocity of drum collector).The spun fibres were collected on the surface of water, in a water bath. The fibres float on water due to their hydro-phobic property of nylon. It has been speculated that, randomly oriented fibres could be aligned into bundled during

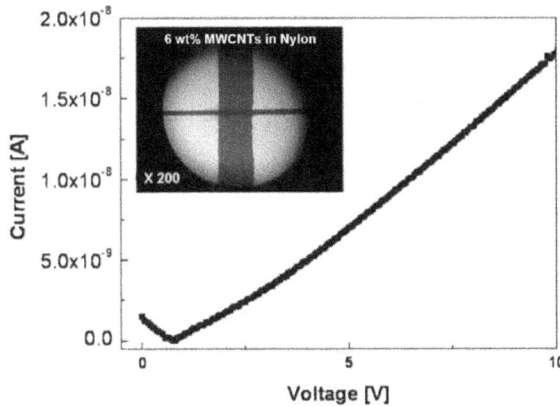

FIGURE 9.23   Current behaviour of electrospun fibre bundle with 6 wt% MWCNTs in nylon. The diameter of fibre bundle is 25 μm approximately. The gap is 230 μm between the electrodes.

their transfer to rotating drum. Figure 9.22e is schematic depiction of fibres transfer from water bath to rotating drum.

The *I–V* measurement was carried out on the fibre bundles of MWCNTs/nylon. The gold electrodes were used to conduct electrical measurements and the gap between the electrodes was kept ~230μm. We observed the conductivity in MWCNTs (6 wt%, 10 wt%, 20 wt%) doped fibre bundles. However, we could not find the systematic relationship as a function of MWCNTs contents because of their different diameters. Moreover, in *I–V* measurement, we have neglected the issues such as, contact resistance, cross section area of bundles. Figure 9.23 shows the *I–V* profile for fibre bundles which contains 6 wt% MWCNTs. The electrical conductivity was found to be improved significantly for MWCNTs/nylon fibre bundles as compared to pure nylon fibre bundles. We suggest that the enhancement in current is due to formation in conduction path due to presence of MWCNTs in the fibre [40]. In addition there are number of other applications, however, due to space crunch we only cited them for interested readers to refer [41–48].

REFERENCES

[1] Bader, S. D. "Methods of Experimental Physics Edited by Robert L. Park and Max G. Lagally (Vol. 22. Solid State Physics: Surfaces, Academic Press, 1985)." *MRS Bulletin* 11, no. 2 (1986): 2–3.

[2] Choi, W. B., D. S. Chung, J. H. Kang, H. Y. Kim, Y. W. Jin, I. T. Han, Y. H. Lee et al. "Fully sealed, high-brightness carbon-nanotube field-emission display." *Applied Physics Letters* 75, no. 20 (1999): 3129–3131.

[3] Spindt, C. A., I. Brodie, L. Humphrey, and E. R. Westerberg. "Physical properties of thin-film field emission cathodes with molybdenum cones." *Journal of Applied Physics* 47, no. 12 (1976): 5248–5263.

[4] Lee, Jong Hak, Seong ManYoo, Dong WookShin, Ji Beom Yoo, Jae. Hong. Park, Yu HeeKim, Prashant S. Alegaonkar et al. "Graphene composite using easy soluble expanded graphite: synthesis and emission parameters." In *2009 22nd International Vacuum Nanoelectronics Conference*, pp. 19–20. IEEE, 2009.

[5] Alegaonkar, P. S., J. H. Park, S. Y. Jeon, J. H. Shin, A. S. Berdinsky, and J. B. Yoo. "Field emission properties of expanded graphite composite." In 한국정보디스플레이학회: 학술대회논문집, pp. 775–777. 2007.

[6] Kolekar, S. K., S. P. Patole, P. S. Alegaonkar, J. B. Yoo, and C. V. Dharmadhikari. "A comparative study of thermionic emission from vertically grown carbon nanotubes and tungsten cathodes." *Applied Surface Science* 257, no. 23 (2011): 10306–10310.

[7] Patole, Shashikant P., Jong Hak Lee, Jae Hong Park, Seong Man Yu, V. G. Makotchenko, A. S. Nazarov, V. E. Fedorov et al. "Field emission properties of a graphene/polymer composite." *Journal of Nanoscience and Nanotechnology* 13, no. 11 (2013): 7689–7694.

[8] Berdinsky, A. S., A. V. Shaporin, J-B. Yoo, J-H. Park, P. S. Alegaonkar, J-H. Han, and G-H. Son. "Field enhancement factor for an array of MWNTs in CNT paste." *Applied Physics A* 83, no. 3 (2006): 377–383.

[9] Jung, Y. J., J. H. Park, S. Y. Jeon, S. J. Park, P. S. Alegaonkar, J. B. Yoo, and C. Y. Park. "Simple fabrication process and characteristic of a screen-printed triode-CNT field emission arrays for the flat lamp application." In 한국정보디스플레이학회: 학술대회논문집, pp. 1214–1218. 2006.

[10] Moon, J. S., P. S. Alegaonkar, J. H. Han, T. Y. Lee, J. B. Yoo, and J. M. Kim. "Enhanced field emission properties of thin-multiwalled carbon nanotubes: role of SiO_x coating." *Journal of Applied Physics* 100, no. 10 (2006): 104303.

[11] Nam, J. W., P. S. Alegaonkar, J. H. Park, J. B. Yoo, D. H. Choe, J. M. Kim, and W. S. Kim. "Field emission properties of plasma treated multiwalled carbon nanotube cathode layers." *Journal of Vacuum Science & Technology B: Microelectronics and Nanometer Structures Processing, Measurement, and Phenomena* 25, no. 2 (2007): 306–311.

[12] Park, J. H., P. S. Alegaonkar, S. Y. Jeon, and J. B. Yoo. "Carbon nanotube composite: dispersion routes and field emission parameters." *Composites Science and Technology* 68, no. 3–4 (2008): 753–759.

[13] Fowler, Ralph Howard, and Lothar Nordheim. "Electron emission in intense electric fields." *Proceedings of the Royal Society of London. Series A, Containing Papers of a Mathematical and Physical Character* 119, no. 781 (1928): 173–181.

[14] Diamond, William T. "New perspectives in vacuum high voltage insulation. I. The transition to field emission." *Journal of Vacuum Science & Technology A: Vacuum, Surfaces, and Films* 16, no. 2 (1998): 707–719.

[15] Forbes, Richard G. "Field emission: new theory for the derivation of emission area from a Fowler–Nordheim plot." *Journal of Vacuum Science & Technology B: Microelectronics and Nanometer Structures Processing, Measurement, and Phenomena* 17, no. 2 (1999): 526–533.

[16] Bonard, Jean-Marc, Christian Klinke, Kenneth A. Dean, and Bernard F. Coll. "Degradation and failure of carbon nanotube field emitters." *Physical Review B* 67, no. 11 (2003): 115406.

[17] Houston, J. M. "The slope of logarithmic plots of the Fowler-Nordheim equation." *Physical Review* 88, no. 2 (1952): 349.

[18] Burgess, R. E., H. Kroemer, and J. M. Houston. "Corrected values of Fowler-Nordheim field emission functions v (y) and s (y)." *Physical Review* 90, no. 4 (1953): 515.

[19] Edgcombe, C. J., and U. Valdre. "Experimental and computational study of field emission characteristics from amorphous carbon single nanotips grown by carbon contamination. I. Experiments and computation." *Philosophical Magazine B* 82, no. 9 (2002): 987–1007.

[20] Edgcombe, C. J., and U. Valdre. "Microscopy and computational modelling to elucidate the enhancement factor for field electron emitters." *Journal of Microscopy* 203, no. 2 (2001): 188–194.

[21] Dyke, W. P., J. K. Trolan, W. W. Dolan, and George Barnes. "The field emitter: fabrication, electron microscopy, and electric field calculations." *Journal of Applied Physics* 24, no. 5 (1953): 570–576.

[22] Gomer, Robert, and Robert Gomer. *Field emission and field ionization*, vol. 34. Harvard University Press, Cambridge, 1961.

[23] Charbonnier, Francis M., William A. Mackie, Robert L. Hartman, and Tianbao Xie. "Robust high current field emitter tips and arrays for vacuum microelectronics devices." *Journal of Vacuum Science & Technology B: Microelectronics and Nanometer Structures Processing, Measurement, and Phenomena* 19, no. 3 (2001): 1064–1072.

[24] Xu, N. S., Y. Chen, S. Z. Deng, J. Chen, X. C. Ma, and E. G. Wang. "Vacuum gap dependence of field electron emission properties of large area multi-walled carbon nanotube films." *Journal of Physics D: Applied Physics* 34, no. 11 (2001): 1597.

[25] Han, Jae Hee, Jae Hong Park, Alexander S. Berdinsky, Prashant S. Alegaonkar, Ji Beom Yoo, Hae Jin Kim, Jin Joo Choi, Joong Woo Nam, and Chun Kyu Lee. "Electron emission reliability of carbon nanotube paste and the diode emission characteristics for a microwave power amplifier." *Electronic Materials Letters* 2, no. 2 (2006): 101–105.

[26] Chalker, Paul R. "Wide bandgap semiconductor materials for high temperature electronics." *Thin Solid Films* 343 (1999): 616–622.

[27] He, Peng, Yong Gao, Jie Lian, Lumin Wang, Dong Qian, Jian Zhao, Wei Wang, Mark J. Schulz, Xing Ping Zhou, and Donglu Shi. "Surface modification and ultrasonication effect on the mechanical properties of carbon nanofiber/polycarbonate composites." *Composites Part A: Applied Science and Manufacturing* 37, no. 9 (2006): 1270–1275.

[28] Liu, C. W., M. H. Lee, M. J. Chen, C. F. Lin, and M. Y. Chern. "Roughness-enhanced electroluminescence from metal oxide silicon tunneling diodes." *IEEE Electron Device Letters* 21 (2000): 601.

[29] Park, J. H., S. Y. Jeon, P. S. Alegaonkar, and J. B. Yoo. "Improvement of emission reliability of carbon nanotube emitters by electrical conditioning." *Thin Solid Films* 516, no. 11 (2008): 3618–3621.

[30] Zhu, Jiang, Haiqing Peng, Fernando Rodriguez-Macias, John L. Margrave, Valery N. Khabashesku, Ashraf M. Imam, Karen Lozano, and Enrique V. Barrera. "Reinforcing epoxy polymer composites through covalent integration of functionalized nanotubes." *Advanced Functional Materials* 14, no. 7 (2004): 643–648.

[31] Li, Jiang, Matthew J. Vergne, Eric D. Mowles, Wei-Hong Zhong, David M. Hercules, and Charles M. Lukehart. "Surface functionalization and characterization of graphitic carbon nanofibers (GCNFs)." *Carbon* 43, no. 14 (2005): 2883–2893.

[32] Sham, Man-Lung, and Jang-Kyo Kim. "Surface functionalities of multi-wall carbon nanotubes after UV/Ozone and TETA treatments." *Carbon* 44, no. 4 (2006): 768–777.

[33] Yu, Guojun, Jinlong Gong, Sen Wang, Dezhang Zhu, Suixia He, and Zhiyuan Zhu. "Etching effects of ethanol on multi-walled carbon nanotubes." *Carbon* 44, no. 7 (2006): 1218–1224.

[34] Nazeeruddin, Mohammad K., and Michael Grätzel. *Semiconductor photochemistry and photophysics/ volume ten*, p. 287. Swiss Federal Institute of Technology, Lausanne, Switzerland, 2003.

[35] Lee, Tae Young, Prashant S. Alegaonkar, and Ji-Beom Yoo. "Fabrication of dye sensitized solar cell using TiO_2 coated carbon nanotubes." *Thin Solid Films* 515, no. 12 (2007): 5131–5135.

[36] Bedja, Idriss, Prashant V. Kamat, Xiao Hua, A. G. Lappin, and Surat Hotchandani. "Photosensitization of nanocrystalline ZnO films by bis (2, 2 '-bipyridine)(2, 2 '-bipyridine-4, 4 '-dicarboxylic acid) ruthenium (II)." *Langmuir* 13, no. 8 (1997): 2398–2403.

[37] O'regan, Brian, and Michael Grätzel. "A low-cost, high-efficiency solar cell based on dye-sensitized colloidal TiO_2 films." *Nature* 353, no. 6346 (1991): 737–740.

[38] Diamant, Yishay, S. G. Chen, Ophira Melamed, and Arie Zaban. "Core– shell nanoporous electrode for dye sensitized solar cells: the effect of the $SrTiO_3$ shell on the electronic properties of the TiO_2 core." *The Journal of Physical Chemistry B* 107, no. 9 (2003): 1977–1981.

[39] Chappel, Shlomit, Si-Guang Chen, and Arie Zaban. "TiO_2-coated nanoporous SnO_2 electrodes for dye-sensitized solar cells." *Langmuir* 18, no. 8 (2002): 3336–3342.

[40] Jeong, Jin Su, Sung Joon Park, Yun Hee Shin, Yong Jun Jung, Prashant Sudhir Alegaonkar, and Ji Beom Yoo. "Fabrication of carbon nanotube embedded nylon nanofiber bundles by electrospinning." In *Solid state phenomena*, vol. 124, pp. 1125–1128. Trans Tech Publications Ltd, 2007.

[41] Joglekar, Shreeram S., Harish M. Gholap, Prashant S. Alegaonkar, and Anup A. Kale. "The interactions between CdTe quantum dots and proteins: understanding nano-bio interface." *AIMS Materials Science* 4, no. 1 (2017): 209–222.

[42] Godbole, Rhushikesh, Vasant P. Godbole, Prashant S. Alegaonkar, and Sunita Bhagwat. "Effect of film thickness on gas sensing properties of sprayed WO_3 thin films." *New Journal of Chemistry* 41, no. 20 (2017): 11807–11816.

[43] Godbole, Rishikesh V., Pratibha Rao, Prashant S. Alegaonkar, and Sunita Bhagwat. "Influence of fuel to oxidizer ratio on LPG sensing performance of $MgFe_2O_4$ nanoparticles." *Materials Chemistry and Physics* 161 (2015): 135–141.

[44] Chinke, Shamal L., Solomon Berhe, and Prashant S. Alegaonkar. "High speed projectile sensor: design, development and system engineering." *IEEE Sensors Journal* 21, no. 23 (2021): 27062–27068.

[45] Alegaonkar, Ashwini P., Satish K. Pardeshi, and Prashant S. Alegaonkar. "Tellurium-reduced graphene oxide two-dimensional (2D) architecture for efficient photo-catalytic effluent: solution for industrial water waste." *Diamond and Related Materials* 108 (2020): 107994.

[46] Gawade, Rohini P., Shamal L. Chinke, and Prashant S. Alegaonkar. "Polymers in cosmetics." In *Polymer science and innovative applications*, pp. 545–565. Elsevier, California 2020.

[47] Joglekar, S. S., P. V. Pimpliskar, V. V. Sirdeshmukh, P. S. Alegaonkar, and A. A. Kale. "FITC embedded ZnO/silica nanocomposites as probe for detection of L-lactate: point-of-care diagnosis." *MRS Advances* 4, no. 46–47 (2019): 2533–2540.

[48] Ghoderao, Prachi, Sanjay Sahare, Prashant Alegaonkar, Anjali A. Kulkarni, and Tejashree Bhave. "Multiwalled carbon nanotubes decorated with Fe_3O_4 nanoparticles for efficacious doxycycline delivery." *ACS Applied Nano Materials* 2, no. 1 (2018): 607–616.

10 Poly-Nanocarbons
Ion-Track Membranes for Devices and Nuclear Radiation-Induced Modifications for Opto-Electronics

10.1 EMERGENCE OF NANO-ION-TRACK MEMBRANE FOR FLAT FLEXIBLE DEVICES

Ion-tracks are the *nano-micro-holes* drilled in a free standing conducting or insulating polymer film of thickness 10–50 µm using the swift heavy ions of energy ranging from 50 MeV to 1 GeV that are obtained from the ion accelerators. Such beams create damaged zones in polymers that can be chemically manipulated to create *holes* through-and-through in the polymeric films to transform into nano-membrane. These holes can be doped, filled, or written with the appropriate materials to create flat, flexible devices termed as the ion-track-based polymeric films [1].

10.1.1 ION-TRACKS: THE BRIEF HISTORY

Ion-tracks in thin polymer films are finding a multitude of new interesting applications not only in electronics but also in a verity of other fields such as medicine or optics, etc. In this chapter, ion-track formation and their applications are described [2]. Furthermore, the design and development of ion-track-based devices, such as micro-magnets, is characterized. The zone of continuous local damage along the trajectories of the swift heavy ions impinging into insulators is denoted as "latent track" (or "nuclear track" or "ion-track"). After removing the modified radiochemical matter, pores emerge, and these pores are called "etched ion-tracks." A literature survey indicates that, so far, not much attention has been paid to the field of latent track applications. Latent track can be studied by, (a) characterizing modified transport properties along tracks, (b) depositing metallic atoms along the tracks, (c) studying chemical changes, and (d) making use of ion-induced phase transitions [3]. There are numerous applications of latent tracks, such as, polyimide films with latent tracks formed there in, can be used as a protective coating against ambient dust and moisture penetration, by maintaining pressure equilibrium and gas exchange with the ambient. In addition, the doped latent tracks can be used for many electronic purpose [2].

Further, the chemical etching of the ion-irradiated polymers is a process that transforms every single latent track into a pore, with a variety of shapes, such as, cylindrical, conical, crater-like, etc., and these shapes depend on the chemical etching conditions of the polymer. From microscopic images of the etched track distributions, one can determine the ion fluence, the composition of the ion flux, the angle of incidence and the energy of the ions by analysing the sizes, shapes and number of the etched ion-tracks.

A considerable amount of work, based on this principle, has been carried out on applications of polymeric track detectors in nuclear physics, radiography, cosmic ray studies, applied radiochemistry, dosimetry, etc. Thus, chemical etching develops "fingerprints" of the ions, which passed through the polymer. Moreover, an ion beam with known parameters in combination with chemical etching, serves as a unique and powerful micro-tool for the creation of a micro- or nanoporous structures [4]. In addition, conducting nano-wires can be formed along tracks, when metals are deposited by

DOI: 10.1201/9781003317258-10

different techniques, such as ELD, galvanic, and evaporation techniques. The SiC needles [5] for conducting cantilevers were studied using AFM microscopy via radiochemical transformation of polysilanes. Moreover, of ion-induced conducting nano-wires in diamond [6] or fullerite [21] are discussed for applications in field emission displays, but this is still in the first experimental stage. The ion-tracks and their combination with other techniques, have a variety of applications in nano-structural devices [7–15].

10.1.1.1 Latent Tracks: The Mechanism of Formation

The latent tracks can be formed, when the incident swift heavy ion, having energy in the range MeV–GeV, deposits a large amount of energy in a medium. However, the energy deposition takes place within an extremely short time (~10^{-17} to 10^{-15} s), and in a very tiny volume (~$10-18$ cm^3), which is called as, ion-track core [16] and reference cited therein. Furthermore, after the ion impact, though the transient breaking of most chemical bonds gets recovered in the ion-track core, during the annealing phase (thermal spike, ~10^{-12} s), a large number of damages occurs permanently [15]. As a result, the number density of radicals increases, carbonaceous clusters and new phases are formed, amorphization sets in, and the free volume is altered, i.e. the physico-chemical properties of the latent tracks are altered. These latent tracks are also called as as-implanted ion-tracks, with modifications in the form of a high density of radicals. Thus, the transient conditions which lead to dramatic modifications of the medium are mainly attributed to the change in the chemical and structural properties of the medium.

10.1.1.2 Track Manipulations: The Chemical Etching Process

The latent tracks which are produced (~300–1000 Å in diameter) in the polymeric matrix can be etched by a suitable chemical reagent, which leads to formation of etched ion-tracks. Moreover, the dimensions of etched tracks that are chemically etched, are orders of magnitude higher than those of latent tracks. The tailoring of ion-tracks of required shapes such as, cylindrical, conical or hyperbolic, transmittent (in thin films) or non-transmittent depends on the choice of projectile, target material, chemical etchant and the thermal conditions during the chemical etching [4]. The latent tracks present in the polymer acts as a structure inhomogeneity, and play an important role in the chemical etching process. Moreover, the accessibility and reactivity of chemically active sites, such as weak bonds, in a latent track lead to the formation of local anisotropy, in the polymer matrix, and as a result, the migration of the etchant is possible in the latent tracks. In addition, structure inhomogeneity such as, regions of glassy state, regions with different molecular mass distributions created during the shear flow, differences in the degree of crosslinking, oriented and not oriented regions, amorphous and crystalline regions, etc., also cause migration of the etchant in the polymer matrix. Furthermore, during the chemical etching of commercially available polymers organic and inorganic admixtures and impurities, such as stabilizers, anti-oxidants, plasticizers, fillers, dyes, traces of catalysts, etc., acts as a structure inhomogeneity in the polymer, which also contribute to the etching process [1]. Furthermore, the availability and distribution of both the intrinsic and the radiation-induced free volume in a polymer determine the migration mechanism of the etchant into the polymer. The migration takes place either by nano-capillarity or via diffusion. In addition, the free volume in amorphous domains gives more degrees of freedom for rearrangements of the molecular chain segments and facilitates every single chain scission event. As a result, selective etching of heavy ion-tracks in polymers is apparent from the properties of latent tracks. One can specify the following typical features, which make the latent tracks preferentially etchable [3]:

1. The material density in latent tracks is lower and, hence, the free volume is larger. (e.g. amorphous tracks in a crystalline or semi-crystalline polymer).
2. The average molecular mass in the track core is lower. Smaller molecules are more mobile and reactive. The chain end, in polymers, requires only one bond cleavage, in absence of a

monomer unit at, whereas two cleavage events are required to release the monomer from the middle of the chain.

3. Molecules in the track core often have polar groups formed in radiation-induced reactions. The polar groups attract ionic etchant and promote penetration of etchant into the track.

4. Apparently, some new chemical bonds and groups may specifically react with a given etchant. In this case the selectivity of the etchant should be especially high.

5. Residual mechanical strain around the ion path causes weak chemical bonds even more susceptible to cleavage. As a result, the etched ion-tracks are produced in the polymer matrix.

A literature survey reveals that, a large number of co-workers have studied the chemical etching of the ion-tracks. It is observed that, polycarbonate (PC) samples irradiated with 450 MeV Kr$^+$ ions can be etched by 6.25 M NaOH solution at a temperature ~50°C for a period of 20 minutes. Polydimethylsiloxane (PDMSiO) can be etched in the solution of 2.5 N NaOH at 115°C for a period of 3 minutes, and polyethylene terephthalate (PET) can be etched in 6.25 N NaOH solution for a period of 45 minutes at a temperature ~50°C. In the present work, the etching of ion-irradiated polyimide samples was carried out by using a 6 M solution of NaOCl at a temperature ~70°C.

10.1.2 Estimation of Etched Track Diameter: Ion Transmission Spectroscopy (ITS) Technique

In the present study, the average diameter and the geometry of the etched ion-tracks were estimated using the ion transmission spectrometry (ITS) technique [1–5]. The experimental set-up of the ITS technique is shown in Figure 10.1. The present ITS system consists of a stainless-steel chamber which is coupled to a vacuum system. In this chamber, one sample at a time could be mounted on a sample holder. In Figure 10.2, the chamber is designated as (a). A 5.5 MeV Am^{241} α-particle source was used for the yield measurements, which is placed at a position (b). The yield of the α-particles

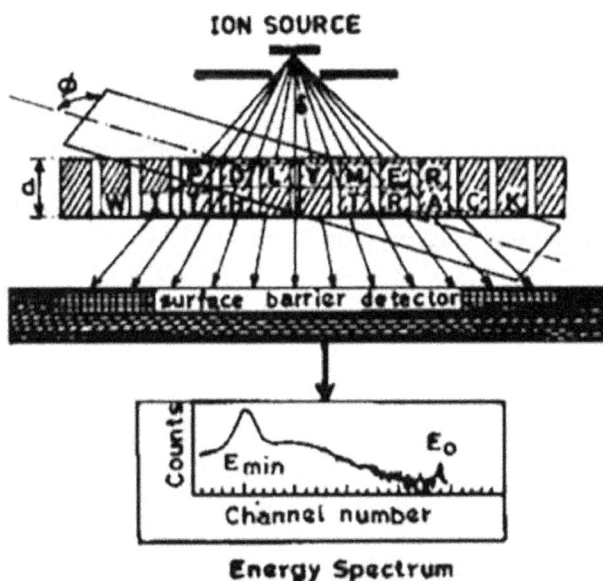

FIGURE 10.1 Schematic diagrams of ion transmission spectrometry (ITS), for the transmission of α-particles through polyimide film, and through etched ion-track.

FIGURE 10.2 Experimental set-up of ion transmission spectrometry @Hanh-Meitner-Inst., Berlin.

was recorded using a silicon surface barrier detector, which is positioned at (c). The detector is coupled to a Multi-channel analyser (MCA), via a preamplifier and amplifier chain (d).

The measurements were carried out at a vacuum level of ~10^{-6} mbar. The initial experiments were conducted for the energy calibration of the system, the recorded the α-particle peak was found to be at ~5500 channel number. Furthermore, a virgin polyimide sample of thickness ~25 µm, was placed in between the detector and the source, mounted on the sample holder, and the spectrum was recorded for a period of ~24 hours. The measurements showed that, the peak position of the recorded α-particle spectrum was shifted from channel number 5500–3750, due to the energy loss suffered by αparticles while passing through the bulk of the polyimide sample. Following the same procedure the virgin polyimide sample was replaced with a 30 minutes. Etched ion-track polyimide sample and the spectrum was recorded for the same period of time. The same experiment was repeated for all the polyimide samples with etching time varying from sample to sample in the range of 30–240 minutes. It was found that, for an etched polyimide sample, a small peak of α-particles emerges at a channel number ~5500 after an etching period of ~120 minutes and further increase in the etching time changes the intensity and the shape of the recorded peak. This change is attributed to the α-particles were passed through the etched ion-tracks present in the polyimide samples. A schematic diagram, describing the principle of the ITS technique is shown in Figure 10.1. The measurements showed that, the intensity and the width of the recorded peak was found to increase with an increasing number of tracks and the track diameter. In the present study, following assumptions were made for the estimation of the average radii of the ion-tracks [1]:

1. The ion-tracks are distributed randomly within a homogeneous polymeric medium, which is attributed to the statistical nature of the ion impact.
2. The radial distribution of the ion-tracks should be symmetric in nature.
3. The trajectories of the incident α-particles should be parallel to each other and to the ion-tracks under examination.

With the above assumptions a formula which correlate transmitted ion yield and the track diameter can be given by the equation: $r_T = \left(\dfrac{T}{\pi \cdot \Phi}\right)^{1/2}$, where T is the ratio of total number of α-particles incident on detector without polyimide samples to with polyimide samples. In case of diverging beam, the ion-track radius is given by the equation: $r_T = \left(\dfrac{T}{\pi \cdot \Phi \cdot \sigma_c}\right)^{1/2}$ where Φ is the number of ion-tracks/ cm², and, σ_c, is the overlapping cross section of the ion-tracks. In addition, one can determine the track diameter directly using Scanning electron microscopy. The experimental results showed the good agreement, in track diameter determination, as estimated by ITS and SEM techniques [1,3].

10.1.3 DESIGN AND DEVELOPMENT OF FLAT, FLEXIBLE PROTOTYPE DEVICES: THE MICRO-TRANSFORMER

In subsequent few sections we would be demonstrating a *proof-of-concept* for ab initio creation of ion-track to device. A polyimide sheet of thickness ~25 µm, was imported form M/s Goodfellow Ltd., Germany. A 450 MeV Kr+ ions beam was used for ion-irradiation of these polyimide ($C_{22}H_{11}N_2O_5$, PMDA-ODA) films. The ion fluence was varied from sample to sample over a range of ~1×10^6–5×10^8 ions/cm². The irradiation was carried out at the Ion Spectrometry Laboratory of the Hahn-Meitner-Institute, Berlin. In this way a large number of ion irradiated polyimide samples were kept ready for the chemical etching process to estimate the number of parameters such as, the variation in the average track diameter, and the etching rate with etching time.

10.1.3.1 Track Formation and Opening: The ITS Analysis

In this, NaOCl solution was used as a chemical etchant for ion irradiated polyimide samples. A freshly prepared NaOCl solution of concentration 6M (with pH ~ 7.3) was prepared and kept ready for the chemical etching process. At all the stages, the chemical etching was carried out at a constant temperature of ~70°C. The NaOCl bath was prepared and the temperature raised and maintained to ~70°C. The ion irradiated polyimide sample was immersed in the NaOCl bath for a period of 30 minutes and after etching the sample was taken out, and cleaned with triply distilled water. This was repeatedly followed in order to avoid over-etching/bulk/etching of the polyimide sample. Following this process, a large number of etched ion-track polyimide samples were prepared with etching time ranging from 30 to 240 minutes. After etching, the samples were cleaned and kept under dry atmosphere. The etched polyimide samples were subjected to the ITS technique to study the variations in the average track diameter with etching time [7,8].

The results showed that, for the polyimide samples, after an etching period of 120 minutes the profile of the α-particle spectrum changes. The variation in the intensity of recorded α-particle with channel numbers as a function of etching time is shown in Figure 10.3. It can be seen from Figure 10.3 that, the counts at higher channel numbers increase with increasing etching time. For sufficiently large etching periods a small peak appears at channel ~5500, which corresponds to the energy of the incident α-particles. This peak at the highest channel number is attributed to the opening of the latent tracks with chemical etching time. Further, the same set of etched polyimide samples were employed to the SEM measurements, to compare the estimated track diameter with the ITS technique. In addition, a few etched polyimide samples were also characterized using optical microscope, AXIO-CAM-CRC, with different magnifications. In similar fashion, same procedure was adopted for the chemical etching of PET and polycarbonate (PC) latent track samples. The result of etched ion-track opening as a function of etching time, for the PET samples, is shown in Figure 10.4a (images for PC not shown).

It is interesting to note that, with gradual increase in the etching time, the position of the peak associated with the bulk polymer region shifts towards higher channel numbers. The shift in peak position, was found to be marginal and attributed to the small amount of bulk etching of the polymer

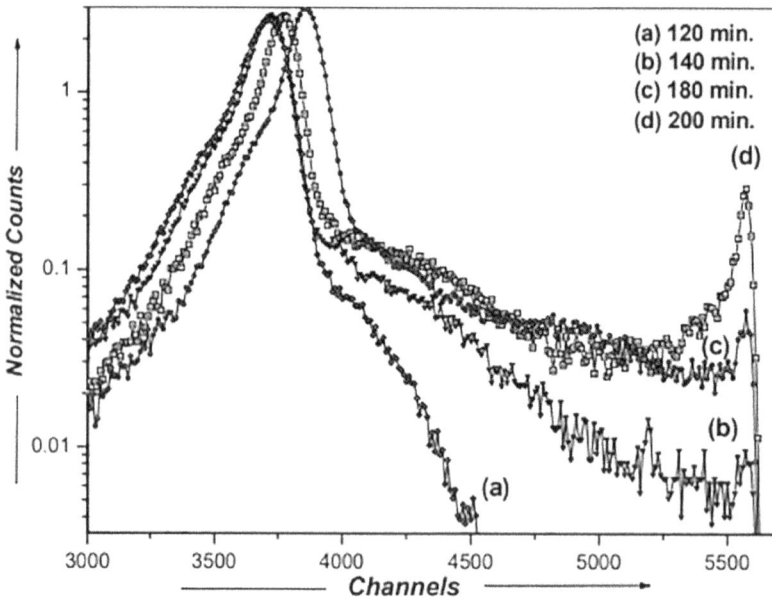

FIGURE 10.3 Recorded ITS spectra for polyimide samples, showing the opening of ion-tracks with etching time ranging from 120 to 195 minutes.

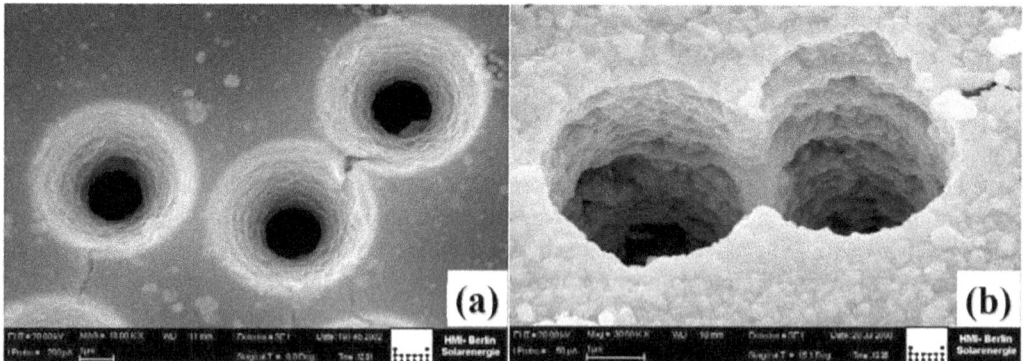

FIGURE 10.4 Surface morphology studied by SEM for (a) PET and (b) polyimide samples.

film, resulting the decrease in the thickness of the polymer sample. The structural and other electronic properties of the etched ion-tracks polymer remain unaltered after chemical etching process. The insignificant shift in the peak position leads to the conclusion that, the bulk etching rate is very small and can be neglected as compared to the etching of the latent track. The surface morphology of the etched ion-tracks (top view) for polyimide is shown in Figure 10.4b.

In another experiment, a systematic study was carried out to observe the shape of the etched tracks. For this purpose, a few etched samples of polyimide and PET were cut into small pieces. These samples were mounted at the edge of the SEM sample holder table in such a way that the charging can be avoided. The sample holder table was tilted at an angle of 90° and positioned with respect to the incident electron beam and in this way a few SEM images were recorded. Figure 10.5a and b shows a cross-sectional photograph of the etched ion-tracks for polyimide and PET samples [3,7,9].

From the measured values of the track diameters, the track radius was estimated and Figure 10.6 shows the plot of track radius as a function of etching time for polyimide samples. It can be seen that

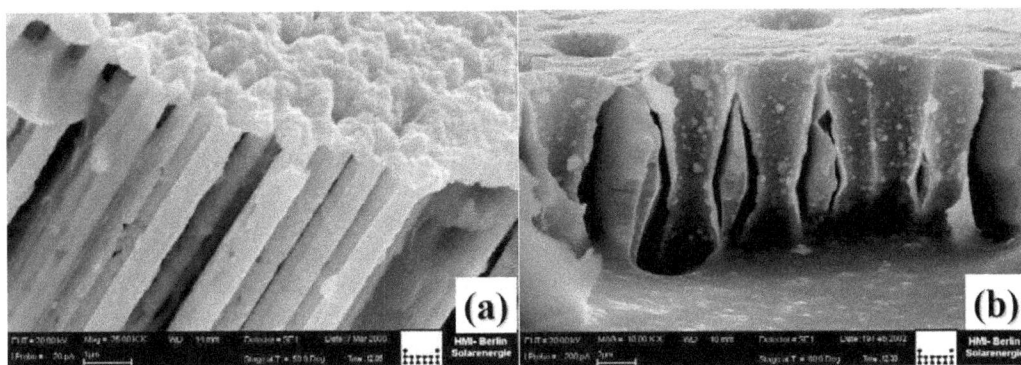

FIGURE 10.5 Cross-sectional view studied by SEM technique to investigate the etched ion-tracks for (a) PET and (b) polyimide samples.

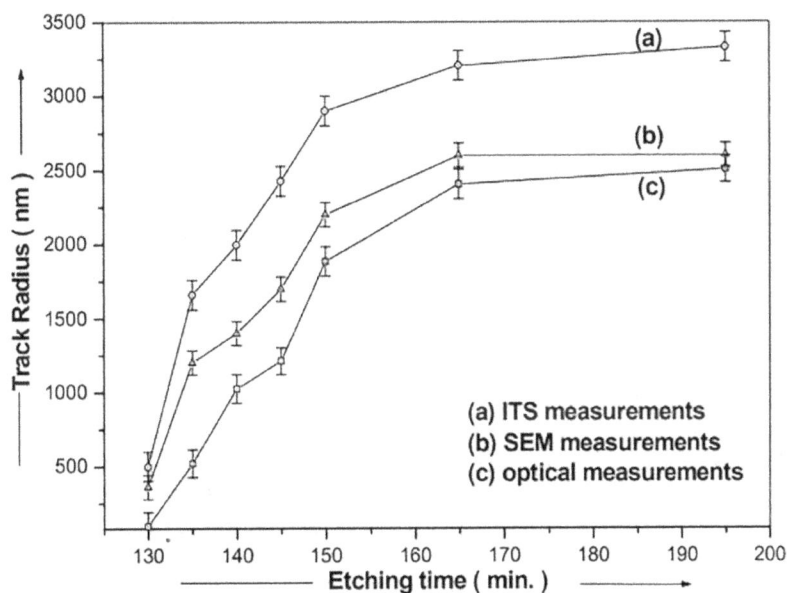

FIGURE 10.6 Comparison of the track radius for polyimide samples, estimated by (a) ITS, (b) optical, and (c) SEM measurements.

the track radius increases markedly during the initial stages of etching, whereas for longer etching times the increase in the track radius is found to be marginal. This shows that, the etching rate has two slopes in nature, initially the rate of etching is high and then for longer etching time the etching rate saturates. These results reveal that, during initial stages of the etching, the radiochemically modified matter in the latent track gets removed and dissolved in the etchant very quickly. With increasing etching time, bulk etching takes place along the ion-track, which is slower as compared to latent track etching. The marginal etching of the bulk polymer shows the proper selection of the etchant.

Under this, an inductor coil was developed, based on the ion-track technique. For this purpose, the metal wires were deposited into the etched ion-tracks using the ELD technique. After metal deposition in tracks using the mask lithography technique the structure of the coil on the surface of the polyimide film was developed. The details of the metal deposition techniques in the etched ion-tracks are described below:

10.1.3.2 Track Deposition

Etched tracks can be filled, in principle, with any material, and the embedded matter can be arranged as either massive wires (also called "fibres, fibrilles") or tubules, or it just can be dispersed discontinuously as small nanoparticles along the track length. The transport of the embedding matter within etched tracks towards the required position is possible by dissolving the material of interest in a suitable liquid (water or some organic solvent). The etched track samples can be immersed in the liquid so that the liquid penetrates into the tracks by capillarity action and can deposit the material of into the track [11,17].

Further, if this liquid is allowed to evaporate within the tracks, then the dissolved matter will supersaturate and precipitate therein as the tubule. The thickness of the tubule, deposited with different materials, depends on the total amount of matter dissolved in the liquid and transported into the tracks. A literature survey reveals that, a considerable amount of work has been done to deposit ion-tracks with materials such as, KCl, C60, PMMA, dyes, etc. This can be achieved by, for example, the electrode-less (or chemical) deposition (ELD) technique [11,17,18].

10.1.3.3 Electrode-Less Deposition

In the ELD technique, a high degree of super-saturation of certain metals like Cu, Ag, Au, Ni, etc. or chalcogenide compounds such as CdS, PbS, CuxS, ZnS, CdSe, ZnSe, CdTe, CuInSe2, etc., takes place by dissolving them in suitable complex agents such as NH3, ethylene diamine (EN), nitrilotri-acetate (NTA), thiourea (TU), and others. The nucleation centre plays an important role in material deposition, which enables the axial tailoring of track structures. The chemical activators such as Sn or Pd ions, have tendency to form bonds with the polymer chains, and hence act as chemical activation centres, whereas physically activated sites can be produced, by laser irradiation or ion irradiation to form damage sites. The range of chemical activation and laser irradiation techniques ranges from ~10 to 50 μm, i.e. over the entire area of the track, whereas ion irradiation activates the etched track only up to the range of ions (typically ~0.1–1 μm for keV and MeV ions). Due to the nucleation centre, the solutions get precipitated heterogeneously, on the inner surfaces of the etched tracks, which leads to the formation of tubules of these materials, e.g. circular contacts for sensor materials embedded in the tracks emerge, which guarantees full accessibility of the sensors to the ambient [3,15].

In contrast, if the nucleation centres are not offered during the ELD process, material deposition will take place only at intrinsic defects sites of the polymer surface, which leads to the growth of only a few dispersed semi-conducting nanocrystals on the walls of the ion-tracks. It has been observed that, the track resistivity depends on the crystallite size and inter-crystallite distance of the material precipitated in the track, thus the resistivity can be tailored to a desire value, in the absence of nucleation centres. Eventually such structures show photo-resistivity [3]. A principal disadvantage of all manipulation techniques applying liquids is that these liquids penetrate to some extent into the used polymers. Furthermore, etched ion-tracks can also be filled with conducting material using a very common technique such as the galvanic deposition technique [3–6].

10.1.3.4 Mask Lithography and Development of Micro-Transformer

Initially, the experiments were carried out on a prototype magnet, which was miniaturized further to a dimension ~200 μm×200 μm×25 μm. A large number of etched polyimide samples were taken for this study. The average track dimensions were estimated using the ITS technique, and the estimated average diameters were re-confirmed by other supporting track characterization techniques such as SEM and optical microscopy. In this the average track diameter was found to be ~6000 nm, with areal density of ion-tracks ~10^8 tracks/cm^2. The design of the prototype inductor coil was made in such a way that metallic wires embedded in etched tracks with suitable metal contact can be coupled to the metal stripes evaporated (or chemically deposited) onto the surface of the polyimide sample [11].

The development of inductor coil structure was carried out, initially, by preparing a set of three prototype masks, with dimensions of ~6 mm×6 mm×0.02 mm using a graphical mapping paper. The details of the masks and production stages are shown in Figure 10.7. During the initial stage of the preparation, two masks (i.e. Mask # 1) were glue pasted on both the sides of the etched polyimide sample. The sample along with the mask was employed to the evaporation technique to

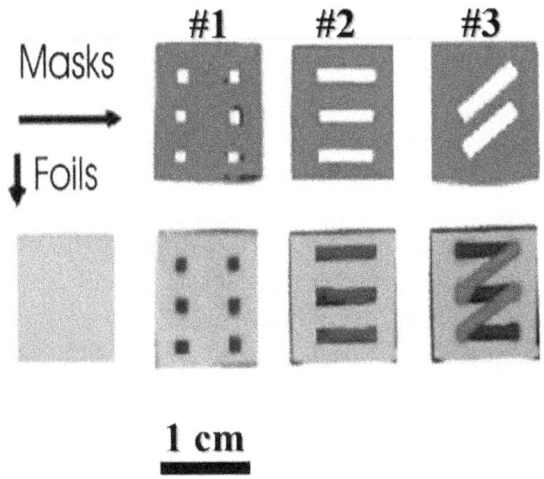

FIGURE 10.7 Ion-track-based prototype inductor coil with dimensions ~6 mm×7 mm×0.02 mm with lithographic masks (top row) and production steps (bottom).

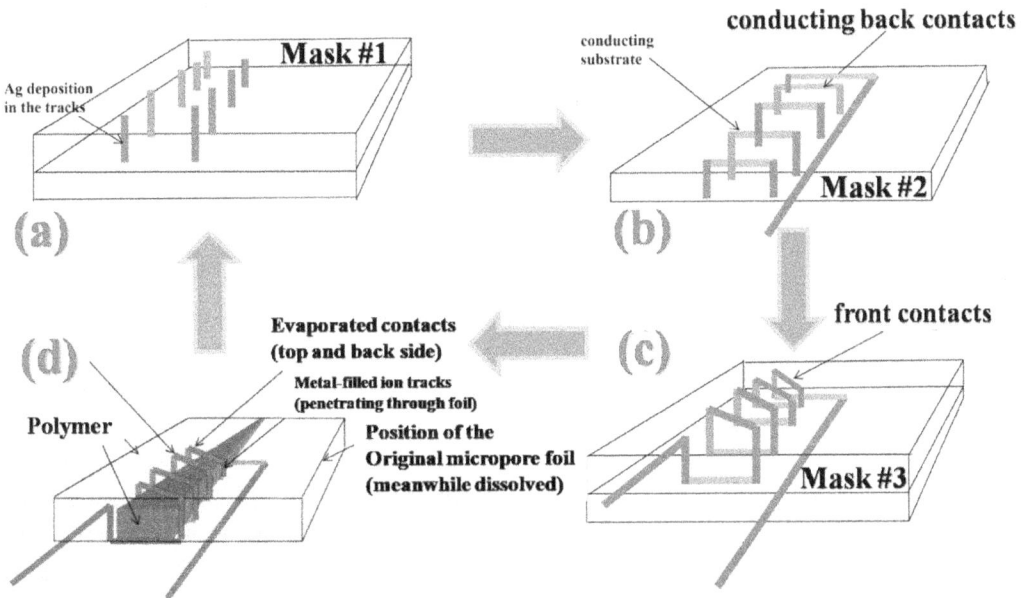

FIGURE 10.8 Production scheme for development of ion-track-based inductor coils. Step (a) deposition of silver in etched ion-track of polyimide using mask #1, (b) fixing silver-doped ion-tracks onto the polyimide surface from the rear side using mask #2, (c) development of contact pads on the front side using mask #3, and (d) integrated view of ion-track-based inductor coils; connecting bar shows the *magnetic core* effect.

deposit the silver in the ion-track. After the silver deposition was over the polyimide sample was immersed in an ethanol bath. During ethanol bathing the glue get dissolved in the ethanol solution and the masks were removed from both the sides from polyimide sample. After the ethanol bathing, the ethanol was found to be chemically inert to interact with deposited silver in the ion-tracks. The structural configuration of the polyimide sample, after gold deposition in the ion-track is shown in Figure 10.7a. Following this process, further masking (i.e. Mask #2 and #3) of polyimide sample was carried out in the identical way, and the configuration of the next step after silver deposition is shown in Figure 10.7b and c. Figure 10.8 describes the corresponding production scheme followed during the silver deposition on the surface and through the ion-tracks of the polyimide samples [11].

10.1.3.5 Device Characteristics and Quality Assessments

The characterization of these ion-track-based coils was made, by measuring the mutual inductance of the coils. For this purpose, by following the same procedure (i.e. masking, deposition etc.), two coils on the same polyimide sample were produced. The measurements of inductance were carried out using an impedance-gain-phase analyser, Model 4192A LF (5 Hz–13 MHz) and Model 4191 (RF 1 MHz–1 GHz), over a frequency range of 10 Hz–1 GHz.

The measurements showed that, test transformer gives good operation response up to ~½ GHz frequency range, with quality factors approaching up to ~10. In addition, the inductance per winding was estimated and found to be 10^{-4} H measured at a frequency 10 Hz and ~2.5×10^{-7} H at frequencies between 10^4 and 5×10^8 Hz. The values of the inductance, L, at a frequency were estimated using a formula: $L = \mu_o \mu_L (A \cdot N^2 / l)$, where μ_o is the permeability of air, μ, is permeability of the medium A is the area of cross section of the coils, N, is the number of turns of the coils, l, is the length of the coil. Figure 10.9, shows the input and output characteristics of the ion-track-based

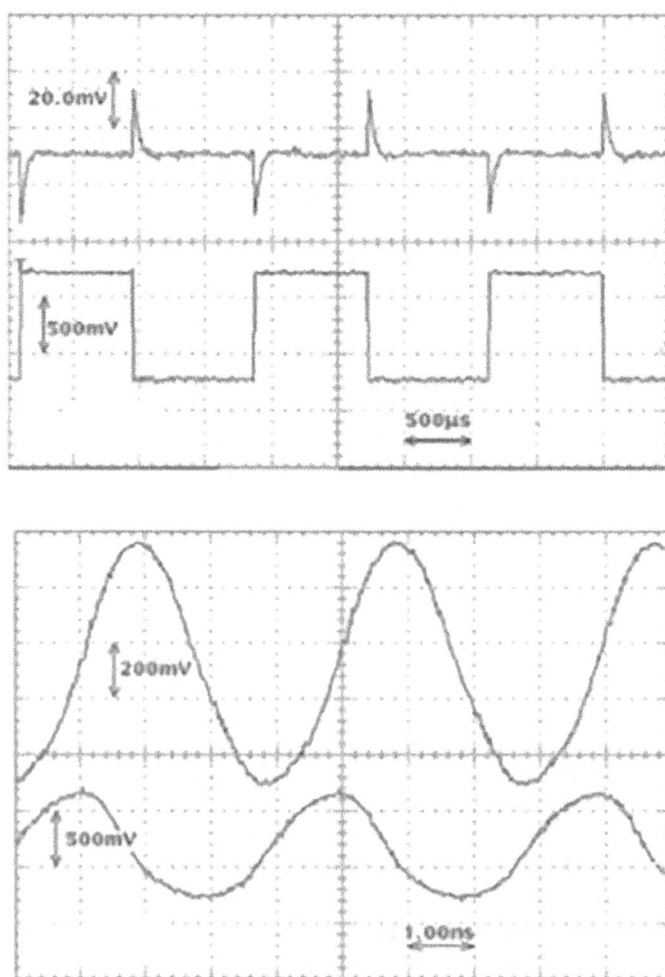

FIGURE 10.9 Response of the ion-track-based micro-transformer. Comparison of input (above) and output (below) signals of the prototype transformer, for rectangular and sinusoidal input at different frequencies. Spectra show the reasonably good response recorded in the frequency range ~ MHz (upper graph). It has been observed that signal transformation is possible only up to the near GHz range when gradually a phase lag emerges between input and output signals.

coils that has been studied at ~MHz and 1 GHz frequency range. These characteristics were studied for rectangular as well as sinusoidal input signals. The measurements showed good response in the MHz range (upper graph), but at GHz frequency range sinusoidal signals starts showing a phase lag between input and output signals [3,11,17].

10.1.3.6 Implementation to Nanostructured Devices: Micro-Transformer

These masks were miniaturized further to the dimension ~200 μm×200 μm×50 μm, by constructing the metallic masks, shown in Figure 10.9. These metallic masks were made using a laser lithographic technique. The masks used in the present study are alkaline resistant, and withstand up to a temperature ~1500°C and their geometry is having an accuracy of ~±30 μm. Following the same procedure two inductor coils were made in the ion-track-based polyimide samples. The gold deposition was carried out using flash evaporation technique. Figure 10.10 shows corresponding gold evaporated inductor coil made using the metallic masks. In a nutshell, polyimide samples of thickness ~25 μm were exposed to 450 MeV Kr^+ - ions. The ion fluence was kept low and varied from samples to sample over the range of ~1×10^{12}–1×10^{13} ions/cm^2. This process led to the formation of latent tracks in the ion exposed polyimide samples [11].

These latent tracks were etched chemically by using NaOCl as an etchant solution. The etching process of carried out at a constant temperature ~70°C. The track dimensions of etched polyimide samples was estimated by ITS technique. The measurements of the ITS techniques showed that the etched ion-tracks have diameters ~6 μm, which was re-confirmed by SEM and optical microscopy (Figure 10.11).

The measurements of the track etching reveal that, for longer etching time the track etching rates get retarded due to the transition from the track etching to the very low bulk etching rate. These polyimide samples were used to design and develop ion-track-based inductor coils. These micro-transformer configuration shows good operation response up to ~1/2 GHz frequency, with a quality factor ~10. Thus etched ion-tracks, in combination with mask lithography, enable many possibilities for creating new micro- and nanostructures within polyimide films that are inaccessible by other techniques. Design and development of an electronic element such as magnet and transformers on the ion-track basis has been demonstrated successfully [3,11–19].

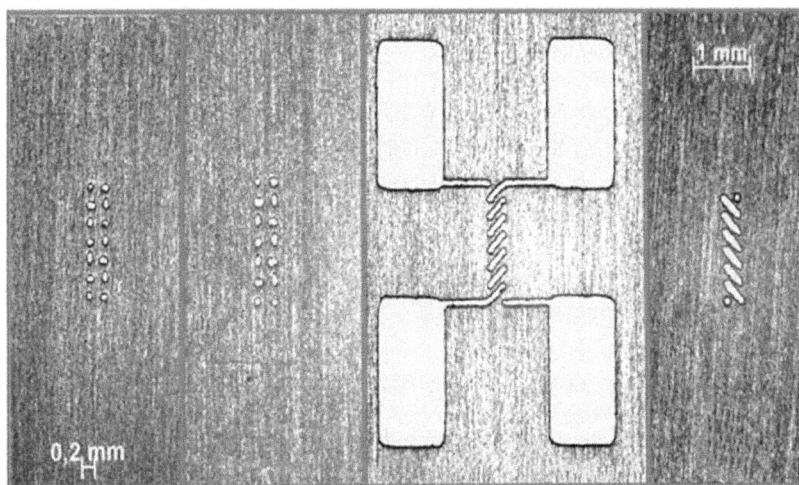

FIGURE 10.10 Metallic masks of miniaturized ion-track-based transformer of dimension ~200 μm×200 μm×50 μm.

FIGURE 10.11 Gold evaporated miniaturized ion-track-based inductor coils of dimension ~200 μm× 200 μm×50 μm, developed in the polyimide samples.

10.2 OPTO-ELECTRONIC PROPERTIES OF RADIATION-INDUCED MODIFIED POLYCARBONS

This subsection gives a general overview and a brief survey of the work carried out on radiation induced modifications polyimide for optoelectronic applications. The effects of different types of radiation, such as, swift heavy ions, electrons, and gamma rays on polymer in general, and polyimide in particular are discussed in brief. The damage induced by energetic radiation in a polymer is usually far more complex than in metals or semiconductors and could be probed by various characterization techniques. It is difficult to probe the overall structure of any damage directly, however, one can study individual effects of radiation in inducing the damage. Further, for the specific applications of the polyimide in space environment, it is desirable to coat the polyimide surface with an element, which can protect the polyimide from the radiations such as atomic oxygen, UV radiations, etc. However, this approach is not useful since the polyimide loses its electrical insulation and becomes conducting. It is therefore desirable to diffuse some low - Z - elements in the surface layer of the polyimide which can protect the surface layer by changing the band energy, keeping the insulating properties unaltered. As such the diffusion of elements in the polyimide at room temperature using the conventional techniques is not possible. In current sub-section, a novel technique has been described, in which either electrons or gamma rays are used for diffusing low - Z - elements such as, B, Li, as well as high - Z elements such as Cu, Ag, in the polyimide surface. The techniques like, Neutron depth profiling, Rutherford Backscattering, X-rays have been used to study the depth distribution of the diffused elements [1].

10.2.1 RADIATION-INDUCED MODIFICATIONS: THE BACKGROUND

Most of the organic polymers are sensitive to high-energy radiations. The neutron affects the nuclei of atoms making up these materials; other energetic radiations affect primarily the electrons. The subsequent alterations in molecular structure govern changes in physical and chemical properties and are called as radiation effects [1,20]. Energetic particles interact with materials to produce displacement of atoms via momentum transfer and to produce ionization in the materials. Displacement disrupts the lattice and produces bulk damage in the solids; ions are far more efficient in producing displacement than electrons. Both electrons and ions produce ionization in materials. In most cases, it is the cumulative effect of ionization, known as radiation dose (a measure of energy deposited per unit mass of material). Insulators and dielectrics are subject to this type of damage, which can degrade electrical and mechanical performance. Degradation of materials due to displacement and ionization process is generally referred to as radiation damage. There is no difference in basic reaction mechanism between the interaction of gamma radiation and the interaction of electrons with organic polymers. However, there is a major difference between radiation damage induced in organic polymers and inorganic materials. In case of inorganic materials the reaction of radiation is the creation of interstitial atoms, vacancies, impurity atoms, and thermal spikes; ionization is less important. In contrast, in case of organic polymer, the radiation effect is largely due to ionization instead of displacement. Both gamma rays and electrons lose most of the energy through ionization and excitation and cause very few displacements by recoils. The electron, however, is completely stopped in the surface of a relatively thick specimen, whereas the photon will penetrate large thickness polymer. The electrons and ions, build up a charge in the polymer which, if not bled off, can produce much more extensive changes than the initial absorption. Heavy ions produce a considerable number of displacements. Furthermore, ionization and excitation rupture the chemical bonds, which lead to the formation of fragments of large polymer molecules, which may retain unpaired electrons from the broken bonds. The free radicals thus produce may react to change the chemical structure of the polymer and alter the physical properties of the polymer. The polymer may undergo cleavage or scission, or crosslinking. When ionizing radiation interact with polymeric materials, mainly following process takes place [1]:

1. Formation of chemical links between macromolecular chains (crosslinking)
2. Degradation and decomposition of macromolecules evidenced by the decrease in the molecular weight.
3. Change in the concentration and the nature of double bonds.
4. Evolution and out diffusion of gases from the polymer matrix.

10.2.1.1   Interaction with Ions

Upon entering a medium a heavy ion loses energy, initially through the process of electronic excitation and ionization, known as electronic energy loss, and later on near the end of its range, through the nuclear elastic collisions, in which atoms of the lattice can be displaced. The later phenomenon of energy loss becomes predominant when the ion energy reduces to ~A keV, where A is the atomic weight of the heavy ion. At this energy the effective charge of the ion is very close to zero and therefore the energy loss through the electronic excitation and ionization becomes negligible. Normally, near the surface, the contribution of the energy deposited through elastic collisions in inducing defects is neglected. This is because the energy loss suffered by a heavy ion through electronic excitation is grater by three orders of magnitude as compared to that through nuclear elastic collisions. However, for a particular target-ion system, if the cross section for the elastic collision (Rutherford scattering) is large, displacement type defects can be induced in the sample from its surface right up to the ion projected range. In addition to this process, the other processes through which the defects at the surface can be produced is Coulomb explosion [1]. When a swift heavy ion passes through the polymers, a tail of radiation damage in the bulk (damage track) [1] as well as material erosion of the surface (electronic sputtering) are observed [1]. Whereas the characteristic of the latent damage tracks has been determined by various physical method ex situ, long after the impacting event [1–3]. Ions with velocities higher than the Bohr velocity (V_B=0.22 cm/nm) deposit their energy in electronic excitation and ionization in a cylindrically symmetric region around the path of the impacting ion, forming the track core. The incident ion can cause ejection of the energetic secondary electrons that transport part of this energy out of the core [1,20,21]. The mean deposited energy density, $e(r)$, after the ion passage is approximated by an $e(r) \propto r^{-2}$ dependence, were, r, is the radial distance from the core of the ion path. Furthermore, the secondary electrons, in turn, either dissipates their energy by collisions with the other electrons or by electron-phonon coupling. Moreover, the energy transfer from ions to macromolecules, results in ionization and electronic excitation. As a result, several processes are induced such as, breaking of original bonds, production of excited and ionised species, of radicals and bond rearrangement, which are responsible of the most observed chemical modifications. In general, the energy deposited by the incoming ion in the polymer, in steady state condition, results in crosslinking or chain scission of the original chain. The prevalence of one effect on the other depends on the structure of the polymer [17]. For ions, the non-relativistic equation for energy loss per unit length of path x, by a particle of mass M, charge z, energy E, and velocity v, in an polymer absorber of atomic number Zeff with N number of atoms per cubic centimetre

is given by equation [1]: $-\dfrac{dE}{dX} = 4\pi e^4 z^2 NZ \bullet \dfrac{1}{mv^2} \bullet \ln\left(\dfrac{2mv^2}{I}\right)$, where e is the charge of electron, m, is the mass of electron, and $<I>$ is the average excitation potential.

10.2.1.2 Electrons

The treatment of electrons differs from that of heavier charged particles, in that the heavier particles are assumed to suffer essentially no deflection in a collision with a bound electron; an impinging electron suffers considerable deflection. Also, after the collision the two electrons, it is impossible to distinguish which has the projectile electron (identity criterion). During irradiation of polymers by electrons cause initially primary reactions leading to a series of secondary reactions. Secondary electrons ejected behave in the same way producing tertiary reaction and so on until all the incident energy is dissipated. The electrons can lose a very large fraction of its initial energy in a single interaction with an electron in the polymer because of its mass equivalence. Hence it undergoes a wide angle scattering so that its path through the polymer is very tortuous. Because of its high speed, the range is much larger even though the specific ionization is much less. The total amount of energy absorbed by an atom or molecule in an encounter with ionizing radiation is 30–60 eV, whereas for ionization of an atom the amount energy required is ~10–30 eV. In case of polymers the bond dissociation energy is of the order of 2–15 eV, hence on exposure to radiations the most probable event for polymer is formation of free radicals by the rupture of the covalent bonds rather than the ionization of the monomer units. The maximum energy, E_{max} (eV), transferred to the molecular chain of mass, M_w, by an electron of energy, E, (MeV) can be given by the following equation: $E_{max}(eV) = \dfrac{2200 \cdot E \cdot (1+E)}{M_W}$, the molecular weight, M_w, (for say polyimide monomer is 341), and estimated value of Emax is ~15 eV. This indicate that, the electrons and the photons of MeV energy are capable of causing chemical changes by direct displacement of the atomic nucleus, rupture of the chemical bonds and subsequent ionization and excitation.

10.2.1.3 Gamma Rays

In a marked contrast to charged particles, a gamma radiation does not have the definite range in matter. This difference results from different types of absorption processes in the two cases. Charged particles results from different types of absorption processes, in the two cases. Charged particles lose their energy in a series of small steps whereas in general gamma rays is either completely removed form a beam by absorption or scattering or continuous on at the same energy. The total process of gamma ray absorption is the sum of each of the three effects: the photoelectric effect, dominant with low energy gamma rays and X-rays; Compton effect, dominant for intermediate energy gamma rays; and pair production, dominant with high energy gamma rays [1].

10.2.1.3.1 Photoelectric Effect

With low energy gamma rays (3 eV–8 keV) the principal contribution to the absorption is made by photoelectric process. Here the gamma rays interact with the bound electrons, ejecting it form the atoms the process is most efficient when the energy of gamma rays is higher than binding energy of the target electrons (i.e. K shell electrons). The interaction cross section for photoelectric effect with energy and type of the absorber is given by the equation:, $\sigma_{P.E.} \approx k \bullet \dfrac{Z^4}{(E_\gamma)^3}$ where $\sigma_{P.E.}$ photoelectric cross section per atom of the absorber, k, is an empirical constant, Z, is the atomic number of the absorber, and E_γ, is the energy of the incident gamma rays. When a photon ejects an electron form a K, L, M, or higher atomic level in the absorber, an electron vacancy results. When this vacancy is filled by an electron from a higher level; characteristic electromagnetic radiations called X–ray fluorescent is emitted. In irradiated samples of appropriate volume, the fluorescent X-rays will be almost totally reabsorbed by photoelectric interaction with electrons in higher electronic levels.

10.2.1.3.2   Compton Effect

This process is responsible for the bulk of effects in polymers with 0.1–2 MeV gamma rays. The gamma ray interacts with an electron and imparts a fraction of its energy to the electron. The gamma ray suffers an energy loss and a change in the wavelength in the process. In order to conserve the momentum and the energy of the system, the direction and the energy of the ejected electron are related to the direction and the energy of the scattered gamma rays and given by the following equation: $E'_\gamma = \hbar v = \dfrac{m_o C^2}{\left((1 - \cos\theta) + \dfrac{m_o C^2}{\hbar v_o} \right)}$, where E'_γ is the energy of the scattered gamma rays, $m_o C^2$ is the rest mass energy of the electron, θ, is the photon scattering angle, $h v_o$, energy of the scattered electron.

10.2.1.3.3   Pair Production

When a photon of energy greater than 1.02 MeV interacts with the field of a nucleus, an electron-positron pair can be created. These particles have same energy. The probability of pair production increases with gamma energy above 2 MeV, it is proportional to the logarithm of the gamma ray energy. Furthermore, the probability of pair production is proportional to Z^2, therefore, the phenomenon of pair production is observed in high - Z - materials as compared to low - Z - materials for the same incident gamma ray energy.

10.2.2   Theory of Radiation-Enhanced/Radiation-Assisted Diffusion

When a polymer is exposed to energetic radiation, the processes of scissioning and crosslinking of the molecular chains are induced [2], into the polymer matrix, due to rupture of covalent bonds [19,20]. Scissioning of molecular chains leads to formation of free radicals at the end groups of the molecular chains that can either react or split off into fragments of neutral molecules or groups of atoms. At a given temperature, these fragments obey Brownian movement [1], which can lead to the formation of free volume (empty sites, vacancies) in the polymer matrix. Free volume, produced due to vacancies, plays a very important role in diffusion processes. The free volume must achieve a configuration, which is energetically favourable for the diffusion of atoms in the polymer matrix. The atoms of an element can diffuse into a polymer only when such fluctuations are sufficiently large to fit the element in the available free volume [1]. Thus, in the energy considerations, external atoms can move to successive potentials of local equilibrium, when a sufficient amount of activation energy has been acquired by the surrounding empty sites. Furthermore, the radicals formed in the matrix during irradiation act as hopping sites for the diffusing atoms. The radiation-assisted diffusion of the atoms of any element occurs in the polymer by means of motion of defects.

During the irradiation the steady state concentration of the vacancies and defects and their build up time depends on the rate of annihilation of vacancies and defects through different processes [1]. In addition, the temperature also plays an important role in changing the vacancy concentration. At a given temperature, the rate at which the vacancy concentration, C_v, increases with time can be related with the rate at which the vacancies are produced, R_v, and annihilated, R_a, by the equation [23]: $\dfrac{dC_v}{dt} = \dfrac{dC_R}{dt} + \dfrac{dC_T}{dt} = (R_V + R_T) + C_V(R_a + R_{aT})$; C_R, is the excess vacancy concentration due to radiation, C_T, is the excess vacancy concentration due to the temperature, T. R_{aT}, is the rate of annihilation of vacancies at temperature, T, and R_T is the rate of vacancy produced due to the temperature, T. During the steady state - concentration of vacancies, C_V, the entity dC_v/dt approaches to zero. In case of a polymer irradiated at room temperature the terms C_T, R_a, and R_T, can be neglected. Following this, above equation reduces to: $C_V = R_V/R_a$. Similarly, the effective diffusion coefficient, D', under the irradiation condition at a temperature, T, can be defined by the relation: $D' = D + \dfrac{R_V}{C_{SR} + C_{ST}}$, where D is the thermal diffusion coefficient, and C_{SR}, is the sink concentration due to radiation and C_{ST}, is the sink concentration due to temperature, T, much above the room temperature. At room temperature, C_{ST}, and D, are negligible. Therefore, equation stated above leads to equation: $D' = \dfrac{R_V}{C_{SR}}$. Under irradiation conditions, the rate of vacancy production, R_v, is proportional to the incident radiation flux.

10.2.3 Theory of Dielectric Function: Tailoring Optoelectronic Properties

10.2.3.1 Molar Polarizability, α

The dielectric constant, ε', density, ρ, and polarizability, α, is related to each other using the molecular Clausius–Mossotti relationship [1] given by: $\left(\dfrac{\varepsilon' - 1}{\varepsilon' + 2}\right) \bullet \dfrac{M_w}{\rho} = \dfrac{N_A \bullet \alpha}{3\varepsilon_o}$, where ε_o is the permittivity of free space (=8.85×10^{-12}F/m), M_w, is the molecular weight of the polymeric chain, and N_A, is Avogadro's number.

It can be observed from the Clausius–Mossotti relationship that the dielectric constant, ε', can be changed with changing the total polarizability, α_T, which originates from three modes of polarization, of the polymeric medium. A brief description [1] of the origin of the three modes of polarization which contribute to the total polarization, α_T, can give us a better understanding of their relationship with chemical changes and with the manner in which the elemental doping can affect each mode of polarization (Figure 10.12).

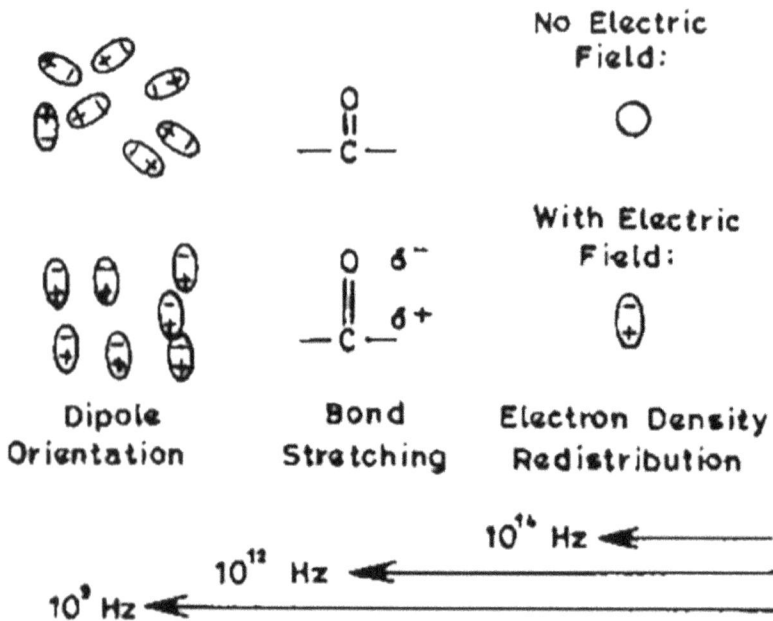

FIGURE 10.12   Polarization phenomenon that influences the dielectric function.

Electronic polarization, α_e, is the slight skewing of the equilibrium electron distribution relative to the positive nuclei to which it is associated (Figure 10.13). Since only the moment of the electrons is involved, this process can occur very rapidly and typically has a time constant of $\sim 10^{-15}$ s, whereas the atomic polarization, α_a, results from the rearrangement of the nuclei in response to an electric field. The positive nuclei are attracted to the negative pole of the applied field. However, the moment of heavy nuclei is more difficult to initiate and reverse than that of the electrons and so cannot follow an oscillating electric field at high a frequency as rapidly as the electron response.

Therefore, the process of atomic polarization, α_a, has a typical time constant $\sim 10^{-12}$ s. Furthermore, the dipole orientation polarization, α_o, results from the redistribution of the charge when a group of atoms with a net permanent dipole moment reorients itself in space in response to the electric field. Since large group masses must reorient, this process is necessarily slower than either atomic and electronic polarization, and even in gas phase will have time constants of the order of $\sim 10^{-9}$ s only, because of the larger inertia that must be overcome to reverse the direction of moment in each cycle of electric field oscillation. In the liquid or solid phases, large intermolecular forces must be overcome, which slows down the process further and decreases the possible polarization under the static condition. The dipole orientation polarization, α_o, is often the dominant mode of polarization contributing to the dielectric constant, ε', in liquid and gases. However, in solids, the dipole moment is usually restricted to the point where this becomes less significant than the electronic mode of polarization. However, in case of polyvinyl fluoride, this is not illustrated well because, enough dipole mobility exits to increase the dielectric constant, ε', dramatically (~ 8). In such polymers, their glass transition temperature, T_g, is near to room temperature (where their dielectric constants were measured), and thus they have substantial chain mobility. In case of polyimide, however, room temperature measurements are hundreds of degrees below the T_g, and molecular motions are severely limited. There are some ways in which the three modes of polarization can interact, but in most of the cases, they act essentially separately and are therefore additive, as shown in equation: $\alpha_T = \alpha_e + \alpha_A + \alpha_O$. Using this equation in the Clausius–Mossotti equation we can see, how each mode of polarization affects the dielectric constant, ε': $\left(\dfrac{\varepsilon'-1}{\varepsilon'+2}\right) \cdot \dfrac{M_w}{\rho} = \dfrac{N_A\left(\alpha_e + \alpha_a + \alpha_o\right)}{3\varepsilon_o}$. There exist a few other models to describe these relationships, but regardless of the exact one used, the dielectric constant, ε', will always scale with the polarizability, α_T. The greater the sum of three modes of polarization, the higher the dielectric constant, ε', will be. Because of the different time

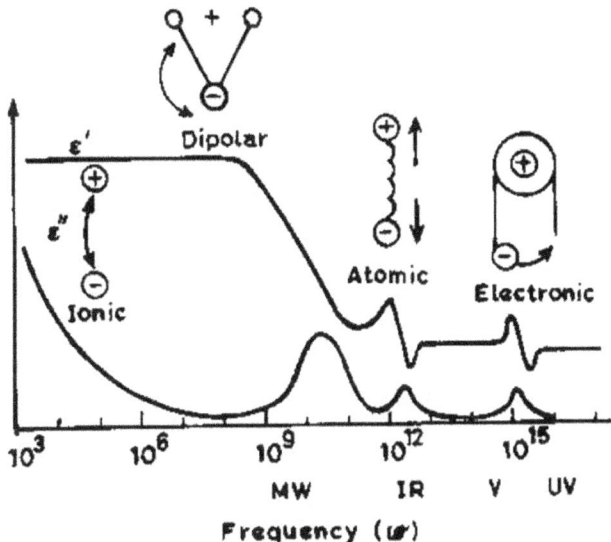

FIGURE 10.13 Frequency response of dielectric mechanism.

constants of three modes of polarization, it follows that dielectric constant, ε', will be frequency dependent. Furthermore, at optical frequencies, where only electronic polarization, α_e, is occurring, the dielectric constant, ε' can be found by the application of the Maxwell's identity, as shown in equation below. The dielectric constant, ε' increment due to electronic polarization, α_e is given by: $\varepsilon'=n^2$ where n is the refractive index of the polymer. Further, since $n^2=\varepsilon_e'$, we can also denote ε_e' as ε' n^2. When at low frequency the value of the dielectric constant, ε', is composed of all three modes of polarization, it is denoted by, $\varepsilon_{o,a,e}'$ and can also be referred as ε'1 kHz (which is in context of this work and will be designed simply as ε'1 kHz measured at a frequency 1 kHz) and is often referred to as the static dielectric constant, or ε's. This equivalency is far from perfect, however, since the materials under study are polymers and are unable to reach their maximum orientation polarization, that static term implies. Following from the additivity of different modes of polarization, the dielectric constant, ε', is assumed also to be essentially the sum of dielectric constant increments due to each mode of polarization separately, as shown in equation: $\varepsilon_{tot}' \approx \varepsilon_e' + \varepsilon_a' + \varepsilon_o'$.

Thus, from above equation, it can be seen that, the change in each component of the dielectric constant, ε', is due to the change in each of the three modes of polarization, that can occur during the chemical synthesis of the polymers. However, the change in the electronic mode is straightforward to determine and tends predictably towards the negative direction. For contrast, the change in the atomic mode is uncertain in magnitude as well as in direction. Further, the orientation polarization has clearly been considered to be small in magnitude. The change in ε_a', had to be determined unambiguously to enable us to find the change in the dipole orientation increment [1].

10.2.3.2 Density Effects (Free Volume Considerations)

A correlation of high free volume and low dielectric constant, ε', has been previously reported for polyimides [1]. In these investigations positron lifetime spectroscopy and group additivity methods were used to quantify the free volume fractions. The introduction of free volume in a polymer decreases the number of polarizable groups per unit volume, and as a result the values of, ε_a', and ε_o' decrease. Furthermore, the doping of pendant groups, flexible bridging units, and bulky groups which limits the chain packing density have all been used to enhance the free volume in polyimides and one can examine their effects on dielectric constant, ε'. In addition, the introduction of free volume in aromatic polyimides can be accomplished by using reactants or charge transfer complexes (CTCs), which result in inefficient chain packing in the polyimide. However, a considerable number of methods is available for achieving a low chain packing density, such as incorporating substituents with o- and m-linkage along the polymer backbone, incorporating flexible bridging units in the back bone and adding pendent group along the polymer backbone [1].

10.2.3.3 Importance of Elemental Incorporation

In addition to polarizability, α, and free volume, the substitution of various elements such as boron, fluorine, silver, copper can also tailor the dielectric constant, ε', of polyimide. However, indiscriminate substitution of them into polyimides may yield undesired effects. Hougham et al. have shown that non-symmetric substitution of fluorine for hydrogen increases the average magnitude of the dielectric constant, ε', by approximately 0.05 per substituted ring [8]. Moreover, symmetric substitution of fluorine does not increase the net dipole moment of the polymer and hence does not increase the dielectric constant, ε'. Furthermore, in the same study Hougham et al. have shown that the dielectric constant, ε', decreases with symmetric fluorine substitution by a combination of lower electronic polarizability, α_e, and larger free volume. In this study, symmetric and non-symmetric fluorinated groups are used to elucidate the influence of fluorine content on dielectric constant, ε'. However, it can be seen that the polarizability, α, is the primary variable influencing the dielectric constant, ε', whereas free volume and fluorine content are secondary variables which can alter the polarizability, α, of a polymer. The enhanced free volume lowers the polarization by decreasing number of polarizable groups per unit volume. Fluorine incorporation increases the free volume, lowers the electronic polarization, α_e, and can either increase or have no effect on dipole polarization depending on whether the fluorine incorporation is symmetric or asymmetric. During the synthesis of a polymer, an appreciable number of defects in the form of crosslinking, discontinuities, bending, etc., may remain in the molecular chains. As a result, dipoles are produced in the polymer matrix, which govern the dielectric properties. This leads to dielectric relaxation, [1, 0, 21] described by the following relation: $\tan\delta = \dfrac{2\pi \cdot f \cdot \tau_o \cdot S_r}{1+\left(2\pi \cdot f \cdot \tau_o\right)^2}$, where S_r is the relaxation strength, which depends on the concentration of dipoles and the square of their dipole moments, τ_o is the relaxation time for the dipole orientation, f is the frequency of the applied electric field, and $\tan\delta$ is the dissipation factor.

10.2.4 METHODOLOGY FOR RADIATION-ASSISTED/RADIATION-ENHANCED DIFFUSION: THE IRRADIATION CONDITIONS

Subsequent work is mainly focused onto manipulating and tailoring dielectric function and allied physical properties of polyimide ($C_{22}H_{10}N_2O_5$, PMDA-ODA, Kapton-H). Number of chemical moieties such as boron, fluorine, phosphorous, silver, copper, etc., were diffused in polyimide, however, discussed only boron, fluorine, and silver cases. The scheme is shown below in Figure 10.14.

For diffusion purpose, polyimide samples, each of size ~12 mm × 12 mm, were obtained by cutting polyimide sheets of thickness ~50 μm. These samples were stored in a dry atmosphere after cleaning. Two numbers of polyimide samples were mounted in a holder such that the spacing between them was ~2 mm. Diffusion experiments were performed by immersing these polyimide samples in an aqueous solution, as stated above kept in a thin walled, radiation resistant PET container, of wall thickness ~80 μm and width ~5 mm. The conFigd system consisting of polyimide samples immersed fully in the AgNO₃ solution was loaded in a Faraday cup cage, and placed near the extraction port of the electron accelerator. A pulsed electron beam of 6-MeV energy was obtained from an electron accelerator, called Race-Track-Microtron of the University of Pune. The electron pulse width was ~1.6 μsec and the pulse repetition rate was 50 pulses per second. In order to have uniform electron intensity on the samples, the electron beam was made to scatter elastically by a thin tungsten foil of thickness ~40 μm.

The system was positioned in such a way that the entire sample area could be exposed to the incident 6-MeV electrons. The intensity of the electrons over a circular area of diameter ~40 mm was

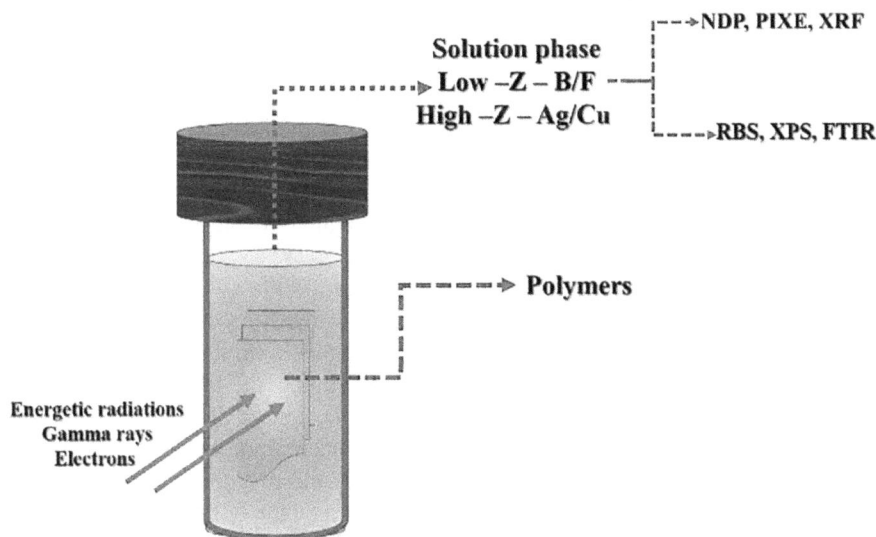

FIGURE 10.14 Schematics of radiation/assisted/enhanced diffusion using 1.6-MeV pulsed electron beam and Co-60 gamma ray source (energy: 1.33 MeV, activity: 10.33 megacurie) to irradiate polyimide and post diagnosis by RBS, XPS, FTIR for high Z elements like P, Ag, Cu and NDP, PIXE, XRF for low Z elements.

found to be uniform within 5%. During irradiation, a current integrator was used for measuring the electron current. Following this process, a set of four polyimide samples was irradiated with electrons and the beam was turned off when the desired electron fluence was received by the samples. In this way different sets of polyimide samples were exposed to electrons and the electron fluence was varied from one set to covering a range of electron fluence from 2×10^{15} to 5×10^{15} e/cm². The electron energy and the flux were kept constant throughout the experiment, and the electron fluence could be therefore changed by varying the time of irradiation. After the irradiation, the samples were removed from the respective solutions, washed with isopropanol, along with distilled water, and dried [20,21].

10.2.5 OPTO-ELECTRONIC PROPERTY INVESTIGATIONS

A large variety of insulating materials are synthesized for scientific and commercial applications. However, the insulating material, having the lowest dielectric constant, ε', is preferred for some special applications such as, to have an inter-level medium in multilayer integrated circuits, etc. Similarly, the insulating material which has a high dielectric constant, ε', is preferred for some other applications, such as, charge storage devices, optical devices, etc. Among commercially available polymers, polyimide is an important class of material for microelectronics and space technology because it has unique properties such as thermal stability, high degree of electrical insulation, etc. The development of polyimide with an increasingly lower dielectric constant, ε', has been the focus of several recent investigations [1]. The dependence of the dielectric constant, ε', and the dielectric loss, ε'', on the film thickness has been reported [20]. Using nano-foam morphology [21] the dielectric constant, ε', of the polyimide has been lowered from 3.2 to 2.5 at ambient, and from 2.9 to 2.3 at 100°C. The refractive index, n, of fluorinated polyimide has been increased [22] by electron irradiation. In most of the cases, studies were carried out on polyimide films of thickness in the range of a few hundreds of nm to a few μm. In general, not much attention has been paid to alter the electrical characteristics of the polyimide of thickness ~50 μm, which is normally used for the technological applications. A literature survey indicates that studies related to the elemental diffusion in polymers are also scares. Diffusion studies at high temperature that are routinely carried out in semiconductors [22] and metals [23], are not possible on polymers because of the relatively low phase transition temperature, $T_g \sim 400$°C. The ion-implantation technique [24] and the thermal diffusion technique (~500°C) [1] have been used to diffuse metallic elements in polymers. However, in these studies the depth of diffusion of elements in polymeric medium was limited from ~μm to nm range [24,25]. A literature survey indicates that, so far no attempt has been made to diffuse metal-atoms, in polymers, at room temperature. This is obvious because of the fact that the diffusion coefficient of metals in polymers is negligible at room temperature. In the present work, elemental silver could be diffused into polyimide up to a depth of ~2.5 μm, at room temperature, by using the radiation-assisted diffusion technique, employing a pulsed electron beam of 6-MeV electrons, for irradiation.

10.2.5.1 Silver Diffusion Using 6-MeV Pulsed Electron Beam

The silver diffused polyimide samples were characterized by the X-ray fluoresce (XRF) technique, by employing Cu K_α, X-rays. A typical XRF spectrum is shown in Figure 10.15A.

Further, for studying the diffusion of silver into the polyimide, the RBS technique is most suitable because the sensitivity of this technique depends on the square of the atomic number of the elements [27], allowing the detection of extremely low concentration of silver atoms in presence of oxygen, nitrogen, carbon, etc., which are constituents of the polyimide matrix. The silver diffused and a few virgin polyimide samples were characterized by the RBS technique, employing 2.4 MeV⁴He particles. The RBS spectra of three polyimide samples, exposed to electron fluences of 2, 3, and 5×10^{15} e/cm² are shown in Figure 10.15B including a virgin polyimide profile. The elements in the polyimide samples were identified by comparing the RBS spectra of the silver diffused polyimide with that of standard RBS spectra. The deconvolution of the RBS spectrum was made using

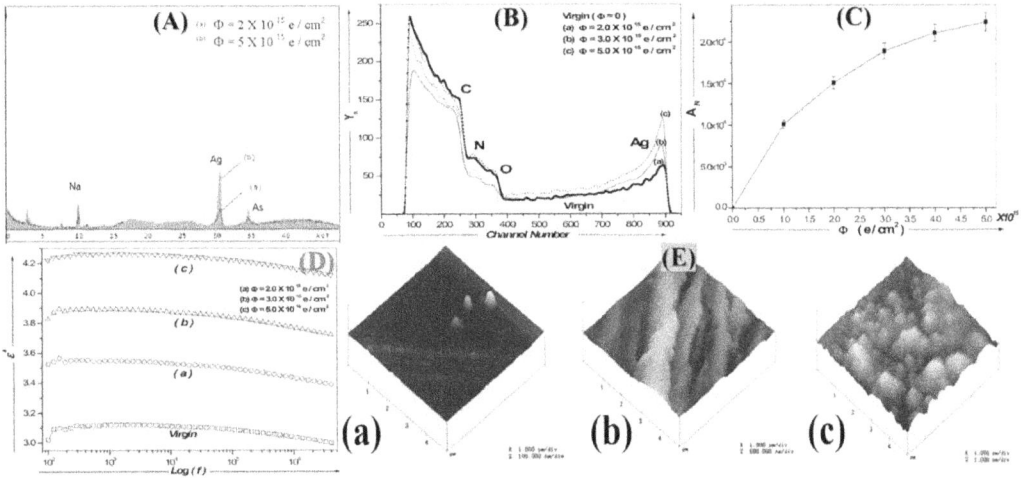

FIGURE 10.15   (A) Recorded XRF spectra to confirm silver, typically, @ a fluence, Φ, (a) 2, and (b) 5×10^{15} e/cm², (B) RBS spectra shown for all fluences including virgin, (C) corresponding area under the peak calculated from RBS for silver, (D) variations in the dielectric function (ε') with Log (f), (E) AFM images recorded typically for (a) virgin, (b) fluence@5×10^{15} e/cm², (c) same fluence in silver solution [24].

standard techniques. The average depth of the diffusion of silver atoms in the polyimide, estimated at ~10% concentration relative to the peak concentration at the surface, was found to be ~2.5 μm. Correspondingly, area under each RBS spectrum was estimated and its variation with electron fluence, Φ, e/cm² is shown in Figure 10.15C. The surface morphology of a few virgin (a), electron-irradiated (b), and silver-diffused polyimide samples (both@5×10^{15} e/cm²) (c) was studied by the Atomic Force Microscopy (AFM) technique, shown in Figure 10.15E. All the silver diffused and a few virgin polyimide samples were also subjected to dielectric relaxation measurements, over a frequency range of 100 Hz–4 MHz, at room temperature. From the measured value of the capacitance, the dielectric constant, ε', at that frequency was estimated. Variations in the dielectric constant, ε', with Log (f) for a virgin and three silver diffused polyimide samples are shown in Figure 10.15D.

It can be seen that, the intensity of the XRF spectra increases with increasing electron irradiation time. This supports the results of RBS indicating that the amount of silver diffused in polyimide samples increases with increasing period of exposure to electrons. A calibration spectrum was used to identify the positions of oxygen, nitrogen, and silver in the recorded RBS spectra of the polyimide samples. In the RBS spectra shown in Figure 10.15(B), the peak with the edge at the channel 908 corresponds to silver. Furthermore, all the three spectra, corresponding to silver, coincide at the channel number 900 and then merge thereafter. This indicates that silver is present at the surface of all the three polyimide samples. The amount of silver diffused in each polyimide sample could not be quantitatively estimated from the recorded RBS spectrum. However, since the experimental conditions were kept identical for all the polyimide samples, the area under the RBS spectrum is considered proportional to the number of silver atoms diffused in the polyimide samples.

It can be observed in Figure 10.15B that, there is no significant change in the intensity of the RBS peak corresponding to oxygen, and nitrogen, as compared to that of the virgin polyimide. This effect may be attributed to degassing of oxygen and nitrogen after dissociation of the liquid during irradiation on the polyimide surface. This result leads to the conclusion that the solution as such did not diffuse into the polyimide during irradiation. It is observed that the number of silver atoms diffused in the polyimide increases with increasing electron fluence, and this process is consistent with equation: $D' = \dfrac{R_V}{C_{SR}}$. The diffusion coefficient, D', for silver atoms in the polyimide at each electron fluence was estimated and found to vary form 1.26×10^{-13} to 9.37×10^{-14} m²/s over the range of electron fluence, 2 to 5×10^{15} e/cm^2. It can be observed the plot in Figure 10.15C that though the area under each RBS spectrum and hence the number of silver atoms diffused in the polyimide sample increases with increasing the electron fluence, however, the linear relationship is not hold good. The concentration of silver atoms tends to saturate at higher electron fluences. It appears that the rate of diffusion of the silver atoms in polyimide decreases with increasing electron fluences. This effect may be attributed to the formation of bridges between the polyimide segments and the silver atoms [25].

A comparison of AFM images shown in Figure 10.15a and b in (E) clearly indicates that, the surface morphology of the polyimide after electron irradiation is modified. The changes in the surface morphology of the electron-irradiated polyimide can be attributed to the degassing of the elements of the polyimide matrix upon electron impact. It can be seen from Figure 10.15E-c that, nano-clustering of silver atoms has occurred on the surface of the polyimide, which was immersed in AgNO$_3$ solution and irradiated with 6-MeV electrons. The surface morphology of this polyimide is distinctly different from that of the virgin and electron-irradiated polyimide. The nano-clustering of silver atoms on the polyimide surface may also contribute to the lowering of the rate of diffusion of silver atoms in the polyimide. The clustering of silver atoms appears to be local phenomena because there was no appreciable change in the surface conductivity of polyimide. Furthermore, though the change in the surface resistivity is not significant, however, the increase in the dielectric constant, ε', was appreciable. It can be seen from Figure 10.15D that the dielectric constant, ε', of polyimide increases with increasing electron fluence and hence with the silver concentration. The silver atoms after diffusing into the polyimide get polarized due to the carbonyl group. This process increases the effective polarization in the polyimide. The observed increase in the dielectric constant, ε', can be attributed to the dipolar interaction between the silver atoms and the carbonyl group present in the polyimide.

The variation in the dielectric constant, ε', over the frequency range 100 Hz–4MHz is negligible. These results reveal that the degree of electronic polarization in the polyimide is unchanged. Further, in accordance with equation $\left(\dfrac{\varepsilon' - 1}{\varepsilon' + 2} \right) \cdot \dfrac{M_w}{\rho} = \dfrac{N_A \left(\alpha_e + \alpha_a + \alpha_o \right)}{3\varepsilon_o}$ [1], the increase in the dielectric constant, ε', is not commensurate with the number of silver atoms diffused in the polyimide. These results reveal that there should be another factor, which opposes the action of the silver atoms in enhancing the dielectric constant, ε'. Furthermore, the effective free volume, which lowers the polarizable units per unit volume also increases gradually along with the silver diffusion in the polyimide. In an earlier study [1], the dielectric constant, ε', of polyimide was found to decrease with increasing electron fluence. These results are important because an insulating material having a large dielectric constant, ε', is required for making charge storage devices and similar applications.

It is interesting to note that, though the total time for which the polyimide samples were irradiated ranges from 20 to 50 minutes, the overall effective time for which the sample was bombarded with electron, was less than a second. In this experiment, electron pulses each of 1.6 µsec duration were obtained from the Race-Track-Microtron. Since the pulse repetition rate was 50 pulses per second, the effective period for which the sample was exposed to electrons was 80 µsec in a second. Even for an exposure period of 50 minutes, the sample was irradiated with electrons, only for 0.240 second. Such a large amount of diffusion of silver atoms in a period of 0.240 second indicates that the electron enhanced diffusion coefficient, D', of silver in the polyimide was enhanced by orders of magnitude, at room temperature, as compared with thermal diffusion.

10.2.5.2 Boron and Fluorine Diffused Polyimide Using Co-60 Gamma Radiations

Figure 10.16A shows that at a given dose, the refractive index, n, of polyimide samples irradiated in BF$_3$ solution, S_S, is smaller than that for the samples, S_A, irradiated in air. These results show that the diffused fluorine atoms might have reduced the degree of electronic polarization because the magnitude of the refractive index, n, at any optical frequency is governed mainly by the degree of electronic polarization. Moreover, the fluorine and boron atoms probably increased the free volume in the surface region, which further led to the lowering of the refractive index, n. The effect of Co-60 gamma rays in the bulk of polyimide irradiated either in air or in BF_3 is expected to be identical for a given dose. However, for the same dose, the refractive index, n, of a polyimide irradiated in BF_3 solution is smaller than that for a polyimide irradiated in air, as seen in Figure 10.16A. This difference in the value of the refractive index, n, can be attributed to the presence of fluorine and boron atoms in the surface region of the polyimide up to a depth of ~3 μm, on both the sides.

One can see in Figure 10.16B that the dielectric loss, ε'', at any frequency is greater for both, S_A and S_S polyimide samples than that for virgin polyimide, S_V. However, the magnitude of the dielectric loss, ε'', is relatively higher for an S_A sample than that for an S_S sample. The dielectric loss, ε'', is attributed to charge-transfer-complexes (CTC) dipoles and hence it appears that the number of CTC dipoles increases with increasing radiation dose. The diffused boron and fluorine atoms must have reduced the number of CTC dipoles in the surface region of the S_S samples. One can see in Figure 10.16B that in each spectrum the magnitude of the first relaxation peak is much smaller than that of the second peak. In addition, the second relaxation loss peak is broad and thus it appears as though the dielectric relaxation period, τ, results from the overlapping of a large number of different dipole relaxation periods [1]. This peak therefore is attributed to an α-relaxation process [26].

Further, the conductivity in a polyimide is governed mostly by the mechanism of hopping charge carriers [27]. Since, the molecular chains have oriented parallel to the surface, the process of hopping of electron from one site to other site can occur mostly in the region containing crosslinked chains. In general, during irradiation the crosslinking of chains can occur, and a large number of molecular chains are broken. This process lead to the formation of amorphous zones. The probability of formation of CTCs is maximum in the amorphous region, mainly because of the fact that in the crystalline region, the molecular chains are oriented parallel to the surface planes. The CTCs in polyimide consists of diamine fragments, acting as electron donors and of dianhydride fragments, acting as electron acceptors. The number of sites available for hopping increases with the gamma

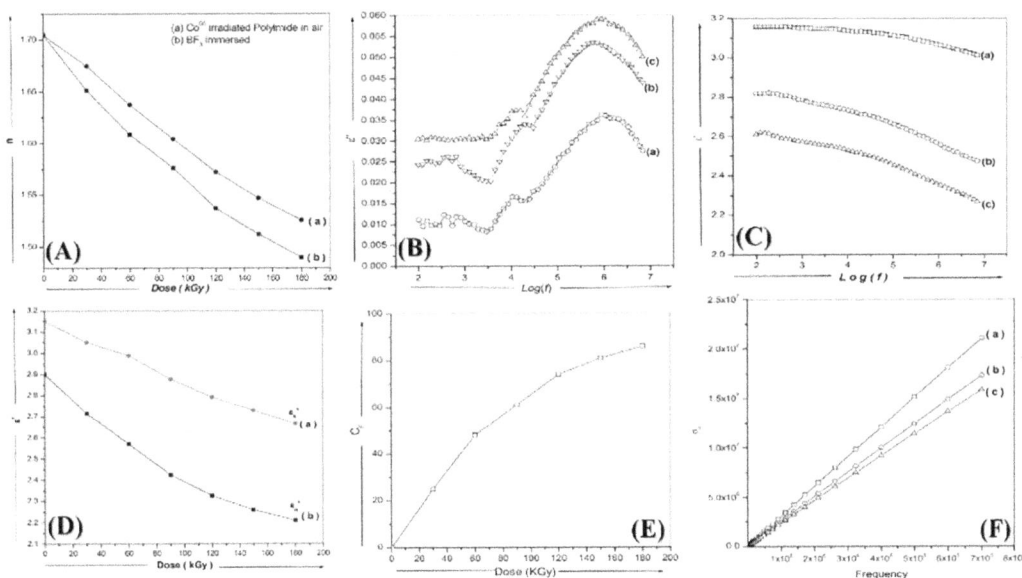

FIGURE 10.16 (A) Variations in calculated refractive index, n, with gamma ray dose: for polyimide samples (a) in air, S_A, and (b) in BF_3 solution, S_S, (B) measured dielectric loss, ε'', for (a) virgin S_V, (b) S_A, and (c) S_S, (C) corresponding, ε', for (a) S_V, (b) S_A, and (c) S_S, (D) comparison of ε_s' @1 kHZ and estimated dielectric constant from Maxwell's identity $\varepsilon = n^2$ for S_S, (E) estimation of relative fluorine concentration, C_F, using XPS for S_S, and (F) variations in a. c. conductivity, $\sigma_0'(\omega)$ with frequency for (a) S_V, (b) S_A, and (c) S_S. All S_A, and S_S@180kGy.

ray dose because a large number of amorphous phase regions can be produced in the crystalline polyimide matrix when exposed to energetic radiation. The number of electrons that can participate in the transition process increases with an increasing dose of Co-60 gamma rays.

As a result, the dielectric loss, ε'', which is associated with the electron transitions, also increases with increasing dose. It is possible that the diffused boron and fluorine atoms have formed bonds with the broken chains in the surface region. This process decreases the number of free electrons, which can participate in the conduction process. As a result, the dielectric loss, ε'', in S_S samples decreases slowly with increasing dose, and hence with boron and fluorine concentrations. Moreover, from the plot of dielectric loss, ε'', vs. frequency, the values of a. c. conductivity, σ' ($\omega\, \varepsilon''$), were calculated for each frequency (Figure 10.16F). The variations for S_V, S_A, and S_S sample were plotted. The variation of a. c. conductivity with frequency was found to obey the following relation: $\sigma'(\omega)\alpha\,\omega^n$; the value of the exponent was found ~1 for S_A and S_S samples but ~0.8 for S_V. Similarly, the a. c. conductivity was found to be orders of magnitude higher than the d. c. conductivity (σ_o). These results yield the conclusion that the hopping mechanism is the dominant mode of polarization. However, the result shows the presence of two components of dielectric polarization. One of these has a frequency dependence of the "universal" type ω^{k-1} with $k=1.0$, independent of film thickness, which is attributed to the electronic hopping. The other component is characterized by a very low loss and depends on chain orientations, and is attributed to the π-electron response of the dielectric lattice.

Figure 10.16C shows that the dielectric constant, ε', decreases slowly with increasing frequency for all three types of samples, S_V, S_A, and S_S. It is interesting to note that the dielectric constant, ε', of a polyimide can be decreased merely by irradiating it with Co-60 gamma rays in air. For the same dose of Co-60 gamma rays, the value of the dielectric constant, ε', can be lowered further by diffusing fluorine and boron into the surface region of the polyimide. This is reconfirmation for results in Figure 10.16D, which shows that the magnitude of both the dielectric constants, ε' (1 kHz) and ε'_n ($\varepsilon'_n=n^2$), decreases with increasing radiation dose. However, the rate of decrease of, ε'_n, is slightly greater than that of, ε'. The observed dielectric constant, ε', at any frequency is the result of the individual contributions of the atomic, dipolar and electronic polarizable units. At an optical frequency corresponding to ~632.8 nm, only the contribution by the electronic mode of polarization is possible, whereas, at 1 kHz, the atomic, dipolar and electronic modes of polarization may contribute to the dielectric constant, ε'.

Further, since the two plots, ε' and ε'_n, are not parallel, these results show that the dielectric constant, ε', changes as the contribution of the electronic mode of polarization decreases with increasing gamma ray dose at a higher rate than that of the atomic and dipolar polarization modes. The degree of electronic polarizability in the surface region decreases by increasing the number of fluorine atoms. In addition, the induced defects in polyimide also act as electronic trapping centres. As a result, the degree of electronic polarizability decreases as the dose of Co-60 gamma rays increases [1].

From Figure 10.16E, the replacement of hydrogen with fluorine is known to decrease the local electron polarization, and it also increases the free volume due to the relatively large volume of fluorine relative to hydrogen. In addition, the diffused boron atoms may also increase the effective free volume. As a result, the number of polarizable groups per unit volume decreases with increasing fluorine and boron concentrations. Also, the electron polarization decreases as the number of fluorine atoms substituted for hydrogen increases. This is mainly because of the lesser electric polarizability of a C–F bond relative to that of a C–H bond [6]. Further, quantitatively the fluorine concentration (C_F in relative %) had been determined by using the relation: $C_F\,(\text{in }\%)=\left\{\dfrac{A_P}{E_P}\right\}\cdot E^{0.71}$, where E is

the kinetic energy of fluorine species, A_p, is the area under the peak, and σ_p, is the photo-ionization cross section. The value of E, was obtained from the incident Mg K_α photon (1253.6 eV) and binding energy E_{BE}, of the fluorine atom in the polyimide matrix (689 eV), which was obtained from the software ESCA-3000, VGX-900. In the spectra recorded for the F1S state with the ESCA technique,

FIGURE 10.17 Recorded SEM micrographs for (a) virgin, S_V, (b) irradiated in air with Co-60 gamma ray to a dose of 180 kGy, S_A, and (c) irradiated in a BF_3 solution with Co-60 gamma ray to a dose of 180 kGy, S_S.

the estimated binding energy of the F- bond was in the range 685 eV to 686 eV, which is 4 eV lower than that of a typical C–F bond (B.E. 689 eV). For a typical C4F bond the binding energy is 687 eV and it decreases as the number of carbon atoms decreases. From the values of the binding energies, it appears that the fluorine atom may be forming a bond of the type $(C–F)_n$ or C_2F_2. From the ESCA measurements, it cannot be predicted whether the fluorine substitution is symmetric or non-symmetrical. The estimated binding energy of boron is 192 eV and so it appears that boron is not bonding with carbon, because the binding energy of boron in B_4C is 186 eV, whereas that in B–F complex is 689.3 eV. The binding energy of boron in B_2O_3 is ~ 192.6 eV, whereas in the present measurements the binding energy of boron is 192 eV. These results indicate that boron is possibly attached to oxygen atoms in the polyimide matrix. Figure 10.16e shows that initially the concentration of fluorine increases rapidly with increasing dose of Co-60 gamma rays but slowly at a later stage. These results, obtained from ESCA measurements, confirm that fluorine atoms could diffuse in the polyimide samples at room temperature under irradiation with Co-60 gamma rays. Similar results were obtained for the boron concentration measured with the neutron depth profile technique, and the depth of diffused boron atoms is ~3 μm. The results for boron are not given here because they are similar to those ones reported earlier for HDPE [1].

The SEM, photographs showed in Figure 10.17 indicates that the degree of surface roughness of the SS sample is smaller than that of the SA sample for the same dose of Co-60 gamma rays. For the polyimide irradiated in BF_3 solution, a large number of the vacancies in the surface region were probably substituted by atoms such as boron, fluorine, hydrogen, oxygen, etc., present in the solution. One can see in micrograph (b) of the SA sample that the polymer chains are parallel to the surface. In Figure 10.17, micrograph (a) shows that the surface is smooth whereas from micrograph (c) one can see that the surface is rough. The bright spots in the image may be attributed to the extrusion of the material and look like corrugated. The separation between the corrugation lines is ~1 μm and the roughness is clearly observable. However, one can see that in micrograph (b) the corrugated surface looks darker whereas in micrograph (c) these lines look brighter. It appears that there is some preferential direction in which the erosion has taken place.

As stated earlier in Section 10.2.3.2, density or free volume effect plays a role in modulating dielectric function of the polyimide. In the following subsection, free volume effects have been quantified by investigating positron annihilation spectroscopic technique.

10.2.6 FREE VOLUME INVESTIGATIONS IN IRRADIATED POLYIMIDE: POSITRON ANNIHILATION

The polymers in general and the polyimide in particular play an important role in the field of space technology. The physical and electrical properties of thin polyimide films are strongly influenced by the structure and stereo-regularity of the molecular chains. During the synthesis an appreciable number of defects in the form of crosslinking, discontinuity, bending, etc., related to the molecular chain remain in the polymers. As a result, isolated amorphous regions with micro-voids in the surface and bulk regions exist in the synthesized polymer film [1]. The free volume of the micro-void decides the physical parameters such as, refractive index, conductivity, dielectric constant, etc. of the polyimide films [2]. The free volume of the micro-voids can be changed by exposing the polyimide film to energetic radiation such as, gamma rays, electrons, heavy ions [3], etc. The existence of a large free volume can influence the other structural characteristics and therefore an enhancement in the diffusivity [4] of certain elements in the irradiated polyimide [5,6] may be achieved. However, not much attention has been paid towards quantitative estimation of the free volume in the polyimide exposed to energetic radiation. In the present work, using the positron annihilation spectroscopy technique, the average free volume, of the micro-void in the electron-irradiated polyimide was estimated and co-related with the electron fluence, Φ. Furthermore, at room temperature the diffusion of silver in the polyimide under electron irradiation was studied and related to the size of micro-voids produced on electron-irradiation. In the polyimide, the life time of the positron trapped in a void of free volume, V_F, is longer than the life time of the positron which is trapped in the dense region. In a medium, the life time of the positron has three components, τ_1, τ_2, τ_3, where τ_1 is the shorter life time, τ_2, is the longer life time and τ_3, corresponds to positronium formation. The longer lifetime, τ_2, corresponds to the situation where a positron is trapped in a micro-voids [7] of free volume, V_F. It is assumed that the micro-void has a volume equivalent to a sphere [8] of radius, R. The life time, τ_2, is related to the radius, R, as per the following relation [9]: $\dfrac{1}{4.5\tau_2} = 1 - \dfrac{R}{R + \Delta R} + \dfrac{1}{2\pi} \sin \dfrac{2\pi R}{R + \Delta R}$. Here, R is in nm and ΔR is a constant ~0.1659 nm for most of the polymeric materials. By measuring the life time, τ_2, the average value of the radius, R, and hence the average value of the free volume, V_F, of the micro-voids in the polyimide can be estimated [24].

10.2.6.1   Positron Annihilation Spectroscopy Technique: The Basics

Polymer contain amorphous domain, which results from fluctuations in the packing density of the macromolecular chains. In amorphous domains, there exist low electron density regions called free volume holes or micro voids. The amount of free volume holes in a polymer is called fractional free volume or free volume content. The exposure of polymers to different energetic radiation may alter the amount of free volume exist in the polymer. The change in the free volume after radiation treatment is called as radiation induced free volume. Nowadays, a sophisticated technique like positron annihilation spectroscopy is available for determining directly the angstrom sized fee volume holes and their concentrations in the polymers [1]. A brief description of the positron annihilation technique is as follows. When an energetic positron forms a radioactive source enters a condensed medium like polymer, it thermalizes by losing its energy in a very short time, then annihilates with an electron of a medium. Annihilation usually takes place from different positron states, viz. free annihilation, or from a localized state (trapped state) or from a bound state called positronium (Ps). Ps can exist in two spin states, a para-positronium (p-Ps spin anti parallel), which annihilates with the life time of 125 ps and ortho-positronium (o-Ps, spin parallel), which annihilates with a life time of 140 ps in free space. In condensed matter, the o-Ps annihilates predominantly via a fast channel with an electron of a surrounding medium possessing an opposite spin; a process called pick off annihilation and the o-Ps life time get reduced to a few ps. Each of these annihilation processes has a characteristic life time. In polymers, the o-Ps lifetime is an important parameter since Ps is a trapped and annihilated in free volume site and hence it is related to the mean size of free volume holes in the polymer matrix [1]. However, in case of polyimide the sizes of the free volume holes or micro-voids are too small for the formation and the localization of positronium atoms [1]. Hence, no evidence of positronium formation can be observed in any of the polyimide films. With this in view, a study has been carried out in electron irradiated and virgin polyimide samples. The Positron annihilation technique was used to measure the size of the micro-voids and its content.

In the present work, the life time of positrons in a few virgin and all the electron-irradiated polyimide samples were measured. These measurements were carried out using a 20 μCi positron source (^{22}Na). The source was prepared by evaporating a water solution of ^{22}NaCl on a rhodium foil of thickness ~8 μm. Positron lifetime spectra were recorded using a fast-fast coincidence system consisting of two 1-inch conical BaF_2 scintillator detector coupled to XP2020Q photomultiplier tubes. The prompt time resolution of the system using a[60]Co source with a[22]Na gate was ~265 ps. The spectrum for each sample was acquired for a period of ~24 hours to get a total number of counts ~10^6. The positron lifetime spectra for these polyimide samples were resolved into three components convoluted by single Gaussian-type time resolution function using the codes RESOLUTION and POSITROFIT. A source component of intensity 10% was removed from all the lifetime spectra. The value of the variances of the fit were keep between 0.99 and 1.08. From these measurements the three components, τ_1, τ_2, τ_3, and the corresponding intensities I_1, I_2, and I_3 were estimated for each of the electron-irradiated and a few virgin polyimide samples. The magnitude of the intensity, I_2, corresponds to the number of annihilation events with the life time, τ_2. The radius, R, was estimated by substituting, τ_2, in equation: $\dfrac{1}{4.5\tau_2} = 1 - \dfrac{R}{R + \Delta R} + \dfrac{1}{2\pi} \sin \dfrac{2\pi R}{R + \Delta R}$. The variations in τ_2, I_2, and, V_F, with electron fluence, Φ, are shown in Figure 10.18a–c, respectively. The variation in, the τ_1, τ_3, I_1, and I_3, with the electron fluence, Φ, was very small and therefore not incorporated in this book.

It can be seen in Figure 10.18a that, the life time, τ_2, in virgin polyimide is ~400 ps. The life time, τ_2, decreases slowly with electron fluence, Φ, and reaches a minimum value of ~376 ps in the polyimide exposed to an electron fluence of 2×10^{15} e/cm^2. With further increase in the electron fluence, Φ, the life time, τ_2, increases and reaches a value of ~437 ps at an electron fluence, Φ, of 5×10^{15} e/cm^2. It is interesting to note that, even though the life time, τ_2, decreased with electron fluence, Φ, however, the intensity, I_2, (Figure 10.18b) did not change much over this range of electron fluence. These results reveal that, the number of positron-trapping voids remains almost the same but the effective free volume of the micro-voids, V_F, decreases with increasing the electron fluence up to 2×10^{15} e/cm^2. It appears that on exposure to electrons, the side groups, stacking conformations and molecular chains undergo partial rupturing. These process leads to a decrease in the intrinsic [15] free volume, V_F, of the irradiated polyimide. The increase in the life time, τ_2, with electron fluence, Φ, above 2×10^{15} e/cm^2 may be attributed to the steep increase in the number of ruptured bonds of oxygen in –ODA and that of nitrogen and carboxyl carbon (-C=O) in - PMDA. As a result, during electron-irradiation the process of evolution of various gas phase species such as, oxygen, C=O, nitrogen, C-O, H-OH, N-H, etc., from the polyimide matrix is accelerated [1].

The results of Residual gas analysis (RGA) results not shown (set-up shown in Figure 10.19) herein confirmed that the gaseous species such as, C=O, N_2, -OH, O_2, etc., evolved from polyimide matrix during electron irradiation. The process of collapsing the molecular chain led to the merging of neighbouring small size voids into a single void with relatively large free volume, V_F. In addition, the passage of gases evolved from bulk to the surface of the polyimide sample may also change the intrinsic free volume, V_F. As a result, the free volume, V_F, of the voids increases but at the same time the number of voids and hence the intensity, I_2, decreases with increasing the electron fluence, Φ, above 2×10^{15} e/cm^2. As a result, voids with large free volume are retained at the surface and therefore the silver atoms could diffuse in the polyimide at room temperature. From Figure 10.18c, it is observed that at an electron fluence of 3×10^{15} e/cm^2, micro-voids with an average free volume,

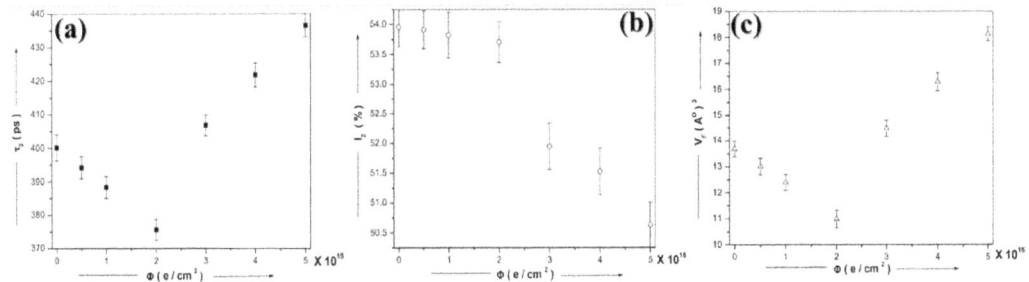

FIGURE 10.18 Recorded variations in τ_2, I_2, and, V_F, with electron fluence, Φ, for polyimide.

FIGURE 10.19 Experimental set-up of the R.G.A. coupled to Race-Track-Microtron 1, 6-MeV pulsed electron accelerator.

V_F, ~ 14.5 $(\text{Å})^3$ and therefore of diameter ~3 Å are created in the polyimide matrix. The average size of these voids ~3 Å appears to be just sufficient to allow the diffusion of silver atoms of atomic diameter ~2.88 Å, into the polyimide matrix. It is interesting to note that silver atoms did not diffuse in the polyimide samples which were initially irradiated under vacuum or atmospheric conditions and then immersed in $AgNO_3$ solution (as discussed in the Section 10.2.5.1. In these polyimide samples either the silver atoms did not diffuse or the number of the silver atoms diffused was below the detection limit of the RBS technique. It appears that, in the surface region of these polyimide samples the processes like segmental relaxation of the dangling bonds and crosslinking were initiated immediately after the turning off of the electron beam. Through these processes the average free volume of the voids in the surface region is reduced below the threshold volume required for the silver diffusion. These processes, however, do not affect the free volume of the voids in the bulk region. The measured average depth of diffusion of silver in polyimide was ~2.5 μm. The enhancement in the diffusivity of silver by orders of magnitude as compared to that in virgin polyimide or polyimide irradiated under vacuum or atmospheric conditions at room temperature can be used for modifying the characteristics of the polyimide [1–27].

REFERENCES

[1] Alegaonkar, Prashant S. "Studies on: radiation induced damage and elemental diffusion in polyimide." (2004). http://hdl.handle.net/10603/126163
[2] Fink, Dietmar, P. S. Alegaonkar, A. V. Petrov, A. S. Berdinsky, V. Rao, M. Müller, K. K. Dwivedi, and L. T. Chadderton. "The emergence of new ion tract applications." *Radiation Measurements* 36, no. 1–6 (2003): 605–609.
[3] Fink, Dietmar, A. V. Petrov, V. Rao, M. Wilhelm, S. Demyanov, P. Szimkowiak, M. Behar, P. S. Alegaonkar, and L. T. Chadderton. "Production parameters for the formation of metallic nanotubules in etched tracks." *Radiation Measurements* 36, no. 1–6 (2003): 751–755.
[4] Martin, Charles R. "Nanomaterials: a membrane-based synthetic approach." *Science* 266, no. 5193 (1994): 1961–1966.

[5] Berdinsky, Alexander S., Dietmar Fink, Alexander V. Petrov, Manfred Müller, Lewis T. Chadderton, Jose F. Chubaci, and Manfredo H. Tabacniks. "Formation and conductive properties of miniaturized fullerite sensors." *MRS Online Proceedings Library (OPL)* 705 (2001): Y4.7.

[6] Biswas, A., D. K. Avasthi, Benoy K. Singh, S. Lotha, J. P. Singh, D. Fink, B. K. Yadav, B. Bhattacharya, and S. K. Bose. "Resonant electron tunneling in single quantum well heterostructure junction of electrodeposited metal semiconductor nanostructures using nuclear track filters." *Nuclear Instruments and Methods in Physics Research Section B: Beam Interactions with Materials and Atoms* 151, no. 1–4 (1999): 84–88.

[7] Berdinsky, Alexander S., D. Fink, Hei-Gon Chun, L. T. Chadderton, V. A. Gridchin, and P. S. Alegaonkar. "Model of conductivity of fullerite tubules in ion tracks of polymer foils." In *7th Korea-Russia International Symposium on Science and Technology, Proceedings KORUS 2003 (IEEE Cat. No. 03EX737)*, vol. 1, pp. 63–69. IEEE, 2003.

[8] Berdinsky, Alexander S., D. Fink, A. V. Petrov, Hei-Gon Chuh, L. T. Chadderton, V. A. Gridchin, and P. S. Alegaonkar. "Pressure dependence of conductivity of fullerite structures." In *7th Korea-Russia International Symposium on Science and Technology, Proceedings KORUS 2003.(IEEE Cat. No. 03EX737)*, vol. 1, pp. 141–146. IEEE, 2003.

[9] Petrov, A. V., D. Fink, G. Richter, P. Szimkowiak, A. Chemseddine, P. S. Alegaonkar, A. S. Berdinsky, L. T. Chadderton, and W. R. Fahrner. "Creation of nanoscale electronic devices by the swift heavy ion technology." In *2003 Siberian Russian Workshop on Electron Devices and Materials. Proceedings. 4th Annual (IEEE Cat. No. 03EX664)*, pp. 40–45. IEEE, 2003.

[10] Berdinsky, A. S., D. Fink, Hui-Gon Chun, Yong-Zoo Yoo, Ji-Beom Yoo, A. V. Petrov, and P. S. Alegaonkar. "Variation of conductivity of fullerite structures under different types of pressure." *Journal of Sensor Science and Technology* 13, no. 5 (2004): 392–398.

[11] Elektronische Nanobauelemente auf der Basis einer einzelnen Ionenspur, D Fink, PS Alegaonkar, AS Petrov, GR Patent 203 20 566.9.

[12] Petrov, A. V., S. E. Demyanov, D. Fink, W. R. Fahrner, A. K. Fedotov, P. S. Alegaonkar, and A. S. Berdinsky. "Novel electronic devices for nanotechnology based on materials with ion tracks." In *Physics, chemistry and application of nanostructures*, edited by V. E. Borisenko, S. V. Gaponenko, and V. S. Gurin, pp. 544–547. Belarusian State University of Informatics and Radioelectronics, Institute of Molecular and Atomic Physics, Belarus. 2005.

[13] Berdinsky, A. S., P. S. Alegaonkar, H. C. Lee, J. S. Jung, J. H. Han, J. B. Yoo, D. Fink, and L. T. Chadderton. "Growth of carbon nanotubes in etched ion tracks in silicon oxide on silicon." *Nano* 2, no. 01 (2007): 59–67.

[14] Fink, D., A. Chandra, P. Alegaonkar, A. Berdinsky, A. Petrov, and D. Sinha. "Nanoclusters and nanotubes for swift ion track technology." *Radiation Effects & Defects in Solids* 162, no. 3–4 (2007): 151–156.

[15] Fink, D., A. Chandra, W. R. Fahrner, K. Hoppe, H. Winkelmann, A. Saad, P. Alegaonkar, A. Berdinsky, D. Grasser, and R. Lorenz. "Ion track-based electronic elements." *Vacuum* 82, no. 9 (2008): 900–905.

[16] Alegaonkar, Prashant S., Vasant N. Bhoraskar, and Sudha V. Bhoraskar. "Polyimide: from radiation-induced degradation stability to flat, flexible devices." In *Polyimide*, edited by B.P. Nandeshwarappa and Sandeep Chandrashekharappa. IntechOpen, California 2021.

[17] Fink, Dietmar, P. S. Alegaonkar, A. V. Petrov, M. Wilhelm, P. Szimkowiak, M. Behar, D. Sinha, W. R. Fahrner, K. Hoppe, and L. T. Chadderton. "High energy ion beam irradiation of polymers for electronic applications." *Nuclear Instruments and Methods in Physics Research Section B: Beam Interactions with Materials and Atoms* 236, no. 1–4 (2005): 11–20.

[18] КАНЮКОВ, ЕЮ, АВ ПЕТРОВ, and СЕ ДЕМЬЯНОВ. "ЮА ИВАНОВА, ДК ИВАНОВ, ЕА СТРЕЛЬЦОВ, АК ФЕДОТОВ." УДК 537.311.322: 621.763

[19] Fink, D., A. Chandra, P. Alegaonkar, A. Berdinsky, A. Petrov, and D. Sinha. "Nanoclusters and-tubes for Swift Ion Track Technology." *SMR* 1758 (2006): 18.

[20] Alegaonkar, P. S., V. N. Bhoraskar, P. Balaya, and P. S. Goyal. "Dielectric properties of 1 MeV electron-irradiated polyimide." *Applied Physics Letters* 80, no. 4 (2002): 640–642.

[21] Alegaonkar, P. S., A. B. Mandale, S. R. Sainkar, and V. N. Bhoraskar. "Refractive index and dielectric constant of the boron and fluorine diffused polyimide." *Nuclear Instruments and Methods in Physics Research Section B: Beam Interactions with Materials and Atoms* 194, no. 3 (2002): 281–288.

[22] Soares, M. R. F., P. Alegaonkar, M. Behar, D. Fink, and M. Müller. "$^6Li^+$ ion implantation into polystyrene." *Nuclear Instruments and Methods in Physics Research Section B: Beam Interactions with Materials and Atoms* 218 (2004): 300–307.

[23] Naddaf, M., C. Balasubramanian, P. S. Alegaonkar, V. N. Bhoraskar, A. B. Mandle, V. Ganeshan, and S. V. Bhoraskar. "Surface interaction of polyimide with oxygen ECR plasma." *Nuclear Instruments and Methods in Physics Research Section B: Beam Interactions with Materials and Atoms* 222, no. 1–2 (2004): 135–144.

[24] Alegaonkar, P. S., and V. N. Bhoraskar. "Effect of MeV electron irradiation on the free volume of polyimide." *Radiation Effects and Defects in Solids* 159, no. 8–9 (2004): 511–516.

[25] Alegaonkar, P. S., and V. N. Bhoraskar. "Characterization of polyimide irradiated in silver solution with a pulsed electron beam." *Nuclear Instruments and Methods in Physics Research Section B: Beam Interactions with Materials and Atoms* 225, no. 3 (2004): 267–274.

[26] Majeed, Riyadh M. A. Abdul, A. Datar, S. V. Bhoraskar, P. S. Alegaonkar, and V. N. Bhoraskar. "Dielectric constant and surface morphology of the elemental diffused polyimide." *Journal of Physics D: Applied Physics* 39, no. 22 (2006): 4855.

[27] Bhave, Tejashree, P. S. Alegaonkar, V. N. Bhoraskar, K. A. Bogle, D. K. Avasthi, and S. V. Bhoraskar. "Electrical characteristics of etched ion-tracks in polyimide filled with silver nanoparticles." *Radiation Effects and Defects in Solids* 173, no. 7–8 (2018): 617–628.

11 Summary and Outlook

To summarize and looking ahead in the area of nanocarbons, the subject is ever fresh. The applications presented herein are typical, representative, and limited to the current interest, but show future landscape of the field. There are always new important applications emerging and enriching the subject.

We began with the status of carbon revealing its importance from minerals, crystals, jewels to exciting synthetic allotropes in the form of buckyballs, nanotubes, graphene, etc. A brief review of special properties of nanocarbons in terms of their superiority over other classes of materials underlined their importance in future applications.

Subsequently, in Chapter 2, a number of carbon systems have been reviewed for their preparation methods. The system involves carbon nanoparticles and carbon nano-spheres and their growth by plasma chemical vapour deposition using porous templates. Sublimation of carbon nano-spheres via a single-step synthesis has been presented, which provided important clues for their applications in energy, sensors, etc. The growth of catalytic carbon nanotubes via plasma, thermal, and rapid thermal CVD deposition has also been presented by clarifying the role and importance of catalytic barrier or buffer layer required for CVD synthesis. The elemental analysis of inter-diffusion of catalytic and barrier layer has been presented. In addition, the super-growth of CNTs by water-assisted chemical deposition is presented. The chapter also presented templated, patterned, and selective deposition of nanotubes useful for sensors, field emission, and other important applications. The synthesis of expanded graphite, disordered graphene, and graphene nano-ribbons, including reduced graphene oxide and CVD graphene, is also presented. The study reveals the synthesis parameters such as method, time, temperature, partial pressure, and precursors involved to obtain controlled, selective, and yield-tuneable growth of such nanocarbons that are subsequently implemented for desired applications. The chapter ends with foundry processing of micrographite to obtain variable density graphite and four-dimensional carbon fabric.

In the next chapter, i.e. Chapter 3, dynamic mechanical properties of typically grown graphene nano-flakes, nano-ribbons, and spheres as well as porous carbons have been presented. We began with the background of dynamic mechanical response connected to blast or explosion to generate an impulse pressure termed as shock wave. They have catastrophic damage to structure. The question addressed is whether such nanocarbons will be useful for the absorption of shock. A brief background of explosion is presented, revealing the characteristics of shock and involved mechanism of energy dissipation into material. A survey of laboratory-level shock synthesis has been presented, including design and development of blast-mitigating architecture of shield. A number of methods have been discussed, such as contact explosion, standoff exposure, shock wave tube, and split Hopkinson pressure bar. A detailed analysis of shock absorption properties of nano-flakes has been presented by analysing the stress–strain curve, establishing a hydrodynamic model to estimate and derive a number of variables such as particle velocity, pressure, and volume by describing its phase state of equation. This has been correlated with fractographic analysis to identify the damage induced in flakes in the form of dislocations such as buckling, gliding, and twining.

For layered flakes, the relevant dynamic force model is presented. The systematic study carried on flakes has been spanned on to other fabricated systems such as ribbons, spheres, and porous nanocarbons revealing and comparing parameters such as elastic limit, compressive strength, and shock damping coefficient. The carbon nano-spheres have shown promising shock mitigation architecture with standing to higher strain rates. The study also reveals signal processing of shock on imprinted on nanocarbons to understand the anatomy of impulse pressure propagated through such structures.

DOI: 10.1201/9781003317258-11

In Chapter 4, radar absorption properties of nanocarbons have been investigated for military purpose applications. Radio-microwaves over frequency range of 0.3–300 GHz is of tactical importance to track and detect the enemy targets. However, the mechanism involved to create clutter in such radar detection can be achieved by designing and developing EMI shielding and coating paints. The chapter begins with the background of radar, its detection range, classification of specific treats, and concerned communication frequency band. The mechanism of scattering by radar waves has been discussed, including simulation of such experiments at laboratory level. Different types of scattering parameter measurement techniques have been presented and discussed. The chapter deals with the fundamental background of wave formalism, estimation of losses, absorption, condition for absorption, minimization of reflection, and enhancement of attenuation performance, including optimization of coating thickness and tailoring reflection law has been discussed. In the chapter, the reader will know how the material architecture can build high-performance shielding coats, and the shielding performance of a number of nanocarbons and their composites have been presented, which include ferrite-based nanotubes, graphene-like nanocarbons, graphene derivatives, hetero-junction-based nanocarbon composites, ferro-nanocarbons, and layered composites. A number of technical parameters and their importance in shield architecture have been discussed, for example representative characteristic impedance, skin depth, surface resistance, standing wave ratio. A rich source of references has been provided.

Subsequently, in Chapter 5, quality assessment and testing of rocket motor grid, and graphitic micrographite nozzles have been investigated. The assessment is presented for thermo-physical properties of variable density graphite. Initially, the importance of micrographitic carbon in missile engineering has been discussed in addition to a brief visit to rocket motor system, its components, and the choice of micrographitic nozzles. The density effect on thermo-physical properties of the nozzles has been discussed. The chapter also presents thermal measurements and estimation of a number of thermal parameters, such as specific heat capacity, thermal conductivity, heat diffusivity, and coefficient of thermal expansion. Its standardization and importance has been presented for crystalline graphite material. The role of physical properties, especially microstructure in the form of green orientating, stacking, packing, intrinsic void, porosity distribution, surfaces/interfaces, degree of graphitization, has been discussed in light of thermal properties of crystalline graphite and micrographite of variable density. Large number of fitment table to the thermal parameters have been presented and discussed for their thermal behaviour. A number of canonical thermodynamic parameters have been estimated, such as Debye temperature, change in entropy, and enthalpy as a function of heat diffusivity, and finally, the suitable density for designing nozzles has been identified. As a continuation of this work, the chosen micrographitic nozzle has been subjected to static field testing. The testing details are presented. For such a nozzle system, the estimation of change in entropy and corresponding enthalpy calculation have been revealed, including the comparison of structural parameters such as lattice distance, crystalline length, in-plane amorphization, and stacking density degree of graphitization.

In Chapter 6, electrochemical and energy storage facet has been discussed for nanocarbons. A few representative systems such as reduced graphene oxide, carbon nano-spheres, and hetero-junction layered materials are presented for their electrochemical performance. The chapter begins with the current scenario of global energy demand, emerging alternative energy storage and energy conversion sources, their performance, and future challenges. The basic formulation of electrochemical parameters has been presented. The preparation and electrochemical performance of tellurium–reduced graphene oxide is presented for their architecting superior electrochemical system. The structure–property relationship has been revealed to underline the optimum composition of tellurium within reduced graphene oxide. Fully sealed device characteristics have been presented for its electrochemical performance. In the same chapter, facile fabrication of flexible and durable supercells made up of carbon nano-spheres has been presented. The pre-analysis of material has been correlated to post-processing after a duty cycle of 20,000 or so. Importantly, the chapter provides an insight into building realistic supercapacitors by choosing materials, electrolyte, window potential, etc., to obtain a high-performance ultra-capacitor that has superior stability and capability and charges swiftly and discharges slowly. Estimations of total power and energy delivered for flat and rolled supercell devices are presented. The chapter ends by providing clues for designing of next-generation ultra-capacitors.

Carbon in its bulk form shows strong diamagnetism; however, in its nano-format, particularly graphene at edges, disordered graphene at defected sites, and ribbons at edges, it shows a strong local magnetic moment induced due to the unsettled electron of dangling state. Magnetism in otherwise non-magnetic carbon is the focus of Chapter 7. The chapter begins by revealing the possible spin transport and magnetic correlations in graphene-like nanocarbons doped with nitrogen molecules. The method of nitrogen doping is described. Relevant magneto-spin investigation carried out using electron spin resonance and spin magnetometry has been presented including details of experimental work performed. In addition, electronic transport process has been revealed in disorder and nitrogenated graphene system. A number of spin transport parameters have been compared for disorder and nitrogenated graphene-like nanocarbons. Some of the important parameters are spin–spin relaxation, spin–lattice relaxation, spin–orbit coupling, momentum relaxation rate, and Pauli-spin susceptibility as a function of the measured temperature over 123–475 K. Magnetic measurement showed saturation level magnetic moment with the presence of corrosion and remanence even at 300 K. The corresponding electron spectroscopic chemical studies are presented to investigate the radical nature of nitrogenated carbon systems. The details of RKKY interaction mechanism are presented with a brief comment on N-GNC qubit. In another study, molecular spin interaction of tellurium adatom in reduced graphene oxide system is presented, including electrolytic conductance measurements. Towards the end of the chapter, tetrakis(dimethylamino)ethylene-induced magnetism in disordered and crystalline graphene has been revealed by investigating the thermomagnetic behaviour simulated through computational techniques. The chapter provides a detailed reference list at the end.

Chapter 8 presents the multifunctional aspect of nanocarbons to demonstrate metamaterials and non-linear optical properties for sensing and mechanically tough fabric applications. With the increase in demand for smart digital home appliance, it has put up a condition to design and architect materials with multifunctional facets. In this, optical gas sensing properties of carbon nanoparticles have been demonstrated by investigating their optical band structure to design and develop sensing characteristics. As a proof of concept, the development of ammonia gas sensor has been presented compounded with measurement details such as sensor transfer function, mechanism of sensing, and molecular-level detection of ammonia. The same material simultaneously showed a high-performance EMI shield architecture as an added feature for multifunctional applications. Coating characteristics, conductivity, and percentage reflection loss have been investigated to reveal shielding effectiveness. In a study shown in this chapter, iron nanoparticles incorporating nanocarbons showed metamaterials aspect to generate X-band electromagnetic cloak. The obtained metamaterials have been investigated for dielectric and magnetic anisotropic measurements to reveal ring resonating characteristics. The modelling and simulation of computational electromagnetic has been demonstrated for Nicolson–Ross–Weir and retrieval methodology to reveal bi-anisotropy of the fabricated metamaterials. In the same chapter, mechanical properties of GNC nanocomposites have been revealed at low weight fraction of filler to exploit tensile, flexural, and fracture toughness as a function of dispersion of GNC in epoxy matrix. The chapter ends with revealing the mechanical properties of electro-spun PVA/CNT composite nanofibers.

In Chapter 9, application engineering of nanocarbon-reinforced composites has been presented with a special emphasis on field emission properties of CNT composites fabricated in the form of paste. Initially, Fowler–Nordheim theory of field emission has been presented, revealing important parameters such as field, geometrical, mean, field enhancement factor including defining turn-on-field, current and current density as a function of applied field, change in inter-electronic distance, activation of cathode layers, etc. Array field emission configuration has been presented, including screen-printed triode carbon nanotube field emission array for flat lamp application. The role and reasons for coating field emitter have been investigated to obtain superior field emission parameters for array field emitter, and plasma treatment on cathode layer has been presented to achieve high emission stability and life time of cathode layers. Moreover, the effect on field emission parameters has been investigated for CNT dispersed via chemical, mechanical, and combinational routes. A comparative study has been presented. In the same chapter, a special emphasis is given to the preparation and production of the TiO_2-coated carbon nanotube dye-sensitized solar cell, including electro-spun fibre mat for mechanically tough and challenging applications.

Chapter 10 presents ion track membranes for devices and nuclear radiation-induced modification in optoelectronics. The chapter begins with the emergence of nano-ion track membrane for flat flexible devices. It includes a brief history of evaluation of ion tracks that had been registered by GeV ions into polymeric foil to create latent tracks which can be manipulated by chemical etching. Various strategies for the choice of swift heavy ions, ion influence, and selection of polymeric foils in combination with recipes for chemical etching have been presented. It followed the estimation of etched ion track dimensions and geometry by the ion transmission spectroscopy technique. Its analysis is presented together with microscopic and optical imaging. The actual device has been designed using mask lithography to fabricate micro-transformer. Quality assessment and quality characteristics, including gain analysis, are presented for the micro-transformer. In the same chapter, radiation-induced modifications in poly-carbon have been presented for optoelectronic properties. A complete background of radiation-induced modifications has been presented, which includes the interaction of ions, electrons, and gamma rays to create modifications and radiation-enhanced diffusion in polymers. A brief theory of radiation-enhanced diffusion is presented, including the theory of molecular dielectric relaxation. A change in refractive index and optical properties of B-, F-, and Si-diffused polymers has been presented. Towards the end of the chapter, radiation-induced free volume analysis is presented. Three appendices (Appendix A–C) are augmented towards the end of this summary and outlook.

Appendix A

CHAPTER 3

In Figure A.3.1a, at HEL, the elastic fraction of the wave is separated from the plastic portion. This elastic fraction, below the HEL, travels at a velocity higher than that of the plastic wave. The effects observed in (a) can thus be explained. Beyond the HEL, the pressure rises continuously to the top due to the non-discontinuous nature of plastic wave front. The rate of increase of this pressure is dictated by the constitutive behaviour of GNF. The involved phase transition cannot be seen clearly in the wave profile. It seems that the wave may be separated into two wave fronts, marginally. At the top of the plot, the pulse duration plateau is seen. When unloading starts in GNF, the profiles suggest that it occurs initially elastically and then plastically. This elasto-plastic transition in unloading GNF leaves the shock signal in an analogous manner to the HEL on loading. The wave on its reflection seems to have fractured the material as seen in spalling *tail* towards the end. A similar trend is seen for $\acute{e}$ vs ε. The other profiles related to $\acute{e}$, force vs time are provided in Chapter 3.

The prepared samples were subjected to high strain rate loading of 1.5 GPa using split Hopkinson pressure bar (sHPB). The strain rate profile and force versus time profile is shown in Figure A.3.2a and b, respectively.

Figure A.3.3 shows images typically recorded for (a) GNF and (b) HSR GNF that showed disintegration of GNF after interaction with impulse pressure.

CHAPTER 5

Figure A.5.1 shows the typical micrographic images recorded using SEM for VDGs at a constant working distance (~6.5 mm). The upper panel of Figure A.5.1a and b shows surface morphology of VDG@1744, respectively, recorded at 15,000 and 30,000 magnifications. The open circle and connecting arrow indicate particular details of a portion. One can see the curvilinear graphitic surfaces upon which a large number of carbonaceous particles were segregated. The particles were found to be random in size and shape compounded with non-uniformity in its spread/distribution over the

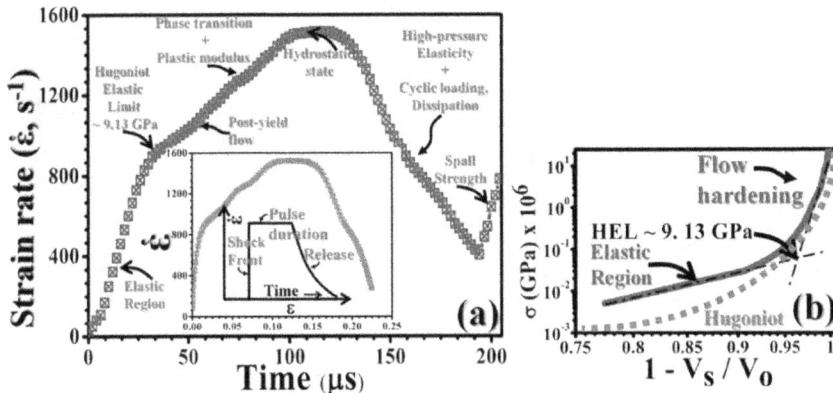

FIGURE A.3.1 (a) Variations in strain rate, $\acute{e}$, with time, imprinting specific shock wave profile onto 2D GNF carbon layers. The curve exhibits a number of stages that are dependent on a peculiar response of GNF to the pressure pulse. Inset shows $\acute{e}$ vs ε, indicating hydrodynamic response of GNF and inside it an idealized shock profile. Plot (b) shock response of GNF (black frozen line).

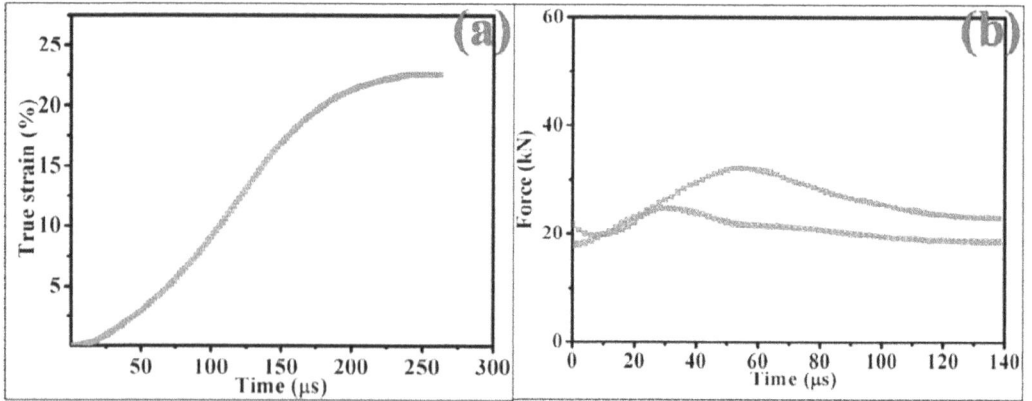

FIGURE A.3.2 GNR subjected to high strain rate of 1.5 GPa; its profiles: (a) true strain as a function of time (μs) and (b) force (kN) v/s time (μs).

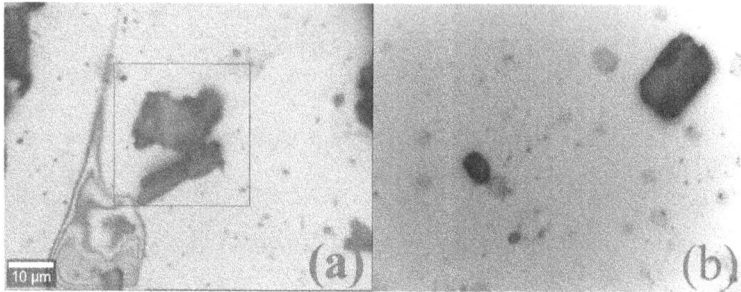

FIGURE A.3.3 Recorded optical micrographic image at 100×.

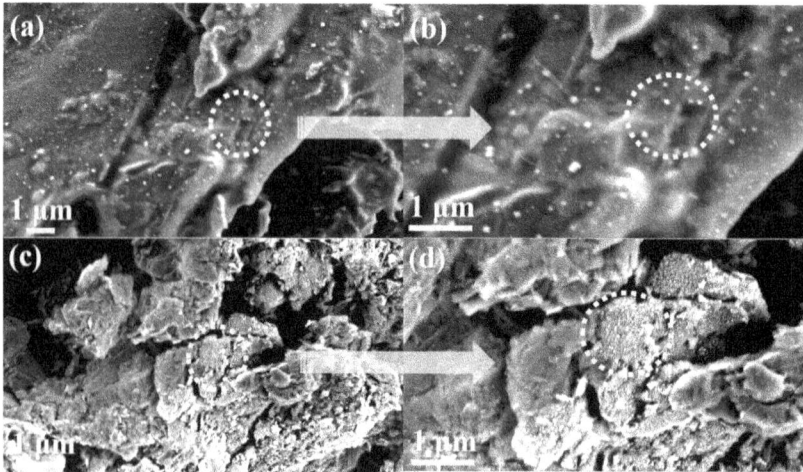

FIGURE A.5.1 Typical SEM images recorded for (a), (b) VDG@1744 and (c), (d) VDG@1811.7 specimens (scale bar: 1 μm). Open circles (left gallery) show images taken at 15,000 magnification; connecting arrows correspondingly indicate zoomed in portion (right gallery) recorded at 30,000 magnification. All images were recorded at the same working distance of ~6.5 mm.

TABLE A.5.1

Fitting Parameters for Milosevic Equation

Density (kg/m³)	Milosevic Fitting Parameters	
	A	n
1611	4201.69 ± 517.32	1.41 ± 0.020
1744	2904.81 ± 602.14	1.30 ± 0.040
1811.7	0971.30 ± 025.73	0.87 ± 0.002
1910.4	2724.49 ± 617.22	1.301 ± 0.036

surface. At some locations, these particles were, apparently, seen to be entrapped within over-layers and wrinkles of graphene. Figure A.5.1c and d is recorded for VDG@1811.7 in an identical fashion as that for the previous specimen. The surface morphology of VDG@1811.7 seems to possess more degree warp compared with VDG@1744. It is a general impression that, for VDG@1811.7, the amount of amorphous carbon deposits was more than that for VDG@1744. However, we were not able to quantify the amount/areal density in both the specimens (Table A.5.1).

Appendix B

CHAPTER 6

Figure B.6.2.1 shows the fitting of (a) C-1s for rGO and Te–rGO, (b) O-1s, and (c) 3d-Te, which were fitted into several components. Figure B.6.2.1a indicates the presence of carbon with five functional components such as (a) non-oxygenated C–C ring (284.6eV), (b) C–N (285.9eV), (c) C in C–O (286.0eV), (d) carbonyl group (287.8eV), and (e) carboxylate carbon, C(O)O (289eV). In Figure B.6.2.1b, the corresponding changes were seen. The decrease in intensity at 286.0 and 287.4eV showed a reduction in oxygen functional groups present in rGO. The peak at 289eV in rGO significantly disappeared in Te–rGO, indicating the substantial removal of C(O)O group. The sharpening and increase in the intensity of C–C peak at 284.6eV indicates the recovery of sp^2 carbon that was ruptured during the oxidation process. Although C1-s in Te–rGO also exhibits these oxygen functional groups, their peak intensities are extremely weaker than those present in rGO. The ratio C-to-O was computed to be 6.06 and 2.36, respectively, for rGO and Te–rGO. This change is indicative of the reduction in oxygen proportion with an increase in sp^2 content.

The deconvoluted spectra of O-1s of Te–rGO and Te-$3d_{5/2}$ are provided in Figure 6.2. In a nutshell, their studies indicated the following: The emerged Te-$3d_{5/2}$ at 568.29eV was having at% 3.64 and deconvoluted into three components. The peak at 568.23eV is attributed to C-Te bonding, which has 54 at% contribution. The other two contributions at 575.63 and 572.22eV respectively, for TeO_2 and Te-unreacted metal fraction. The 3d component corresponds to the spin–orbit splitting character of the bond. For pure Te atoms, a doublet of $3d_{5/2}$–$3d_{3/2}$ is reported with binding energy separation ~10.4eV; however, in our case, no such splitting was observed. It seems that the bonding between Te and C quenched the coupling character, significantly (Figures B.6.2.2 and B.6.8.1).

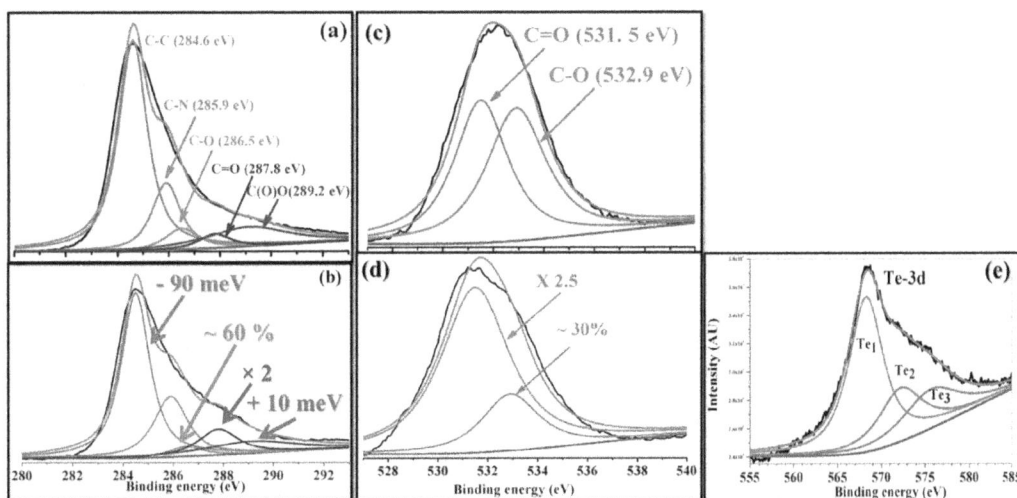

FIGURE B.6.2.1 Recorded ESCA for (a), (b) C-1s and (c), (d) O-1s, respectively, for rGO and 1 wt% Te–rGO; (d) 3d-Te elemental scan in 1 wt% rGO.

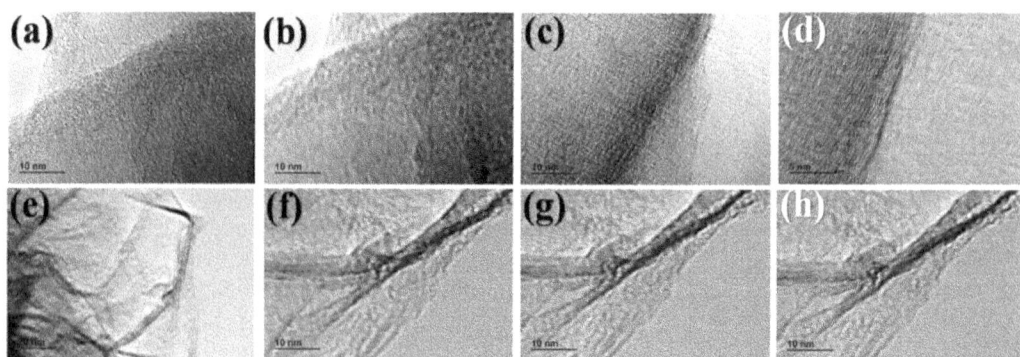

FIGURE B.6.2.2   Recorded HRTEM images: (a–d) rGO and (e–h) Te–rGO.

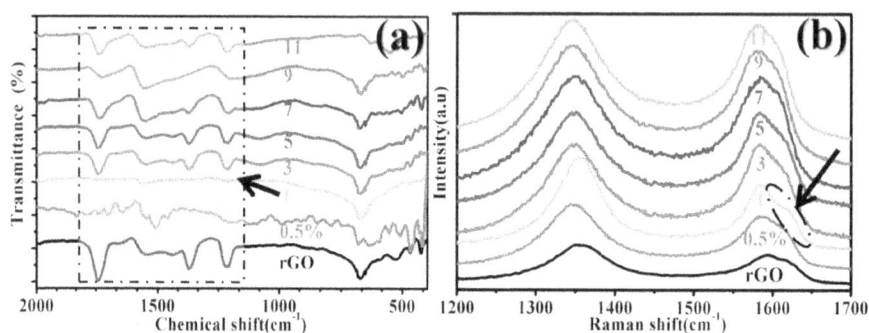

FIGURE B.6.8.1   Recorded (a) FTIR and (b) Raman spectra of rGO and 0.5–11 wt% Te–rGO.

From FTIR and Raman study (Figure B.6.8.1), we observed a drastic change in 1 wt% Te–rGO and rGO. The relevant discussion is also presented in Figure 6.8. For higher wt%, the peaks at 1210, 1370, 1440, 1525, 1730, 2312, and 2660 cm^{-1} remain almost unaltered when compared to rGO samples. Hence, the skeletal molecular vibrations of rGO are dramatically varied by dilute presence of Te within rGO network. It may be possible that, at higher concentrations, Te–Te molecular interactions are more dominant than their interaction with rGO framework. There may be clustering of Te having weaker interaction with rGO. Due to this, we mainly focused on these two systems only (B.6.8.1).

The XPS analysis of O-1s is as follows: The presence of oxygen-containing functional groups in carbon nano-spheres (CNS) is confirmed by the O-1s spectrum shown in Figure B.6.9.1. The peaks at 531.0, 532.4, 533.9, and 547.9 eV are ascribed to the carbonyl (10%), hydroxyl (33%), etheric oxygen (25%), and carboxyl (37%), respectively. Among them, the presence of carbonyl oxygen is relatively low compared to others. It may be possible that carboxyl and hydroxyl moieties may participate in charge and discharge processes, readily over other oxygen impurities in CNS [1] (Figures B.6.12.1–B.6.12.4 and Tables B.6.1–B.6.3).

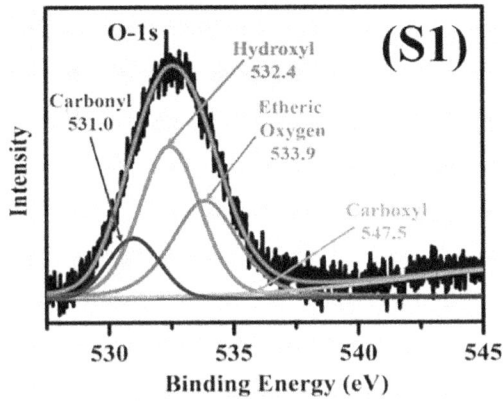

FIGURE B.6.9.1   Recorded O-1s spectrum for CNS.

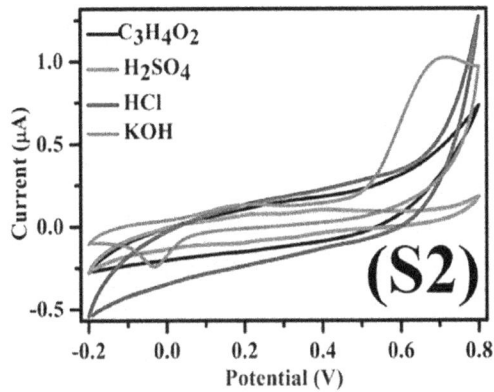

FIGURE B.6.12.1   Two-electrode cyclic voltammetry (CV) curves @10 mV/s for different electrolytes.

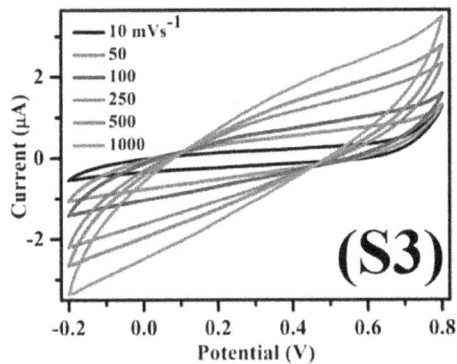

FIGURE B.6.12.2   Two-electrode CV curves, at different scan rates, for 1 M HCl.

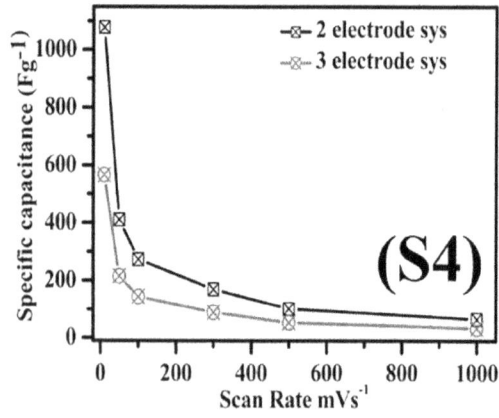

FIGURE B.6.12.3   Specific capacitance (CSP) vs scan rate for both electrode systems, in 1 M HCl.

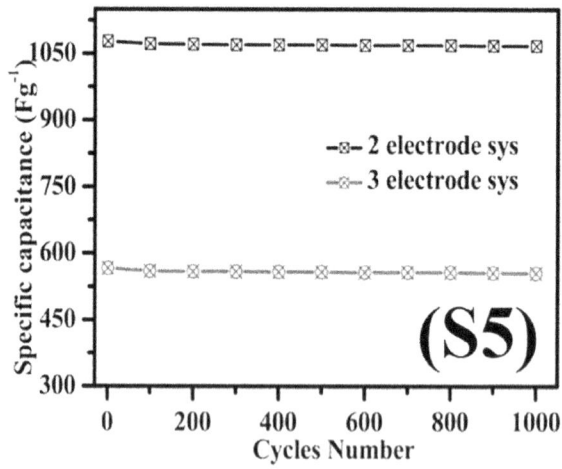

FIGURE B.6.12.4   Stability curve for both electrode systems @ 10 mV/s, in 1 M HCl.

TABLE B.6.1
Electrochemical Parameters for Nanocarbon Electrodes

Additives and Treatments	S_A (m²/g)	C_{sp} (Fg⁻¹)	E_D (Wh/kg)	P_D (kW/kg)	Cyclic Stability	Measurements done at	Electrolyte
Graphene nano-sheet/MWCNTs	169.2	274.0	86.4	0.7	93% (10,000)	1 A/g	1 M Et₄NBF₄
Polymer (P123)	30.0	484.0	–	–	76% (1000)	2 A/g	1 M H₂SO₄
Activated with KOH	79.6	77.0	1.4	0.6	100,000	1 mA/cm²	0.1 M H₂SO₄
Methanol, NiO nanowires	106.0	1950.0	83.0	75.0	85%–90% (2000)	100 mV/s	1 M NaOH
Reduced graphene oxide	376.0	438.0	–	–	76% (1000)	2 A/g	1 M H₂SO₄
MnO₂ powder, methanol	40.0	450.0	26.0	6.0	88% (1000)	0.5 mA	0.1 M KOH
Carbon xerogels, nickel oxide, activated with phosphoric acid	215.0	151.0	–	–	–	2.5 A/g	1 M H₂SO₄
Ferrocene	40.8	1.0	0.3	–	83%	100 mVs⁻¹	Na₂SO₄
Methanol, lithium nickel manganese oxide powders	30.0	154.0	–	–	83% (200)	–	–
MnO₂ nanowires, titanium foils	50.0	483.0	96.0	32.0	10,000	1 mA	0.1 M KOH
Methanol, lithium titanate (LTO) spine	–	200.0	0.3	2.8	4000	0.5 mA	1 M LiPF₆
Current work	791.0	560.0	78.0	2.8	97% (1000)	10 mV/s	1 M HCl

TABLE B.6.2
Elemental Composition of CNS

Orbital	Binding Energy (eV)	Peak Area (Counts-eV/sec)	Sensitivity Factor	Atomic Concentration (Atomic %)
C-1s	284.4	15,562.0	0.296	88.0
sp²	282.9	04248.0		~27.3[a]
sp³	284.6	07146.0		~45.9[a]
Oxidized carbon (C=O)	287.1	02111.0		~13.6[a]
Oxidized carbon (C(O)O)	289.6	01184.0		~07.6[a]
π–π* transition	292.5	00873.0		~05.6[a]
O-1s	532.3	05055.0	0.711	12.0
Carbonyl	531.0	00513.0		~10[a]
Hydroxyl	532.4	01670.0		~33[a]
Etheric oxygen	531.9	01277.0		~25[a]
Carboxyl	547.9	01595.0		~37[a]

[a] Indicates the estimated % of subcomponents in intensity-weighted fractions.

TABLE B.6.3
Performance Characteristics of CNS in Aqueous and Non-Aqueous Electrolytes, Estimated from CV Curves

Electrolyte (1 M)	C_{SP} Fg^{-1} @ 10 mV/s	
	Two Electrodes	Three Electrodes
$C_3H_4O_2$[a]	772.0	400.0
H_2SO_4	338.0	178.0
KOH	832.0	437.0
HCl	1080.0	570.0

[a] Acrylic acid is reported for the first time in camphoric nanocarbon electrodes.

REFERENCE

[1] Xu, Shunjian, Can Liu, and Jörg Wiezorek. "20 renewable biowastes derived carbon materials as green counter electrodes for dye-sensitized solar cells." *Materials Chemistry and Physics* 204 (2018): 294–304.

Appendix C

CHAPTER 7

Analytical calculations of spin transport data for GNCs, N-GNCs, graphene, and N-graphene
The optimization of spin relaxation time and the understanding of relaxation mechanism is crucial. Due to external perturbations, the deformed spin system regains the state of equilibrium "up" or "down" over the characteristic time scale and is termed as spin relaxation time. And one of the most prominent modes of spin relaxation is SO interaction [1]. The quantification of SO could be done using the correlation: $\Delta g = \alpha \left(\dfrac{Li}{\Delta} \right)$, where α is the band structure-dependent constant ≈ 1, and L_i is the spin–orbit coupling constant (SOCC). The value of $\Delta g = 5.25 \times 10^{-3}$, and the magnitude of π–π^* bandwidth, Δ, is computed to be ~4.06 eV for GNCs. The L_i is found to be ~22.12 meV for GNCs. The spin relaxation rate, Γ_{spin}, is related to T_{sl} via relation: $\Gamma_{\text{spin}} = \hbar/T_{\text{sl}}$, where $\hbar = 6.59 \times 10^{-16}$ eV–s and is found to be ~7.81×10^{-8} eV. Further, the relation between momentum relaxation rate, Γ, and Γ_{spin} is given by: $\Gamma_{\text{spin}} = \infty \cdot \left(\dfrac{L_i}{\Delta} \right)^2 \cdot \Gamma$ and is found to be 2.63 meV for GNCs. The Γ_{spin} has functional dependence on pseudo-chemical potential, $\widetilde{\mu}$, and is given by relation (C.7.1):

$$\widetilde{\mu} \left(\mu, \Gamma_{\text{spin}} \right) = \frac{\Gamma_{\text{spin}}}{\pi} \cdot \left(\frac{\mu^2 + \Gamma^2_{\text{spin}}}{D^2} \right) + \frac{\Delta}{2} \left(1 - \frac{2}{\pi} \arctan \frac{2\Gamma_{\text{spin}}}{\Delta} \right) \tag{C.7.1}$$

where μ is the chemical potential and D, the continuum cut-off parameter, is ≈ 3.00 eV. Since $\mu = \Delta/2$, the magnitude of $\widetilde{\mu}$ is estimated to be ~1.94 eV for GNCs. The pseudo-chemical potential, $\widetilde{\mu}$ appeared in the expression of density of states ρ ($\widetilde{\mu}, \Gamma_{\text{spin}}$), with finite values of μ, Γ_{spin}, as computed above. The ρ ($\widetilde{\mu}, \Gamma_{\text{spin}}$) is measured in units of states/eV–atom and given by:

$$\rho(\widetilde{\mu}, \Gamma_{\text{spin}}) = \frac{2A_c \; \widetilde{\mu} \left(\mu, \Gamma_{\text{spin}} \right)}{\hbar^2 v_f^2} = 0.0385\widetilde{\mu} \tag{C.7.2}$$

where $A_c = \dfrac{\left(5.24 \; \text{A}^0 \right)^2}{2}$ per atoms in unit cell for graphene and v_f is the Fermi velocity of carriers at Fermi level. The value of ρ ($\widetilde{\mu}, \Gamma_{\text{spin}}$) is found to be 0.07469 states/eV–atom, which is used in calculating the volume Pauli-spin susceptibility, χ_{spin}, of GNCs using the following relation:

$$\chi_{\text{spin}} = \mu_O \; \mu_B^2 \; \rho\left(\widetilde{\mu}, \Gamma_{\text{spin}} \right) \frac{N}{A} \cdot \left(k_B T \right)^{-1} \tag{C.7.3}$$

where μ_O is the permeability of vacuum, μ_B is the Bohr magneton, N/A is the number of carbon atoms per unit area ~3.81×10^{15} atoms/cm^2, k_B is Boltzmann's constant, and T is the temperature. The value is estimated to be ~4.80×10^{-7} for GNCs at 300 K. In general, the performance of ESR is given by limit of detection (L_D), which is the number concentration of non-degenerated spin $\frac{1}{2}$ particles at 300 K that gives a signal-to-noise ratio, S/N, of 20 for 0.05 mT line width. The value of L_D for our system is 5×10^9 spins/mT. This gives L_D^{GNCs} for GNCs as:

$$L_D^{GNCs} = L_D' \cdot \frac{f\left(\Delta B\right)}{25 \text{ meV } \left(\tilde{\mu}, \Gamma_{\text{spin}}\right)} \tag{C.7.4}$$

where ΔB is 0.05 mT and $f\left(\Delta B\right) = 1$ for values less than 1 mT. Note that $\tilde{\mu} \gg \Gamma_{\text{spin}}$ and, under this condition, DOS is approximately given by $\rho\left(\mu\right) = 0.0385\,\mu$; one can get $L_D^{GNCs} = 514.81 \cdot L_D'$ (Figure C.7.2.1 and Tables C.7.1.1–C.7.1.4).

TABLE C.7.1.1

Dispersion Line Width, ΔH_{pp}, and Spin Anisotropy, Δg, in the Range 123–473 K

	Graphene		N-graphene		GNCs		N-GNCs	
Temp (in K)	ΔH_{PP} (in mT)	Δg	ΔH_{PP} (in mT)	Δg	ΔH_{PP} (in mT)	Δg	ΔH_{PP} (in mT)	Δg
123	0.64390	0.00495	0.72591	0.00518	0.73626	0.00460	0.38851	0.00434
173	0.65778	0.00506	0.60557	0.00509	0.71285	0.00467	0.48593	0.00425
223	0.69512	0.00493	0.63612	0.00514	0.67046	0.00463	0.43882	0.00433
273	0.69522	0.00499	0.58783	0.00562	0.71434	0.00448	0.44144	0.00416
298	0.65098	0.00578	0.59264	0.00584	0.67123	0.00458	0.45361	0.00439
323	0.69534	0.00506	0.63001	0.00616	0.70144	0.00445	0.43450	0.00438
373	0.67963	0.00521	0.62687	0.00600	0.70391	0.00448	0.42721	0.00430
423	0.66392	0.00535	0.62372	0.00583	0.70637	0.00445	0.42953	0.00421
473	0.81821	0.00505	0.76981	0.00604	0.73002	0.00457	0.41454	0.00415
$<A>_T$	0.68678 ± 0.05321	0.00516 ± 2.1×10^{-4}	0.64131 ± 0.06356	0.00563 ± 4.0×10^{-4}	0.70428 ± 0.0299	0.00455 ± 7.86×10^{-5}	0.43470 ± 0.0268	0.00427 ± 9.48×10^{-5}

TABLE C.7.1.2

Spin–Orbit Coupling Constant, L_i, and Momentum Relaxation Rate, Γ_{spin}, in the Measured Temperature Range 123–473 K

	Graphene		N-graphene		GNCs		N-GNCs	
Temp (in K)	L_i (in meV)	$\Gamma_{\text{spin}} 10^{-8}$ (eV)	L_i (in meV)	$\Gamma_{\text{spin}} 10^{-8}$ (eV)	L_i (in meV)	$\Gamma_{\text{spin}} 10^{-8}$ (eV)	L_i (in meV)	$\Gamma_{\text{spin}} 10^{-8}$ (eV)
123	20.63	9.66	27.26	10.89	18.68	11.05	16.45	5.83
173	21.09	7.02	26.79	6.46	18.96	7.60	16.11	5.18
223	20.55	5.76	27.05	5.26	18.80	5.55	16.42	3.63
273	20.81	4.69	29.58	3.97	18.19	4.83	15.77	2.98
298	20.84	4.03	30.73	3.67	18.59	4.15	16.64	2.81
323	21.09	3.97	32.42	3.60	18.07	4.01	16.61	2.48
373	21.70	3.47	31.55	3.16	18.10	3.55	16.29	2.18
423	22.31	2.96	30.68	2.72	18.07	3.08	15.96	1.88
473	21.05	3.19	31.79	3.01	18.55	2.85	15.73	1.61
$<A>_T$	21.271 ± 0.86	4.98 ± 2.19	29.63 ± 0.21	4.75 ± 2.59	18.49 ± 0.32	5.19 ± 2.63	16.17 ± 0.37	3.17 ± 1.45

TABLE C.7.1.3
Spin–Spin Relaxation, T_{ss}, and Effective Magnetic Moment, μ_{eff}, in 123–473 K

Temp (in K)	Graphene		N-graphene		GNCs		N-GNCs	
	T_{ss} (in ps)	μ_{eff} (in A-m^2)	T_{ss} (in ps)	μ_{eff} (in A-m^2)	T_{ss} (in ps)	μ_{eff} (in A-m^2)	T_{ss} (in ps)	μ_{eff} (in A-m^2)
123	8.81	0.0704	7.82	0.0661	7.71	0.075	14.60	0.055
173	8.63	0.0712	9.37	0.0681	7.97	0.074	11.60	0.062
223	8.17	0.0692	8.92	0.0666	8.47	0.072	12.90	0.058
273	8.17	0.0709	9.66	0.0673	7.95	0.074	12.80	0.058
298	8.73	0.0707	9.58	0.0676	8.46	0.073	12.50	0.059
323	8.16	0.0732	9.01	0.0697	8.09	0.074	13.00	0.058
373	8.58	0.0724	9.46	0.0696	8.17	0.073	13.30	0.057
423	8.55	0.0716	9.10	0.0694	8.03	0.074	13.20	0.058
473	6.94	0.0794	7.37	0.0771	7.78	0.075	13.70	0.053
$<A>_T$	8.32 ± 0.580	0.0722 ± 0.0026	8.93 ± 0.801	0.0691 ± 0.0033	8.08 ± 0.265	0.0738 ± 9.7 × 10^{-5}	13.10 ± 0.819	0.0576 ± 0.00251

TABLE C.7.1.4
DOS Variations Over 123–473 K for the Systems

Temperature (in K)	Density of States (in Stat/eV-Atom)			
	Graphene	N-graphene	GNCs	N-GNCs
123	0.07486	0.09585	0.07178	0.06490
173	0.07656	0.09803	0.07342	0.06721
223	0.07706	0.09873	0.07469	0.06896
273	0.07774	0.09976	0.07493	0.06920
298	0.07813	0.10001	0.07554	0.07008
323	0.07822	0.10019	0.07548	0.07019
423	0.07897	0.10041	0.07613	0.07067
473	0.07864	0.10037	0.07641	0.07079

Analysis of N-1s and O-1s for GNCs and N-GNCs using ESCA

The measurement details of ESCA are provide in the experimental section. Table C.7.4.1 shows the comparison of C-1s for GNCs and N-GNCs (Figure C.7.4.1).

The at% of nitrogen in GNCs is ~2.19 at%, and after doping with nitrogen, it is increased to 9.20 at%. For GNCs, the main peak of nitrogen N-1s appeared at 400.45 eV along with a satellite feature N_1 at 405.06 eV. After treating GNCs with TDAE, the main peak is shifted to lower binding energy at 399.89 eV and the feature of satellite disappeared. The negative chemical shift in the main peak of nitrogen is ~0.45 eV is due to the transfer of charge from GNCs to TDAE. It clearly indicates p-type dopant received from GNCs. GNCs are electron-rich due to the delocalization of π-electrons. This could also be reconfirmed by comparing the peak intensity for GNCs with N-GNCs. After nitrogen doping, the peak intensity of nitrogen is decreased. This indicates that the charge is transferred from GNCs to nitrogen. The N_1 peak is mainly due to shake-up effect, which is not observed in GNCs treated with TDAE. This could be explained as follows: (a) In the TDAE molecules, the binding energy of four nitrogen atoms bonded to carbon atoms is larger than that of residual nitrogen that exists in GNCs. This residual nitrogen shows satellite feature. (b) TDAE is a strong reducing agent that lowers the band gap (see Figure C.7.4.3; UV–visible spectra and relevant analysis). Downshift in band gap is in favour of producing secondary electrons whose kinetic energy is less than that of the photoelectrons main peak. Thus, shake-up-induced peak asymmetry could have taken place in

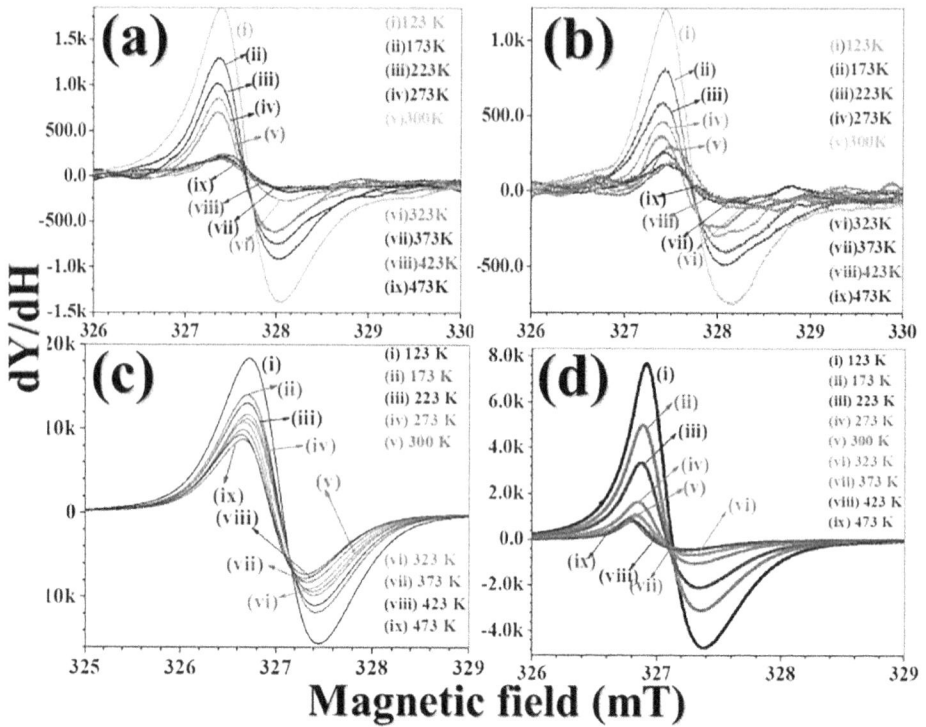

FIGURE C.7.2.1 Recorded ESR spectra for the systems. (a) Graphene, (b) N-graphene, (c) GNCs, and (d) N-GNCs.

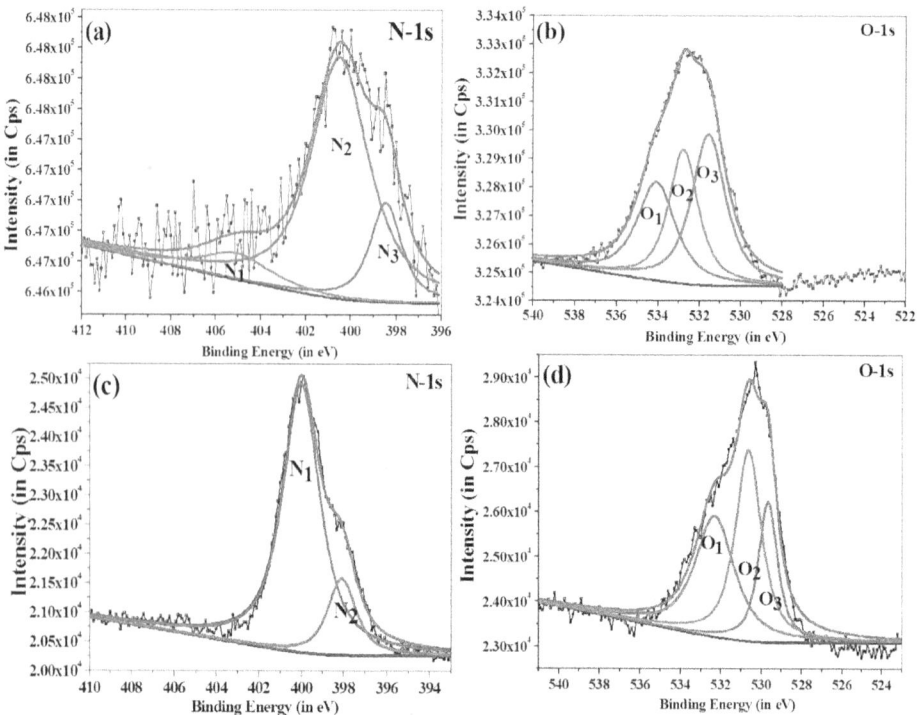

FIGURE C.7.4.1 Recorded ESCA spectra of N-1s and O-1s. Upper panel: data for GNCs, and lower panel: data for N-GNCs.

TABLE C.7.4.1

Details of Convoluted C-1s, N-1s, and O-1s Spectra for GNCs and N-GNCs

Sample	Main Peak	Deconvoluted Peak(s)	Binding Energy Position (in eV)	Atomic %	FWHM	Species
GNCs	C-1s	–	285.33	**93.00**	–	–
		C_1	289.20	17.76	4.19	COOR or C-C=O or HOPG
		C_2	286.23	40.45	2.25	C=O, C–R, C–N (residual nitrogen in GNCs)
		C_3	285.07	43.23	1.74	C=C, C-1s
N-GNCs	C-1s	–	284.66	**93.15**	–	–
		C_1	286.21	31.32	1.83	HOPG-like
		C_2	284.99	40.21	1.33	C=O, C–R, C–N
		C_3	284.04	32.09	1.19	C=N, C=N$^+$,C–OH
GNCs	N-1s	–	400.45	**02.19**		–
		N_1	405.06	–	3.62	Satellite peak
		N_2	400.48	–	3.02	Main N-1s
		N_3	398.46	–	1.67	N-1s
N-GNCs	N-1s	–	399.89	**09.20**	–	–
		N_1	400.00	–	2.12	N-1s
		N_2	398.08	–	1.57	C=N
GNCs	O-1s	–	532.32	**03.25**	–	–
		O_1	534.0789	–	2.24	O$^-_2$ related/C-OH
		O_2	532.77	–	1.73	C-OH
		O_3	531.55	–	1.97	C=O/O-1s
N-GNCs	O-1s	–	532.60	**01.45**		–
		O_1	531.13	–	2.82	COO⁻C=O
		O_2	532.90	–	1.90	C-OH
		O_3	532.1	–	1.31	O-1s

Bold value indicate total composition of an element

the TDAE-treated GNCs. Hence, we do not see shake-up feature [2]. It is interesting to note that the residual nitrogen is in single valance state in case of N-GNCs and in mixed state in GNCs. Other details are provided in Table C.7.4.1.

After doing of TDAE to GNCs, the total oxygen content is decreased from 3.25 to 1.45 at%. The deceased oxygen may be utilized for fragmentation/radicalization of TDAE. FTIR supports this finding. The intensities of peaks related to C–O groups are found to be converted into C–N after doping nitrogen in GNCs. This indicates that doping not only introduced TDAE and its fragments into the host disordered graphene, but also decreased the content of native oxygen.

Analysis of GNCs and N-GNCs using Fourier Transform Infrared Spectroscopy

Figure C.7.4.2 shows the recorded FTIR spectrum for (a) GNCs and (b) N-GNCs. In profile (a), mainly three bands were observed, below 3000 cm^{-1}. The absence of major bands above 3000 cm^{-1} indicates that GNCs contain no hydroxyl (–OH) and amino (–NH$_2$) group. Moreover, the set of bands between 1500 and 1600 cm^{-1} could be attributed to the conjugated double bond C=C. The recorded FTIR profile (b), for N-GNCs, is distinctly different than that of profile (a). TDAE has

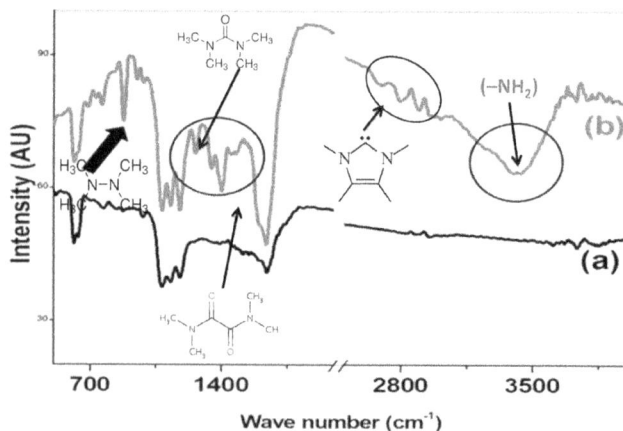

FIGURE C.7.4.2 FTIR spectrum recorded for (a) GNCs and (b) N-GNCs.

108 fundamental vibration modes, more than half of which consist of C–H bond stretching and bending within eight methyl fragments. The complete assignment of them is a difficult task; therefore, we have focused our attention mainly on C–N bond stretching modes where different charge states are expected from different fragments [3]. The emergence of broad band, for N-GNCs, at ~3433–3335 cm^{-1} (indicated by the open circle) is assigned for –N–H amines/amides, associated with one of the fragments of the TDAE that is dimethylamine (DMA). The small bands appeared between 3000 and 2650 cm^{-1} is –C–H stretch, whereas the bands at 3350–3310 cm^{-1} are for N-heterocyclic compounds. This could easily be reconfirmed with –C=N–H mode that appeared at ~1641 cm^{-1}. It also explains R–NH$_2$ and Ar–NH$_2$ in-plane bending between 1599 and 1641 cm^{-1} and could be attributed to the fragment of tetramethylurea (TMU). The small peak appearing at ~983 cm^{-1} (indicated by an arrow) could be attributed to the out-of-plane bending corresponding to CH$_2$ twisting. The shake-up feature that appeared at ~1599 cm^{-1}, as a shoulder to the main band, is attributed to the fragment of the tetramethyloxamide (TMO). This feature explains the Ar–NH– bonding with aromatic ring and is confirmed by the band appearing at ~1599 cm^{-1}.

The bands appearing at 1450–1350 cm^{-1} is the stretching of C–N bonds and N$_2$C=CN$_2$ fragments, while peaks at 1372–1382 cm^{-1} contain the C–N bonds stretching within the N–CH$_3$ fragments and could be attributed to the existence of fragments of tetrakisdimethylamino-1,2-dioxetane (TDMD) and tetramethylhydrazine (THM). Further, the aromatic C–N stretching could also be explained using the stretching bands that appeared between 1350 and 1265 cm^{-1} (indicated by a circle in Figure C.7.4.2). These are due to the C–C character with C–N stretch. The di-substitution of amines to the sp^2 bonded carbon network could be confirmed by observing two bands closely appearing at 1641 and 1599 cm^{-1}. Moreover, the bands that appeared at 1448 and 1408 cm^{-1} are stretch vibrations assigned to the geminal dimethyl group. The band appearing at 1350 cm^{-1} is for –CO–CH$_3$ on the zigzag carbon of graphene, which is attributed to the asymmetric bending mode that appeared at 1448 cm^{-1} attached with the bis(dimethylamino)methane (BMAM). Thus, from the above band assignments, it is revealed that the fragments of TDAE get attached to the zigzag chains of graphene edges or the honeycomb network. Our further investigations revealed that these radicals played an important role in charge transfer process, which could generate reduced exchange ordering in N-GNCs [4].

UV–visible spectra for GNCs and N-GNCs

Figure C.7.4.3 shows the UV–visible spectrum recorded for (a) GNCs and (b) N-GNCs samples. One can see from profile (a) that the absorption peak for GNCs appeared at wavelength, λ_{max}, 245 nm. On the contrary, in profile (b), the peak is found to be shifted to, λ_{max}, 230 nm.

Thus, a blueshift is observed after incorporating TDAE into GNCs. The observed peak, in profile (b), is broad and extended up to ~200 nm, i.e. towards UV region. The observed blueshift indicates that the charge transfer occurred from non-bonded TDAE electron to π^*-electron energy level of GNCs [5].

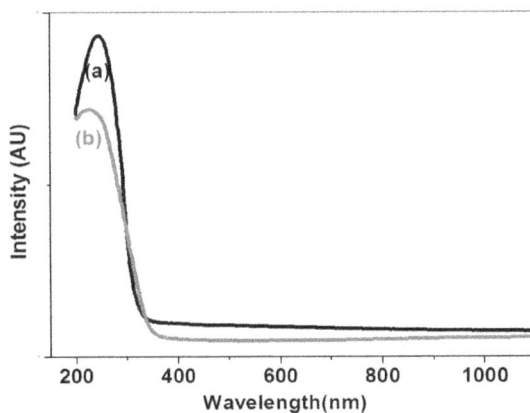

FIGURE C.7.4.3 UV–visible spectrum recorded for (a) GNCs and (b) N-GNCs.

CHAPTER 8

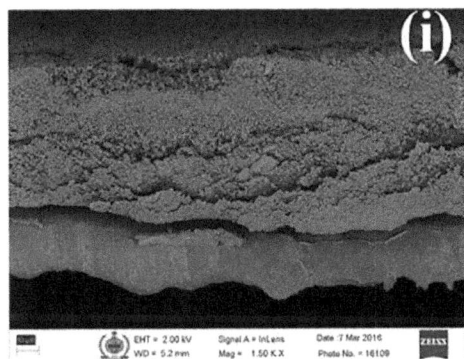

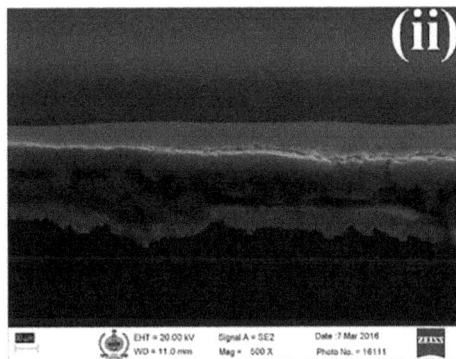

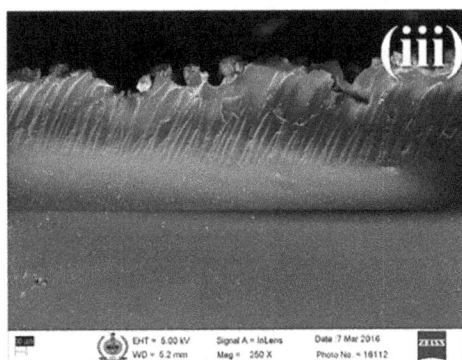

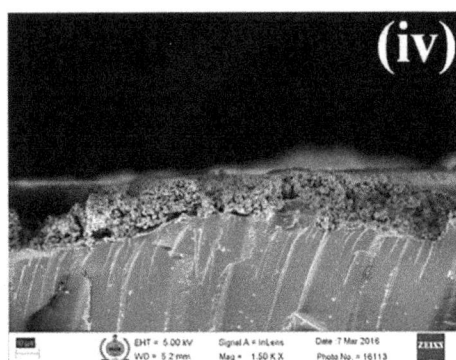

FIGURE C.8.7.1 Recorded cross-sectional FESEM images of nanocarbons grown on copper and glass substrates showing peeling off effect.

REFERENCES

[1] Dóra, Balázs, Ferenc Murányi, and Ferenc Simon. "Electron spin dynamics and electron spin resonance in graphene." *EPL (Europhysics Letters)* 92, no. 1 (2010): 17002.

[2] Svensson, Johannes, Yu Tarakanov, DongSu Lee, Jari M. Kinaret, YungWoo Park, and Eleanor E. B. Campbell. "A carbon nanotube gated carbon nanotube transistor with 5 ps gate delay." *Nanotechnology* 19, no. 32 (2008): 325201.

[3] Pokhodnia, K. I., Georgios Papavassiliou, Polona Umek, A. Omerzu, and D. Mihailovič. "A structural and infrared study of the charge states of tetrakis (dimethylamino) ethylene (TDAE) in TDAE-C_{60} and $(TDAE)(Cl)_2$." *The Journal of Chemical Physics* 110, no. 7 (1999): 3606–3611.

[4] Nakanishi, Koji. *Infrared absorption spectroscopy, practical.* 1964.

[5] Lai, Qi, Shifu Zhu, Xueping Luo, Min Zou, and Shuanghua Huang. "Ultraviolet-visible spectroscopy of graphene oxides." *AIP Advances* 2, no. 3 (2012): 032146.

Index

For Product Safety Concerns and Information please contact our EU
representative GPSR@taylorandfrancis.com
Taylor & Francis Verlag GmbH, Kaufingerstraße 24, 80331 München, Germany